D0151298

CAUSALITY

Models, Reasoning, and Inference
Second Edition

Written by one of the preeminent researchers in the field, this book provides a comprehensive exposition of modern analysis of causation. It shows how causality has grown from a nebulous concept into a mathematical theory with significant applications in the fields of statistics, artificial intelligence, economics, philosophy, cognitive science, and the health and social sciences.

Judea Pearl presents a comprehensive theory of causality which unifies the probabilistic, manipulative, counterfactual, and structural approaches to causation and offers simple mathematical tools for studying the relationships between causal connections and statistical associations. The book opens the way for including causal analysis in the standard curricula of statistics, artificial intelligence, business, epidemiology, social sciences, and economics. Students in these fields will find natural models, simple inferential procedures, and precise mathematical definitions of causal concepts that traditional texts have evaded or made unduly complicated.

The first edition of *Causality* has led to a paradigmatic change in the way that causality is treated in statistics, philosophy, computer science, social science, and economics. Cited in more than 2,800 scientific publications, it continues to liberate scientists from the traditional molds of statistical thinking. In this revised edition, Pearl elucidates thorny issues, answers readers' questions, and offers a panoramic view of recent advances in this field of research.

Causality will be of interest to students and professionals in a wide variety of fields. Anyone who wishes to elucidate meaningful relationships from data, predict effects of actions and policies, assess explanations of reported events, or form theories of causal understanding and causal speech will find this book stimulating and invaluable.

Judea Pearl is professor of computer science and statistics at the University of California, Los Angeles, where he directs the Cognitive Systems Laboratory and conducts research in artificial intelligence, human reasoning, and philosophy of science. The author of *Heuristics* and *Probabilistic Reasoning*, he is a member of the National Academy of Engineering and a Founding Fellow of the American Association for Artificial Intelligence. Dr. Pearl is the recipient of the IJCAI Research Excellence Award for 1999, the London School of Economics Lakatos Award for 2001, and the ACM Alan Newell Award for 2004. In 2008, he received the Benjamin Franklin Medal for computer and cognitive science from the Franklin Institute.

Commendation for the First Edition

"Judea Pearl's previous book, *Probabilistic Reasoning in Intelligent Systems,* was arguably the most influential book in Artificial Intelligence in the past decade, setting the stage for much of the current activity in probabilistic reasoning. In this book, Pearl turns his attention to causality, boldly arguing for the primacy of a notion long ignored in statistics and misunderstood and mistrusted in other disciplines, from physics to economics. He demystifies the notion, clarifies the basic concepts in terms of graphical models, and explains the source of many misunderstandings. This book should prove invaluable to researchers in artificial intelligence, statistics, economics, epidemiology, and philosophy, and, indeed, all those interested in the fundamental notion of causality. It may well prove to be one of the most influential books of the next decade."

> – Joseph Halpern, Computer Science Department, Cornell University

"This lucidly written book is full of inspiration and novel ideas that bring clarity to areas where confusion has prevailed, in particular concerning causal interpretation of structural equation systems, but also on concepts such as counterfactual reasoning and the general relation between causal thinking and graphical models. Finally the world can get a coherent exposition of these ideas that Judea Pearl has developed over a number of years and presented in a flurry of controversial yet illuminating articles."

> – Steffen L. Lauritzen, Department of Mathematics, Aalborg University

"Judea Pearl's new book, *Causality: Models, Reasoning, and Inference,* is an outstanding contribution to the causality literature. It will be especially useful to students and practitioners of economics interested in policy analysis."

> – Halbert White, Professor of Economics, University of California, San Diego

"This book fulfills a long-standing need for a rigorous yet accessible treatise on the mathematics of causal inference. Judea Pearl has done a masterful job of describing the most important approaches and displaying their underlying logical unity. The book deserves to be read by all statisticians and scientists who use nonexperimental data to study causation, and would serve well as a graduate or advanced undergraduate course text."

> – Sander Greenland, School of Public Health, University of California, Los Angeles

"Judea Pearl has written an account of recent advances in the modeling of probability and cause, substantial parts of which are due to him and his co-workers. This is essential reading for anyone interested in causality."

> – Brian Skyrms, Department of Philosophy, University of California, Irvine

CAUSALITY

Models, Reasoning, and Inference
Second Edition

Judea Pearl

University of California, Los Angeles

CAMBRIDGE
UNIVERSITY PRESS

CAMBRIDGE UNIVERSITY PRESS
Cambridge, New York, Melbourne, Madrid, Cape Town, Singapore,
São Paulo, Delhi, Dubai, Tokyo

Cambridge University Press
32 Avenue of the Americas, New York, NY 10013-2473, USA

www.cambridge.org
Information on this title: www.cambridge.org/9780521895606

First published 2000
8th printing 2008
Second edition 2009

Printed in the United States of America

A catalog record for this publication is available from the British Library.

The Library of Congress has cataloged the first edition as follows:

Pearl, Judea
Causality : models, reasoning, and inference / Judea Pearl.
p. cm.
ISBN 0-521-77362-8 (hardback)
1. Causation. 2. Probabilities. I. Title.
BD541.P43   2000
122 – dc21   99-042108

ISBN 978-0-521-89560-6 Hardback

Development of Western science is based on two great achievements: the invention of the formal logical system (in Euclidean geometry) by the Greek philosophers, and the discovery of the possibility to find out causal relationships by systematic experiment (during the Renaissance).

Albert Einstein (1953)

TO DANNY
AND THE GLOWING AUDACITY OF GOODNESS

Contents

Preface to the First Edition

The central aim of many studies in the physical, behavioral, social, and biological sciences is the elucidation of cause–effect relationships among variables or events. However, the appropriate methodology for extracting such relationships from data – or even from theories – has been fiercely debated.

The two fundamental questions of causality are: (1) What empirical evidence is required for legitimate inference of cause–effect relationships? (2) Given that we are willing to accept causal information about a phenomenon, what inferences can we draw from such information, and how? These questions have been without satisfactory answers in part because we have not had a clear semantics for causal claims and in part because we have not had effective mathematical tools for casting causal questions or deriving causal answers.

In the last decade, owing partly to advances in graphical models, causality has undergone a major transformation: from a concept shrouded in mystery into a mathematical object with well-defined semantics and well-founded logic. Paradoxes and controversies have been resolved, slippery concepts have been explicated, and practical problems relying on causal information that long were regarded as either metaphysical or unmanageable can now be solved using elementary mathematics. Put simply, causality has been mathematized.

This book provides a systematic account of this causal transformation, addressed primarily to readers in the fields of statistics, artificial intelligence, philosophy, cognitive science, and the health and social sciences. Following a description of the conceptual and mathematical advances in causal inference, the book emphasizes practical methods for elucidating potentially causal relationships from data, deriving causal relationships from combinations of knowledge and data, predicting the effects of actions and policies, evaluating explanations for observed events and scenarios, and – more generally – identifying and explicating the assumptions needed for substantiating causal claims.

Ten years ago, when I began writing *Probabilistic Reasoning in Intelligent Systems* (1988), I was working within the empiricist tradition. In this tradition, probabilistic relationships constitute the foundations of human knowledge, whereas causality simply provides useful ways of abbreviating and organizing intricate patterns of probabilistic relationships. Today, my view is quite different. I now take causal relationships to be the fundamental building blocks both of physical reality and of human understanding of that reality, and I regard probabilistic relationships as but the surface phenomena of the causal machinery that underlies and propels our understanding of the world.

Accordingly, I see no greater impediment to scientific progress than the prevailing practice of focusing all of our mathematical resources on probabilistic and statistical inferences while leaving causal considerations to the mercy of intuition and good judgment. Thus I have tried in this book to present mathematical tools that handle causal relationships side by side with probabilistic relationships. The prerequisites are startlingly simple, the results embarrassingly straightforward. No more than basic skills in probability theory and some familiarity with graphs are needed for the reader to begin solving causal problems that are too complex for the unaided intellect. Using simple extensions of probability calculus, the reader will be able to determine mathematically what effects an intervention might have, what measurements are appropriate for control of confounding, how to exploit measurements that lie on the causal pathways, how to trade one set of measurements for another, and how to estimate the probability that one event was the actual cause of another.

Expert knowledge of logic and probability is nowhere assumed in this book, but some general knowledge in these areas is beneficial. Thus, Chapter 1 includes a summary of the elementary background in probability theory and graph notation needed for the understanding of this book, together with an outline of the developments of the last decade in graphical models and causal diagrams. This chapter describes the basic paradigms, defines the major problems, and points readers to the chapters that provide solutions to those problems.

Subsequent chapters include introductions that serve both to orient the reader and to facilitate skipping; they indicate safe detours around mathematically advanced topics, specific applications, and other explorations of interest primarily to the specialist.

The sequence of discussion follows more or less the chronological order by which our team at UCLA has tackled these topics, thus re-creating for the reader some of our excitement that accompanied these developments. Following the introductory chapter (Chapter 1), we start with the hardest questions of how one can go about discovering cause–effect relationships in raw data (Chapter 2) and what guarantees one can give to ensure the validity of the relationships thus discovered. We then proceed to questions of identifiability – namely, predicting the direct and indirect effects of actions and policies from a combination of data and fragmentary knowledge of where causal relationships might operate (Chapters 3 and 4). The implications of these findings for the social and health sciences are then discussed in Chapters 5 and 6 (respectively), where we examine the concepts of structural equations and confounding. Chapter 7 offers a formal theory of counterfactuals and structural models, followed by a discussion and a unification of related approaches in philosophy, statistics, and economics. The applications of counterfactual analysis are then pursued in Chapters 8–10, where we develop methods of bounding causal relationships and illustrate applications to imperfect experiments, legal responsibility, and the probability of necessary, sufficient, and single-event causation. We end this book (Epilogue) with a transcript of a public lecture that I presented at UCLA, which provides a gentle introduction to the historical and conceptual aspects of causation.

Readers who wish to be first introduced to the nonmathematical aspects of causation are advised to start with the Epilogue and then to sweep through the other historical/conceptual parts of the book: Sections 1.1.1, 3.3.3, 4.5.3, 5.1, 5.4.1, 6.1, 7.2, 7.4, 7.5, 8.3, 9.1, 9.3, and 10.1. More formally driven readers, who may be anxious to delve directly into the mathematical aspects and computational tools, are advised to start with Section 7.1 and then to proceed as follows for tool building: Section 1.2, Chapter 3, Sections 4.2–4.4, Sections 5.2–5.3, Sections 6.2–6.3, Section 7.3, and Chapters 8–10.

I owe a great debt to many people who assisted me with this work. First, I would like to thank the members of the Cognitive Systems Laboratory at UCLA, whose work and ideas formed the basis of many of these sections: Alex Balke, Blai Bonet, David Chickering, Adnan Darwiche, Rina Dechter, David Galles, Hector Geffner, Dan Geiger, Moisés Goldszmidt, Jin Kim, Jin Tian, and Thomas Verma. Tom and Dan have proven some of the most basic theorems in causal graphs; Hector, Adnan, and Moisés were responsible for keeping me in line with the logicist approach to actions and change; and Alex and David have taught me that counterfactuals are simpler than the name may imply.

My academic and professional colleagues have been very generous with their time and ideas as I began ploughing the peaceful territories of statistics, economics, epidemiology, philosophy, and the social sciences. My mentors–listeners in statistics have been Phil Dawid, Steffen Lauritzen, Don Rubin, Art Dempster, David Freedman, and David Cox. In economics, I have benefited from many discussions with John Aldrich, Kevin Hoover, James Heckman, Ed Learner, and Herbert Simon. My forays into epidemiology resulted in a most fortunate and productive collaboration with Sander Greenland and James Robins. Philosophical debates with James Woodward, Nancy Cartwright, Brian Skyrms, Clark Glymour, and Peter Spirtes have sharpened my thinking of causality in and outside philosophy. Finally, in artificial intelligence, I have benefited from discussions with and the encouragement of Nils Nilsson, Ray Reiter, Don Michie, Joe Halpern, and David Heckerman.

The National Science Foundation deserves acknowledgment for consistently and faithfully sponsoring the research that led to these results, with special thanks to H. Moraff, Y. T. Chien, and Larry Reeker. Other sponsors include Abraham Waksman of the Air Force Office of Scientific Research, Michael Shneier of the Office of Naval Research, the California MICRO Program, Northrop Corporation, Rockwell International, Hewlett-Packard, and Microsoft.

I would like to thank Academic Press and Morgan Kaufmann Publishers for their kind permission to reprint selected portions of previously published material. Chapter 3 includes material reprinted from *Biometrika,* vol. 82, Judea Pearl, "Causal Diagrams for Empirical Research," pp. 669–710, Copyright 1995, with permission from Oxford University Press. Chapter 5 includes material reprinted from *Sociological Methods and Research,* vol. 27, Judea Pearl, "Graphs, Causality, and Structural Equation Models," pp. 226–84, Copyright 1998, with permission from Sage Publications, Inc. Chapter 7 includes material reprinted from *Foundations of Science,* vol. 1, David Galles and Judea Pearl, "An Axiomatic Characterization of Causal Counterfactuals," pp. 151–82, Copyright 1998, with permission from Kluwer Academic Publishers. Chapter 7 also includes material reprinted from *Artificial Intelligence,* vol. 97, David Galles and Judea Pearl, "Axioms of Causal Relevance," pp. 9–43, Copyright 1997, with permission from Elsevier Science. Chapter 8 includes material modified from *Journal of the American Statistical Association*, vol. 92, Alexander Balke and Judea Pearl, "Bounds on Treatment Effects from Studies with Imperfect Compliance," pp. 1171–6, Copyright 1997, with permission from the American Statistical Association.

The manuscript was most diligently typed, processed, and illustrated by Kaoru Mulvihill. Jin Tian and Blai Bonet helped in proofing selected chapters. Matt Darnell did a masterful job of copyediting these pages. Alan Harvey has been my consoling ombudsman and virtual editor throughout the production process.

Finally, my humor and endurance through the writing of this book owe a great debt to my family – to Tammy, Danny, Michelle, and Leora for filling my heart with their smiles, and to my wife Ruth for surrounding me with so much love, support, and meaning.

J. P.
Los Angeles
August 1999

Preface to the Second Edition

It has been more than eight years since the first edition of this book presented readers with the friendly face of causation and her mathematical artistry. The popular reception of the book and the rapid expansion of the structural theory of causation call for a new edition to assist causation through her second transformation – from a demystified wonder to a commonplace tool in research and education. This edition (1) provides technical corrections, updates, and clarifications in all ten chapters of the original book, (2) adds summaries of new developments and annotated bibliographical references at the end of each chapter, and (3) elucidates subtle issues that readers and reviewers have found perplexing, objectionable, or in need of elaboration. These are assembled into an entirely new chapter (11) which, I sincerely hope, clears the province of causal thinking from the last traces of controversy.

Teachers who have taught from this book before should find the revised edition more lucid and palatable, while those who have waited for scouts to carve the path will find the road paved and tested. Supplementary educational material, slides, tutorials, and homework can be found on my website, http://www.cs.ucla.edu/~judea/.

My main audience remain the students: students of statistics who wonder why instructors are reluctant to discuss causality in class; students of epidemiology who wonder why elementary concepts such as confounding are so hard to define mathematically; students of economics and social science who question the meaning of the parameters they estimate; and, naturally, students of artificial intelligence and cognitive science, who write programs and theories for knowledge discovery, causal explanations, and causal speech.

I hope that each of these groups will find the unified theory of causation presented in this book to be both inspirational and instrumental in tackling new challenges in their respective fields.

J. P.
Los Angeles
July 2008

CHAPTER ONE

Introduction to Probabilities, Graphs, and Causal Models

Chance gives rise to thoughts,
and chance removes them.
Pascal (1670)

1.1 INTRODUCTION TO PROBABILITY THEORY

1.1.1 Why Probabilities?

Causality connotes lawlike necessity, whereas probabilities connote exceptionality, doubt, and lack of regularity. Still, there are two compelling reasons for starting with, and in fact stressing, probabilistic analysis of causality; one is fairly straightforward, the other more subtle.

The simple reason rests on the observation that causal utterances are often used in situations that are plagued with uncertainty. We say, for example, "reckless driving causes accidents" or "you will fail the course because of your laziness" (Suppes 1970), knowing quite well that the antecedents merely tend to make the consequences more likely, not absolutely certain. Any theory of causality that aims at accommodating such utterances must therefore be cast in a language that distinguishes various shades of likelihood – namely, the language of probabilities. Connected with this observation, we note that probability theory is currently the official mathematical language of most disciplines that use causal modeling, including economics, epidemiology, sociology, and psychology. In these disciplines, investigators are concerned not merely with the presence or absence of causal connections but also with the relative strengths of those connections and with ways of inferring those connections from noisy observations. Probability theory, aided by methods of statistical analysis, provides both the principles and the means of coping with – and drawing inferences from – such observations.

The more subtle reason concerns the fact that even the most assertive causal expressions in natural language are subject to exceptions, and those exceptions may cause major difficulties if processed by standard rules of deterministic logic. Consider, for example, the two plausible premises:

1. My neighbor's roof gets wet whenever mine does.
2. If I hose my roof it will get wet.

Taken literally, these two premises imply the implausible conclusion that my neighbor's roof gets wet whenever I hose mine.

1

Such paradoxical conclusions are normally attributed to the finite granularity of our language, as manifested in the many exceptions that are implicit in premise 1. Indeed, the paradox disappears once we take the trouble of explicating those exceptions and write, for instance:

1*.   My neighbor's roof gets wet whenever mine does, except when it is covered with plastic, or when my roof is hosed, etc.

Probability theory, by virtue of being especially equipped to tolerate unexplicated exceptions, allows us to focus on the main issues of causality without having to cope with paradoxes of this kind.

As we shall see in subsequent chapters, tolerating exceptions solves only some of the problems associated with causality. The remaining problems – including issues of inference, interventions, identification, ramification, confounding, counterfactuals, and explanation – will be the main topic of this book. By portraying those problems in the language of probabilities, we emphasize their universality across languages. Chapter 7 will recast these problems in the language of deterministic logic and will introduce probabilities merely as a way to express uncertainty about unobserved facts.

1.1.2   Basic Concepts in Probability Theory

The bulk of the discussion in this book will focus on systems with a finite number of discrete variables and thus will require only rudimentary notation and elementary concepts in probability theory. Extensions to continuous variables will be outlined but not elaborated in full generality. Readers who want additional mathematical machinery are invited to study the many excellent textbooks on the subject – for example, Feller (1950), Hoel et al. (1971), or the appendix to Suppes (1970). This section provides a brief summary of elementary probability concepts, based largely on Pearl (1988b), with special emphasis on Bayesian inference and its connection to the psychology of human reasoning under uncertainty. Such emphasis is generally missing from standard textbooks.

We will adhere to the Bayesian interpretation of probability, according to which probabilities encode degrees of belief about events in the world and data are used to strengthen, update, or weaken those degrees of belief. In this formalism, degrees of belief are assigned to propositions (sentences that take on true or false values) in some language, and those degrees of belief are combined and manipulated according to the rules of probability calculus. We will make no distinction between sentential propositions and the actual events represented by those propositions. For example, if A stands for the statement "Ted Kennedy will seek the nomination for president in year 2012," then $P(A \mid K)$ stands for a person's subjective belief in the event described by A given a body of knowledge K, which might include that person's assumptions about American politics, specific proclamations made by Kennedy, and an assessment of Kennedy's age and personality. In defining probability expressions, we often simply write $P(A)$, leaving out the symbol K. However, when the background information undergoes changes, we need to identify specifically the assumptions that account for our beliefs and explicitly articulate K (or some of its elements).

In the Bayesian formalism, belief measures obey the three basic axioms of probability calculus:

$$0 \leq P(A) \leq 1, \tag{1.1}$$

$$P \text{ (sure proposition) } = 1, \tag{1.2}$$

$$P(A \text{ or } B) = P(A) + P(B) \text{ if } A \text{ and } B \text{ are mutually exclusive.} \tag{1.3}$$

The third axiom states that the belief assigned to any set of events is the sum of the beliefs assigned to its nonintersecting components. Because any event A can be written as the union of the joint events $(A \wedge B)$ and $(A \wedge \neg B)$, their associated probabilities are given by[1]

$$P(A) = P(A, B) + P(A, \neg B), \tag{1.4}$$

where $P(A, B)$ is short for $P(A \wedge B)$. More generally, if B_i, $i = 1, 2,\ldots, n$, is a set of exhaustive and mutually exclusive propositions (called a *partition* or a *variable*), then $P(A)$ can be computed from $P(A, B_i)$, $i = 1, 2,\ldots, n$, by using the sum

$$P(A) = \sum_i P(A, B_i), \tag{1.5}$$

which has come to be known as the "law of *total* probability." The operation of summing up probabilities over all B_i is also called "marginalizing over B"; and the resulting probability, $P(A)$, is called the *marginal* probability of A. For example, the probability of A, "The outcomes of two dice are equal," can be computed by summing over the joint events $(A \wedge B_i)$, $i = 1, 2,\ldots, 6$, where B_i stands for the proposition "The outcome of the first die is i." This yields

$$P(A) = \sum_i P(A, B_i) = 6 \times \frac{1}{36} = \frac{1}{6}. \tag{1.6}$$

A direct consequence of (1.2) and (1.4) is that a proposition and its negation must be assigned a total belief of unity,

$$P(A) + P(\neg A) = 1, \tag{1.7}$$

because one of the two statements is certain to be true.

The basic expressions in the Bayesian formalism are statements about *conditional probabilities* – for example, $P(A \mid B)$ – which specify the belief in A under the assumption that B is known with absolute certainty. If $P(A \mid B) = P(A)$, we say that A and B are *independent*, since our belief in A remains unchanged upon learning the truth of B. If $P(A \mid B, C) = P(A \mid C)$, we say that A and B are *conditionally independent* given C; that is, once we know C, learning B would not change our belief in A.

Contrary to the traditional practice of defining conditional probabilities in terms of joint events,

$$P(A \mid B) = \frac{P(A, B)}{P(B)}, \tag{1.8}$$

---

[1] The symbols $\wedge$, $\vee$, $\neg$, $\Rightarrow$ denote the logical connectives *and, or, not,* and *implies,* respectively.

Bayesian philosophers see the conditional relationship as more basic than that of joint events – that is, more compatible with the organization of human knowledge. In this view, B serves as a pointer to a context or frame of knowledge, and $A \mid B$ stands for an event A in the context specified by B (e.g., a symptom A in the context of a disease B). Consequently, empirical knowledge invariably will be encoded in conditional probability statements, whereas belief in joint events (if it is ever needed) will be computed from those statements via the product

$$P(A, B) = P(A \mid B) \, P(B),\tag{1.9}$$

which is equivalent to (1.8). For example, it was somewhat unnatural to assess

$$P(A, B_i) = \frac{1}{36}$$

directly in (1.6). The mental process underlying such assessment presumes that the two outcomes are independent, so to make this assumption explicit the probability of the joint event (equality, B_i) should be assessed from the conditional event (equality $\mid B_i$) via the product

$$P(\text{equality} \mid B_i) \, P(B_i) = P(\text{outcome of second die is } i \mid B_i) P(B_i)$$

$$= \frac{1}{6} \times \frac{1}{6} = \frac{1}{36}.$$

As in (1.5), the probability of any event A can be computed by conditioning it on any set of exhaustive and mutually exclusive events B_i, $i = 1, 2, \ldots, n$, and then summing:

$$P(A) = \sum_i P(A \mid B_i) P(B_i).\tag{1.10}$$

This decomposition provides the basis for hypothetical or "assumption-based" reasoning. It states that the belief in any event A is a weighted sum over the beliefs in all the distinct ways that A might be realized. For example, if we wish to calculate the probability that the outcome X of the first die will be greater than the outcome Y of the second, we can condition the event $A : X > Y$ on all possible values of X and obtain

$$P(A) = \sum_{i=1}^{6} P(Y < X \mid X = i) P(X = i)$$

$$= \sum_{i=1}^{6} P(Y < i) \frac{1}{6} = \sum_{i=1}^{6} \sum_{j=1}^{i-1} P(Y = j) \frac{1}{6}$$

$$= \frac{1}{6} \sum_{i=2}^{6} \frac{i-1}{6} = \frac{5}{12}.$$

It is worth reemphasizing that formulas like (1.10) are always understood to apply in some larger context K, which defines the assumptions taken as common knowledge (e.g., the fairness of dice rolling). Equation (1.10) is really a shorthand notation for the statement

$$P(A \mid K) = \sum_{i} P(A \mid B_i, K)P(B_i \mid K). \tag{1.11}$$

This equation follows from the fact that every conditional probability $P(A \mid K)$ is itself a genuine probability function; hence it satisfies (1.10).

Another useful generalization of the product rule (equation (1.9)) is the *chain rule* formula. It states that if we have a set of n events, $E_1, E_2, \ldots, E_n$, then the probability of the joint event $(E_1, E_2, \ldots, E_n)$ can be written as a product of n conditional probabilities:

$$P(E_1, E_2, \ldots, E_n) = P(E_n \mid E_{n-1}, \ldots, E_2, E_1) \ldots P(E_2 \mid E_1) P(E_1). \tag{1.12}$$

This product can be derived by repeated application of (1.9) in any convenient order.

The heart of Bayesian inference lies in the celebrated inversion formula,

$$P(H \mid e) = \frac{P(e \mid H)P(H)}{P(e)}, \tag{1.13}$$

which states that the belief we accord a hypothesis H upon obtaining evidence e can be computed by multiplying our previous belief $P(H)$ by the likelihood $P(e \mid H)$ that e will materialize if H is true. This $P(H \mid e)$ is sometimes called the posterior probability (or simply *posterior*), and $P(H)$ is called the prior probability (or *prior*). The denominator $P(e)$ of (1.13) hardly enters into consideration because it is merely a normalizing constant $P(e) = P(e \mid H)P(H) + P(e \mid \neg H)P(\neg H)$, which can be computed by requiring that $P(H \mid e)$ and $P(\neg H \mid e)$ sum to unity.

Whereas formally (1.13) might be dismissed as a tautology stemming from the definition of conditional probabilities,

$$P(A \mid B) = \frac{P(A, B)}{P(B)} \quad \text{and} \quad P(B \mid A) = \frac{P(A, B)}{P(A)}, \tag{1.14}$$

the Bayesian subjectivist regards (1.13) as a normative rule for updating beliefs in response to evidence. In other words, although conditional probabilities can be viewed as purely mathematical constructs (as in (1.14)), the Bayes adherent views them as primitives of the language and as faithful translations of the English expression "..., given that I know A." Accordingly, (1.14) is not a definition but rather an empirically verifiable relationship between English expressions. It asserts, among other things, that the belief a person attributes to B after discovering A is never lower than that attributed to $A \wedge B$ before discovering A. Also, the ratio between these two beliefs will increase proportionally with the degree of surprise $[P(A)]^{-1}$ one associates with the discovery of A.

The importance of (1.13) is that it expresses a quantity $P(H \mid e)$ – which people often find hard to assess – in terms of quantities that often can be drawn directly from our experiential knowledge. For example, if a person at the next gambling table declares the outcome "twelve," and we wish to know whether he was rolling a pair of dice or spinning a roulette wheel, our models of the gambling devices readily yield the quantities $P(\text{twelve} \mid \text{dice})$ and $P(\text{twelve} \mid \text{roulette})$: 1/36 for the former and 1/38 for the latter. Similarly, we can judge the prior probabilities $P(\text{dice})$ and $P(\text{roulette})$ by estimating the number of roulette wheels and dice tables at the casino. Issuing a direct judgment of

$P(\text{dice} \mid \text{twelve})$ would have been much more difficult; only a specialist in such judgments, trained at the very same casino, could do it reliably.

In order to complete this brief introduction, we must discuss the notion of *probabilistic model* (also called *probability space*). A probabilistic model is an encoding of information that permits us to compute the probability of every well-formed sentence S in accordance with the axioms of (1.1)–(1.3). Starting with a set of atomic propositions $A, B, C,\ldots$, the set of well-formed sentences consists of all Boolean formulas involving these propositions, for example, $S = (A \wedge B) \vee \neg C$. The traditional method of specifying probabilistic models employs a *joint distribution function*, which is a function that assigns nonnegative weights to every *elementary event* in the language (an elementary event being a conjunction in which every atomic proposition or its negation appears once) such that the sum of the weights adds up to 1. For example, if we have three atomic propositions, A, B, and C, then a joint distribution function should assign nonnegative weights to all eight combinations – $(A \wedge B \wedge C), (A \wedge B \neg C), \ldots, (\neg A \wedge \neg B \wedge \neg C)$ – such that the eight weights sum to 1.

The reader may recognize the set of elementary events as the *sample space* in probability textbooks. For example, if A, B, and C correspond to the propositions that coins 1, 2, and 3 will come up heads, then the sample space will consist of the set {HHH, HHT, HTH,…, TTT}. Indeed, it is sometimes convenient to view the conjunctive formulas corresponding to elementary events as *points* (or *worlds* or *configurations*), and to regard other formulas as *sets* made up of these points. Since every Boolean formula can be expressed as a disjunction of elementary events, and since the elementary events are mutually exclusive, we can always compute $P(S)$ using the additivity axiom (equation (1.3)). Conditional probabilities can be computed the same way, using (1.14). Thus, any joint probability function represents a complete probabilistic model.

Joint distribution functions are mathematical constructs of great importance. They allow us to determine quickly whether we have sufficient information to specify a complete probabilistic model, whether the information we have is consistent, and at what point additional information is needed. The criteria are simply to check (i) whether the information available is sufficient for uniquely determining the probability of every elementary event in the domain and (ii) whether the probabilities add up to 1.

In practice, however, joint distribution functions are rarely specified explicitly. In the analysis of continuous random variables, the distribution functions are given by algebraic expressions such as those describing normal or exponential distributions; for discrete variables, indirect representation methods have been developed where the overall distribution is inferred from local relationships among small groups of variables. Graphical models, the most popular of these representations, provide the basis of discussion throughout this book. Their use and formal characterization will be discussed in the next few sections.

1.1.3 Combining Predictive and Diagnostic Supports

The essence of Bayes's rule (equation 1.13)) is conveniently portrayed using the *odds* and *likelihood ratio* parameters. Dividing (1.13) by the complementary form for $P(\neg H \mid e)$, we obtain

$$\frac{P(H \mid e)}{P(\neg H \mid e)} = \frac{P(e \mid H)}{P(e \mid \neg H)} \frac{P(H)}{P(\neg H)}. \tag{1.15}$$

Defining the *prior odds* on H as

$$O(H) = \frac{P(H)}{P(\neg H)} = \frac{P(H)}{1 - P(H)} \tag{1.16}$$

and the *likelihood ratio* as

$$L(e \mid H) = \frac{P(e \mid H)}{P(e \mid \neg H)}, \tag{1.17}$$

the *posterior odds*

$$O(H \mid e) = \frac{P(H \mid e)}{P(\neg H \mid e)} \tag{1.18}$$

are given by the product

$$O(H \mid e) = L(e \mid H)O(H). \tag{1.19}$$

Thus, Bayes's rule dictates that the overall strength of belief in a hypothesis H, based on both our previous knowledge K and the observed evidence e, should be the product of two factors: the prior odds $O(H)$ and the likelihood ratio $L(e \mid H)$. The first factor measures the *predictive* or *prospective* support accorded to H by the background knowledge alone, while the second represents the *diagnostic* or *retrospective* support given to H by the evidence actually observed.[2]

Strictly speaking, the likelihood ratio $L(e \mid H)$ might depend on the content of the tacit knowledge base K. However, the power of Bayesian techniques comes primarily from the fact that, in causal reasoning, the relationship $P(e \mid H)$ is fairly local: given that H is true, the probability of e can be estimated naturally since it is usually not dependent on many other propositions in the knowledge base. For example, once we establish that a patient suffers from a given disease H, it is natural to estimate the probability that she will develop a certain symptom e. The organization of medical knowledge rests on the paradigm that a symptom is a stable characteristic of the disease and should therefore be fairly independent of other factors, such as epidemic conditions, previous diseases, and faulty diagnostic equipment. For this reason the conditional probabilities $P(e \mid H)$, as opposed to $P(H \mid e)$, are the atomic relationships in Bayesian analysis. The former possess modularity features similar to logical rules. They convey a degree of confidence in rules such as "If H then e," a confidence that persists regardless of what other rules or facts reside in the knowledge base.

Example 1.1.1   Imagine being awakened one night by the shrill sound of your burglar alarm. What is your degree of belief that a burglary attempt has taken place? For

---

[2] In epidemiology, if H stands for exposure and e stands for disease, then the likelihood ratio L is called the "risk ratio" (Rothman and Greenland 1998, p. 50). Equation (1.18) would then give the odds that a person with disease e had been exposed to H.

illustrative purposes we make the following judgments: (a) There is a 95% chance that an attempted burglary will trigger the alarm system – $P(\text{alarm} \mid \text{burglary}) = 0.95$; (b) based on previous false alarms, there is a slight (1%) chance that the alarm will be triggered by a mechanism other than an attempted burglary – $P(\text{alarm} \mid \text{no burglary}) = 0.01$; (c) previous crime patterns indicate that there is a one in ten thousand chance that a given house will be burglarized on a given night – $P(\text{burglary}) = 10^{-4}$.

Putting these assumptions together using (1.19), we obtain

$$O(\text{burglary} \mid \text{alarm}) = L(\text{alarm} \mid \text{burglary})O(\text{burglary})$$

$$= \frac{0.95}{0.01} \frac{10^{-4}}{1 - 10^{-4}} = 0.0095.$$

So, from

$$P(A) = \frac{O(A)}{1 + O(A)} \tag{1.20}$$

we have

$$P(\text{burglary} \mid \text{alarm}) = \frac{0.0095}{1 + 0.0095} = 0.00941.$$

Thus, the retrospective support imparted to the burglary hypothesis by the alarm evidence has increased its degree of belief almost a hundredfold, from one in ten thousand to 94.1 in ten thousand. The fact that the belief in burglary is still below 1% should not be surprising, given that the system produces a false alarm almost once every three months. Notice that it was not necessary to estimate the absolute values of the probabilities $P(\text{alarm} \mid \text{burglary})$ and $P(\text{alarm} \mid \text{no burglary})$. Only their ratio enters the calculation, so a direct estimate of this ratio could have been used instead.

1.1.4   Random Variables and Expectations

By a *variable* we will mean an attribute, measurement or inquiry that may take on one of several possible outcomes, or *values*, from a specified domain. If we have beliefs (i.e., probabilities) attached to the possible values that a variable may attain, we will call that variable a *random variable*.[3] For example, the color of the shoes that I will wear tomorrow is a random variable named "color," and the values it may take come from the domain {yellow, green, red,...}.

Most of our analysis will concern a finite set V of random variables (also called *partitions*) where each variable $X \in V$ may take on values from a finite domain D_X. We will use capital letters (e.g., X, Y, Z) for variable names and lowercase letters (x, y, z)

---

[3] This is a minor generalization of the textbook definition, according to which a random variable is a mapping from the sample space (e.g., the set of elementary events) to the real line. In our definition, the mapping is from the sample space to any set of objects called "values," which may or may not be ordered.

as generic symbols for specific values taken by the corresponding variables. For example, if X stands for the color of an object, then x will designate any possible choice of an element from the set {yellow, green, red,...}. Clearly, the proposition $X = $ yellow describes an *event*, namely, a subset of possible states of affair that satisfy the proposition "the color of the object is yellow." Likewise, each variable X can be viewed as a partition of the states of the world, since the statement $X = x$ defines a set of exhaustive and mutually exclusive sets of states, one for each value of x.

In most of our discussions, we will not make notational distinction between variables and sets of variables, because a set of variables essentially defines a compound variable whose domain is the Cartesian product of the domains of the individual constituents in the set. Thus, if Z stands for the set {X, Y}, then z stands for pairs (x, y) such that $x \in D_X$ and $y \in D_Y$. When the distinction between variables and sets of variables requires special emphasis, indexed letters (say, $X_1, X_2,..., X_n$ or $V_1, V_2,..., V_n$) will be used to represent individual variables.

We shall consistently use the abbreviation $P(x)$ for the probabilities $P(X = x)$, $x \in D_x$. Likewise, if Z stands for the set {X, Y}, then $P(z)$ will be defined as

$$P(z) \triangleq P(Z = z) = P(X = x, Y = y), \quad x \in D_X, \quad y \in D_Y.$$

When the values of a random variable X are real numbers, X is called a *real* random variable; one can then define the *mean* or *expected value* of X as

$$E(X) \triangleq \sum_x xP(x) \tag{1.21}$$

and the *conditional mean* of X, given event $Y = y$, as

$$E(X \mid y) \triangleq \sum_x xP(x \mid y). \tag{1.22}$$

The expectation of any function g of X is defined as

$$E[g(X)] \triangleq \sum_x g(x)P(x). \tag{1.23}$$

In particular, the function $g(X) = (X - E(X))^2$ has received much attention; its expectation is called the *variance* of X, denoted σ_X^2;

$$\sigma_X^2 \triangleq E[(X - E(X))^2].$$

The conditional mean $E(X \mid Y = y)$ is the *best estimate* of X, given the observation $Y = y$, in the sense of minimizing the expected square error $\sum_x (x - x')^2 P(x \mid y)$ over all possible x'.

The expectation of a function $g(X, Y)$ of two variables, X and Y, requires the joint probability $P(x, y)$ and is defined as

$$E[g(X, Y)] \triangleq \sum_{x, y} g(x, y)P(x, y)$$

(cf. equation (1.23)). Of special importance is the expectation of the product $(g(X, Y) = (X - E(X))(Y - E(Y)))$, which is known as the *covariance* of X and Y,

$$\sigma_{XY} \triangleq E\left[(X - E(X))(Y - E(Y))\right],$$

and which is often normalized to yield the *correlation coefficient*

$$\rho_{XY} = \frac{\sigma_{XY}}{\sigma_X \sigma_Y}$$

and the *regression coefficient*

$$r_{XY} \triangleq \rho_{XY}\frac{\sigma_X}{\sigma_Y} = \frac{\sigma_{XY}}{\sigma_Y^2}.$$

The *conditional* variance, covariance, and correlation coefficient, given $Z = z$, are defined in a similar manner, using the conditional distribution $P(x, y \mid z)$ in taking expectations. In particular, the *conditional correlation coefficient*, given $Z = z$, is defined as

$$\rho_{XY|z} = \frac{\sigma_{XY|z}}{\sigma_{X|z}\,\sigma_{Y|z}}. \tag{1.24}$$

Additional properties, specific to normal distributions, will be reviewed in Chapter 5 (Section 5.2.1).

The foregoing definitions apply to discrete random variables – that is, variables that take on finite or denumerable sets of values on the real line. The treatment of expectation and correlation is more often applied to continuous random variables, which are characterized by a *density function* $f(x)$ defined as follows:

$$P(a \leq X \leq b) = \int_a^b f(x)\,dx$$

for any two real numbers a and b with $a < b$. If X is discrete, then $f(x)$ coincides with the probability function $P(x)$, once we interpret the integral through the translation

$$\int_{-\infty}^{\infty} f(x)dx \iff \sum_x P(x). \tag{1.25}$$

Readers accustomed to continuous analysis should bear this translation in mind whenever summation is used in this book. For example, the expected value of a continuous random variable X can be obtained from (1.21), to read

$$E(X) = \int_{-\infty}^{\infty} xf(x)\,dx,$$

with analogous translations for the variance, correlation, and so forth.

We now turn to define *conditional independence* relationships among variables, a central notion in causal modelling.

1.1.5   Conditional Independence and Graphoids

Definition 1.1.2 (Conditional Independence)

Let $V = \{V_1, V_2,...\}$ be a finite set of variables. Let $P(\cdot)$ be a joint probability function over the variables in V, and let X, Y, Z stand for any three subsets of variables in V. The sets X and Y are said to be conditionally independent given Z *if*

$$P(x \mid y, z) = P(x \mid z) \quad \text{whenever} \quad P(y, z) > 0. \tag{1.26}$$

In words, learning the value of Y does not provide additional information about X, once we know Z. (Metaphorically, Z "screens off" X from Y.)

Equation (1.26) is a terse way of saying the following: For any configuration x of the variables in the set X and for any configurations y and z of the variables in Y and Z satisfying $P(Y = y, Z = z) > 0$, we have

$$P(X = x \mid Y = y, Z = z) = P(X = x \mid Z = z). \tag{1.27}$$

We will use Dawid's (1979) notation $(X \perp\!\!\!\perp Y \mid Z)_P$ or simply $(X \perp\!\!\!\perp Y \mid Z)$ to denote the conditional independence of X and Y given Z; thus,

$$(X \perp\!\!\!\perp Y \mid Z)_P \quad \text{iff} \quad P(x \mid y, z) = P(x \mid z) \tag{1.28}$$

for all values x, y, z such that $P(y, z) > 0$. Unconditional independence (also called *marginal independence*) will be denoted by $(X \perp\!\!\!\perp Y \mid \emptyset)$; that is,

$$(X \perp\!\!\!\perp Y \mid \emptyset) \text{ iff } P(x \mid y) = P(x) \quad \text{whenever} \quad P(y) > 0 \tag{1.29}$$

("iff" is shorthand for "if and only if"). Note that $(X \perp\!\!\!\perp Y \mid Z)$ implies the conditional independence of all pairs of variables $V_i \in X$ and $V_j \in Y$, but the converse is not necessarily true.

The following is a (partial) list of properties satisfied by the conditional independence relation $(x \perp\!\!\!\perp y \mid Z)$.

Symmetry: $(X \perp\!\!\!\perp Y \mid Z) \Longrightarrow (Y \perp\!\!\!\perp X \mid Z).$

Decomposition: $(X \perp\!\!\!\perp YW \mid Z) \Longrightarrow (X \perp\!\!\!\perp Y \mid Z).$

Weak union: $(X \perp\!\!\!\perp YW \mid Z) \Longrightarrow (X \perp\!\!\!\perp Y \mid ZW).$

Contraction: $(X \perp\!\!\!\perp Y \mid Z) \ \& \ (X \perp\!\!\!\perp W \mid ZY) \Longrightarrow (X \perp\!\!\!\perp YW \mid Z).$

Intersection: $(X \perp\!\!\!\perp W \mid ZY) \ \& \ (X \perp\!\!\!\perp Y \mid ZW) \Longrightarrow (X \perp\!\!\!\perp YW \mid Z).$

(Intersection is valid in strictly positive probability distributions.)

The proof of these properties can be derived by elementary means from (1.28) and the basic axioms of probability theory.[4] These properties were called *graphoid axioms* by

---

[4] These properties were first introduced by Dawid (1979) and Spohn (1980) in a slightly different form, and were independently proposed by Pearl and Paz (1987) to characterize the relationships between graphs and informational relevance. Geiger and Pearl (1993) present an in-depth analysis.

Pearl and Paz (1987) and Geiger et al. (1990) and have been shown to govern the concept of informational relevance in a wide variety of interpretations (Pearl 1988b). In graphs, for example, these properties are satisfied if we interpret $(X \perp\!\!\!\perp Y \mid Z)$ to mean "all paths from a subset X of nodes to a subset Y of nodes are intercepted by a subset Z of nodes."

The intuitive interpretation of the graphoid axioms is as follows (Pearl 1988b, p. 85). The *symmetry* axiom states that, in any state of knowledge Z, if Y tells us nothing new about X, then X tells us nothing new about Y. The *decomposition* axiom asserts that if two combined items of information are judged irrelevant to X, then each separate item is irrelevant as well. The *weak union* axiom states that learning irrelevant information W cannot help the irrelevant information Y become relevant to X. The *contraction* axiom states that if we judge W irrelevant to X after learning some irrelevant information Y, then W must have been irrelevant before we learned Y. Together, the weak union and contraction properties mean that irrelevant information should not alter the relevance status of other propositions in the system; what was relevant remains relevant, and what was irrelevant remains irrelevant. The *intersection* axiom states that if Y is irrelevant to X when we know W and if W is irrelevant to X when we know Y, then neither W nor Y (nor their combination) is relevant to X.

1.2   GRAPHS AND PROBABILITIES

1.2.1   Graphical Notation and Terminology

A graph consists of a set V of *vertices* (or *nodes*) and a set E of *edges* (or *links*) that connect some pairs of vertices. The vertices in our graphs will correspond to variables (whence the common symbol V), and the edges will denote a certain relationship that holds in pairs of variables, the interpretation of which will vary with the application. Two variables connected by an edge are called *adjacent*.

Each edge in a graph can be either directed (marked by a single arrowhead on the edge), or undirected (unmarked links). In some applications we will also use "bidirected" edges to denote the existence of unobserved common causes (sometimes called *confounders*). These edges will be marked as dotted curved arcs with two arrowheads (see Figure 1.1(a)). If all edges are directed (see Figure 1.1(b)), we then have a *directed graph*. If we strip away all arrowheads from the edges in a graph G, the resultant undirected graph is called the *skeleton* of G. A *path* in a graph is a sequence of edges (e.g., $((W, Z), (Z, Y), (Y, X), (X, Z))$ in Figure 1.1(a)) such that each edge starts with the vertex ending the preceding edge. In other words, a path is any unbroken, nonintersecting route traced out along the edges in a graph, which may go either along or against the arrows. If every edge in a path is an arrow that points from the first to the second vertex of the pair, we have a *directed path*. In Figure 1.1(a), for example, the path $((W, Z), (Z, Y))$ is directed, but the paths $((W, Z), (Z, Y), (Y, X))$ and $((W, Z), (Z, X))$ are not. If there exists a path between two vertices in a graph, then the two vertices are said to be *connected*; else they are *disconnected*.

Directed graphs may include directed cycles (e.g., $X \rightarrow Y, Y \rightarrow X$), representing mutual causation or feedback processes, but not self-loops (e.g., $X \rightarrow X$). A graph (like the two in Figure 1.1) that contains no directed cycles is called *acyclic*. A graph that is

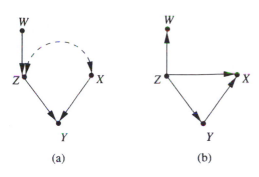

Figure 1.1   (a) A graph containing both directed and bidirected edges. (b) A directed acyclic graph (DAG) with the same skeleton as (a).

both directed and acyclic (Figure 1.1(b)) is called a *directed acyclic graph* (DAG), and such graphs will occupy much of our discussion of causality. We make free use of the terminology of kinship (e.g., *parents, children, descendants, ancestors, spouses*) to denote various relationships in a graph. These kinship relations are defined along the full arrows in the graph, including arrows that form directed cycles but ignoring bidirected and undirected edges. In Figure 1.1(a), for example, Y has two parents (X and Z), three ancestors (X, Z, and W), and no children, while X has no parents (hence, no ancestors), one spouse (Z), and one child (Y). A *family* in a graph is a set of nodes containing a node and all its parents. For example, $\{W\}$, $\{Z, W\}$, $\{X\}$, and $\{Y, Z, X\}$ are the families in the graph of Figure 1.1(a).

A node in a directed graph is called a *root* if it has no parents and a *sink* if it has no children. Every DAG has at least one root and at least one sink. A connected DAG in which every node has at most one parent is called a *tree*, and a tree in which every node has at most one child is called a *chain*. A graph in which every pair of nodes is connected by an edge is called *complete*. The graph in Figure 1.1(a), for instance, is connected but not complete, because the pairs (W, X) and (W, Y) are not adjacent.

1.2.2   Bayesian Networks

The role of graphs in probabilistic and statistical modeling is threefold:

1. to provide convenient means of expressing substantive assumptions;

2. to facilitate economical representation of joint probability functions; and

3. to facilitate efficient inferences from observations.

We will begin our discussion with item 2.

Consider the task of specifying an arbitrary joint distribution, $P(x_1,\ldots, x_n)$, for n dichotomous variables. To store $P(x_1,\ldots, x_n)$ explicitly would require a table with 2^n entries, an unthinkably large number by any standard. Substantial economy can be achieved when each variable depends on just a small subset of other variables. Such dependence information permits us to decompose large distribution functions into several small distributions – each involving a small subset of variables – and then to piece them together coherently to answer questions of a global nature. Graphs play an essential role in such decomposition, for they provide a vivid representation of the sets of variables that are relevant to each other in any given state of knowledge.

Both directed and undirected graphs have been used by researchers to facilitate such decomposition. Undirected graphs, sometimes called *Markov networks* (Pearl 1988b), are used primarily to represent symmetrical spatial relationships (Isham 1981; Cox and Wermuth 1996; Lauritzen 1996). Directed graphs, especially DAGs, have been used to represent causal or temporal relationships (Lauritzen 1982; Wermuth and Lauritzen 1983; Kiiveri et al. 1984) and came to be known as *Bayesian networks*, a term coined in Pearl (1985) to emphasize three aspects: (1) the subjective nature of the input information; (2) the reliance on Bayes's conditioning as the basis for updating information; and (3) the distinction between causal and evidential modes of reasoning, a distinction that underscores Thomas Bayes's paper of 1763. Hybrid graphs (involving both directed and undirected edges) have also been proposed for statistical modeling (Wermuth and Lauritzen 1990), but in this book our main interest will focus on directed acyclic graphs, with occasional use of directed cyclic graphs to represent feedback cycles.

The basic decomposition scheme offered by directed acyclic graphs can be illustrated as follows. Suppose we have a distribution P defined on n discrete variables, which we may order arbitrarily as $X_1, X_2,..., X_n$. The chain rule of probability calculus (equation (1.12)) always permits us to decompose P as a product of n conditional distributions:

$$P(x_1,\ldots,x_n) = \prod_j P(x_j \mid x_1,\ldots,x_{j-1}). \tag{1.30}$$

Now suppose that the conditional probability of some variable X_j is not sensitive to all the predecessors of X_j but only to a small subset of those predecessors. In other words, suppose that X_j is independent of all other predecessors, once we know the value of a select group of predecessors called PA_j. We can then write

$$P(x_j \mid x_1,\ldots,x_{j-1}) = P(x_j \mid pa_j) \tag{1.31}$$

in the product of (1.30), which will considerably simplify the input information required. Instead of specifying the probability of X_j conditional on all possible realizations of its predecessors $X_1,..., X_{j-1}$, we need only concern ourselves with the possible realizations of the set PA_j. The set PA_j is called the *Markovian parents* of X_j, or *parents* for short. The reason for the name becomes clear when we build graphs around this concept.

Definition 1.2.1 (Markovian Parents)
Let $V = \{X_1,..., X_n\}$ be an ordered set of variables, and let $P(v)$ be the joint probability distribution on these variables. A set of variables PA_j is said to be Markovian parents *of X_j if PA_j is a minimal set of predecessors of X_j that renders X_j independent of all its other predecessors. In other words, PA_j is any subset of $\{X_1,..., X_{j-1}\}$ satisfying*

$$P(x_j \mid pa_j) = P(x_j \mid x_1,\ldots,x_{j-1}) \tag{1.32}$$

and such that no proper subset of PA_j satisfies (1.32).[5]

---

[5] Lowercase symbols (e.g., x_j, pa_j) denote particular realizations of the corresponding variables (e.g., X_j, PA_j).

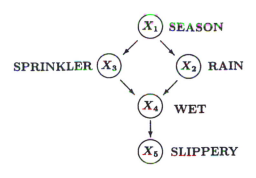

Figure 1.2 A Bayesian network representing dependencies among five variables.

Definition 1.2.1 assigns to each variable X_j a select set PA_j of preceding variables that are sufficient for determining the probability of X_j; knowing the values of other preceding variables is redundant once we know the values pa_j of the parent set PA_j. This assignment can be represented in the form of a DAG in which variables are represented by nodes and arrows are drawn from each node of the parent set PA_j toward the child node X_j. Definition 1.2.1 also suggests a simple recursive method for constructing such a DAG: Starting with the pair (X_1, X_2), we draw an arrow from X_1 to X_2 if and only if the two variables are dependent. Continuing to X_3, we draw no arrow in case X_3 is independent of $\{X_1, X_2\}$; otherwise, we examine whether X_2 screens off X_3 from X_1 or X_1 screens off X_3 from X_2. In the first case, we draw an arrow from X_2 to X_3; in the second, we draw an arrow from X_1 to X_3. If no screening condition is found, we draw arrows to X_3 from both X_1 and X_2. In general: at the jth stage of the construction, we select any minimal set of X_j's predecessors that screens off X_j from its other predecessors (as in equation (1.32)), call this set PA_j and draw an arrow from each member in PA_j to X_j. The result is a directed acyclic graph, called a Bayesian network, in which an arrow from X_i to X_j assigns X_i, as a Markovian parent of X_j, consistent with Definition 1.2.1.

It can be shown (Pearl 1988b) that the set PA_j is unique whenever the distribution $P(v)$ is strictly positive (i.e., involving no logical or definitional constraints), so that every configuration v of variables, no matter how unlikely, has some finite probability of occurring. Under such conditions, the Bayesian network associated with $P(v)$ is unique, given the ordering of the variables.

Figure 1.2 illustrates a simple yet typical Bayesian network. It describes relationships among the season of the year (X_1), whether rain falls (X_2), whether the sprinkler is on (X_3), whether the pavement would get wet (X_4), and whether the pavement would be slippery (X_5). All variables in this figure are binary (taking a value of either true or false) except for the root variable X_1, which can take one of four values: spring, summer, fall, or winter. The network was constructed in accordance with Definition 1.2.1, using causal intuition as a guide. The absence of a direct link between X_1 and X_5, for example, captures our understanding that the influence of seasonal variations on the slipperiness of the pavement is mediated by other conditions (e.g., the wetness of the pavement). This intuition coincides with the independence condition of (1.32), since knowing X_4 renders X_5 independent of $\{X_1, X_2, X_3\}$.

The construction implied by Definition 1.2.1 defines a Bayesian network as a carrier of conditional independence relationships along the order of construction. Clearly, every distribution satisfying (1.32) must decompose (using the chain rule of (1.30)) into the product

$$P(x_1, \ldots, x_n) = \prod_i P(x_i \mid pa_i). \tag{1.33}$$

For example, the DAG in Figure 1.2 induces the decomposition

$$P(x_1, x_2, x_3, x_4, x_5) = P(x_1)P(x_2 \mid x_1)P(x_3 \mid x_1)P(x_4 \mid x_2, x_3)P(x_5 \mid x_4). \tag{1.34}$$

The product decomposition in (1.33) is no longer order-specific since, given P and G, we can test whether P decomposes into the product given by (1.33) without making any reference to variable ordering. We therefore conclude that a necessary condition for a DAG G to be a Bayesian network of probability distribution P is for P to admit the product decomposition dictated by G, as given in (1.33).

Definition 1.2.2 (Markov Compatibility)
If a probability function P admits the factorization of (1.33) relative to DAG G, we say that G represents P, that G and P are compatible, *or that P is* Markov relative *to G.*[6]

Ascertaining compatibility between DAGs and probabilities is important in statistical modeling primarily because compatibility is a necessary and sufficient condition for a DAG G *to explain* a body of empirical data represented by P, that is, to describe a stochastic process capable of *generating P* (e.g., Pearl 1988b, pp. 210–23). If the value of each variable X_i is chosen at random with some probability $P_i(x_i \mid pa_i)$, based solely on the values pa_i previously chosen for PA_i, then the overall distribution P of the generated instances $x_1, x_2, \ldots, x_n$ will be Markov relative to G. Conversely, if P is Markov relative to G, then there exists a set of probabilities $P_i(x_i \mid pa_i)$ according to which we can choose the value of each variable X_i such that the distribution of the generated instances $x_1, x_2, \ldots, x_n$ will be equal to P. (In fact, the correct choice of $P_i(x_i \mid pa_i)$ would be simply $P(x_i \mid pa_i)$.)

A convenient way of characterizing the set of distributions compatible with a DAG G is to list the set of (conditional) independencies that each such distribution must satisfy. These independencies can be read off the DAG by using a graphical criterion called *d-separation* (Pearl 1988b; the d denotes *directional*), which will play a major role in many discussions in this book.

1.2.3 The *d*-Separation Criterion

Consider three disjoint sets of variables, X, Y, and Z, which are represented as nodes in a directed acyclic graph G. To test whether X is independent of Y given Z in any distribution compatible with G, we need to test whether the nodes corresponding to variables Z "block" all paths from nodes in X to nodes in Y. By *path* we mean a sequence of consecutive edges (of any directionality) in the graph, and blocking is to be interpreted as stopping the flow of information (or of dependency) between the variables that are connected by such paths, as defined next.

Definition 1.2.3 (*d*-Separation)
A path p is said to be d-separated (or blocked) by a set of nodes Z if and only if

---

[6] The latter expression seems to gain strength in recent literature (e.g., Spirtes et al. 1993; Lauritzen 1996). Pearl (1988b, p. 116) used "G is an *I-map* of *P*."

1. *p contains a chain $i \rightarrow m \rightarrow j$ or a fork $i \leftarrow m \rightarrow j$ such that the middle node m is in Z, or*

2. *p contains an inverted fork (or* collider*) $i \rightarrow m \leftarrow j$ such that the middle node m is not in Z and such that no descendant of m is in Z.*

A set Z is said to d-separate X from Y if and only if Z blocks every path from a node in X to a node in Y.

The intuition behind *d*-separation is simple and can best be recognized if we attribute causal meaning to the arrows in the graph. In causal chains $i \rightarrow m \rightarrow j$ and causal forks $i \leftarrow m \rightarrow j$, the two extreme variables are marginally dependent but become independent of each other (i.e., blocked) once we condition on (i.e., know the value of) the middle variable. Figuratively, conditioning on *m* appears to "block" the flow of information along the path, since learning about *i* has no effect on the probability of *j*, given *m*. Inverted forks $i \rightarrow m \leftarrow j$, representing two causes having a common effect, act the opposite way; if the two extreme variables are (marginally) independent, they will become dependent (i.e., connected through unblocked path) once we condition on the middle variable (i.e., the common effect) or any of its descendants. This can be confirmed in the context of Figure 1.2. Once we know the season, X_3 and X_2 are independent (assuming that sprinklers are set in advance, according to the season); whereas finding that the pavement is wet or slippery renders X_2 and X_3 dependent, because refuting one of these explanations increases the probability of the other.

In Figure 1.2, $X = \{X_2\}$ and $Y = \{X_3\}$ are *d*-separated by $Z = \{X_1\}$, because both paths connecting X_2 and X_3 are blocked by *Z*. The path $X_2 \leftarrow X_1 \rightarrow X_3$ is blocked because it is a fork in which the middle node X_1 is in *Z*, while the path $X_2 \rightarrow X_4 \leftarrow X_3$ is blocked because it is an inverted fork in which the middle node X_4 and all its descendants are outside *Z*. However, *X* and *Y* are not *d*-separated by the set $Z' = \{X_1, X_5\}$: the path $X_2 \rightarrow X_4 \leftarrow X_3$ (an inverted fork) is not blocked by *Z'*, since X_5, a descendant of the middle node X_4, is in *Z'*. Metaphorically, learning the value of the consequence X_5 renders its causes X_2 and X_3 dependent, as if a pathway were opened along the arrows converging at X_4.

At first glance, readers might find it a bit odd that conditioning on a node not lying on a blocked path may unblock the path. However, this corresponds to a general pattern of causal relationships: observations on a common consequence of two independent causes tend to render those causes dependent, because information about one of the causes tends to make the other more or less likely, given that the consequence has occurred. This pattern is known as *selection bias* or *Berkson's paradox* in the statistical literature (Berkson 1946) and as the *explaining away effect* in artificial intelligence (Kim and Pearl 1983). For example, if the admission criteria to a certain graduate school call for either high grades as an undergraduate or special musical talents, then these two attributes will be found to be correlated (negatively) in the student population of that school, even if these attributes are uncorrelated in the population at large. Indeed, students with low grades are likely to be exceptionally gifted in music, which explains their admission to the graduate school.

Figure 1.3 illustrates more elaborate examples of *d*-separation: example (a) contains a bidirected arc $Z_1 \blacktriangleleft - - \blacktriangleright Z_3$, and (b) involves a directed cycle $X \rightarrow Z_2 \rightarrow Z_1 \rightarrow X$. In

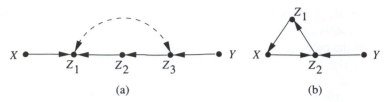

Figure 1.3 Graphs illustrating d-separation. In (a), X and Y are d-separated given Z_2 and d-connected given Z_1. In (b), X and Y cannot be d-separated by any set of nodes.

Figure 1.3(a), the two paths between X and Y are blocked when none of $\{Z_1, Z_2, Z_3\}$ is measured. However, the path $X \rightarrow Z_1 \blacktriangleleft - - \blacktriangleright Z_3 \leftarrow Y$ becomes unblocked when Z_1 is measured. This is so because Z_1 unblocks the "colliders" at both Z_1 and Z_3; the first because Z_1 is the collision node of the collider, the second because Z_1 is a descendant of the collision node Z_3 through the path $Z_1 \leftarrow Z_2 \leftarrow Z_3$. In Figure 1.3(b), X and Y cannot be d-separated by any set of nodes, including the empty set. If we condition on Z_2, we block the path $X \leftarrow Z_1 \leftarrow Z_2 \leftarrow Y$ yet unblock the path $X \rightarrow Z_2 \leftarrow Y$. If we condition on Z_1, we again block the path $X \leftarrow Z_1 \leftarrow Z_2 \leftarrow Y$ and unblock the path $X \rightarrow Z_2 \leftarrow Y$, because Z_1 is a descendant of the collision node Z_2.

The connection between d-separation and conditional independence is established through the following theorem due to Verma and Pearl (1988; see also Geiger et al. 1990).

Theorem 1.2.4 (Probabilistic Implications of d-Separation)
If sets X and Y are d-separated by Z in a DAG G, then X is independent of Y conditional on Z in every distribution compatible with G. Conversely, if X and Y are not d-separated by Z in a DAG G, then X and Y are dependent conditional on Z in at least one distribution compatible with G.

The converse part of Theorem 1.2.4 is in fact much stronger – the absence of d-separation implies dependence in *almost all* distributions compatible with G. The reason is that a precise tuning of parameters is required to generate independency along an unblocked path in the diagram, and such tuning is unlikely to occur in practice (see Spirtes et al. 1993 and Sections 2.4 and 2.9.1).

In order to distinguish between the probabilistic notion of conditional independence $(X \perp\!\!\!\perp Y \mid Z)_P$ and the graphical notion of d-separation, for the latter we will use the notation $(X \perp\!\!\!\perp Y \mid Z)_G$. We can thereby express Theorem 1.2.4 more succinctly as follows.

Theorem 1.2.5
For any three disjoint subsets of nodes (X, Y, Z) in a DAG G and for all probability functions P, we have:

(i) *$(X \perp\!\!\!\perp Y \mid Z)_G \Longrightarrow (X \perp\!\!\!\perp Y \mid Z)_P$ whenever G and P are compatible; and*

(ii) *if $(X \perp\!\!\!\perp Y \mid Z)_P$ holds in all distributions compatible with G, it follows that $(X \perp\!\!\!\perp Y \mid Z)_G$.*

An alternative test for d-separation has been devised by Lauritzen et al. (1990), based on the notion of ancestral graphs. To test for $(X \perp\!\!\!\perp Y \mid Z)_G$, delete from G all nodes except those in $\{X, Y, Z\}$ and their ancestors, connect by an edge every pair of nodes that share a common child, and remove all arrows from the arcs. Then $(X \perp\!\!\!\perp Y \mid Z)_G$ holds if and only if Z intercepts all paths between X and Y in the resulting undirected graph.

Note that the ordering with which the graph was constructed does not enter into the d-separation criterion; it is only the topology of the resulting graph that determines the set of independencies that the probability P must satisfy. Indeed, the following theorem can be proven (Pearl 1988b, p. 120).

Theorem 1.2.6 (Ordered Markov Condition)
A necessary and sufficient condition for a probability distribution P to be Markov relative a DAG G is that, conditional on its parents in G, each variable be independent of all its predecessors in some ordering of the variables that agrees with the arrows of G.

A consequence of this theorem is an order-independent criterion for determining whether a given probability P is Markov relative to a given DAG G.

Theorem 1.2.7 (Parental Markov Condition)
A necessary and sufficient condition for a probability distribution P to be Markov relative a DAG G is that every variable be independent of all its nondescendants (in G), conditional on its parents. (We exclude X_i when speaking of its "nondescendants.")

This condition, which Kiiveri et al. (1984) and Lauritzen (1996) called the "local" Markov condition, is sometimes taken as the definition of Bayesian networks (Howard and Matheson 1981). In practice, however, the ordered Markov condition is easier to use.

Another important property that follows from d-separation is a criterion for determining whether two given DAGs are observationally equivalent – that is, whether every probability distribution that is compatible with one of the DAGs is also compatible with the other.

Theorem 1.2.8 (Observational Equivalence)
Two DAGs are observationally equivalent if and only if they have the same skeletons and the same sets of v-structures, that is, two converging arrows whose tails are not connected by an arrow (Verma and Pearl 1990).[7]

Observational equivalence places a limit on our ability to infer directionality from probabilities alone. Two networks that are observationally equivalent cannot be distinguished without resorting to manipulative experimentation or temporal information. For example, reversing the direction of the arrow between X_1 and X_2 in Figure 1.2 would neither introduce nor destroy a v-structure. Therefore, this reversal yields an observationally equivalent network, and the directionality of the link $X_1 \rightarrow X_2$ cannot be determined from probabilistic information. The arrows $X_2 \rightarrow X_4$ and $X_4 \rightarrow X_5$, however, are of different nature; there is no way of reversing their directionality without creating a new v-structure. Thus, we see that some probability functions P (such as the one responsible for the construction of the Bayesian network in Figure 1.2), when unaccompanied

---

[7] An identical criterion was independently derived by Frydenberg (1990) in the context of chain graphs, where strict positivity is assumed.

by temporal information, can constrain the directionality of some arrows in the graph. The precise meaning of such directionality constraints – and the possibility of using these constraints for inferring causal relationships from data – will be formalized in Chapter 2.

1.2.4 Inference with Bayesian Networks

Bayesian networks were developed in the early 1980s to facilitate the tasks of prediction and "abduction" in artificial intelligence (AI) systems. In these tasks, it is necessary to find a coherent interpretation of incoming observations that is consistent with both the observations and the prior information at hand. Mathematically, the task boils down to the computation of $P(y \mid x)$, where X is a set of observations and Y is a set of variables that are deemed important for prediction or diagnosis.

Given a joint distribution P, the computation of $P(y \mid x)$ is conceptually trivial and invokes straightforward application of Bayes's rule to yield

$$P(y \mid x) = \frac{\sum_s P(y, x, s)}{\sum_{y,s} P(y, x, s)}, \tag{1.35}$$

where S stands for the set of all variables *excluding* X and Y. Because every Bayesian network defines a joint probability P (given by the product in (1.33)), it is clear that $P(y \mid x)$ can be computed from a DAG G and the conditional probabilities $P(x_i \mid pa_i)$ defined on the families of G.

The challenge, however, lies in performing these computations efficiently and within the representation level provided by the network topology. The latter is important in systems that generate explanations for their reasoning processes. Although such inference techniques are not essential to our discussion of causality, we will nevertheless survey them briefly, for they demonstrate (i) the effectiveness of organizing probabilistic knowledge in the form of graphs and (ii) the feasibility of performing coherent probabilistic calculations (and approximations thereof) on such organization. Details can be found in the references cited.

The first algorithms proposed for probabilistic calculations in Bayesian networks used message-passing architecture and were limited to trees (Pearl 1982; Kim and Pearl 1983). With this technique, each variable is assigned a simple processor and permitted to pass messages asynchronously to its neighbors until equilibrium is achieved (in a finite number of steps). Methods have since been developed that extend this tree propagation (and some of its synchronous variants) to general networks. Among the most popular are Lauritzen and Spiegelhalter's (1988) method of join-tree propagation and the method of cut-set conditioning (Pearl 1988b, pp. 204–10; Jensen 1996). In the join-tree method, we decompose the network into clusters (e.g., cliques) that form tree structures and then treat the set variables in each cluster as a compound variable that is capable of passing messages to its neighbors (which are also compound variables). For example, the network of Figure 1.2 can be structured as a Markov-compatible chain of three clusters:

$$\{X_1, X_2, X_3\} \rightarrow \{X_2, X_3, X_4\} \rightarrow \{X_4, X_5\}.$$

In the cut-set conditioning method, a set of variables is instantiated (given specific values) such that the remaining network forms a tree. The propagation is then performed on that tree, and a new instantiation chosen, until all instantiations have been exhausted; the results are then averaged. In Figure 1.2, for example, if we instantiate X_1 to any specific value (say, X_1 = summer), then we break the pathway between X_2 and X_3 and the remaining network becomes tree-structured. The main advantage of the cut-set conditioning method is that its storage-space requirement is minimal (linear in the size of the network), whereas that of the join-tree method might be exponential. Hybrid combinations of these two basic algorithms have also been proposed (Shachter et al. 1994; Dechter 1996) to allow flexible trade-off of storage versus time (Darwiche 2009).

Whereas inference in general networks is "NP-hard" (Cooper 1990), the computational complexity for each of the methods cited here can be estimated prior to actual processing. When the estimates exceed reasonable bounds, an approximation method such as stochastic simulation (Pearl 1988b, pp. 210–23) can be used instead. This method exploits the topology of the network to perform Gibbs sampling on local subsets of variables, sequentially as well as concurrently.

Additional properties of DAGs and their applications to evidential reasoning in expert systems are discussed in Pearl (1988b), Lauritzen and Spiegelhalter (1988), Pearl (1993a), Spiegelhalter et al. (1993), Heckerman et al. (1995), and Darwiche (2009).

1.3   CAUSAL BAYESIAN NETWORKS

The interpretation of direct acyclic graphs as carriers of independence assumptions does not necessarily imply causation; in fact, it will be valid for any set of recursive independencies along any ordering of the variables, not necessarily causal or chronological. However, the ubiquity of DAG models in statistical and AI applications stems (often unwittingly) primarily from their causal interpretation – that is, as a system of processes, one per family, that could account for the generation of the observed data. It is this causal interpretation that explains why DAG models are rarely used in any variable ordering other than those which respect the direction of time and causation.

The advantages of building DAG models around causal rather than associational information are several. First, the judgments required in the construction of the model are more meaningful, more accessible and hence more reliable. The reader may appreciate this point by attempting to construct a DAG representation for the associations in Figure 1.2 along the ordering $(X_5, X_1, X_3, X_2, X_4)$. Such exercises illustrate not only that some independencies are more vividly accessible to the mind than others but also that conditional independence judgments are accessible (hence reliable) only when they are anchored onto more fundamental building blocks of our knowledge, such as causal relationships. In the example of Figure 1.2, our willingness to assert that X_5 is independent of X_2 and X_3 once we know X_4 (i.e., whether the pavement is wet) is defensible because we can easily translate the assertion into one involving causal relationships: that the *influence* of rain and sprinkler on slipperiness is *mediated* by the wetness of the pavement. Dependencies that are not supported by causal links are considered odd or spurious and are even branded "paradoxical" (see the discussion of Berkson's paradox, Section 1.2.3).

We will have several opportunities throughout this book to demonstrate the primacy of causal over associational knowledge. In extreme cases, we will see that people tend to ignore probabilistic information altogether and attend to causal information instead (see Section 6.1.4).[8] This puts into question the ruling paradigm of graphical models in statistics (Wermuth and Lauritzen 1990; Cox and Wermuth 1996), according to which conditional independence assumptions are the primary vehicle for expressing substantive knowledge.[9] It seems that if conditional independence judgments are by-products of stored causal relationships, then tapping and representing those relationships directly would be a more natural and more reliable way of expressing what we know or believe about the world. This is indeed the philosophy behind causal Bayesian networks.

The second advantage of building Bayesian networks on causal relationships – one that is basic to the understanding of causal organizations – is the ability to represent and respond to external or spontaneous *changes*. Any local reconfiguration of the mechanisms in the environment can be translated, with only minor modification, into an isomorphic reconfiguration of the network topology. For example, to represent a disabled sprinkler in the story of Figure 1.2, we simply delete from the network all links incident to the node Sprinkler. To represent the policy of turning the sprinkler off if it rains, we simply add a link between Rain and Sprinkler and revise $P(x_3 \mid x_1, x_2)$. Such changes would require much greater remodeling efforts if the network were not constructed along the causal direction but instead along (say) the order $(X_5, X_1, X_3, X_2, X_4)$. This remodeling flexibility may well be cited as the ingredient that marks the division between deliberative and reactive agents and that enables the former to manage novel situations instantaneously, without requiring training or adaptation.

1.3.1 Causal Networks as Oracles for Interventions

The source of this flexibility rests on the assumption that each parent–child relationship in the network represents a stable and autonomous physical mechanism – in other words, that it is conceivable to change one such relationship *without* changing the others. Organizing one's knowledge in such *modular* configurations permits one to predict the effect of external interventions with a minimum of extra information. Indeed, causal models (assuming they are valid) are much more informative than probability models. A joint distribution tells us how probable events are and how probabilities would change with subsequent observations, but a causal model also tells us how these probabilities would change as a result of external interventions – such as those encountered in policy analysis, treatment management, or planning everyday activity. Such changes cannot be deduced from a joint distribution, even if fully specified.

The connection between modularity and interventions is as follows. Instead of specifying a new probability function for each of the many possible interventions, we specify

---

[8] The Tversky and Kahneman (1980) experiments with causal biases in probability judgment constitute another body of evidence supporting this observation. For example, most people believe that it is more likely for a girl to have blue eyes, given that her mother has blue eyes, than the other way around; the two probabilities are in fact equal.

[9] The author was as guilty of advocating the centrality of conditional independence as were his colleagues in statistics; see Pearl (1988b, p. 79).

Figure 1.4   Network representation of the action "turning the sprinkler On."

merely the immediate change implied by the intervention and, by virtue of autonomy, we assume that the change is local, and does not spread over to mechanisms other than those specified. Once we know the identity of the mechanism altered by an intervention and the nature of the alteration, the overall effect of an intervention can be predicted by modifying the corresponding factors in (1.33) and using the modified product to compute a new probability function. For example, to represent the action "turning the sprinkler On" in the network of Figure 1.2, we delete the link $X_1 \to X_3$ and assign X_3 the value On. The graph resulting from this operation is shown in Figure 1.4, and the resulting joint distribution on the remaining variables will be

$$P_{X_3 = \text{On}}(x_1, x_2, x_4, x_5) = P(x_1)\, P(x_2 \mid x_1)\, P(x_4 \mid x_2, X_3 = \text{On})\, P(x_5 \mid x_4), \qquad (1.36)$$

in which all the factors on the right-hand side (r.h.s.), by virtue of autonomy, are the same as in (1.34).

The deletion of the factor $P(x_3 \mid x_1)$ represents the understanding that, whatever relationship existed between seasons and sprinklers prior to the action, that relationship is no longer in effect while we perform the action. Once we physically turn the sprinkler on and keep it on, a new mechanism (in which the season has no say) determines the state of the sprinkler.

Note the difference between the action $do(X_3 = \text{On})$ and the observation $X_3 = \text{On}$. The effect of the latter is obtained by ordinary Bayesian conditioning, that is, $P(x_1, x_2, x_4, x_5 \mid X_3 = \text{On})$, while that of the former by conditioning a mutilated graph, with the link $X_1 \to X_3$ removed. This indeed mirrors the difference between seeing and doing: after *observing* that the sprinkler is on, we wish to infer that the season is dry, that it probably did not rain, and so on; no such inferences should be drawn in evaluating the effects of a contemplated *action* "turning the sprinkler On."

The ability of causal networks to predict the effects of actions of course requires a stronger set of assumptions in the construction of those networks, assumptions that rest on causal (not merely associational) knowledge and that ensure the system would respond to interventions in accordance with the principle of autonomy. These assumptions are encapsulated in the following definition of causal Bayesian networks.

Definition 1.3.1 (Causal Bayesian Network)
Let $P(v)$ be a probability distribution on a set V of variables, and let $P_x(v)$ denote the distribution resulting from the intervention $do(X = x)$ that sets a subset X of variables

to constants x.[10] *Denote by* P_* *the set of all interventional distributions* $P_x(v), X \subseteq V$, *including* $P(v)$, *which represents no intervention (i.e.,* $X = \phi$). *A DAG G is said to be a* causal Bayesian network *compatible with* P_* *if and only if the following three conditions hold for every* $P_x \in P_*$:

(i) $P_x(v)$ *is Markov relative to G;*

(ii) $P_x(v_i) = 1$ *for all* $V_i \in X$ *whenever* v_i *is consistent with* $X = x$;

(iii) $P_x(v_i | pa_i) = P(v_i | pa_i)$ *for all* $V_i \notin X$ *whenever* pa_i *is consistent with* $X = x$, *i.e., each* $P(v_i | pa_i)$ *remains invariant to interventions not involving* V_i.

Definition 1.3.1 imposes constraints on the interventional space P_* that permit us to encode this vast space economically, in the form of a single Bayesian network G. These constraints enable us to compute the distribution $P_x(v)$ resulting from any intervention $do(X = x)$ as a *truncated factorization*

$$P_x(v) = \prod_{\{i \,|\, V_i \notin X\}} P(v_i | pa_i) \qquad \text{for all } v \text{ consistent with } x, \qquad (1.37)$$

which follows from Definition 1.3.1 and justifies the family deletion procedure on G, as in (1.36). It is not hard to show that, whenever G is a causal Bayes network with respect to P_*, the following two properties must hold.

Property 1
For all i,

$$P(v_i | pa_i) = P_{pa_i}(v_i). \qquad (1.38)$$

Property 2
For all i and for every subset S of variables disjoint of $\{V_i, PA_i\}$, *we have*

$$P_{pa_i, s}(v_i) = P_{pa_i}(v_i). \qquad (1.39)$$

Property 1 renders every parent set PA_i *exogenous* relative to its child V_i, ensuring that the conditional probability $P(v_i | pa_i)$ coincides with the effect (on V_i) of setting PA_i to pa_i by external control. Property 2 expresses the notion of invariance; once we control its direct causes PA_i, no other interventions will affect the probability of V_i.

1.3.2 Causal Relationships and Their Stability

This mechanism-based conception of interventions provides a semantical basis for notions such as "causal effects" or "causal influence," to be defined formally and analyzed in Chapters 3 and 4. For example, to test whether a variable X_i has a causal influence on another variable X_j, we compute (using the truncated factorization formula of (1.37)) the (marginal) distribution of X_j under the actions $do(X_i = x_i)$ – namely, $P_{x_i}(x_j)$ for all

---

[10] The notation $P_x(v)$ will be replaced in subsequent chapters with $P(v \,|\, do(x))$ and $P(v \,|\, \hat{x})$ to facilitate algebraic manipulations.

values x_i of X_i – and test whether that distribution is sensitive to x_i. It is easy to see from our previous examples that only variables that are descendants of X_i in the causal network can be influenced by X_i; deleting the factor $P(x_i | pa_i)$ from the joint distribution turns X_i into a root node in the mutilated graph, and root variables (as the d-separation criterion dictates) are independent of all other variables except their descendants.

This understanding of causal influence permits us to see precisely why, and in what way, causal relationships are more "stable" than probabilistic relationships. We expect such difference in stability because causal relationships are *ontological*, describing objective physical constraints in our world, whereas probabilistic relationships are *epistemic*, reflecting what we know or believe about the world. Therefore, causal relationships should remain unaltered as long as no change has taken place in the environment, even when our knowledge about the environment undergoes changes. To demonstrate, consider the causal relationship S_1, "Turning the sprinkler on would not affect the rain," and compare it to its probabilistic counterpart S_2, "The state of the sprinkler is independent of (or unassociated with) the state of the rain." Figure 1.2 illustrates two obvious ways in which S_2 will change while S_1 remains intact. First, S_2 changes from false to true when we learn what season it is (X_1). Second, given that we know the season, S_2 changes from true to false once we observe that the pavement is wet ($X_4 =$ true). On the other hand, S_1 remains true regardless of what we learn or know about the season or about the pavement.

The example reveals a stronger sense in which causal relationships are more stable than the corresponding probabilistic relationships, a sense that goes beyond their basic ontological–epistemological difference. The relationship S_1 will remain invariant to changes in the mechanism that regulates how seasons affect sprinklers. In fact, it remains invariant to changes in *all* mechanisms shown in this causal graph. We thus see that causal relationships exhibit greater robustness to ontological changes as well; they are sensitive to a smaller set of mechanisms. More specifically, and in marked contrast to probabilistic relationships, causal relationships remain invariant to changes in the mechanism that governs the causal variables (X_3 in our example).

In view of this stability, it is no wonder that people prefer to encode knowledge in causal rather than probabilistic structures. Probabilistic relationships, such as marginal and conditional independencies, may be helpful in hypothesizing initial causal structures from uncontrolled observations. However, once knowledge is cast in causal structure, those probabilistic relationships tend to be forgotten; whatever judgments people express about conditional independencies in a given domain are derived from the causal structure acquired. This explains why people feel confident asserting certain conditional independencies (e.g., that the price of beans in China is independent of the traffic in Los Angeles) having no idea whatsoever about the numerical probabilities involved (e.g., whether the price of beans will exceed $10 per bushel).

The element of stability (of mechanisms) is also at the heart of the so-called explanatory accounts of causality, according to which causal models need not encode behavior under intervention but instead aim primarily to provide an "explanation" or "understanding" of how data are generated.[11] Regardless of what use is eventually made

---

[11] Elements of this explanatory account can be found in the writings of Dempster (1990), Cox (1992), and Shafer (1996); see also King et al. (1994, p. 75).

of our "understanding" of things, we surely would prefer an understanding in terms of durable relationships, transportable across situations, over those based on transitory relationships. The sense of "comprehensibility" that accompanies an adequate explanation is a natural by-product of the transportability of (and hence of our familiarity with) the causal relationships used in the explanation. It is for reasons of stability that we regard the falling barometer as predicting but not explaining the rain; those predictions are not transportable to situations where the pressure surrounding the barometer is controlled by artificial means. True understanding enables predictions in such novel situations, where some mechanisms change and others are added. It thus seems reasonable to suggest that, in the final analysis, the explanatory account of causation is merely a variant of the manipulative account, albeit one where interventions are dormant. Accordingly, we may as well view our unsatiated quest for understanding "how data is generated" or "how things work" as a quest for acquiring the ability to make predictions under a wider range of circumstances, including circumstances in which things are taken apart, reconfigured, or undergo spontaneous change.

1.4 FUNCTIONAL CAUSAL MODELS

The way we have introduced the causal interpretation of Bayesian networks represents a fundamental departure from the way causal models (and causal graphs) were first introduced into genetics (Wright 1921), econometrics (Haavelmo 1943), and the social sciences (Duncan 1975), as well as from the way causal models are used routinely in physics and engineering. In those models, causal relationships are expressed in the form of deterministic, *functional* equations, and probabilities are introduced through the assumption that certain variables in the equations are unobserved. This reflects Laplace's (1814) conception of natural phenomena, according to which nature's laws are deterministic and randomness surfaces owing merely to our ignorance of the underlying boundary conditions. In contrast, all relationships in the definition of causal Bayesian networks were assumed to be inherently stochastic and thus appeal to the modern (i.e., quantum mechanical) conception of physics, according to which all nature's laws are inherently probabilistic and determinism is but a convenient approximation.

   In this book, we shall express preference toward Laplace's quasi-deterministic conception of causality and will use it, often contrasted with the stochastic conception, to define and analyze most of the causal entities that we study. This preference is based on three considerations. First, the Laplacian conception is more general. Every stochastic model can be emulated by many functional relationships (with stochastic inputs), but not the other way around; functional relationships can only be approximated, as a limiting case, using stochastic models. Second, the Laplacian conception is more in tune with human intuition. The few esoteric quantum mechanical experiments that conflict with the predictions of the Laplacian conception evoke surprise and disbelief, and they demand that physicists give up deeply entrenched intuitions about locality and causality (Maudlin 1994). Our objective is to preserve, explicate, and satisfy – not destroy – those intuitions.[12]

---

[12] The often heard argument that human intuitions belong in psychology and not in science or philosophy is inapplicable when it comes to causal intuition – the original authors of causal thoughts

Finally, certain concepts that are ubiquitous in human discourse can be defined only in the Laplacian framework. We shall see, for example, that such simple concepts as "the probability that event B occured *because* of event A" and "the probability that event B would have been *different* if it were not for event A" cannot be defined in terms of purely stochastic models. These so-called *counterfactual* concepts will require a synthesis of the deterministic and probabilistic components embodied in the Laplacian model.

1.4.1 Structural Equations

In its general form, a functional causal model consists of a set of equations of the form

$$x_i = f_i(pa_i, u_i), \quad i = 1, \ldots, n, \tag{1.40}$$

where pa_i (connoting *parents*) stands for the set of variables that directly determine the value of X_i and where the U_i represent errors (or "disturbances") due to omitted factors. Equation (1.40) is a nonlinear, nonparametric generalization of the linear structural equation models (SEMs)

$$x_i = \sum_{k \neq 1} \alpha_{ik} x_k + u_i, \quad i = 1, \ldots, n, \tag{1.41}$$

which have become a standard tool in economics and social science (see Chapter 5 for a detailed exposition of this enterprise). In linear models, pa_i corresponds to those variables on the r.h.s. of (1.41) that have nonzero coefficients.

The interpretation of the functional relationship in (1.40) is the standard interpretation that functions carry in physics and the natural sciences; it is a recipe, a strategy, or a *law* specifying what value nature would assign to X_i in response to every possible value combination that (PA_i, U_i) might take on. A set of equations in the form of (1.40) and in which each equation represents an autonomous mechanism is called a *structural model*; if each variable has a distinct equation in which it appears on the left-hand side (called the *dependent* variable), then the model is called a *structural causal model* or a *causal model* for short.[13] Mathematically, the distinction between structural and algebraic equations is that any subset of structural equations is, in itself, a valid structural model – one that represents conditions under some set of interventions.

To illustrate, Figure 1.5 depicts a canonical econometric model relating price and demand through the equations

$$q = b_1 p + d_1 i + u_1, \tag{1.42}$$

$$p = b_2 q + d_2 w + u_2, \tag{1.43}$$

where Q is the quantity of household demand for a product A, P is the unit price of product A, I is household income, W is the wage rate for producing product A, and U_1 and

---

cannot be ignored when the meaning of the concept is in question. Indeed, compliance with human intuition has been the ultimate criterion of adequacy in every philosophical study of causation, and the proper incorporation of background information into statistical studies likewise relies on accurate interpretation of causal judgment.

[13] Formal treatment of causal models, structural equations, and error terms are given in Chapter 5 (Section 5.4.1) and Chapter 7 (Sections 7.1 and 7.2.5).

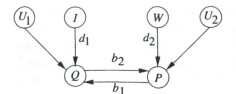

Figure 1.5 Causal diagram illustrating the relationship between price (P), demand (Q), income (I), and wages (W).

U_2 represent error terms – unmodeled factors that affect quantity and price, respectively (Goldberger 1992). The graph associated with this model is cyclic, and the vertices associated with the variables U_1, U_2, I, and W are root nodes, conveying the assumption of mutual independence. The idea of *autonomy* (Aldrich 1989), in this context, means that the two equations represent two loosely coupled segments of the economy, consumers and producers. Equation (1.42) describes how consumers decide what quantity Q to buy, and (1.43) describes how manufacturers decide what price P to charge. Like all feedback systems, this too represents implicit dynamics; today's prices are determined on the basis of yesterday's demand, and these prices will determine the demand in the next period of transactions. The solution to such equations represents a long-term equilibrium under the assumption that the background quantities, U_1 and U_2, remain constant.

The two equations are considered to be "autonomous" relative to the dynamics of changes in the sense that external changes affecting one equation do not imply changes to the others. For example, if government decides on price control and sets the price P at p_0, then (1.43) will be modified to read $p = p_0$ but the relationships in (1.42) will remain intact, yielding $q = b_1 p_0 + d_1 i + u_1$. We thus see that b_1, the "demand elasticity," should be interpreted as the rate of change of Q per unit *controlled* change in P. This is different, of course, from the rate of change of Q per unit *observed* change in P (under uncontrolled conditions), which, besides b_1, is also affected by the parameters of (1.43) (see Section 7.2.1, equation (7.14)). The difference between controlled and observed changes is essential for the correct interpretation of structural equation models in social science and economics, and it will be discussed at length in Chapter 5. If we have reasons to believe that consumer behavior will also change under a price control policy, then this modified behavior would need to be modeled explicitly – for example, by treating the coefficients b_1 and d_1 as dependent variables in auxiliary equations involving P.[14] Section 7.2.1 will present an analysis of policy-related problems using this model.

To illustrate the workings of nonlinear functional models, consider again the causal relationships depicted in Figure 1.2. The causal model associated with these relationships will consist of five functions, each representing an autonomous mechanism governing one variable:

$$x_1 = u_1,$$

$$x_2 = f_2(x_1, u_2),$$

---

[14] Indeed, consumers normally react to price fixing by hoarding goods in anticipation of shortages (Lucas 1976). Such phenomena are not foreign to structural models, though; they simply call for more elaborate equations to capture consumers' expectations.

$$x_3 = f_3(x_1, u_3),\tag{1.44}$$

$$x_4 = f_4(x_3, x_2, u_4),$$

$$x_5 = f_5(x_4, u_5).$$

The error variables $U_1,\ldots, U_5$ are not shown explicitly in the graph; by convention, this implies that they are assumed to be mutually independent. When some disturbances are judged to be dependent, it is customary to encode such dependencies by augmenting the graph with double-headed arrows, as shown in Figure 1.1(a).

A typical specification of the functions $\{f_1,\ldots,f_5\}$ and the disturbance terms is given by the following Boolean model:

$$x_2 = [(X_1 = \text{winter}) \lor (X_1 = \text{fall}) \lor u_2] \land \lnot u'_2,$$

$$x_3 = [(X_1 = \text{summer}) \lor (X_1 = \text{spring}) \lor u_3] \land \lnot u'_3,$$

$$x_4 = (x_2 \lor x_3 \lor u_4) \land \lnot u'_4,\tag{1.45}$$

$$x_5 = (x_4 \lor u_5) \land \lnot u'_5,$$

where x_i stands for $X_i = $ true and where u_i and u'_i stand for triggering and inhibiting abnormalities, respectively. For example, u_4 stands for (unspecified) events that might cause the pavement to get wet (X_4) when the sprinkler is off ($\lnot x_3$) and it does not rain ($\lnot x_2$) (e.g., a broken water pipe), while u'_4 stands for (unspecified) events that would keep the pavement dry in spite of the rain (x_2), the sprinkler (x_3), and u_4 (e.g., pavement covered with a plastic sheet).

It is important to emphasize that, in the two models just described, the variables placed on the left-hand side of the equality sign (the dependent or output variables) act distinctly from the other variables in each equation. The role of this distinction becomes clear when we discuss interventions, since it is only through this distinction that we can identify which equation ought to be modified under local interventions of the type "fix the price at p_0" ($do(P = p_0)$) or "turn the sprinkler On" ($do(X_3 = \text{true})$).[15]

We now compare the features of functional models as defined in (1.40) with those of causal Bayesian networks defined in Section 1.3. Toward this end, we will consider the processing of three types of queries:

predictions (e.g., would the pavement be slippery if we *find* the sprinkler off?);

interventions (e.g., would the pavement be slippery if we *make sure* that the sprinkler is off?); and

counterfactuals (e.g., would the pavement be slippery *had* the sprinkler been off, given that the pavement is in fact not slippery and the sprinkler is on?).

We shall see that these three types of queries represent a hierarchy of three fundamentally different types of problems, demanding knowledge with increasing levels of detail.

---

[15] Economists who write the supply–demand equations as $\{q = ap + u_1, q = bp + u_2\}$, with q appearing on the l.h.s. of both equations, are giving up the option of analyzing price control policies unless additional symbolic machinery is used to identify which equation will be modified by the $do(P = p_0)$ operator.

1.4.2 Probabilistic Predictions in Causal Models

Given a causal model (equation (1.40)), if we draw an arrow from each member of PA_i toward X_i, then the resulting graph G will be called a *causal diagram*. If the causal diagram is acyclic, then the corresponding model is called *semi-Markovian* and the values of the X variables will be uniquely determined by those of the U variables. Under such conditions, the joint distribution $P(x_1,..., x_n)$ is determined uniquely by the distribution $P(u)$ of the error variables. If, in addition to acyclicity, the error terms are jointly independent, the model is called *Markovian*.

A fundamental theorem about Markovian models establishes a connection between causation and probabilities via the parental Markov condition of Theorem 1.2.7.

Theorem 1.4.1 (Causal Markov Condition)

Every Markovian causal model M induces a distribution $P(x_1,..., x_n)$ that satisfies the parental Markov condition relative the causal diagram G associated with M; that is, each variable X_i is independent of all its nondescendants, given its parents PA_i in G (Pearl and Verma 1991).[16]

The proof is immediate. Considering that the set $\{PA_i, U_i\}$ determines one unique value of X_i, the distribution $P(x_1,..., x_n, u_1,..., u_n)$ is certainly Markov relative the augmented DAG $G(X, U)$, in which the U variables are represented explicitly. The required Markov condition of the marginal distribution $P(x_1,..., x_n)$ follows by d-separation in $G(X, U)$.

Theorem 1.4.1 shows that the parental Markov condition of Theorem 1.2.7 follows from two causal assumptions: (1) our commitment to include in the model (not in the background) every variable that is a cause of two or more other variables; and (2) Reichenbach's (1956) common-cause assumption, also known as "no correlation without causation," stating that, if any two variables are dependent, then one is a cause of the other *or* there is a third variable causing both. These two assumptions imply that the background factors in U are mutually independent and hence that the causal model is Markovian. Theorem 1.4.1 explains both why Markovian models are so frequently assumed in causal analysis and why the parental Markov condition (Theorem 1.2.7) is so often regarded as an inherent feature of causal models (see, e.g., Kiiveri et al. 1984; Spirtes et al. 1993).[17]

The causal Markov condition implies that characterizing each child–parent relationship as a deterministic function, instead of the usual conditional probability $P(x_i \mid pa_i)$, imposes equivalent independence constraints on the resulting distribution and leads to the same recursive decomposition that characterizes Bayesian networks (see equation (1.33)). More significantly, this holds regardless of the choice of functions $\{f_i\}$ and regardless

---

[16] Considering its generality and transparency, I would not be surprised if some version of this theorem has appeared earlier in the literature, but I am not aware of any nonparametric version.

[17] Kiiveri et al.'s (1984) paper, entitled "Recursive Causal Models," provides the first proof (for strictly positive distributions) that the parental Markov condition of Theorem 1.2.7 follows from the factorization of (1.33). This implication, however, is purely probabilistic and invokes no aspect of causation. In order to establish a connection between causation and probability we must first devise a model for causation, either in terms of manipulations (as in Definition 1.3.1) or in terms of functional relationships in structural equations (as in Theorem 1.4.1).

of the error distributions $P(u_i)$. Thus, we need not specify in advance the functional form of $\{f_i\}$ or the distributions $P(u_i)$; once we measure (or estimate) $P(x_i \mid pa_i)$, all probabilistic properties of a Markovian causal model are determined, regardless of the mechanism that actually generates those conditional probabilities. Druzdzel and Simon (1993) showed that, for every Bayesian network G characterized by a distribution P (as in (1.33)), there exists a functional model (as in (1.40)) that generates a distribution identical to P.[18] It follows that in all probabilistic applications of Bayesian networks – including statistical estimation, prediction, and diagnosis – we can use an equivalent functional model as specified in (1.40), and we can regard functional models as just another way of encoding joint distribution functions.

Nonetheless, the causal–functional specification has several advantages over the probabilistic specification, even in purely predictive (i.e., nonmanipulative) tasks. First and foremost, all the conditional independencies that are displayed by the causal diagram G are guaranteed to be *stable* – that is, invariant to parametric changes in the mechanisms represented by the functions f_i and the distributions $P(u_i)$. This means that agents who choose to organize knowledge using Markovian causal models can make reliable assertions about conditional independence relations without assessing numerical probabilities – a common ability among humanoids[19] and a useful feature for inference. Second, the functional specification is often more meaningful and natural, and it yields a small number of parameters. Typical examples are the linear structural equations used in social science and economics (see Chapter 5) and the "noisy OR gate" that has become quite popular in modeling the effect of multiple dichotomous causes (Pearl 1988b, p. 184). Third (and perhaps hardest for an empiricist to accept), judgmental assumptions of conditional independence among observable quantities are simplified and made more reliable in functional models, because such assumptions are cast directly as judgments about the presence or absence of *unobserved* common causes (e.g., why is the price of beans in China judged to be independent of the traffic in Los Angeles?). In the construction of Bayesian networks, for example, instead of judging whether each variable is independent of all its nondescendants (given its parents), we need to judge whether the parent set contains *all* relevant immediate causes – in particular, whether no factor omitted from the parent set is a cause of another observed variable. Such judgments are more natural because they are discernible directly from a qualitative causal structure, the very structure that our mind has selected for storing stable aspects of experience.

Finally, there is an additional advantage to basing prediction models on causal mechanisms that stems from considerations of stability (Section 1.3.2). When some conditions in the environment undergo change, it is usually only a few causal mechanisms that are affected by the change; the rest remain unaltered. It is simpler then to reassess (judgmentally) or reestimate (statistically) the model parameters knowing that

---

[18] In Chapter 9 we will show that, except in some pathological cases, there actually exist an infinite number of functional models with this property.

[19] Statisticians who are reluctant to discuss causality yet have no hesitation expressing background information in the form of conditional independence statements would probably be shocked to realize that such statements acquire their validity from none other than the *causal* Markov condition (Theorem 1.4.1). See note 9.

the corresponding symbolic change is also local, involving just a few parameters, than to reestimate the entire model from scratch.[20]

1.4.3 Interventions and Causal Effects in Functional Models

The functional characterization $x_i = f_i(pa_i, u_i)$, like its stochastic counterpart, provides a convenient language for specifying how the resulting distribution would change in response to external interventions. This is accomplished by encoding each intervention as an alteration on a select set of functions instead of a select set of conditional probabilities. The overall effect of the intervention can then be predicted by modifying the corresponding equations in the model and using the modified model to compute a new probability function. Thus, all features of causal Bayesian networks (Section 1.3) can be emulated in Markovian functional models.

For example, to represent the action "turning the sprinkler On" in the model of (1.44), we delete the equation $x_3 = f_3(x_1, u_3)$ and replace it with $x_3 = $ On. The modified model will contain all the information needed for computing the effect of the action on other variables. For example, the probability function induced by the modified model will be equal to that given by (1.36), and the modified diagram will coincide with that of Figure 1.4.

More generally, when an intervention forces a subset X of variables to attain fixed values x, then a subset of equations is to be pruned from the model in (1.40), one for each member of X, thus defining a new distribution over the remaining variables that characterizes the effect of the intervention and coincides with the truncated factorization obtained by pruning families from a causal Bayesian network (equation (1.37)).[21]

The functional model's representation of interventions offers greater flexibility and generality than that of a stochastic model. First, the analysis of interventions can be extended to cyclic models, like the one in Figure 1.5, so as to answer policy-related questions[22] (e.g.: What would the demand quantity be if we control the price at p_0?). Second, interventions involving the modification of equational parameters (like b_1 and d_1 in (1.42)) are more readily comprehended than those described as modifiers of conditional probabilities, perhaps because stable physical mechanisms are normally associated with equations and not with conditional probabilities. Conditional probabilities are perceived to be derivable from, not generators of, joint distributions. Third, the analysis of causal effects in non-Markovian models will be greatly simplified using functional models. The reason is: there are infinitely many conditional probabilities $P(x_i | pa_i)$ but only a finite number of functions $x_i = f_i(pa_i, u_i)$ among discrete variables X_i and PA_i. This fact will enable us in Chapter 8 (Section 8.2.2) to use linear-programming techniques to obtain sharp bounds on causal effects in studies involving noncompliance.

---

[20] To the best of my knowledge, this aspect of causal models has not been studied formally; it is suggested here as a research topic for students of adaptive systems.

[21] An explicit translation of interventions to "wiping out" equations from the model was first proposed by Strotz and Wold (1960) and later used in Fisher (1970) and Sobel (1990). More elaborate types of interventions, involving conditional actions and stochastic strategies, will be formulated in Chapter 4.

[22] Such questions, especially those involving the control of endogenous variables, are conspicuously absent from econometric textbooks (see Chapter 5).

Finally, functional models permit the analysis of context-specific actions and policies. The notion of causal effect as defined so far is of only minor use in practical policy making. The reason is that causal effects tell us the general tendency of an action to bring about a response (as with the tendency of a drug to enhance recovery in the overall population) but are not specific to actions in a given situation characterized by a set of particular observations that may themselves be affected by the action. A physician is usually concerned with the effect of a treatment on a patient who has already been examined and found to have certain symptoms. Some of those symptoms will themselves be affected by the treatment. Likewise, an economist is concerned with the effect of taxation in a given economic context characterized by various economical indicators, which (again) will be affected by taxation if applied. Such context-specific causal effects cannot be computed by simulating an intervention in a static Bayesian network, because the context itself varies with the intervention and so the conditional probabilities $P(x_i \mid pa_i)$ are altered in the process. However, the functional relationships $x_i = f_i(pa_i, u_i)$ remain invariant, which enables us to compute context-specific causal effects as outlined in the next section (see Sections 7.2.1, 8.3, and 9.3.4 for full details).

1.4.4   Counterfactuals in Functional Models

We now turn to the most distinctive characteristic of functional models – the analysis of *counterfactuals*. Certain counterfactual sentences, as we remarked before, cannot be defined in the framework of stochastic causal networks. To see the difficulties, let us consider the simplest possible causal Bayesian network consisting of a pair of independent (hence unconnected) binary variables X and Y. Such a network ensues, for example, in a controlled (i.e., randomized) clinical trial when we find that a treatment X has no effect on the distribution of subjects' response Y, which may stand for either recovery ($Y = 0$) or death ($Y = 1$). Assume that a given subject, Joe, has taken the treatment and died; we ask whether Joe's death occurred *because of* the treatment, *despite* the treatment, or *regardless of* the treatment. In other words, we ask for the probability Q that Joe would have died had he not been treated.

To highlight the difficulty in answering such counterfactual questions, let us take an extreme case where 50% of the patients recover and 50% die in both the treatment and the control groups; assume further that the sample size approaches infinity, thus yielding

$$P(y \mid x) = 1/2 \quad \text{for all } x \text{ and } y. \tag{1.46}$$

Readers versed in statistical testing will recognize immediately the impossibility of answering the counterfactual question from the available data, noting that Joe, who took the treatment and died, was never tested under the no-treatment condition. Moreover, the difficulty does not stem from addressing the question to a particular individual, Joe, for whom we have only one data point. Rephrasing the question in terms of population frequencies – asking what percentage Q of subjects who died under treatment would have recovered had they not taken the treatment – will encounter the same difficulties because none of those subjects was tested under the no-treatment condition. Such difficulties have prompted some statisticians to dismiss counterfactual questions as metaphysical and to advocate the restriction of statistical analysis to only those questions that can be answered by direct tests (Dawid 2000).

However, that our scientific, legal, and ordinary languages are loaded with counter-factual utterances indicates clearly that counterfactuals are far from being metaphysical; they must have definite testable implications and must carry valuable substantive infor-mation. The analysis of counterfactuals therefore represents an opportunity to anyone who shares the aims of this book: integrating substantive knowledge with statistical data so as to refine the former and interpret the latter. Within this framework, the counterfac-tual issue demands answers to tough, yet manageable technical questions: What is the empirical content of counterfactual queries? What knowledge is required to answer those queries? How can this knowledge be represented mathematically? Given such represen-tation, what mathematical machinery is needed for deriving the answers?

Chapter 7 (Section 7.2.2) presents an empirical explication of counterfactuals as claims about the temporal persistence of certain mechanisms. In our example, the response to treatment of each (surviving) patient is assumed to be persistent. If the outcome Y were a reversible condition, rather than death, then the counterfactual claim would translate di-rectly into predictions about response to future treatments. But even in the case of death, the counterfactual quantity Q implies not merely a speculation about the hypothetical be-havior of subjects who died but also a testable claim about surviving untreated subjects under subsequent treatment. We leave it as an exercise for the reader to prove that, based on (1.46) and barring sampling variations, the percentage Q of deceased subjects from the treatment group who would have recovered had they not taken the treatment precisely equals the percentage Q' of surviving subjects in the nontreatment group who will die if given treatment.[23] Whereas Q is hypothetical, Q' is unquestionably testable.

Having sketched the empirical interpretation of counterfactuals, our next step in this introductory chapter is the question of representation: What knowledge is required to an-swer questions about counterfactuals? And how should this knowledge be formulated so that counterfactual queries can be answered quickly and reliably? That such representation exists is evident by the swiftness and consistency with which people distinguish plausi-ble from implausible counterfactual statements. Most people would agree that President Clinton's place in history would be different had he not met Monica Lewinsky, but only a few would assert that his place in history would change had he not eaten breakfast yes-terday. In the cognitive sciences, such consistency of opinion is as close as one can get to a proof that an effective machinery for representing and manipulating counterfactuals re-sides someplace in the human mind. What then are the building blocks of that machinery?

A straightforward representational scheme would (i) store counterfactual knowledge in the form of counterfactual premises and (ii) derive answers to counterfactual queries using some logical rules of inference capable of taking us from premises to conclusions. This approach has indeed been taken by the philosophers Robert Stalnaker (1968) and David Lewis (1973a,b), who constructed logics of counterfactuals using closest-world

---

[23] For example, if Q equals 100% (i.e., all those who took the treatment and died would have recov-ered had they not taken the treatment), then all surviving subjects from the nontreatment group will die if given treatment (again, barring sampling variations). Such exercises will become rou-tine when we develop the mathematical machinery for analyzing probabilities of causes (see Chapter 9, Theorem 9.2.12, equations (9.11)–(9.12)).

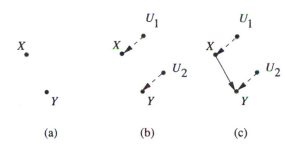

Figure 1.6 (a) A causal Bayesian network that represents the distribution of (1.47). (b) A causal diagram representing the process generating the distribution in (a), according to model 1. (c) Same, according to model 2. (Both U_1 and U_2 are unobserved.)

semantics (i.e., "B would be true if it were A" just in case B is true in the closest possible world (to ours) in which A is true). However, the closest-world semantics still leaves two questions unanswered. (1) What choice of distance measure would make counterfactual reasoning compatible with ordinary conceptions of cause and effect? (2) What mental representation of interworld distances would render the computation of counterfactuals manageable and practical (for both humans and machines)? These two questions are answered by the structural model approach expanded in Chapter 7.

An approach similar to Lewis's (though somewhat less formal) has been pursued by statisticians in the potential-outcome framework (Rubin 1974; Robins 1986; Holland 1988). Here, substantive knowledge is expressed in terms of probabilistic relationships (e.g., independence) among counterfactual variables and then used in the estimation of causal effects. The question of representation shifts from the closest-world to the potential-outcome approach: How are probabilistic relationships among counterfactuals stored or inferred in the investigator's mind? In Chapter 7 (see also Section 3.6.3) we provide an analysis of the closest-world and potential-outcome approaches and compare them to the structural model approach, to be outlined next, in which counterfactuals are *derived* from (and in fact defined by) a functional causal model (equation (1.40)).

In order to see the connection between counterfactuals and structural equations, we should first examine why the information encoded in a Bayesian network, even in its causal interpretation, is insufficient to answer counterfactual queries. Consider again our example of the controlled randomized experiment (equation (1.46)), which corresponds to an edgeless Bayesian network (Figure 1.6(a)) with two independent binary variables and a joint probability:

$$P(y, x) = 0.25 \quad \text{for all } x \text{ and } y. \tag{1.47}$$

We now present two functional models, each generating the joint probability of (1.47) yet each giving a different value to the quantity of interest, Q = the probability that a subject who died under treatment ($x = 1, y = 1$) would have recovered ($y = 0$) had he or she not been treated ($x = 0$).

Model 1 (Figure 1.6(b))
Let

$$x = u_1,$$

$$y = u_2,$$

where U_1 and U_2 are two independent binary variables with $P(u_1 = 1) = P(u_2 = 1) = \frac{1}{2}$ (e.g., random coins).

Model 1	$u_2 = 0$		$u_2 = 1$		Marginal	
	$x = 1$	$x = 0$	$x = 1$	$x = 0$	$x = 1$	$x = 0$
$y = 1$ (death)	0	0	0.25	0.25	0.25	0.25
$y = 0$ (recovery)	0.25	0.25	0	0	0.25	0.25

Model 2	$u_2 = 0$		$u_2 = 1$		Marginal	
	$x = 1$	$x = 0$	$x = 1$	$x = 0$	$x = 1$	$x = 0$
$y = 1$ (death)	0	0.25	0.25	0	0.25	0.25
$y = 0$ (recovery)	0.25	0	0	0.25	0.25	0.25

Figure 1.7 Contingency tables showing the distributions $P(x, y, u_2)$ and $P(x, y)$ for the two models discussed in the text.

Model 2 (Figure 1.6(c))
Let

$$x = u_1,$$

$$y = xu_2 + (1 - x)(1 - u_2), \tag{1.48}$$

where, as before, U_1 and U_2 are two independent binary variables.

Model 1 corresponds to treatment (X) that has no effect on any of the subjects; in model 2, every subject is affected by treatment. The reason that the two models yield the same distribution is that model 2 describes a mixture of two subpopulations. In one ($u_2 = 1$), each subject dies ($y = 1$) if and only if treated; in the other ($u_2 = 0$), each subject recovers ($y = 0$) if and only if treated. The distributions $P(x, y, u_2)$ and $P(x, y)$ corresponding to these two models are shown in the tables of Figure 1.7.

The value of Q differs in these two models. In model 1, Q evaluates to zero, because subjects who died correspond to $u_2 = 1$ and, since the treatment has no effect on y, changing X from 1 to 0 would still yield $y = 1$. In model 2, however, Q evaluates to unity, because subjects who died under treatment must correspond to $u_2 = 1$ (i.e., those who die if treated), meaning they would recover if and only if not treated.

The first lesson of this example is that stochastic causal models are insufficient for computing probabilities of counterfactuals; knowledge of the actual process behind $P(y \mid x)$ is needed for the computation.[24] A second lesson is that a functional causal model constitutes a mathematical object sufficient for the computation (and definition) of such probabilities. Consider, for example, model 2 of (1.48). The way we concluded that a deceased treated subject ($y = 1, x = 1$) would have recovered if not treated involved three mental steps. First, we applied the evidence at hand, $e : \{y = 1, x = 1\}$, to the model and concluded that e is compatible with only one realization of U_1 and U_2 – namely, $\{u_1 = 1,$

---

[24] In the potential-outcome framework (Sections 3.6.3 and 7.4.4), such knowledge obtains stochastic appearance by defining distributions over *counterfactual variables* Y_1 and Y_0, which stand for the potential response of an individual to treatment and no treatment, respectively. These hypothetical variables play a role similar to the functions $f_i(pa_i, u_i)$ in our model; they represent the deterministic assumption that every individual possesses a definite response to treatment, regardless of whether that treatment was realized.

$u_2 = 1$}. Second, to simulate the hypothetical condition "had he or she not been treated," we substituted $x = 0$ into (1.48) while ignoring the first equation $x = u_1$. Finally, we solved (1.48) for y (assuming $x = 0$ and $u_2 = 1$) and obtained $y = 0$, from which we concluded that the probability of recovery ($y = 0$) is unity under the hypothetical condition considered.

These three steps can be generalized to any causal model M as follows. Given evidence e, to compute the probability of $Y = y$ under the hypothetical condition $X = x$ (where X is a subset of variables), apply the following three steps to M.

> **Step 1 (abduction):** Update the probability $P(u)$ to obtain $P(u \mid e)$.

> **Step 2 (action):** Replace the equations corresponding to variables in set X by the equations $X = x$.

> **Step 3 (prediction):** Use the modified model to compute the probability of $Y = y$.

In temporal metaphors, this three-step procedure can be interpreted as follows. Step 1 explains the past (U) in light of the current evidence e; step 2 bends the course of history (minimally) to comply with the hypothetical condition $X = x$; finally, step 3 predicts the future (Y) based on our new understanding of the past and our newly established condition, $X = x$.

Recalling that for each value u of U there is a unique solution for Y, it is clear that step 3 always gives a unique solution for the needed probability; we simply sum up the probabilities $P(u \mid e)$ assigned to all those u that yield $Y = y$ as a solution. Chapter 7 develops effective procedures for computing probabilities of counterfactuals, procedures that are based on probability propagation in "twin" networks (Balke and Pearl 1995a): one network represents the actual world; the other, the counterfactual world.

Note that the hypothetical condition $X = x$ always stands in contradiction to the prevailing values u of U in the model considered (else $X = x$ would actually be realized and thus would not be considered hypothetical). It is for this reason that we invoke (in step 2) an external intervention (alternatively, a "theory change" or a "miracle"; Lewis 1973b), which modifies the model and thus explains the contradiction away. In Chapter 7 we extend this structural–interventional model to give a full semantical and axiomatic account for both counterfactuals and the probability of counterfactuals. In contrast with Lewis's theory, this account is not based on an abstract notion of similarity among hypothetical worlds; rather, it rests on the actual mechanisms involved in the production of the hypothetical worlds considered. Likewise, in contrast with the potential-outcome framework, counterfactuals in the structural account are not treated as undefined primitives but rather as quantities to be derived from the more fundamental concepts of causal mechanisms and their structure.

The three-step model of counterfactual reasoning also uncovers the real reason why stochastic causal models are insufficient for computing probabilities of counterfactuals. Because the U variables do not appear explicitly in stochastic models, we cannot apply step 1 so as to update $P(u)$ with the evidence e at hand. This implies that several ubiquitous notions based on counterfactuals – including probabilities of causes (given the effects), probabilities of explanations, and context-dependent causal effect – cannot be defined in such models. For these, we must make some assumptions about the form of the functions f_i and the probabilities of the error terms. For example, the assumptions of linearity, normality, and error independence are sufficient for computing all counterfactual queries in the model of Figure 1.5 (see Section 7.2.1). In Chapter 9, we will present conditions under which counterfactual queries concerning probability of causation can be inferred from data when f_i and $P(u)$ are unknown, and only general features (e.g., monotonicity) of these entities are assumed. Likewise, Chapter 8 (Section 8.3) will present methods of *bounding* probabilities of counterfactuals when only stochastic models are available.

The preceding considerations further imply that the three tasks listed in the beginning of this section – prediction, intervention, and counterfactuals – form a natural hierarchy of causal reasoning tasks, with increasing levels of refinement and increasing demands on the knowledge required for accomplishing these tasks. Prediction is the simplest of the three, requiring only a specification of a joint distribution function. The analysis of interventions requires a causal structure in addition to a joint distribution. Finally, processing counterfactuals is the hardest task because it requires some information about the functional relationships and/or the distribution of the omitted factors.

This hierarchy also defines a natural partitioning of the chapters in this book. Chapter 2 will deal primarily with the probabilistic aspects of causal Bayesian networks (though the underlying causal structure will serve as a conceptual guide). Chapters 3–6 will deal exclusively with the interventional aspects of causal models, including the identification of causal effects, the clarification of structural equation models, and the relationships between confounding and collapsibility. Chapters 7–10 will deal with counterfactual analysis, including axiomatic foundation, applications to policy analysis, the bounding of counterfactual queries, the identification of probabilities of causes, and the explication of single-event causation.

I wish the reader a smooth and rewarding journey through these chapters. But first, an important stop for terminological distinctions.

1.5 CAUSAL VERSUS STATISTICAL TERMINOLOGY

This section defines fundamental terms and concepts that will be used throughout this book. These definitions may not agree with those given in standard sources, so it is important to refer to this section in case of doubts regarding the interpretation of these terms.

A *probabilistic parameter* is any quantity that is defined in terms[25] of a joint probability function. Examples are the quantities defined in Sections 1.1 and 1.2.

A *statistical parameter* is any quantity that is defined in terms of a joint probability distribution of observed variables, making no assumption whatsoever regarding the existence or nonexistence of unobserved variables.

Examples: the conditional expectation $E(Y\,|\,x)$,
the regression coefficient r_{YX},
the value of the density function at $y = 0, x = 1$.

A *causal parameter* is any quantity that is defined in terms of a causal model (as in (1.40)) and is not a statistical parameter.

---

[25] A quantity Q is said to be *defined in terms of* an object of class C if Q can be computed uniquely from the description of any object in class C (i.e., if Q is defined by a functional mapping from C to the domain of Q).

Examples: the coefficients a_{ik} in (1.41),
whether X_9 has influence on X_3 for some u,
the expected value of Y under the intervention $do(X = 0)$,
the number of parents of variable X_7.

Remark: The exclusion of unmeasured variables from the definition of statistical parameters is devised to prevent one from hiding causal assumptions under the guise of latent variables. Such constructions, if permitted, would qualify any quantity as statistical and would thus obscure the important distinction between quantities that can be estimated from statistical data alone, and those that require additional assumptions beyond the data.

A ***statistical assumption*** is any constraint on a joint distribution of observed variable; for example, that f is multivariate normal or that P is Markov relative to a given DAG D.

A ***causal assumption*** is any constraint on a causal model that cannot be realized by imposing statistical assumptions; for example, that f_i is linear, that U_i and U_j (unobserved) are uncorrelated, or that x_3 does not appear in $f_4(pa_4, u_4)$. Causal assumptions may or may not have statistical implications. In the former case we say that the assumption is "testable" or "falsifiable." Often, though not always, causal assumptions can be falsified from experimental studies, in which case we say that they are "experimentally testable." For example, the assumption that X has no effect on $E(Y)$ in model 2 of Figure 1.6 is empirically testable, but the assumption that X may cure a given subject in the population is not.

Remark: The distinction between causal and statistical parameters is crisp and fundamental – the two do not mix. Causal parameters cannot be discerned from statistical parameters unless causal assumptions are invoked. The formulation and simplification of these assumptions will occupy a major part of this book.

Remark: Temporal precedence among variables may furnish some information about (the absence of) causal relationships – a later event cannot be the cause of an earlier event. Temporally indexed distributions such as $P(y_t \mid y_{t-1}, x_t)$, $t = 1,\ldots$, which are used routinely in economic analysis, may therefore be regarded as borderline cases between statistical and causal models. We shall nevertheless classify those models as statistical because the great majority of policy-related questions *cannot* be discerned from such distributions, given our commitment to making no assumption regarding the presence or absence of unmeasured variables. Consequently, econometric concepts such as "Granger causality" (Granger 1969) and "strong exogeneity" (Engle et al. 1983) will be classified as statistical rather than causal.[26]

Remark: The terms "theoretical" and "structural" are often used interchangeably with "causal"; we will use the latter two, keeping in mind that some structural models may not be causal (see Section 7.2.5).

[26] Caution must also be exercised in labeling as a "data-generating model" the probabilistic sequence $P(y_t|y_{t-1} x_t)$, $t = 1,\ldots$ (e.g., Davidson and MacKinnon 1993, p. 53; Hendry 1995). Such sequences are statistical in nature and, unless causal assumptions of the type developed in Chapter 2 (see Definitions 2.4.1 and 2.7.4) are invoked, they cannot be applied to policy-evaluation tasks.

Causal versus Statistical Concepts

The demarcation line between causal and statistical parameters extends as well to general concepts and will be supported by terminological distinction. Examples of *statistical* concepts are: correlation, regression, conditional independence, association, likelihood, collapsibility, risk ratio, odds ratio, propensity score, Granger's causality, and so on. Examples of *causal* concepts are: randomization, influence, effect, confounding, exogeneity, ignorability, disturbance (e.g., (1.40)), spurious correlation, path coefficients, instrumental variables, intervention, explanation, and so on. The purpose of this demarcation line is not to exclude causal concepts from the province of statistical analysis but, rather, to encourage investigators to treat nonstatistical concepts with the proper set of tools.

Some readers may be surprised by the idea that textbook concepts such as randomization, confounding, spurious correlation, and effects are nonstatistical. Others may be shocked at the idea that controversial concepts such as exogeneity, confounding, and counterfactuals *can* be defined in terms of causal models. This book is written with these readers in mind, and the coming pages will demonstrate that the distinctions just made between causal and statistical concepts are essential for clarifying both.

Two Mental Barriers to Causal Analysis

The sharp distinction between statistical and causal concepts can be translated into a useful principle: behind every causal claim there must lie some causal assumption that is not discernable from the joint distribution and, hence, not testable in observational studies. Such assumptions are usually provided by humans, resting on expert *judgment*. Thus, the way humans organize and communicate experiential knowledge becomes an integral part of the study, for it determines the veracity of the judgments experts are requested to articulate.

Another ramification of this causal–statistical distinction is that any mathematical approach to causal analysis must acquire *new notation*. The vocabulary of probability calculus, with its powerful operators of expectation, conditionalization, and marginalization, is defined strictly in terms of distribution functions and is therefore insufficient for expressing causal assumptions or causal claims. To illustrate, the syntax of probability calculus does not permit us to express the simple fact that "symptoms do not cause diseases," let alone draw mathematical conclusions from such facts. All we can say is that two events are dependent – meaning that if we find one, we can expect to encounter the other, but we cannot distinguish statistical dependence, quantified by the conditional probability $P(disease \mid symptom)$, from causal dependence, for which we have no expression in standard probability calculus.

The preceding two requirements: (1) to commence causal analysis with untested, judgmental assumptions, and (2) to extend the syntax of probability calculus, constitute the two main obstacles to the acceptance of causal analysis among professionals with traditional training in statistics (Pearl 2003c, also sections 11.1.1 and 11.6.4). This book helps overcome the two barriers through an effective and friendly notational system based on symbiosis of graphical and algebraic approaches.

CHAPTER TWO

A Theory of Inferred Causation

*I would rather discover one causal law
than be King of Persia.*
                    Democritus (460–370 B.C.)

Preface

The possibility of learning causal relationships from raw data has been on philosophers'
dream lists since the time of Hume (1711–1776). That possibility entered the realm of formal treatment and feasible computation in the mid-1980s, when the mathematical relationships between graphs and probabilistic dependencies came to light. The approach
described herein is an outgrowth of Rebane and Pearl (1987) and Pearl (1988b, Chap. 8),
which describes how causal relationships can be inferred from nontemporal statistical data
if one makes certain assumptions about the underlying process of data generation (e.g., that
it has a tree structure). The prospect of inferring causal relationships from weaker structural assumptions (e.g., general directed acyclic graphs) has motivated parallel research
efforts at three universities: UCLA, Carnegie Mellon University (CMU), and Stanford. The
UCLA and CMU teams pursued an approach based on searching the data for patterns of conditional independencies that reveal fragments of the underlying structure and then piec-
ing those fragments together to form a coherent causal model (or a set of such models). The
Stanford group pursued a Bayesian approach, where data are used to update prior probabilities assigned to candidate causal structures (Cooper and Herskovits 1991). The UCLA
and CMU efforts have led to similar theories and almost identical discovery algorithms,
which were implemented in the TETRAD II program (Spirtes et al. 1993). The Bayesian
approach has since been pursued by a number of research teams (Singh and Valtorta 1995;
Heckerman et al. 1994) and now serves as the basis for several graph-based learning methods (Jordan 1998). This chapter describes the approach pursued by Tom Verma and me in
the period 1988–1992, and it briefly summarizes related extensions, refinements, and
improvements that have been advanced by the CMU team and others. Some of the philosophical rationale behind this development, primarily the assumption of minimality, are
implicit in the Bayesian approach as well (Section 2.9.1).

The basic idea of automating the discovery of causes – and the specific implementa-
tion of this idea in computer programs – came under fierce debate in a number of forums
(Cartwright 1995a; Humphreys and Freedman 1996; Cartwright 1999; Korb and Wallace
1997; McKim and Turner 1997; Robins and Wasserman 1999). Selected aspects of this
debate will be addressed in the discussion section at the end of this chapter (Section 2.9.1).

Acknowledging that statistical associations do not *logically* imply causation, this chapter asks whether weaker relationships exist between the two. In particular, we ask:

1. What clues prompt people to perceive causal relationships in uncontrolled observations?

2. What assumptions would allow us to infer causal models from these clues?

3. Would the models inferred tell us anything useful about the causal mechanisms that underly the observations?

In Section 2.2 we define the notions of causal models and causal structures and then describe the task of causal discovery as an inductive game that scientists play against Nature. In Section 2.3 we formalize the inductive game by introducing "minimal model" semantics – the semantical version of Occam's razor – and exemplify how, contrary to common folklore, causal relationships can be distinguished from spurious covariations following this standard norm of inductive reasoning. Section 2.4 identifies a condition, called *stability* (or *faithfulness*), under which effective algorithms exist that uncover structures of casual influences as defined here. One such algorithm (called IC), introduced in Section 2.5, uncovers the set of all causal models compatible with the data, assuming all variables are observed. Another algorithm (IC*), described in Section 2.6, is shown to uncover many (though not all) valid causal relationships when some variables are *not* observable. In Section 2.7 we extract from the IC* algorithm the essential conditions under which causal influences are inferred, and we offer these as independent definitions of genuine influences and spurious associations, with and without temporal information. Section 2.8 offers an explanation for the puzzling yet universal agreement between the temporal and statistical aspects of causation. Finally, Section 2.9 summarizes the claims made in this chapter, re-explicates the assumptions that lead to these claims, and offers new justifications of these assumption in light of ongoing debates.

2.1 INTRODUCTION – THE BASIC INTUITIONS

An autonomous intelligent system attempting to build a workable model of its environment cannot rely exclusively on preprogrammed causal knowledge; rather, it must be able to translate direct observations to cause-and-effect relationships. However, given that statistical analysis is driven by covariation, not causation, and assuming that the bulk of human knowledge derives from passive observations, we must still identify the clues that prompt people to perceive causal relationships in the data. We must also find a computational model that emulates this perception.

Temporal precedence is normally assumed to be essential for defining causation, and it is undoubtedly one of the most important clues that people use to distinguish causal from other types of associations. Accordingly, most theories of causation invoke an explicit requirement that a cause precede its effect in time (Reichenbach 1956; Good 1961; Suppes 1970; Shoham 1988). Yet temporal information alone cannot distinguish genuine causation from spurious associations caused by unknown factors – the barometer falls before it rains yet does not cause the rain. In fact, the statistical and philosophical literature has adamantly warned analysts that, unless one knows in advance all causally relevant factors or unless one can carefully manipulate some variables, no genuine causal inferences are possible (Fisher 1935; Skyrms 1980; Cliff 1983; Eells and Sober 1983; Holland 1986; Gardenfors 1988; Cartwright 1989).[1] Neither condition is realizable in normal learning environments, and the question remains how causal knowledge is ever acquired from experience.

The clues that we explore in this chapter come from certain patterns of statistical associations that are characteristic of causal organizations – patterns that, in fact, can be given meaningful interpretation only in terms of causal directionality. Consider, for example, the following *intransitive* pattern of dependencies among three events: A and B are dependent, B and C are dependent, yet A and C are independent. If you ask a person to supply an example of three such events, the example would invariably portray A and C as two independent causes and B as their common effect, namely, $A \rightarrow B \leftarrow C$. (In my favorite example, A and C are the outcomes of two fair coins, and B represents a bell that rings whenever either coin comes up heads.) Fitting this dependence pattern with a scenario in which B is the cause and A and C are the effects is mathematically feasible but very unnatural (the reader is encouraged to try this exercise).

Such thought experiments tell us that certain patterns of dependency, void of temporal information, are conceptually characteristic of certain causal directionalities and not others. Reichenbach (1956), who was the first to wonder about the origin of those patterns, suggested that they are characteristic of Nature, reflective of the second law of thermodynamics. Rebane and Pearl (1987) posed the question in reverse, and asked whether the distinctions among the dependencies associated with the three basic causal substructures: $X \rightarrow Y \rightarrow Z$, $X \leftarrow Y \rightarrow Z$, and $X \rightarrow Y \leftarrow Z$ can be used to uncover genuine causal influences in the underlying data-generating process. They quickly realized that the key to determining the direction of the causal relationship between X and Y lies in "the presence of a third variable Z that correlates with Y but not with X," as in the collider $X \rightarrow Y \leftarrow Z$, and developed an algorithm that recovers both the edges and directionalities in the class of causal graphs that they considered (i.e., a polytrees).

The investigation in this chapter formalizes these intuitions and extends the Rebane-Pearl recovery algorithm to general graphs, including graphs with unobserved variables.

2.2 THE CAUSAL DISCOVERY FRAMEWORK

We view the task of causal discovery as an induction game that scientists play against Nature. Nature possesses stable causal mechanisms that, on a detailed level of descriptions, are deterministic functional relationships between variables, some of which are unobservable. These mechanisms are organized in the form of an acyclic structure, which the scientist attempts to identify from the available observations.

---

[1] Some of the popular quotes are: "No causation without manipulation" (Holland 1986), "No causes in, no causes out" (Cartwright 1989), "No computer program can take account of variables that are not in the analysis" (Cliff 1983). My favorite: "Causation first, Manipulation second."

Definition 2.2.1. (Causal Structure)
A causal structure of a set of variables V is a directed acyclic graph (DAG) in which each node corresponds to a distinct element of V, and each link represents a direct functional relationship among the corresponding variables.

A causal structure serves as a blueprint for forming a "causal model" – a precise speci-fication of *how* each variable is influenced by its parents in the DAG, as in the structural equation model of (1.40). Here we assume that Nature is at liberty to impose arbitrary functional relationships between each effect and its causes and then to perturb these rela-tionships by introducing arbitrary (yet mutually independent) disturbances. These distur-bances reflect "hidden" or unmeasurable conditions that Nature governs by some undis-closed probability function.

Definition 2.2.2 (Causal Model)
A causal model is a pair $M = \langle D, \Theta_D \rangle$ consisting of a causal structure D and a set of pa-rameters Θ_D compatible with D. The parameters Θ_D assign a function $x_i = f_i(pa_i, u_i)$ to each $X_i \in V$ and a probability measure $P(u_i)$ to each u_i, where PA_i are the parents of X_i in D and where each U_i is a random disturbance distributed according to $P(u_i)$, independently of all other u.

As we have seen in Chapter 1 (Theorem 1.4.1), the assumption of independent distur-bances renders the model *Markovian* in the sense that each variable is independent of all its nondescendants, conditional on its parents in D. The ubiquity of the Markov assump-tion in human discourse may be reflective of the granularity of the models we deem use-ful for understanding Nature. We can start in the deterministic extreme, where all vari-ables are explicated in microscopic detail and where the Markov condition certainly holds. As we move up to macroscopic abstractions by aggregating variables and intro-ducing probabilities to summarize omitted variables, we need to decide at what stage the abstraction has gone too far and where useful properties of causation are lost. Evidently, the Markov condition has been recognized by our ancestors (the authors of our causal thoughts) as a property worth protecting in this abstraction; correlations that are not explained by common causes are considered spurious, and models containing such cor-relations are considered incomplete. The Markov condition guides us in deciding when a set of parents PA_i is considered complete in the sense that it includes *all* the relevant immediate causes of variable X_i. It permits us to leave some of these causes out of PA_i (to be summarized by probabilities), but not if they also affect other variables modeled in the system. If a set PA_i in a model is too narrow, there will be disturbance terms that influence several variables simultaneously, and the Markov property will be lost. Such disturbances will be treated explicitly as "latent" variables (see Definition 2.3.2). Once we acknowledge the existence of latent variables and represent their existence explicitly as nodes in a graph, the Markov property is restored.

Once a causal model M is formed, it defines a joint probability distribution $P(M)$ over the variables in the system. This distribution reflects some features of the causal structure (e.g., each variable must be independent of its grandparents, given the values of its parents). Nature then permits the scientist to inspect a select subset $O \subseteq V$ of "observed" variables and to ask questions about $P_{[O]}$, the probability distribution over the observables, but it hides the underlying causal model as well as the causal structure. We investigate the feasibility of recovering the topology D of the DAG from features of the probability distribution $P_{[O]}$.[2]

2.3   MODEL PREFERENCE (OCCAM'S RAZOR)

In principle, since V is unknown, there is an unbounded number of models that would fit a given distribution, each invoking a different set of "hidden" variables and each connecting the observed variables through different causal relationships. Therefore, with no restriction on the type of models considered, the scientist is unable to make any meaningful assertions about the structure underlying the phenomena. For example, every probability distribution $P_{[O]}$ can be generated by a structure in which no observed variable is a cause of another but instead all variables are consequences of one latent common cause, U.[3] Likewise, assuming $V = O$ but lacking temporal information, the scientist can never rule out the possibility that the underlying structure is a complete, acyclic, and arbitrarily ordered graph – a structure that (with the right choice of parameters) can *mimic* the behavior of any model, regardless of the variable ordering. However, following standard norms of scientific induction, it is reasonable to rule out any theory for which we find a simpler, less elaborate theory that is equally consistent with the data (see Definition 2.3.5). Theories that survive this selection process are called *minimal*. With this notion, we can construct our (preliminary) definition of inferred causation as follows.

Definition 2.3.1 (Inferred Causation (Preliminary))
A variable X is said to have a causal influence *on a variable Y if a directed path from X to Y exists in every minimal structure consistent with the data.*

Here we equate a causal structure with a scientific theory, since both contain a set of free parameters that can be adjusted to fit the data. We regard Definition 2.3.1 as preliminary because it assumes that all variables are observed. The next few definitions generalize the concept of minimality to structures with unobserved variables.

Definition 2.3.2 (Latent Structure)
A latent structure is a pair $L = \langle D, O \rangle$, where D is a causal structure over V and where $O \subseteq V$ is a set of observed variables.

Definition 2.3.3 (Structure Preference)
One latent structure $L = \langle D, O \rangle$ is preferred to another $L' = \langle D', O \rangle$ (written $L \preceq L'$) if and only if D' can mimic D over O – that is, if and only if for every Θ_D there exists a

---

[2] This formulation invokes several idealizations of the actual task of scientific discovery. It assumes, for example, that the scientist obtains the distribution directly, rather than events sampled from the distribution. Additionally, we assume that the observed variables actually appear in the original causal model and are not some aggregate thereof. Aggregation might result in feedback loops, which we do not discuss in this chapter.

[3] This can be realized by letting U have as many states as O, assigning to U the prior distribution $P(u) = P(o(u))$ (where $o(u)$ is the cell of O corresponding to state u), and letting each observed variable O_i take on its corresponding value in $o(u)$.

$\Theta'_{D'}$ *such that* $P_{[O]}(\langle D', \Theta'_{D'} \rangle) = P_{[O]}(\langle D, \Theta_D \rangle).$ *Two latent structures are* equivalent, *written* $L' \equiv L$, *if and only if* $L \preceq L'$ *and* $L \succeq L'.$[4]

Note that the preference for simplicity imposed by Definition 2.3.3 is gauged by the expressive power of a structure, not by its syntactic description. For example, one latent structure L_1 may invoke many more parameters than L_2 and still be preferred if L_2 can accommodate a richer set of probability distributions over the observables. One reason scientists prefer simpler theories is that such theories are more constraining and thus more falsifiable; they provide the scientist with less opportunities to overfit the data "hindsightedly" and therefore command greater credibility if a fit is found (Popper 1959; Pearl 1978; Blumer et al. 1987).

We also note that the set of independencies entailed by a causal structure imposes limits on its expressive power, that is, its power to mimic other structures. Indeed, L_1 cannot be preferred to L_2 if there is even one observable dependency that is permitted by L_1 and forbidden by L_2. Thus, tests for preference and equivalence can sometimes be reduced to tests of induced dependencies, which in turn can be determined directly from the topology of the DAGs without ever concerning ourselves with the set of parameters. This is the case in the absence of hidden variables (see Theorem 1.2.8) but does not hold generally in all latent structures. Verma and Pearl (1990) showed that some latent structures impose numerical rather than independence constraints on the observed distribution (see, e.g., Section 8.4, equations (8.21)–(8.23)); this makes the task of verifying model preference complicated but does still permit us to extend the semantical definition of inferred causation (Definition 2.3.1) to latent structures.

Definition 2.3.4 (Minimality)

A latent structure L is minimal *with respect to a class $\mathcal{L}$ of latent structures if and only if there is no member of $\mathcal{L}$ that is strictly preferred to L – that is, if and only if for every $L' \in \mathcal{L}$ we have $L \equiv L'$ whenever $L' \preceq L$.*

Definition 2.3.5 (Consistency)

A latent structure $L = |D, O|$ is consistent *with a distribution $\hat{P}$ over O if D can accommodate some model that generates $\hat{P}$ – that is, if there exists a parameterization Θ_D such that $P_{[O]}(\langle D, \Theta_D \rangle) = \hat{P}$.*

Clearly, a necessary (and sometimes sufficient) condition for L to be consistent with $\hat{P}$ is that L can account for all the dependencies embodied in $\hat{P}$

Definition 2.3.6 (Inferred Causation)

Given $\hat{P}$, a variable C has a causal influence *on variable E if and only if there exists a directed path from C to E in every minimal latent structure consistent with $\hat{P}$.*

We view this definition as normative because it is based on one of the least disputed norms of scientific investigation: Occam's razor in its semantical casting. However, as with any

---

[4] We use the succinct term "preferred to" to mean "preferred or equivalent to," a relation that has also been named "a submodel of."

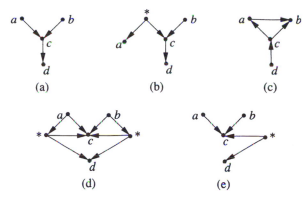

Figure 2.1 Causal structures illustrating the minimality of (a) and (b) and the justification for inferring the relationship $c \rightarrow d$. The node ($*$) represents a hidden variable with any number of states.

scientific inquiry, we make no claims that this definition is guaranteed to always identify stable physical mechanisms in Nature. It identifies the mechanisms we can plausibly infer from nonexperimental data; moreover, it guarantees that any alternative mechanism will be less trustworthy than the one inferred because the alternative would require more contrived, hindsighted adjustment of parameters (i.e., functions) to fit the data.

As an example of a causal relation that is identified by Definition 2.3.6, imagine that observations taken over four variables $\{a, b, c, d\}$ reveal two independencies: "a is independent of b" and "d is independent of $\{a, b\}$ given c." Assume further that the data reveals *no other* independence besides those that logically follow from these two. This dependence pattern would be typical, for example, of the following variables: a = having a cold, b = having hay fever, c = having to sneeze, d = having to wipe one's nose. It is not hard to see that structures (a) and (b) in Figure 2.1 are minimal, for they entail the observed independencies and none other.[5] Furthermore, any structure that explains the observed dependence between c and d by an arrow from d to c, or by a hidden common cause ($*$) between the two, cannot be minimal, because any such structure would be able to "out-mimic" the one shown in Figure 2.1(a) (or the one in Figure 2.1(b)), which reflects all observed independencies. For example, the structure of Figure 2.1(c), unlike that of Figure 2.1(a), accommodates distributions with arbitrary relations between a and b. Similarly, Figure 2.1(d) is not minimal because it fails to impose the conditional independence between d and $\{a, b\}$ given c and will therefore accommodate distributions in which d and $\{a, b\}$ are dependent given c. In contrast, Figure 2.1(e) is not consistent with the data, since it imposes an unobserved marginal independence between $\{a, b\}$ and d.

This example (taken from Pearl and Verma 1991) illustrates a remarkable connection between causality and probability: certain patterns of probabilistic dependencies (in our case, all dependencies except ($a \perp\!\!\!\perp b$) and ($d \perp\!\!\!\perp \{a, b\} \mid c$)) imply unambiguous *causal* dependencies (in our case, $c \rightarrow d$) without making any assumption about the presence

---

[5] To verify that (a) and (b) are equivalent, we note that (b) can mimic (a) if we let the link $a \leftarrow *$ impose equality between the two variables. Conversely, (a) can mimic (b), since it is capable of generating every distribution that possesses the independencies entailed by (b). (For theory and methods of "reading off" conditional independencies from graphs, see Section 1.2.3 or Pearl 1988b.)

or absence of latent variables.[6] The only assumption invoked in this implication is minimality – models that overfit the data are ruled out.

2.4 STABLE DISTRIBUTIONS

Although the minimality principle is sufficient for forming a normative theory of inferred causation, it does not guarantee that the structure of the actual data-generating model would be minimal, or that the search through the vast space of minimal structures would be computationally practical. Some structures may admit peculiar parameterizations that would render them indistinguishable from many other minimal models that have totally disparate structures. For example, consider a binary variable C that takes the value 1 whenever the outcomes of two fair coins (A and B) are the same and takes the value 0 otherwise. In the trivariate distribution generated by this parameterization, each pair of variables is marginally independent yet is dependent conditional on the third variable. Such a dependence pattern may in fact be generated by three minimal causal structures, each depicting one of the variables as causally dependent on the other two, but there is no way to decide among the three. In order to rule out such "pathological" parameterizations, we impose a restriction on the distribution called *stability*, also known as DAG-isomorphism or perfect-mapness (Pearl 1988b, p. 128) and faithfulness (Spirtes et al. 1993). This restriction conveys the assumption that all the independencies embedded in P are stable; that is, they are entailed by the structure of the model D and hence remain invariant to any change in the parameters Θ_D. In our example, only the correct structure (namely, $A \rightarrow C \leftarrow B$) will retain its independence pattern in the face of changing parameterizations – say, when the coins become slightly biased.

Definition 2.4.1 (Stability)
Let $I(P)$ denote the set of all conditional independence relationships embodied in P. A causal model $M = \langle D, \Theta_D \rangle$ generates a stable *distribution if and only if $P(\langle D, \Theta_D \rangle)$ contains no extraneous independences – that is, if and only if $I(P(\langle D, \Theta_D \rangle)) \subseteq I(P(\langle D, \Theta'_D \rangle))$ for any set of parameters Θ'_D.*

The stability condition states that, as we vary the parameters from Θ to Θ', no independence in P can be destroyed; hence the name "stability." Succinctly, P is a stable distribution of M if it "maps" the structure D of M, that is, $(X \perp\!\!\!\perp Y \mid Z)_P \Longleftrightarrow (X \perp\!\!\!\perp Y \mid Z)_D$ for any three sets of variables $X, Y,$ and Z (see Theorem 1.2.5).

The relationship between minimality and stability can be illustrated using the following analogy. Suppose we see a picture of a chair and that we need to decide between two theories as follows.

T_1:   The object in the picture is a chair.
T_2:   The object in the picture is either a chair or two chairs positioned such that one hides the other.

---

[6] Standard probabilistic definitions of causality (e.g., Suppes 1970; Eells 1991) invariably require knowledge of all relevant factors that may influence the observed variables (see Section 7.5.3).

Our preference for T_1 over T_2 can be justified on two principles, one based on minimality and the other on stability. The minimality principle argues that T_1 is preferred to T_2 because the set of scenes composed of single objects is a proper subset of scenes composed of two or fewer objects and, unless we have evidence to the contrary, we should prefer the more specific theory. The stability principle rules out T_2 a priori, arguing that it would be rather unlikely for two objects to align themselves so as to have one perfectly hide the other. Such an alignment would be *unstable* relative to slight changes in environmental conditions or viewing angle.

The analogy with independencies is clear. Some independencies are *structural*, that is, they would persist for every functional–distributional parameterization of the graph. Others are sensitive to the precise numerical values of the functions and distributions. For example, in the structure $Z \leftarrow X \rightarrow Y$, which stands for the relations

$$z = f_1(x, u_1), \quad y = f_2(x, u_2), \tag{2.1}$$

the variables Z and Y will be independent, conditional on X, for all functions f_1 and f_2. In contrast, if we add an arrow $Z \rightarrow Y$ to the structure and use a linear model

$$z = \gamma x + u_1, \quad y = \alpha x + \beta z + u_2, \tag{2.2}$$

with $\alpha = -\beta\gamma$, then Y and X will be independent. However, the independence between Y and X is unstable because it disappears as soon as the equality $\alpha = -\beta\gamma$ is violated. The stability assumption presumes that this type of independence is unlikely to occur in the data, that all independencies are structural.

To further illustrate the relations between stability and minimality, consider the causal structure depicted in Figure 2.1(c). The minimality principle rejects this structure on the ground that it fits a broader set of distributions than those fitted by structure (a). The stability principle rejects this structure on the ground that, in order to fit the data (specifically, the independence $(a \perp\!\!\!\perp b)$), the association produced by the arrow $a \rightarrow b$ must cancel precisely the one produced by the path $a \leftarrow c \rightarrow b$. Such precise cancelation cannot be stable, for it cannot be sustained for all functions connecting variables a, b, and c. In structure (a), by contrast, the independence $(a \perp\!\!\!\perp b)$ is stable.

2.5 RECOVERING DAG STRUCTURES

With the added assumption of stability, every distribution has a unique minimal causal structure (up to d-separation equivalence), as long as there are no hidden variables. This uniqueness follows from Theorem 1.2.8, which states that two causal structures are equivalent (i.e., they can mimic each other) if and only if they relay the same dependency information – namely, they have the same skeleton and same set of v-structures.

In the absence of unmeasured variables, the search for the minimal model then boils down to reconstructing the structure of a DAG D from queries about conditional independencies, assuming that those independencies reflect d-separation conditions in some undisclosed underlying DAG D_0. Naturally, since D_0 may have equivalent structures, the reconstructed DAG will not be unique, and the best we can do is to find a graphical representation for the equivalence class of D_0. Such graphical representation was introduced in Verma and Pearl (1990) under the name *pattern*. A pattern is a partially directed

DAG, in particular, a graph in which some edges are directed and some are nondirected. The directed edges represent arrows that are common to every member in the equivalence class of D_0, while the undirected edges represent ambivalence; they are directed one way in some equivalent structures and another way in others.

The following algorithm, introduced in Verma and Pearl (1990), takes as input a stable probability distribution $\hat{P}$ generated by some underlying DAG D_0 and outputs a pattern that represents the equivalence class of D_0.[7]

IC Algorithm (**Inductive Causation**)

Input: $\hat{P}$, a stable distribution on a set V of variables.

Output: a pattern $H(\hat{P})$ compatible with $\hat{P}$.

1. For each pair of variables a and b in V, search for a set S_{ab} such that $(a \perp\!\!\!\perp b \mid S_{ab})$ holds in $\hat{P}$ – in other words, a and b should be independent in $\hat{P}$, conditioned on S_{ab}. Construct an undirected graph G such that vertices a and b are connected with an edge if and only if no set S_{ab} can be found.

2. For each pair of nonadjacent variables a and b with a common neighbor c, check if $c \in S_{ab}$.
   If it is, then continue.
   If it is not, then add arrowheads pointing at c (i.e., $a \rightarrow c \leftarrow b$).

3. In the partially directed graph that results, orient as many of the undirected edges as possible subject to two conditions: (i) Any alternative orientation would yield a new v-structure; or (ii) Any alternative orientation would yield a directed cycle.

The IC algorithm leaves the details of steps 1 and 3 unspecified, and several refinements have been proposed for optimizing these two steps. Verma and Pearl (1990) noted that, in sparse graphs, the search can be trimmed substantially if commenced with the Markov network of P, namely, the undirected graph formed by linking only pairs that are dependent conditionally on all other variables. In linear Gaussian models, the Markov network can be found in polynomial time, through matrix inversion, by assigning edges to pairs that correspond to the nonzero entries of the inverse covariance matrix. Spirtes and Glymour (1991) proposed a general systematic way of searching for the sets S_{ab} in step 1. Starting with sets S_{ab} of cardinality 0, then cardinality 1, and so on, edges are recursively removed from a complete graph as soon as separation is found. This refinement, called the PC algorithm (after its authors, Peter and Clark), enjoys polynomial time in graphs of finite degree because, at every stage, the search for a separating set S_{ab} can be limited to nodes that are adjacent to a and b.

Step 3 of the IC algorithm can be systematized in several ways. Verma and Pearl (1992) showed that, starting with any pattern, the following four rules are required for obtaining a maximally oriented pattern.

---

[7] The IC algorithm, as introduced in Verma and Pearl (1990), was designed to operate on latent structures. For clarity, we here present the algorithm in two separate parts, IC and IC*, with IC restricted to DAGs and IC* operating on latent structures.

R_1: Orient $b - c$ into $b \to c$ whenever there is an arrow $a \to b$ such that a and c are nonadjacent.

R_2: Orient $a - b$ into $a \to b$ whenever there is chain $a \to c \to b$.

R_3: Orient $a - b$ into $a \to b$ whenever there are two chains $a - c \to b$ and $a - d \to b$ such that c and d are nonadjacent.

R_4: Orient $a - b$ into $a \to b$ whenever there are two chains $a - c \to d$ and $c \to d \to b$ such that c and b are nonadjacent and a and d are adjacent.

Meek (1995) showed that these four rules are also sufficient, so that repeated application will eventually orient *all* arrows that are common to the equivalence class of D_0. Moreover, R_4 is not required if the starting orientation is limited to v-structures.

Another systematization is offered by an algorithm due to Dor and Tarsi (1992) that tests (in polynomial time) if a given partially oriented acyclic graph can be fully oriented without creating a new v-structure or a directed cycle. The test is based on recursively removing any vertex v that has the following two properties:

1. no edge is directed outward from v;
2. every neighbor of v that is connected to v through an undirected edge is also adjacent to all the other neighbors of v.

A partially oriented acyclic graph has an admissible extension in a DAG if and only if all its vertices can be removed in this fashion. Thus, to find the maximally oriented pattern, we can (i) separately try the two orientations, $a \to b$ and $a \leftarrow b$, for every undirected edge $a - b$, and (ii) test whether both orientations, or just one, have extensions. The set of uniquely orientable arrows constitutes the desired maximally oriented pattern. Additional refinements can be found in Chickering (1995), Andersson et al. (1997), and Moole (1997).

Latent structures, however, require special treatment, because the constraints that a latent structure imposes upon the distribution cannot be completely characterized by any set of conditional independence statements. Fortunately, certain sets of those independence constraints can be identified (Verma and Pearl 1990); this permits us to recover valid fragments of latent structures.

2.6 RECOVERING LATENT STRUCTURES

When Nature decides to "hide" some variables, the observed distribution $\hat{P}$ need no longer be stable relative to the observable set O. That is, we are no longer guaranteed that, among the minimal latent structures compatible with $\hat{P}$, there exists one that has a DAG structure. Fortunately, rather then having to search through this unbounded space of latent structures, the search can be confined to graphs with finite and well-defined structures. For every latent structure L, there is a dependency-equivalent latent structure (the projection) of L on O in which every unobserved node is a root node with exactly two observed children. We characterize this notion explicitly as follows.

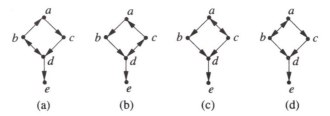

Figure 2.4 Latent structures equivalent to those of Figure 2.3(a).

$S_{bc} = \{a\}$, $S_{be} = \{d\}$, and $S_{ce} = \{d\}$. Thus, step 1 of IC* yields the undirected graph of Figure 2.3(b).

2. The triplet (b, d, c) is the only one that satisfies the condition of step 2, since d is not in S_{bc}. Accordingly, we obtain the partially directed graph of Figure 2.3(c).

3. Rule R_1 of step 3 is applicable to the triplet (b, d, e) (and to (c, d, e)), since b and e are nonadjacent and there is an arrowhead at d from b but not from e. We therefore add an arrowhead at e, and mark the link, to obtain Figure 2.3(d). This is also the final output of IC*, because R_1 and R_2 are no longer applicable.

The absence of arrowheads on $a - b$ and $a - c$, and the absence of markings on $b \rightarrow d$ and $c \rightarrow d$, correctly represent the ambiguities presented by $\hat{P}$. Indeed, each of the latent structures shown in Figure 2.4 is observationally equivalent to that of Figure 2.3(a). Marking the link $d \rightarrow e$ in Figure 2.3(d) advertises the existence of a directed link $d \rightarrow e$ in each and every latent structure that is independence-equivalent to the one in Figure 2.3(a).

2.7 LOCAL CRITERIA FOR INFERRING CAUSAL RELATIONS

The IC* algorithm takes a distribution $\hat{P}$ and outputs a partially directed graph. Some of the links are marked unidirectional (denoting genuine causation), some are *un*marked unidirectional (denoting potential causation), some are bidirectional (denoting spurious association), and some are undirected (denoting relationships that remain undetermined). The conditions that give rise to these labelings can be taken as definitions for the various kinds of causal relationships. In this section we present explicit definitions of potential and genuine causation as they emerge from the IC* algorithm. Note that, in all these definitions, the criterion for causation between two variables (X and Y) will require that a third variable Z exhibit a specific pattern of dependency with X and Y. This is not surprising, since the essence of causal claims is to stipulate the behavior of X and Y under the influence of a third variable, one that corresponds to an external control of X (or Y) – as echoed in the paradigm of "no causation without manipulation" (Holland 1986). The difference is only that the variable Z, acting as a virtual control, must be identified within the data itself, as if Nature had performed the experiment. The IC* algorithm can be regarded as offering a systematic way of searching for variables Z that qualify as virtual controls, given the assumption of stability.

Definition 2.7.1 (Potential Cause)
A variable X has a potential causal influence on another variable Y (that is inferable from $\hat{P}$) if the following conditions hold.

1. *X and Y are dependent in every context.*
2. *There exists a variable Z and a context S such that*
    (i)   *X and Z are independent given S (i.e., $X \perp\!\!\!\perp Z \mid S$) and*
    (ii) *Z and Y are dependent given S (i.e., $Z \not\!\perp\!\!\!\perp Y \mid S$).*

By "context" we mean a set of variables tied to specific values. In Figure 2.3(a), for example, variable b qualifies as a potential cause of d by virtue of variable $Z = c$ being dependent on d and independent of b in context $S = a$. Likewise, c qualifies as a potential cause of d (with $Z = b$ and $S = a$). Neither b nor c qualifies as a genuine cause of d, because this pattern of dependencies is also compatible with a latent common cause, shown as bidirected arcs in Figures 2.4(a)–(b). However, Definition 2.7.1 disqualifies d as a cause of b (or c), and this leads to the classification of d as a *genuine* cause of e, as formulated in Definition 2.7.2.[9] Note that Definition 2.7.1 precludes a variable X from being a potential cause of itself or of any other variable that functionally determines X.

Definition 2.7.2 (Genuine Cause)
A variable X has a genuine causal influence on another variable Y if there exists a variable Z such that either:

1. *X and Y are dependent in any context and there exists a context S satisfying*
    (i)   *Z is a potential cause of X (per Definition 2.7.1),*
    (ii) *Z and Y are dependent given S (i.e., $Z \not\!\perp\!\!\!\perp Y \mid S$), and*
    (iii) *Z and Y are independent given $S \cup X$ (i.e., $Z \perp\!\!\!\perp Y \mid S \cup X$);*
    or
2. *X and Y are in the transitive closure of the relation defined in criterion 1.*

Conditions (i)–(iii) are illustrated in Figure 2.3(a) with $X = d$, $Y = e$, $Z = b$, and $S = \emptyset$. The destruction of the dependence between b and e through conditioning on d cannot be attributed to spurious association between d and e; genuine causal influence is the only explanation, as shown in the structures of Figure 2.4.

Definition 2.7.3 (Spurious Association)
Two variables X and Y are spuriously associated if they are dependent in some context and there exist two other variables (Z_1 and Z_2) and two contexts (S_1 and S_2) such that:

---

[9] Definition 2.7.1 was formulated in Pearl (1990) as a relation between events (rather than variables) with the added condition $P(Y \mid X) > P(Y)$ (in the spirit of Reichenbach 1956, Good 1961, and Suppes 1970). This refinement is applicable to any of the definitions in this section, but it will not be formulated explicitly.

patterns of dependencies that should be sufficient to uncover genuine causal relationships. These relationships cannot be attributed to hidden causes lest we violate one of the basic maxims of scientific methodology: the semantical version of Occam's razor. Adherence to this maxim may explain why humans reach consensus regarding the directionality and nonspuriousness of causal relationships in the face of opposing alternatives that are perfectly consistent with observation. Echoing Cartwright (1989), we summarize our claim with the slogan "No causes in – No causes out; Occam's razor in – Some causes out."

How safe are the causal relationships inferred by the IC algorithm – or by the TETRAD program of Spirtes et al. (1993) or the Bayesian methods of Cooper and Herskovits (1991) or Heckerman et al. (1994)?

Recasting this question in the context of visual perception, we may equally well ask: How safe are our predictions when we recognize three-dimensional objects from their two-dimensional shadows, or from the two-dimensional pictures that objects reflect on our retinas? The answer is: Not absolutely safe, but good enough to tell a tree from a house and good enough to make useful inferences without having to touch every physical object that we see. Returning to causal inference, our question then amounts to assessing whether there are enough discriminating clues in a typical learning environment (say, in skill acquisition tasks or in epidemiological studies) to allow us to make reliable discriminations between cause and effect. Rephrased as a logical guarantee, we can categorically assert that the IC* algorithm will never label an arrow $a \rightarrow b$ as genuine if in fact a has no causal influence on b and if the observed distribution is stable relative to its underlying causal model.

On the practical side, we have shown that the assumption of model minimality, together with that of "stability" (no accidental independencies) lead to an effective algorithm for structuring candidate causal models capable of generating the data, transparent as well as latent. Simulation studies conducted at our laboratory in 1990 showed that networks containing tens of variables require fewer than 5,000 samples to have their structure recovered by the algorithm. For example, 1,000 samples taken from (a binary version of) the process shown in (2.3), each containing ten successive X, Y pairs, were sufficient to recover its double-chain structure (and the correct direction of time). The greater the noise, the quicker the recovery (up to a point). In testing this modeling scheme on real-life data, we have examined the observations reported in Sewal Wright's seminal paper "Corn and Hog Correlations" (Wright 1925). As expected, corn price (X) can clearly be identified as a cause of hog price (Y) but not the other way around. The reason lies in the existence of the variable corn crop (Z), which satisfies the conditions of Definition 2.7.2 (with $S = \emptyset$). Several applications of the principles and algorithms discussed in this chapter are described in Glymour and Cooper (1999, pp. 441–541).

It should be natural to ask how the new criteria for causation could benefit current research in machine learning and data mining. In some sense, our method resembles a standard, machine-learning search through a space of hypotheses (Mitchell 1982) where each hypothesis stands for a causal model. Unfortunately, this is where the resemblance ends. The prevailing paradigm in the machine-learning literature has been to define each hypothesis (or theory, or concept) as a subset of observable instances; once we observe the entire extension of this subset, the hypothesis is defined unambiguously. This is not the case in causal discovery. Even if the training sample exhausts the hypothesis subset (in our case, this corresponds to observing P precisely), we are still left with a vast number of equivalent causal theories, each stipulating a drastically different set of causal claims. Therefore, *fitness to data is an insufficient criterion for validating causal theories.* Whereas in traditional learning tasks we attempt to generalize from one set of instances to another, the causal modeling task is to generalize from behavior under one set of conditions to behavior under another set. Causal models should therefore be chosen by a criterion that challenges their stability against changing conditions, and this is indeed what scientists attempt to accomplish through controlled experimentation. Absent such experimentation, the best one can do is to rely on virtual control variables, like those revealed by Nature through the dependence patterns of Definitions 2.7.1–2.7.4.

2.9.1 On Minimality, Markov, and Stability

The idea of inferring causation from association cannot be expected to go unchallenged by scientists trained along the lines of traditional doctrines. Naturally, the assumptions underlying the theory described in this chapter – minimality and stability – come under attack from statisticians and philosophers. This section contains additional thoughts in defense of these assumptions.

Although few have challenged the principle of minimality (to do so would amount to challenging scientific induction), objections have been voiced against the way we defined the objects of minimization – namely, causal models. Definition 2.2.2 assumes that the stochastic terms u_i are mutually independent, an assumption that endows each model with the Markov property: conditioned on its parents (direct causes), each variable is independent of its nondescendants. This implies, among the other ramifications of d-separation, several familiar relationships between causation and association that are usually associated with Reichenbach's (1956) principle of common cause – for example, "no correlation without causation," "causes screen off their effects," "no action at a distance."

The Markovian assumption, as explained in our discussion of Definition 2.2.2, is a matter of convention, to distinguish complete from incomplete models.[14] By building the Markovian assumption into the definition of complete causal models (Definition 2.2.2) and then relaxing the assumption through latent structures (Definition 2.3.2), we declare our preparedness to miss the discovery of non-Markovian causal models that cannot be described as latent structures. I do not consider this loss to be very serious, because such models – even if any exist in the macroscopic world – would have limited utility as guides to decisions. For example, it is not clear how one would predict the effects of interventions from such a model, save for explicitly listing the effect of every conceivable intervention in advance.

---

[14] Discovery algorithms for certain non-Markovian models, involving cycles and selection bias, have been reported in Spirtes et al. (1995) and Richardson (1996).

over the space of possible structures to seek the one(s) with the highest posterior score. Methods based on this approach have the advantage of operating well under small-sample conditions, but they encounter difficulties in coping with hidden variables. The assumption of parameter independence, which is made in all practical implementations of the Bayesian approach, induces preferences toward models with fewer parameters and hence toward minimality. Likewise, parameter independence can be justified only when the parameters represent mechanisms that are free to change independently of one another – that is, when the system is autonomous and hence stable.

Postscript for the Second Edition

Work on causal discovery has been pursued vigorously by the TETRAD group at Carengie Mellon University and reported in Spirtes et al. (2000), Robins et al. (2003), Scheines (2002), and Moneta and Spirtes (2006).

Applications of causal discovery in economics are reported in Bessler (2002), Swanson and Granger (1997), and Demiralp and Hoover (2003). Gopnik et al. (2004) applied causal Bayesian networks to explain how children acquire causal knowledge from observations and actions (see also Glymour 2001).

Hoyer et al. (2006) and Shimizu et al. (2005, 2006) have proposed a new scheme of discovering causal directionality, based not on conditional independence but on functional composition. The idea is that in a linear model $X \rightarrow Y$ with non-Gaussian noise, variable Y is a linear combination of two independent noise terms. As a consequence, $P(y)$ is a convolution of two non-Gaussian distributions and would be, figuratively speaking, "more Gaussian" than $P(x)$. The relation of "more Gaussian than" can be given precise numerical measure and used to infer directionality of certain arrows.

Tian and Pearl (2001a,b) developed yet another method of causal discovery based on the detection of "shocks," or spontaneous local changes in the environment which act like "Nature's interventions," and unveil causal directionality toward the consequences of those shocks.

Verma and Pearl (1990) noted that two latent structures may entail the same set of conditional independencies and yet impose different equality constraints on the joint distributions. These constraints, dubbed "dormant independencies," were characterized systematically in Tian and Pearl (2002b) and Shpitser and Pearl (2008); they promise to provide a powerful new discovery tool for structure learning.

A program of benchmarks of causal discovery algorithms, named "Causality Workbench," has been reported by Guyon et al. (2008a,b; http://clopinet.com/causality). Regular contests are organized in which participants are given real data or data generated by a concealed causal model, and the challenge is to predict the outcome of a select set of interventions.

CHAPTER THREE

Causal Diagrams and the Identification of Causal Effects

> *The eye obeys exactly*
> *the action of the mind.*
> Emerson (1860)

Preface

In the previous chapter we dealt with ways of learning causal relationships from raw data. In this chapter we explore the ways of inferring such relationships from a combination of data and qualitative causal assumptions that are deemed plausible in a given domain. More broadly, this chapter aims to help researchers communicate qualitative assumptions about cause–effect relationships, elucidate the ramifications of such assumptions, and derive causal inferences from a combination of assumptions, experiments, and data. Our major task will be to decide whether the assumptions given are sufficient for assessing the strength of causal effects from nonexperimental data.

Causal effects permit us to predict how systems would respond to hypothetical interventions – for example, policy decisions or actions performed in everyday activity. As we have seen in Chapter 1 (Section 1.3), such predictions are the hallmark of the empirical sciences and are not discernible from probabilistic information alone; they rest on – and, in fact, define – causal relationships. This chapter uses causal diagrams to give formal semantics to the notion of *intervention*, and it provides explicit formulas for postintervention probabilities in terms of preintervention probabilities. The implication is that the effects of every intervention can be estimated from nonexperimental data, provided the data is supplemented with a causal diagram that is both acyclic and contains no latent variables.

If some variables are not measured then the question of identifiability arises, and this chapter develops a nonparametric framework for analyzing the identification of causal relationships in general and causal effects in particular. We will see that causal diagrams provide a powerful mathematical tool in this analysis; they can be queried, using extremely simple tests, to determine if the assumptions available are sufficient for identifying causal effects. If so, the diagrams produce mathematical expressions for causal effects in terms of observed distributions; otherwise, the diagrams can be queried to suggest additional observations or auxiliary experiments from which the desired inferences can be obtained.

Another tool that emerges from the graphical analysis of causal effects is a *calculus of interventions* – a set of inference rules by which sentences involving interventions and observations can be transformed into other such sentences, thus providing a syntactic method of deriving (or verifying) claims about interventions and the way they interact with observations. With the help of this calculus the reader will be able to (i) determine mathematically whether a given set of covariates is appropriate for control of confounding, (ii) deal with measurements that lie on the causal pathways, and (iii) trade one set of measurements for another.

Finally, we will show how the new calculus disambiguates concepts that have triggered controversy and miscommunication among philosophers, statisticians, economists, and psychologists. These include distinctions between structural and regression equations, definitions of direct and indirect effects, and relationships between structural equations and Neyman-Rubin models.

3.1 INTRODUCTION

The problems addressed in this chapter can best be illustrated through a classical example due to Cochran (see Wainer 1989). Consider an experiment in which soil fumigants (X) are used to increase oat crop yields (Y) by controlling the eelworm population (Z); the fumigants may also have direct effects (both beneficial and adverse) on yields besides the control of eelworms. We wish to assess the total effect of the fumigants on yields when this typical study is complicated by several factors. First, controlled randomized experiments are unfeasible – farmers insist on deciding for themselves which plots are to be fumigated. Second, farmers' choice of treatment depends on last year's eelworm population (Z_0), an unknown quantity that is strongly correlated with this year's population. Thus we have a classical case of confounding bias that interferes with the assessment of treatment effects regardless of sample size. Fortunately, through laboratory analysis of soil samples, we can determine the eelworm populations before and after the treatment; furthermore, because the fumigants are known to be active for a short period only, we can safely assume that they do not affect the growth of eelworms surviving the treatment. Instead, eelworms' growth depends on the population of birds (and other predators), which is correlated with last year's eelworm population and hence with the treatment itself.

The method developed in this chapter permits the investigator to translate complex considerations of this sort into a formal language and thereby facilitate the following tasks:

1. explicating the assumptions that underlie the model;

2. deciding whether the assumptions are sufficient to obtain consistent estimates of the target quantity – the total effect of the fumigants on yields;

3. providing (if the answer to item 2 is affirmative) a closed-form expression for the target quantity in terms of distributions of observed quantities; and

4. suggesting (if the answer to item 2 is negative) a set of observations and experiments that, if performed, would render a consistent estimate feasible.

The first step in this analysis is to construct a causal diagram like the one given in Figure 3.1, which represents the investigator's understanding of the major causal influences among measurable quantities in the domain. For example, the quantities Z_1, Z_2, Z_3 represent the eelworm population before treatment, after treatment, and at the end of the season, respectively. The Z_0 term represents last year's eelworm population; because it is an unknown quantity, it is denoted by a hollow circle, as is the quantity B, the

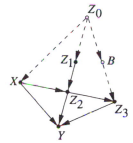

Figure 3.1 A causal diagram representing the effect of fumigants (X) on yields (Y)

population of birds and other predators. Links in the diagram are of two kinds: those emanating from unmeasured quantities are designated by dashed arrows, those connecting measured quantities by solid arrows. The substantive assumptions embodied in the diagram are negative causal assertions which are conveyed through the links *missing* from the diagram. For example, the missing arrow between Z_1 and Y signifies the investigator's understanding that pretreatment eelworms cannot affect oat plants directly; their entire influence on oat yields is mediated by the posttreatment conditions, Z_2 and Z_3. Our purpose is not to validate or repudiate such domain-specific assumptions but rather to test whether a given set of assumptions is sufficient for quantifying causal effects from nonexperimental data – here, estimating the total effect of fumigants on yields.

The causal diagram in Figure 3.1 is similar in many respects to the path diagrams devised by Wright (1921); both reflect the investigator's subjective and qualitative knowledge of causal influences in the domain, both employ directed acyclic graphs, and both allow for the incorporation of latent or unmeasured quantities. The major differences lie in the method of analysis. First, whereas path diagrams have been analyzed mostly in the context of linear models with Gaussian noise, causal diagrams permit arbitrary nonlinear interactions. In fact, our analysis of causal effects will be entirely nonparametric, entailing no commitment to a particular functional form for equations and distributions. Second, causal diagrams will be used not only as a passive language to communicate assumptions but also as an active computational device through which the desired quantities are derived. For example, the method to be described allows an investigator to inspect the diagram of Figure 3.1 and make the following immediate conclusions.

1. The total effect of X on Y can be estimated consistently from the observed distribution of X, Z_1, Z_2, Z_3, and Y.

2. The total effect of X on Y (assuming discrete variables throughout) is given by the formula[1]

$$P(y \mid \hat{x}) = \sum_{z_1} \sum_{z_2} \sum_{z_3} P(y \mid z_2, z_3, x) \, P(z_2 \mid z_1, x)$$
$$\times \sum_{x'} P(z_3 \mid z_1, z_2, x') \, P(z_1, x'), \qquad (3.1)$$

---

[1] The notation $P_x(y)$ was used in Chapter 1; it is changed henceforth to $P(y \mid \hat{x})$ or $P(y \mid do(x))$ because of the inconvenience in handling subscripts. The reader need not be intimidated if, at this point, (3.1) appears unfamiliar. After reading Section 3.4, the reader should be able to derive such formulas with greater ease than solving algebraic equations. Note that x' is merely an index of summation that ranges over the values of X.

of the intervention can be predicted by modifying the corresponding equations in the model and using the modified model to compute a new probability function.

The simplest type of external intervention is one in which a single variable, say X_i, is forced to take on some fixed value x_i. Such an intervention, which we call "atomic," amounts to lifting X_i from the influence of the old functional mechanism $x_i = f_i(pa_i, u_i)$ and placing it under the influence of a new mechanism that sets the value x_i while keeping all other mechanisms unperturbed. Formally, this atomic intervention, which we denote by $do(X_i = x_i)$, or $do(x_i)$ for short,[2] amounts to removing the equation $x_i = f_i(pa_i, u_i)$ from the model and substituting $X_i = x_i$ in the remaining equations. The new model thus created represents the system's behavior under the intervention $do(X_i = x_i)$ and, when solved for the distribution of X_j, yields the causal effect of X_i on X_j, which is denoted $P(x_j \mid \hat{x}_i)$. More generally, when an intervention forces a subset X of variables to attain fixed values x, then a subset of equations is to be pruned from the model given in (3.4), one for each member of X, thus defining a new distribution over the remaining variables that completely characterizes the effect of the intervention.[3]

Definition 3.2.1 (Causal Effect)
Given two disjoint sets of variables, X and Y, the causal effect of X on Y, denoted either as $P(y \mid \hat{x})$ or as $P(y \mid do(x))$, is a function from X to the space of probability distributions on Y. For each realization x of X, $P(y \mid \hat{x})$ gives the probability of Y = y induced by deleting from the model of (3.4) all equations corresponding to variables in X and substituting X = x in the remaining equations.

Clearly, the graph corresponding to the reduced set of equations is a subgraph of G from which all arrows entering X have been pruned (Spirtes et al. 1993). The difference $E(Y \mid do(x')) - E(Y \mid do(x''))$ is sometimes taken as the definition of "causal effect" (Rosenbaum and Rubin 1983), where x' and x'' are two distinct realizations of X. This difference can always be computed from the general function $P(y \mid do(x))$, which is defined for every level x of X and provides a more refined characterization of the effect of interventions.

3.2.2 Interventions as Variables

An alternative (but sometimes more appealing) account of intervention treats the force responsible for the intervention as a variable within the system (Pearl 1993b). This is

---

[2] An equivalent notation, using *set* (*x*) instead of *do*(*x*), was used in Pearl (1995a). The *do*(*x*) notation was first used in Goldszmidt and Pearl (1992) and is gaining in popular support. Lauritzen (2001) used $P(y \mid X \leftarrow x)$. The expression $P(y \mid do(x))$ is equivalent in intent to $P(Y_x = y)$ in the potential-outcome model introduced by Neyman (1923) and Rubin (1974) and to the expression $P[(X = x) \; \square\!\!\rightarrow (Y = y)]$ in the counter-factual theory of Lewis (1973b). The semantical bases of these notions are discussed in Section 3.6.3 and in Chapter 7.

[3] The basic view of interventions as equation modifiers originates with Marschak (1950) and Simon (1953). An explicit translation of interventions to "wiping out" equations from the model was first proposed by Strotz and Wold (1960) and later used in Fisher (1970) and Sobel (1990). Graphical ramifications of this translation were explicated first in Spirtes et al. (1993) and later in Pearl (1993b). Balke and Pearl (1994a,b) proposed it as the basis for defining counterfactuals; see (3.51).

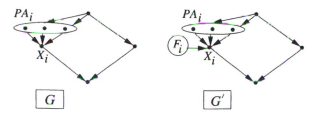

Figure 3.2 Representing external intervention F_i by an augmented network $G' = G \cup \{F_i \rightarrow X_i\}$.

facilitated by representing the function f_i itself as a value of a variable F_i and then writing (3.2) as

$$x_i = I(pa_i, f_i, u_i), \tag{3.7}$$

where I is a three-argument function satisfying

$$I(a, b, c) = f_i(a, c) \qquad \text{whenever } b = f_i.$$

This amounts to conceptualizing the intervention as an external force F_i that alters the function f_i between X_i and its parents. Graphically, we can represent F_i as an added parent node of X_i, and the effect of such an intervention can be analyzed by standard conditionalization – that is, by conditioning our probability on the event that variable F_i attains the value f_i.

The effect of an atomic intervention $do(X_i = x'_i)$ is encoded by adding to G a link $F_i \rightarrow X_i$ (see Figure 3.2), where F_i is a new variable taking values in $\{do(x'_i), \text{idle}\}$, x'_i ranges over the domain of X_i, and "idle" represents no intervention. Thus, the new parent set of X_i in the augmented network is $PA'_i = PA_i \cup \{F_i\}$, and it is related to X_i by the conditional probability

$$P(x_i \mid pa'_i) = \begin{cases} P(x_i \mid pa_i) & \text{if } F_1 = \text{idle}, \\ 0 & \text{if } F_i = do(x'_i) \text{ and } x_i \neq x'_i, \\ 1 & \text{if } F_i = do(x'_i) \text{ and } x_i = x'_i. \end{cases} \tag{3.8}$$

The effect of the intervention $do(x'_i)$ is to transform the original probability function $P(x_1, \ldots, x_n)$ into a new probability function $P(x_1, \ldots, x_n \mid \hat{x}'_i)$, given by

$$P(x_1, \ldots, x_n \mid \hat{x}'_i) = P'(x_1, \ldots, x_n \mid F_i = do(x'_i)), \tag{3.9}$$

where P' is the distribution specified by the augmented network $G' = G \cup \{F_i \rightarrow X_i\}$ and (3.8), with an arbitrary prior distribution on F_i. In general, by adding a hypothetical intervention link $F_i \rightarrow X_i$ to each node in G, we can construct an augmented probability function $P'(x_1, \ldots, x_n; F_1, \ldots, F_n)$ that contains information about richer types of interventions. Multiple interventions would be represented by conditioning P' on a subset of the F_i (taking values in their respective $do(x'_i)$ domains), and the preintervention probability function P would be viewed as the posterior distribution induced by conditioning each F_i in P' on the value "idle."

One advantage of the augmented network representation is that it is applicable to *any* change in the functional relationship f_i and not merely to the replacement of f_i by a constant. It also displays clearly the ramifications of spontaneous changes in f_i, unmediated by external control. Figure 3.2 predicts, for example, that only descendants of X_i would be effected by changes in f_i and hence the marginal probability $P(z)$ will remain unaltered for every set Z of nondescendants of X_i. Likewise, Figure 3.2 dictates that the conditional probability $P(y \mid x_i)$ remains invariant to changes in f_i for any set Y of descendants of X_i, provided X_i d-separates F_i from Y. Kevin Hoover (1990, 2001) used such invariant features to determine the direction of causal influences among economic variables (e.g., employment and money supply) by observing the changes induced by sudden modifications in the processes that govern these variables (e.g., tax reform, labor dispute). Indeed, whenever we obtain reliable information (e.g., from historical or institutional knowledge) that an abrupt local change has taken place in a specific mechanism f_i that constrains a given family (X_i, PA_i) of variables, we can use the observed changes in the marginal and conditional probabilities surrounding those variables to determine whether X_i is indeed the child (or dependent variable) of that family, thus determining the structure of causal influences in the domain (Tian and Pearl, 2001a). The statistical features that remain invariant under such changes, as well as the causal assumptions underlying this invariance, are displayed in the augmented network G'.

3.2.3   Computing the Effect of Interventions

Regardless of whether we represent interventions as a modification of an existing model (Definition 3.2.1) or as a conditionalization in an augmented model (equation (3.9)), the result is a well-defined transformation between the preintervention and postintervention distributions. In the case of an atomic intervention $do(X_i = x'_i)$, this transformation can be expressed in a simple *truncated factorization* formula that follows immediately from (3.2) and Definition 3.2.1:[4]

$$P(x_1, \ldots, x_n \mid \hat{x}'_i) = \begin{cases} \prod_{j \neq i} P(x_j \mid pa_j) & \text{if } x_i = x'_i, \\ 0 & \text{if } x_i \neq x'_i. \end{cases} \tag{3.10}$$

Equation (3.10) reflects the removal of the term $P(x_i \mid pa_i)$ from the product of (3.5), since pa_i no longer influences X_i. For example, the intervention $do(X = x')$ will transform the preintervention distribution given in (3.6) into the product

$$P(z_0, z_1, b, z_2, z_3, y \mid \hat{x}') = P(z_0)\, P(z_1 \mid z_0)\, P(b \mid z_0)$$
$$\times\ P(z_2 \mid x', z_1) P(z_3 \mid z_2, b) P(y \mid x', z_2, z_3).$$

Graphically, the removal of the term $P(x_i \mid pa_i)$ is equivalent to removing the links between PA_i and X_i while keeping the rest of the network intact. Clearly, the transformation defined in (3.10) satisfies the condition of Definition 1.3.1 as well as the properties of (1.38)–(1.39).

---

[4] Equation (3.10) can also be obtained from the G-computation formula of Robins (1986, p. 1423; see also Section 3.6.4) and the manipulation theorem of Spirtes et al. (1993) (according to this source, this formula was "independently conjectured by Fienberg in a seminar in 1991"). Additional properties of the transformation defined in (3.10) and (3.11) are given in Goldszmidt and Pearl (1992) and Pearl (1993b).

Multiplying and dividing (3.10) by $P(x'_i | pa_i)$, the relationship to the preintervention distribution becomes more transparent:

$$P(x_1, \ldots, x_n | \hat{x}'_i) = \begin{cases} \dfrac{P(x_1, \ldots, x_n)}{P(x'_i | pa_i)} & \text{if } x_i = x'_i, \\ 0 & \text{if } x_i \neq x'_i. \end{cases} \tag{3.11}$$

If we regard a joint distribution as an assignment of mass to a collection of abstract points $(x_1, \ldots, x_n)$, each representing a possible state of the world, then the transformation described in (3.11) reveals some interesting properties of the change in mass distribution that take place as a result of an intervention $do(X_i = x'_i)$ (Goldszmidt and Pearl 1992). Each point $(x_1, \ldots, x_n)$ is seen to increase its mass by a factor equal to the inverse of the conditional probability $P(x'_i | pa_i)$ corresponding to that point. Points for which this conditional probability is low would boost their mass value substantially, while those possessing a pa_i value that anticipates a natural (noninterventional) realization of x'_i (i.e., $P(x'_i | pa_i) \approx 1$) will keep their mass unaltered. In standard Bayes conditionalization, each excluded point $(x_i \neq x'_i)$ transfers its mass to the entire set of preserved points through a renormalization constant. However, (3.11) describes a different transformation: each excluded point $(x_i \neq x'_i)$ transfers its mass to a select set of points that share the same value of pa_i. This can be seen from the constancy of both the total mass assigned to each stratum pa_i and the relative masses of points within each such stratum:

$$P(pa_i | do(x'_i)) = P(pa_i);$$

$$\frac{P(s_i, pa_i, x'_i | do(x'_i))}{P(s'_i, pa_i, x'_i | do(x'_i))} = \frac{P(s_i, pa_i, x'_i)}{P(s'_i, pa_i, x'_i)}.$$

Here S_i denotes the set of all variables excluding $\{PA_i \cup X_i\}$. This select set of mass-receiving points can be regarded as "closest" to the point excluded by virtue of sharing the same history, as summarized by pa_i (see Sections 4.1.3 and 7.4.3).

Another interesting form of (3.11) obtains when we interpret the division by $P(x'_i | pa_i)$ as conditionalization on x'_i and pa_i:

$$P(x_1, \ldots, x_n | \hat{x}'_i) = \begin{cases} P(x_1, \ldots, x_n | x'_i, pa_i) \, P(pa_i) & \text{if } x_i = x'_i, \\ 0 & \text{if } x_i \neq x'_i. \end{cases} \tag{3.12}$$

This formula becomes familiar when used to compute the effect of an intervention $do(X_i = x'_i)$ on a set of variables Y disjoint from $(X_i \cup PA_i)$. Summing (3.12) over all variables except $Y \cup X_i$ yields the following theorem.

Theorem 3.2.2 (Adjustment for Direct Causes)
Let PA_i denote the set of direct causes of variable X_i and let Y be any set of variables disjoint of $\{X_i \cup PA_i\}$. The effect of the intervention $do(X_i = x'_i)$ on Y is given by

$$P(y | \hat{x}'_i) = \sum_{pa_i} P(y | x'_i, pa_i) \, P(pa_i), \tag{3.13}$$

where $P(y | x'_i, pa_i)$ and $P(pa_i)$ represent preintervention probabilities.

In the special case of a strategy $S*$ composed of elementary actions $do(X_k = x_k)$, the function g degenerates into a constant, x_k, and we obtain

$$P*(y) = P(y \mid \hat{x}_1, \hat{x}_2, \dots, \hat{x}_n)$$

$$= \sum_{z_1, \dots, z_n} P(y \mid z_1, z_2, \dots, z_n, x_1, x_2, \dots, x_n) \prod_k P(z_k \mid z_{k-1}, x_{k-1}), \qquad (3.18)$$

which can also be obtained from (3.14).

The planning problem illustrated by this example is typical of Markov decision processes (MDPs) (Howard 1960; Dean and Wellman 1991; Bertsekas and Tsitsiklis 1996), where the target of analysis is finding the best next action $do(X_k = x_k)$, given the current state Z_k and past actions. In MDPs, we are normally given the transition functions $P(z_{k+1} \mid z_k, \hat{x}_k)$ and the cost function to be minimized. In the problem we have just analyzed, neither function is given; instead, they must be learned from data gathered under past (presumably suboptimal) strategies. Fortunately, because all variables in the model were measured, both functions were identifiable and could be estimated directly from the corresponding conditional probabilities as follows:

$$P(z_{k+1} \mid z_k, \hat{x}_k) = P(z_{k+1} \mid z_k, x_k);$$

$$P(y \mid z_1, z_2, \dots, z_n, \hat{x}_1, \hat{x}_2, \dots, \hat{x}_n) = P(y \mid z_1, \dots z_2, \dots, z_n, x_1, x_2, \dots, x_n).$$

In Chapter 4 (Section 4.4) we will deal with partially observable Markov decision processes (POMDPs), where some states Z_k are unobserved; learning the transition and cost functions in those problems will require a more intricate method of identification.

It is worth noting that, in this example, to predict the effect of a new strategy it is necessary first to measure variables (Z_k) that are affected by some control variables (X_{k-1}). Such measurements are generally shunned in the classical literature on experimental design (Cox 1958, p. 48), because they lie on the causal pathways between treatment and outcome and thus tend to confound the desired effect estimate. However, our analysis shows that, when properly processed, such measurements may be indispensable in predicting the effect of dynamic control programs. This will be especially true in semi-Markovian models (i.e., DAGs involving unmeasured variables), which are analyzed in Section 3.3.2.

Summary

The immediate implication of the analysis provided in this section is that – given a causal diagram in which all direct causes (i.e., parents) of intervened variables are observable – one can infer postintervention distributions from preintervention distributions; hence, under such assumptions we can estimate the effects of interventions from passive (i.e., nonexperimental) observations, using the truncated factorization formula of (3.14). Yet the more challenging problem is to derive causal effects in situations like Figure 3.1, where some members of PA_i are unobservable and so prevent estimation of $P(x'_i \mid pa_i)$. In Sections 3.3 and 3.4 we provide simple graphical tests for deciding when $P(x_j \mid \hat{x}_i)$ is estimable in such models. But first we need to define more formally what it means for a causal quantity Q to be estimable from passive observations, a question that falls under the technical term *identification*.

3.2.4   Identification of Causal Quantities

Causal quantities, unlike statistical parameters, are defined relative to a causal model M and not relative to a joint distribution $P_M(v)$ over the set V of observed variables. Since nonexperimental data provides information about $P_M(v)$ alone, and since several models can generate the same distribution, the danger exists that the desired quantity will not be discernible unambiguously from the data – even when infinitely many samples are taken. Identifiability ensures that the added assumptions conveyed by M (e.g., the causal graph or the zero coefficients in structural equations) will supply the missing information without explicating M in full detail.

Definition 3.2.3 (Identifiability)

Let $Q(M)$ be any computable quantity of a model M. We say that Q is identifiable in a class M of models if, for any pairs of models M_1 and M_2 from M, $Q(M_1) = Q(M_2)$ whenever $P_{M_1}(v) = P_{M_2}(v)$. If our observations are limited and permit only a partial set F_M of features (of $P_M(v)$) to be estimated, we define Q to be identifiable from F_M if $Q(M_1) = Q(M_2)$ whenever $F_{M1} = F_{M2}$.

Identifiability is essential for integrating statistical data (summarized by $P(v)$) with incomplete causal knowledge of $\{f_i\}$, as it enables us to estimate quantities Q consistently from large samples of $P(v)$ without specifying the details of M; the general characteristics of the class M suffice. For the purpose of our analysis, the quantity Q of interest is the causal effect $P_M(y \mid \hat{x})$, which is certainly computable from a given model M (using Definition 3.2.1) but which we often need to compute from an incomplete specification of M – in the form of qualitative features portrayed in the graph G associated with M. We will therefore consider a class M of models that have the following characteristics in common:

(i) they share the same parent–child families (i.e., the same causal graph G); and

(ii) they induce positive distributions on the observed variables (i.e., $P(v) > 0$).

Relative to such classes, we now have the following.

Definition 3.2.4 (Causal Effect Identifiability)

The causal effect *of X on Y is* identifiable *from a graph G if the quantity $P(y \mid \hat{x})$ can be computed uniquely from any positive probability of the observed variables – that is, if $P_{M_1}(y \mid \hat{x}) = P_{M_2}(y \mid \hat{x})$ for every pair of models M_1 and M_2 with $P_{M_1}(v) = P_{M_2}(v) > 0$ and $G(M_1) = G(M_2) = G$.*

The identifiability of $P(y \mid \hat{x})$ ensures that it is possible to infer the effect of action $do(X = x)$ on Y from two sources of information:

(i) passive observations, as summarized by the probability function $P(v)$; and

(ii) the causal graph G, which specifies (qualitatively) which variables make up the stable mechanisms in the domain or, alternatively, which variables participate in the determination of each variable in the domain.

Restricting identifiability to positive distributions assures us that the condition $X = x'$ is represented in the data in the appropriate context, thus avoiding a zero denominator in (3.11). It would be impossible to infer the effect of action $do(X = x')$ from data in which X never attains the value x' in the context wherein the action is applied. Extensions to some nonpositive distributions are feasible but will not be treated here. Note that, to prove nonidentifiability, it is sufficient to present two sets of structural equations that induce identical distributions over observed variables but have different causal effects.

Using the concept of identifiability, we can now summarize the results of Section 3.2.3 in the following theorem.

Theorem 3.2.5

Given a causal diagram G of any Markovian model in which a subset V of variables are measured, the causal effect $P(y \mid \hat{x})$ is identifiable whenever $\{X \cup Y \cup PA_X\} \subseteq V$, that is, whenever X, Y, and all parents of variables in X are measured. The expression for $P(y \mid \hat{x})$ is then obtained by adjusting for PA_x, as in (3.13).

A special case of Theorem 3.2.5 holds when *all* variables are assumed to be observed.

Corollary 3.2.6

Given the causal diagram G of any Markovian model in which all variables are measured, the causal effect $P(y \mid \hat{x})$ is identifiable for every two subsets of variables X and Y and is obtained from the truncated factorization of (3.14).

We now turn our attention to identification problems in semi-Markovian models.

3.3   CONTROLLING CONFOUNDING BIAS

Whenever we undertake to evaluate the effect of one factor (X) on another (Y), the question arises as to whether we should *adjust* (or "standardize") our measurements for possible variations in some other factors (Z), otherwise known as "covariates," "concomitants," or "confounders" (Cox 1958, p. 48). Adjustment amounts to partitioning the population into groups that are homogeneous relative to Z, assessing the effect of X on Y in each homogeneous group, and then averaging the results (as in (3.13)). The illusive nature of such adjustment was recognized as early as 1899, when Karl Pearson discovered what is now called *Simpson's paradox* (see Section 6.1): Any statistical relationship between two variables may be reversed by including additional factors in the analysis. For example, we may find that students who smoke obtain higher grades than those who do not smoke but, adjusting for age, smokers obtain lower grades in every age group and, further adjusting for family income, smokers again obtain higher grades than nonsmokers in every income–age group, and so on.

Despite a century of analysis, Simpson's reversal continues to "trap the unwary" (Dawid 1979), and the practical question that it poses – whether an adjustment for a given covariate is appropriate – has resisted mathematical treatment. Epidemiologists, for example, are still debating the meaning of "confounding" (Grayson 1987; Shapiro 1997) and often adjust for wrong sets of covariates (Weinberg 1993; see also Chapter 6). The potential-outcome analyses of Rosenbaum and Rubin (1983) and Pratt and Schlaifer

(1988) have led to a concept named "ignorability," which recasts the confounding problem in counterfactual vocabulary but falls short of providing researchers with a workable criterion to guide the choice of covariates (see Section 11.3.2). Ignorability reads: "Z is an admissible set of covariates if, given Z, the value that Y would obtain had X been x is independent of X." Since counterfactuals are not observable, and since judgments about conditional independence of counterfactuals are not readily assertable from common scientific knowledge, the question has remained open: What criterion should one use to decide which variables are appropriate for adjustment?

Section 3.3.1 presents a general and formal solution of the adjustment problem using the friendly language of causal graphs. In Section 3.3.2 we extend this result to nonstandard covariates that are affected by X and hence require several steps of adjustment. Finally, Section 3.3.3 illustrates the use of these criteria in an example.

3.3.1 The Back-Door Criterion

Assume we are given a causal diagram G, together with nonexperimental data on a subset V of observed variables in G, and suppose we wish to estimate what effect the interventions $do(X = x)$ would have on a set of response variables Y, where X and Y are two subsets of V. In other words, we seek to estimate $P(y \mid \hat{x})$ from a sample estimate of $P(v)$, given the assumptions encoded in G.

We show that there exists a simple graphical test, named the "back-door criterion" in Pearl (1993b), that can be applied directly to the causal diagram in order to test if a set $Z \subseteq V$ of variables is sufficient for identifying $P(y \mid \hat{x})$.[5]

Definition 3.3.1 (Back-Door)
A set of variables Z satisfies the back-door *criterion relative to an ordered pair of variables (X_i, X_j) in a DAG G if:*

(i) *no node in Z is a descendant of X_i; and*

(ii) *Z blocks every path between X_i and X_j that contains an arrow into X_i.*

Similarly, if X and Y are two disjoint subsets of nodes in G, then Z is said to satisfy the back-door criterion relative to (X, Y) if it satisfies the criterion relative to any pair (X_i, X_j) such that $X_i \in X$ and $X_j \in Y$.

The name "back-door" echoes condition (ii), which requires that only paths with arrows pointing at X_i be blocked; these paths can be viewed as entering X_i through the back door. In Figure 3.4, for example, the sets $Z_1 = \{X_3, X_4\}$ and $Z_2 = \{X_4, X_5\}$ meet the back-door criterion, but $Z_3 = \{X_4\}$ does not because X_4 does not block the path $(X_i, X_3, X_1, X_4, X_2, X_5, X_j)$. Chapter 11 (Section 11.3.1) gives the intuition for (i) and (ii).

Theorem 3.3.2 (Back-Door Adjustment)
If a set of variables Z satisfies the back-door criterion relative to (X, Y), then the causal effect of X on Y is identifiable and is given by the formula

---

[5] This criterion may also be obtained from Theorem 7.1 of Spirtes et al. (1993). An alternative criterion, using a single d-separation test, is established in Section 3.4 (see (3.37)).

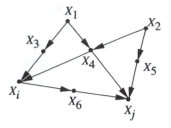

Figure 3.4 A diagram representing the back-door criterion; adjusting for variables $\{X_3, X_4\}$ (or $\{X_4, X_5\}$) yields a consistent estimate of $P(x_j \mid \hat{x}_i)$. Adjusting for $\{X_4\}$ or $\{X_6\}$ would yield a biased estimate.

$$P(y \mid \hat{x}) = \sum_z P(y \mid x, z)P(z). \tag{3.19}$$

The summation in (3.19) represents the standard formula obtained under adjustment for Z; variables X for which the equality in (3.19) is valid were named "conditionally ignorable given Z" in Rosenbaum and Rubin (1983). Reducing ignorability conditions to the graphical criterion of Definition 3.3.1 replaces judgments about counterfactual dependencies with ordinary judgments about cause-effect relationships, as represented in the diagram. The graphical criterion can be tested by systematic procedures that are applicable to diagrams of any size and shape. The criterion also enables the analyst to search for an optimal set of covariates – namely, a set Z that minimizes measurement cost or sampling variability (Tian et al. 1998). The use of a similar graphical criterion for identifying path coefficients in linear structural equations is demonstrated in Chapter 5. Applications to epidemiological research are given in Greenland et al. (1999a) and Glymour and Greenland (2008), where the set Z is called "sufficient set"; *admissible* or *deconfounding* set is a better term.

Proof of Theorem 3.3.2

The proof originally offered in Pearl (1993b) was based on the observation that, when Z blocks all back-door paths from X to Y, setting $(X = x)$ or conditioning on $X = x$ has the same effect on Y. This can best be seen from the augmented diagram G' of Figure 3.2, to which the intervention arcs $F_X \rightarrow X$ were added. If all back-door paths from X to Y are blocked, then all paths from F_X to Y must go through the children of X, and those would be blocked if we condition on X. The implication is that Y is independent of F_X given X,

$$P(y \mid x, F_X = do(x)) = P(y \mid x, F_X = \text{idle}) = P(y \mid x), \tag{3.20}$$

which means that the observation $X = x$ cannot be distinguished from the intervention $F_X = do(x)$.

Formally, we can prove this observation by writing $P(y \mid \hat{x})$ in terms of the augmented probability function P' in accordance with (3.9) and conditioning on Z to obtain

$$P(y \mid \hat{x}) = P'(y \mid F_x) = \sum_z P'(y \mid z, F_x)\, P'(z \mid F_x)$$

$$= \sum_z P'(y \mid z, x, F_x)\, P'(z \mid F_x). \tag{3.21}$$

The addition of x to the last expression is licensed by the implication $F_x \Longrightarrow X = x$. To eliminate F_x from the two terms on the right-hand side of (3.21), we invoke the two

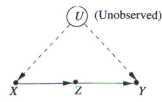

Figure 3.5 A diagram representing the front-door criterion. A two-step adjustment for Z yields a consistent estimate of $P(y \mid \hat{x})$.

conditions of Definition 3.3.1. Since F_x consists of root nodes with children restricted to X, it must be independent of all nondescendants of X, including Z. Thus, condition (i) yields

$$P'(z \mid F_x) = P'(z) = P(z).$$

Invoking now the back-door condition (ii), together with (3.20), permits us to eliminate F_x from (3.21), thus proving (3.19). Section 11.3.3 provides an alternative proof.   □

3.3.2 The Front-Door Criterion

Condition (i) of Definition 3.3.1 reflects the prevailing practice that "the concomitant observations should be quite unaffected by the treatment" (Cox 1958, p. 48). This section demonstrates how concomitants that *are* affected by the treatment can be used to facilitate causal inference. The emerging criterion, named the front-door criterion in Pearl (1995a), will constitute the second building block of the general test for identifying causal effects (Section 3.4 and Theorem 3.6.1).

Consider the diagram in Figure 3.5, which represents the model of Figure 3.4 when the variables $X_1, \dots, X_5$ are unobserved and $\{X_i, X_6, X_j\}$ are relabeled $\{X, Z, Y\}$, respectively. Although Z does not satisfy any of the back-door conditions, measurements of Z can nevertheless enable consistent estimation of $P(y \mid \hat{x})$. This will be shown by reducing the expression for $P(y \mid \hat{x})$ to formulas that are computable from the observed distribution function $P(x, y, z)$.

The joint distribution associated with Figure 3.5 can be decomposed (equation (3.5)) into

$$P(x, y, z, u) = P(u) P(x \mid u) P(z \mid x) P(y \mid z, u). \tag{3.22}$$

From (3.10), the intervention $do(x)$ removes the factor $P(x \mid u)$ and induces the post-intervention distribution

$$P(y, z, u \mid \hat{x}) = P(y \mid z, u) P(z \mid x) P(u). \tag{3.23}$$

Summing over z and u then gives

$$P(y \mid \hat{x}) = \sum_z P(z \mid x) \sum_u P(y \mid z, u) P(u). \tag{3.24}$$

In order to eliminate u from the r.h.s. of (3.24), we use the two conditional independence assumptions encoded in the graph of Figure 3.5:

$$P(u \mid z, x) = P(u \mid x),\tag{3.25}$$

$$P(y \mid x, z, u) = P(y \mid z, u).\tag{3.26}$$

This yields the equalities

$$\sum_u P(y \mid z, u)\, P(u) = \sum_x \sum_u P(y \mid z, u)\, P(u \mid x)\, P(x)$$

$$= \sum_x \sum_u P(y \mid x, z, u)\, P(u \mid x, z)\, P(x)$$

$$= \sum_x P(y \mid x, z)\, P(x)\tag{3.27}$$

and allows the reduction of (3.24) to a form involving only observed quantities:

$$P(y \mid \hat{x}) = \sum_z P(z \mid x) \sum_{x'} P(y \mid x', z)\, P(x').\tag{3.28}$$

All factors on the r.h.s. of (3.28) are consistently estimable from nonexperimental data, so it follows that $P(y \mid \hat{x})$ is estimable as well. Thus, we are in possession of an unbiased nonparametric estimand for the causal effect of X on Y whenever we can find a mediating variable Z that meets the conditions of (3.25) and (3.26).

Equation (3.28) can be interpreted as a two-step application of the back-door formula. In the first step, we find the causal effect of X on Z; since there is no unblocked back-door path from X to Z in Figure 3.5, we simply have

$$P(z \mid \hat{x}) = P(z \mid x).$$

Next, we compute the causal effect of Z on Y, which we can no longer equate with the conditional probability $P(y \mid z)$ because there is a back-door path $Z \leftarrow X \leftarrow U \rightarrow Y$ from Z to Y. However, since X blocks (d-separates) this path, X can play the role of a concomitant in the back-door criterion, which allows us to compute the causal effect of Z on Y in accordance with (3.19), giving $P(y \mid \hat{z}) = \sum_{x'} P(y \mid x', z)\, P(x')$. Finally, we combine the two causal effects via

$$P(y \mid \hat{x}) = \sum_z P(y \mid \hat{z})\, P(z \mid \hat{x}),$$

which reduces to (3.28).

We summarize this result by a theorem after formally defining the assumptions.

Definition 3.3.3 (Front-Door)

A set of variables Z is said to satisfy the front-door *criterion relative to an ordered pair of variables (X, Y) if:*

    (i) *Z intercepts all directed paths from X to Y;*

    (ii) *there is no unblocked back-door path from X to Z; and*

    (iii) *all back-door paths from Z to Y are blocked by X.*

Theorem 3.3.4 (Front-Door Adjustment)
If Z satisfies the front-door criterion relative to (X, Y) and if $P(x, z) > 0$, then the causal effect of X on Y is identifiable and is given by the formula

$$P(y \mid \hat{x}) = \sum_z P(z \mid x) \sum_{x'} P(y \mid x', z) P(x'). \qquad (3.29)$$

The conditions stated in Definition 3.3.3 are overly restrictive; some of the back-door paths excluded by conditions (ii) and (iii) can actually be allowed provided they are blocked by some concomitants. For example, the variable Z_2 in Figure 3.1 satisfies a front-door-like criterion relative to (X, Z_3) by virtue of Z_1 blocking all back-door paths from X to Z_2 as well as those from Z_2 to Z_3. To allow the analysis of such intricate structures, including nested combinations of back-door and front-door conditions, a more powerful symbolic machinery will be introduced in Section 3.4, one that will sidestep algebraic manipulations such as those used in the derivation of (3.28). But first let us look at an example illustrating possible applications of the front-door condition.

3.3.3   Example: Smoking and the Genotype Theory

Consider the century-old debate on the relation between smoking (X) and lung cancer (Y) (Sprites et al. 1993, pp. 291–302). According to many, the tobacco industry has managed to forestall antismoking legislation by arguing that the observed correlation between smoking and lung cancer could be explained by some sort of carcinogenic genotype (U) that involves inborn craving for nicotine.

The amount of tar (Z) deposited in a person's lungs is a variable that promises to meet the conditions listed in Definition 3.3.3, thus fitting the structure of Figure 3.5. To meet condition (i), we must assume that smoking cigarettes has no effect on the production of lung cancer except as mediated through tar deposits. To meet conditions (ii) and (iii), we must assume that, even if a genotype is aggravating the production of lung cancer, it nevertheless has no effect on the amount of tar in the lungs except indirectly (through cigarette smoking). Likewise, we must assume that no other factor that affects tar deposit has any influence on smoking. Finally, condition $P(x, z) > 0$ of Theorem 3.3.4 requires that high levels of tar in the lungs be the result not only of cigarette smoking but also of other factors (e.g., exposure to environmental pollutants) and that tar may be absent in some smokers (owing perhaps to an extremely efficient tar-rejecting mechanism). Satisfaction of this last condition can be tested in the data.

To demonstrate how we can assess the degree to which cigarette smoking increases (or decreases) lung-cancer risk, we will assume a hypothetical study in which the three variables X, Y, Z were measured simultaneously on a large, randomly selected sample of the population. To simplify the exposition, we will further assume that all three variables are binary, taking on true (1) or false (0) values. A hypothetical data set from a study on the relations among tar, cancer, and cigarette smoking is presented in Table 3.1. It shows that 95% of smokers and 5% of nonsmokers have developed high levels of tar in their lungs. Moreover, 81% of subjects with tar deposits have developed lung cancer, compared to only 14% among those with no tar deposits. Finally, within each of these two groups (tar and no-tar), smokers show a much higher percentage of cancer than nonsmokers.

Rule 3 (*Insertion/deletion of actions*):

$$P(y \mid \hat{x}, \hat{z}, w) = P(y \mid \hat{x}, w) \; if \; (Y \perp\!\!\!\perp Z \mid X, W)_{G_{\overline{X}, \overline{Z(W)}}}, \tag{3.33}$$

where Z(W) is the set of Z-nodes that are not ancestors of any W-node in $G_{\overline{X}}$.

Each of these inference rules follows from the basic interpretation of the "hat" $\hat{x}$ operator as a replacement of the causal mechanism that connects X to its preaction parents by a new mechanism $X = x$ introduced by the intervening force. The result is a *submodel* characterized by the subgraph $G_{\overline{X}}$ (named "manipulated graph" in Spirtes et al. 1993).

Rule 1 reaffirms d-separation as a valid test for conditional independence in the distribution resulting from the intervention $do(X = x)$, hence the graph $G_{\overline{X}}$. This rule follows from the fact that deleting equations from the system does not introduce any dependencies among the remaining disturbance terms (see (3.2)).

Rule 2 provides a condition for an external intervention $do(Z = z)$ to have the same effect on Y as the passive observation $Z = z$. The condition amounts to $\{X \cup W\}$ blocking all back-door paths from Z to Y (in $G_{\overline{X}}$), since $G_{\overline{X}\underline{Z}}$ retains all (and only) such paths.

Rule 3 provides conditions for introducing (or deleting) an external intervention $do(Z = z)$ without affecting the probability of $Y = y$. The validity of this rule stems, again, from simulating the intervention $do(Z = z)$ by the deletion of all equations corresponding to the variables in Z (hence the graph $G_{\overline{X}\overline{Z}}$). The reason for limiting the deletion to nonancestors of W-nodes is provided with the proofs of Rules 1–3 in Pearl (1995a).

Corollary 3.4.2

A causal effect $q = P(y_1, \dots, y_k \mid \hat{x}_1, \dots, \hat{x}_m)$ *is identifiable in a model characterized by a graph G if there exists a finite sequence of transformations, each conforming to one of the inference rules in Theorem 3.4.1, that reduces q into a standard (i.e., "hat"-free) probability expression involving observed quantities.*

Rules 1–3 have been shown to be *complete*, namely, sufficient for deriving all identifiable causal effects (Shpitser and Pearl 2006a; Huang and Valtorta 2006). Moreover, as illustrated in Section 3.4.3, symbolic derivations using the hat notation are more convenient than algebraic derivations that aim at eliminating latent variables from standard probability expressions (as in Section 3.3.2, equation (3.24)). However, the task of deciding whether a sequence of rules exists for reducing an arbitrary causal effect expression has not been systematized, and direct graphical criteria for identification are therefore more desirable. These will be developed in Chapter 4.

3.4.3 Symbolic Derivation of Causal Effects: An Example

We will now demonstrate how Rules 1–3 can be used to derive all causal effect estimands in the structure of Figure 3.5. Figure 3.6 displays the subgraphs that will be needed for the derivations that follow.

Task 1: Compute $P(z \mid \hat{x})$

This task can be accomplished in one step, since G satisfies the applicability condition for Rule 2. That is, $X \perp\!\!\!\perp Z$ in $G_{\underline{X}}$ (because the path $X \leftarrow U \rightarrow Y \leftarrow Z$ is blocked by the converging arrows at Y), and we can write

$$P(z \mid \hat{x}) = P(z \mid x). \tag{3.34}$$

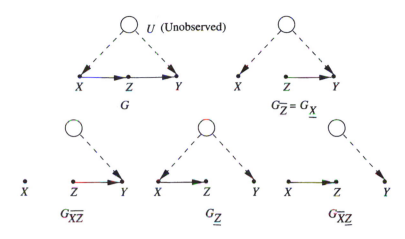

Figure 3.6 Subgraphs of G used in the derivation of causal effects.

Task 2: Compute $P(y \mid \hat{z})$

Here we cannot apply Rule 2 to exchange $\hat{z}$ with z because $G_{\underline{Z}}$ contains a back-door path from Z to Y: $Z \leftarrow X \leftarrow U \rightarrow Y$. Naturally, we would like to block this path by measuring variables (such as X) that reside on that path. This involves conditioning and summing over all values of X:

$$P(y \mid \hat{z}) = \sum_x P(y \mid x, \hat{z}) \, P(x \mid \hat{z}). \tag{3.35}$$

We now have to deal with two terms involving $\hat{z}$, $P(y \mid x, \hat{z})$ and $P(x \mid \hat{z})$. The latter can be readily computed by applying Rule 3 for action deletion:

$$P(x \mid \hat{z}) = P(x) \quad \text{if} \quad (Z \perp\!\!\!\perp X)_{G_{\overline{Z}}}, \tag{3.36}$$

since X and Z are d-separated in $G_{\overline{Z}}$. (Intuitively, manipulating Z should have no effect on X, because Z is a descendant of X in G.) To reduce the former term, $P(y \mid x, \hat{z})$, we consult Rule 2:

$$P(y \mid x, \hat{z}) = P(y \mid x, z) \quad \text{if} \quad (Z \perp\!\!\!\perp Y \mid X)_{G_{\underline{Z}}}, \tag{3.37}$$

noting that X d-separates Z from Y in $G_{\underline{Z}}$. This allows us to write (3.35) as

$$P(y \mid \hat{z}) = \sum_x P(y \mid x, z) \, P(x) = E_x P(y \mid x, z), \tag{3.38}$$

which is a special case of the back-door formula (equation (3.19)). The legitimizing condition, $(Z \perp\!\!\!\perp Y \mid X)_{G_{\underline{Z}}}$, offers yet another graphical test for a set X to be sufficient for control of confounding (between Y and Z) that is equivalent to the opaque "ignorability" condition of Rosenbaum and Rubin (1983).

Task 3: Compute $P(y \mid \hat{x})$

Writing

$$P(y \mid \hat{x}) = \sum_z P(y \mid z, \hat{x}) \, P(z \mid \hat{x}), \tag{3.39}$$

we see that the term $P(z \mid \hat{x})$ was reduced in (3.34) but that no rule can be applied to eliminate the hat symbol $^\wedge$ from the term $P(y \mid z, \hat{x})$. However, we can legitimately add this symbol via Rule 2:

$$P(y \mid z, \hat{x}) = P(y \mid \hat{z}, \hat{x}), \tag{3.40}$$

since the applicability condition $(Y \perp\!\!\!\perp Z \mid X)_{G_{\overline{XZ}}}$ holds (see Figure 3.6). We can now delete the action $\hat{x}$ from $P(y \mid \hat{z}, \hat{x})$ using Rule 3, since $Y \perp\!\!\!\perp X \mid Z$ holds in $G_{\overline{XZ}}$. Thus, we have

$$P(y \mid z, \hat{x}) = P(y \mid \hat{z}), \tag{3.41}$$

which was calculated in (3.38). Substituting (3.38), (3.41), and (3.34) back into (3.39) finally yields

$$P(y \mid \hat{x}) = \sum_{z} P(z \mid x) \sum_{x'} P(y \mid x', z) P(x'), \tag{3.42}$$

which is identical to the front-door formula of (3.28).

Task 4: Compute $P(y, z \mid \hat{x})$

We have

$$P(y, z \mid \hat{x}) = P(y \mid z, \hat{x}) P(z \mid \hat{x}').$$

The two terms on the r.h.s. were derived before in (3.34) and (3.41), from which we obtain

$$P(y, z \mid \hat{x}) = P(y \mid \hat{z}) P(z \mid x)$$
$$= P(z \mid x) \sum_{x'} P(y \mid x', z) P(x'). \tag{3.43}$$

Task 5: Compute $P(x, y \mid \hat{z})$

We have

$$P(x, y \mid \hat{z}) = P(y \mid x, \hat{z}) P(x \mid \hat{z})$$
$$= P(y \mid x, z) P(x). \tag{3.44}$$

The first term on the r.h.s. is obtained by Rule 2 (licensed by $G_{\underline{Z}}$) and the second term by Rule 3 (as in (3.36)).

Note that, in all the derivations, the graph G has provided both the license for applying the inference rules and the guidance for choosing the right rule to apply.

3.4.4 Causal Inference by Surrogate Experiments

Suppose we wish to learn the causal effect of X on Y when $P(y \mid \hat{x})$ is not identifiable and, for practical reasons of cost or ethics, we cannot control X by randomized experiment. The question arises of whether $P(y \mid \hat{x})$ can be identified by randomizing a surrogate variable Z that is easier to control than X. For example, if we are interested in assessing the effect of cholesterol levels (X) on heart disease (Y), a reasonable experiment to conduct would be to control subjects' diet (Z), rather than exercising direct control over cholesterol levels in subjects' blood.

Formally, this problem amounts to transforming $P(y \mid \hat{x})$ into expressions in which only members of Z obtain the hat symbol. Using Theorem 3.4.1, it can be shown that the following conditions are sufficient for admitting a surrogate variable Z:

(i) X intercepts all directed paths from Z to Y; and

(ii) $P(y \mid \hat{x})$ is identifiable in $G_{\bar{Z}}$.

Indeed, if condition (i) holds, then we can write $P(y \mid \hat{x}) = P(y \mid \hat{x}, \hat{z})$, because $(Y \perp\!\!\!\perp Z \mid X)_{G_{\overline{XZ}}}$. But $P(y \mid \hat{x}, \hat{z})$ stands for the causal effect of X on Y in a model governed by $G_{\bar{z}}$, which – by condition (ii) – is identifiable. Translated to our cholesterol example, these conditions require that there be no direct effect of diet on heart conditions and no confounding of cholesterol levels and heart disease, unless we can neutralize such confounding by additional measurements.

Figures 3.9(e) and 3.9(h) (in Section 3.5.2) illustrate models in which both conditions hold. With Figure 3.9(e), for example, we obtain this estimand

$$P(y \mid \hat{x}) = P(y \mid x, \hat{z}) = \frac{P(y, x \mid \hat{z})}{P(x \mid \hat{z})}. \tag{3.45}$$

This can be established directly by first applying Rule 3 to add $\hat{z}$,

$$P(y \mid \hat{x}) = P(y \mid \hat{x}, \hat{z}) \quad \text{because} \ (Y \perp\!\!\!\perp Z \mid X)_{G_{\overline{XZ}}},$$

and then applying Rule 2 to exchange $\hat{x}$ with x:

$$P(y \mid \hat{x}, \hat{z}) = P(y \mid x, \hat{z}) \quad \text{because} \ (Y \perp\!\!\!\perp X \mid Z)_{G_{X\overline{Z}}}.$$

According to (3.45), only one level of Z suffices for the identification of $P(y \mid \hat{x})$ for any values of y and x. In other words, Z need not be varied at all; it can simply be held constant by external means and, if the assumptions embodied in G are valid, the r.h.s. of (3.45) should attain the same value regardless of the (constant) level at which Z is being held. In practice, however, several levels of Z will be needed to ensure that enough samples are obtained for each desired value of X. For example, if we are interested in the difference $E(Y \mid \hat{x}) - E(Y \mid \hat{x}')$, where x and x' are two treatment levels, then we should choose two values z and z' of Z that maximize the number of samples in x and x' (respectively) and then estimate

$$E(Y \mid \hat{x}) - E(Y \mid \hat{x}') = E(Y \mid x, \hat{z}) - E(Y \mid x', \hat{z}').$$

3.5   GRAPHICAL TESTS OF IDENTIFIABILITY

Figure 3.7 shows simple diagrams in which $P(y \mid \hat{x})$ cannot be identified owing to the presence of a "bow" pattern – a confounding arc (dashed) embracing a causal link between X and Y. A confounding arc represents the existence in the diagram of a back-door

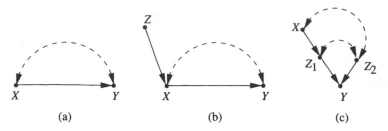

Figure 3.7 (a) A bow pattern: a confounding arc embracing a causal link $X \rightarrow Y$, thus preventing the identification of $P(y \mid \hat{x})$ even in the presence of an instrumental variable Z, as in (b). (c) A bowless graph that still prohibits the identification of $P(y \mid \hat{x})$.

path that contains only unobserved variables and has no converging arrows. For example, the path X, Z_0, B, Z_3 in Figure 3.1 can be represented as a confounding arc between X and Z_3. A bow pattern represents an equation $y = f_Y(x, u, \varepsilon_Y)$, where U is unobserved and dependent on X. Such an equation does not permit the identification of causal effects, since any portion of the observed dependence between X and Y may always be attributed to spurious dependencies mediated by U.

The presence of a bow pattern prevents the identification of $P(y \mid \hat{x})$ even when it is found in the context of a larger graph, as in Figure 3.7(b). This is in contrast to linear models, where the addition of an arc to a bow pattern can render $P(y \mid \hat{x})$ identifiable (see Chapter 5, Figure 5.9). For example, if Y is related to X via a linear relation $y = bx + u$, where U is an unobserved disturbance possibly correlated with X, then $b = \frac{\partial}{\partial x} E(Y \mid \hat{x})$ is not identifiable. However, adding an arc $Z \rightarrow X$ to the structure (i.e., finding a variable Z that is correlated with X but not with U) would facilitate the computation of $E(Y \mid \hat{x})$ via the instrumental variable formula (Bowden and Turkington 1984; see also Chapter 5):

$$b \triangleq \frac{\partial}{\partial x} E(Y \mid \hat{x}) = \frac{E(Y \mid z)}{E(X \mid z)} = \frac{r_{YZ}}{r_{XZ}}. \tag{3.46}$$

In nonparametric models, adding an instrumental variable Z to a bow pattern (Figure 3.7(b)) does not permit the identification of $P(y \mid \hat{x})$. This is a familiar problem in the analysis of clinical trials in which treatment assignment (Z) is randomized (hence, no link enters Z) but compliance is imperfect (see Chapter 8). The confounding arc between X and Y in Figure 3.7(b) represents unmeasurable factors that influence subjects' choice of treatment (X) as well as subjects' response to treatment (Y). In such trials, it is not possible to obtain an unbiased estimate of the treatment effect $P(y \mid \hat{x})$ without making additional assumptions on the nature of the interactions between compliance and response (as is done, for example, in the analyses of Imbens and Angrist 1994 and Angrist et al. 1996). Although the added arc $Z \rightarrow X$ permits us to calculate bounds on $P(y \mid \hat{x})$ (Robins 1989, sec. lg; Manski 1990; Balke and Pearl 1997) and the upper and lower bounds may even coincide for certain types of distributions $P(x, y, z)$ (Section 8.2.4), there is no way of computing $P(y \mid \hat{x})$ for *every* positive distribution $P(x, y, z)$, as required by Definition 3.2.4.

In general, the addition of arcs to a causal diagram can impede, but never assist, the identification of causal effects in nonparametric models. This is because such addition reduces the set of d-separation conditions carried by the diagram; hence, if a causal effect derivation fails in the original diagram, it is bound to fail in the augmented diagram as well. Conversely, any causal effect derivation that succeeds in the augmented diagram (by a sequence of symbolic transformations, as in Corollary 3.4.2) would succeed in the original diagram.

Our ability to compute $P(y_1 \mid \hat{x})$ and $P(y_2 \mid \hat{x})$ for pairs (Y_1, Y_2) of singleton variables does not ensure our ability to compute joint distributions, such as $P(y_1, y_2 \mid \hat{x})$. Figure 3.7(c), for example, shows a causal diagram where both $P(z_1 \mid \hat{x})$ and $P(z_2 \mid \hat{x})$ are computable yet $P(z_1, z_2 \mid \hat{x})$ is not. Consequently, we cannot compute $P(y \mid \hat{x})$. It is interesting to note that this diagram is the smallest graph that does not contain a bow pattern and still presents an uncomputable causal effect.

Another interesting feature demonstrated by Figure 3.7(c) is that computing the effect of a joint intervention is often easier than computing the effects of its constituent singleton interventions.[6] Here, it is possible to compute $P(y \mid \hat{x}, \hat{z}_2)$ and $P(y \mid \hat{x}, \hat{z}_1)$, yet there is no way of computing $P(y \mid \hat{x})$. For example, the former can be evaluated by invoking Rule 2 in $G_{\overline{X}\underline{Z}_2}$, giving

$$P(y \mid \hat{x}, \hat{z}_2) = \sum_{z_1} P(y \mid z_1, \hat{x}, \hat{z}_2) \, P(z_1 \mid \hat{x}, \hat{z}_2)$$

$$= \sum_{z_1} P(y \mid z_1, x, z_2) \, P(z_1 \mid x). \tag{3.47}$$

However, Rule 2 cannot be used to convert $P(z_1 \mid \hat{x}, z_2)$ into $P(z_1 \mid x, z_2)$ because, when conditioned on Z_2, X and Z_1 are d-connected in $G_{\underline{X}}$ (through the dashed lines). A general approach to computing the effect of joint and sequential interventions was developed by Pearl and Robins (1995) and is described in Chapter 4 (Section 4.4).

3.5.1 Identifying Models

Figure 3.8 shows simple diagrams in which the causal effect of X on Y is identifiable (where X and Y are single variables). Such models are called "identifying" because their structures communicate a sufficient number of assumptions (missing links) to permit the identification of the target quantity $P(y \mid \hat{x})$. Latent variables are not shown explicitly in these diagrams; rather, such variables are implicit in the confounding arcs (dashed). Every causal diagram with latent variables can be converted to an equivalent diagram involving measured variables interconnected by arrows and confounding arcs. This conversion corresponds to substituting out all latent variables from the structural equations of (3.2) and then constructing a new diagram by connecting any two variables X_i and X_j by (i) an arrow from X_j to X_i whenever X_j appears in the equation for X_i, and (ii) a confounding arc whenever the same ε term appears in both f_i and f_j. The result is a diagram in which all unmeasured variables are exogenous and mutually independent.

Several features should be noted from examining the diagrams in Figure 3.8.

---

[6] This was brought to my attention by James Robins, who has worked out many of these computations in the context of sequential treatment management (Robins 1986, p. 1423).

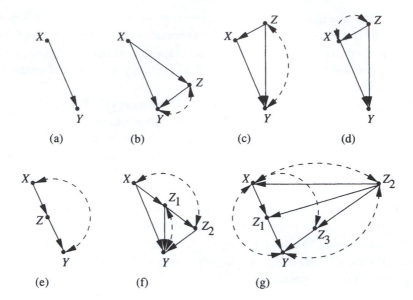

Figure 3.8 Typical models in which the effect of X on Y is identifiable. Dashed arcs represent confounding paths, and Z represents observed covariates.

1. Since the removal of any arc or arrow from a causal diagram can only assist the identifiability of causal effects, $P(y \mid \hat{x})$ will still be identified in any edge subgraph of the diagrams shown in Figure 3.8. Likewise, the introduction of mediating observed variables onto any edge in a causal graph can assist, but never impede, the identifiability of any causal effect. Therefore, $P(y \mid \hat{x})$ will still be identified from any graph obtained by adding mediating nodes to the diagrams shown in Figure 3.8.

2. The diagrams in Figure 3.8 are maximal in the sense that the introduction of any additional arc or arrow onto an existing pair of nodes would render $P(y \mid \hat{x})$ no longer identifiable. Note the conformity with Theorem 3.6.1, p. 105.

3. Although most of the diagrams in Figure 3.8 contain bow patterns, none of these patterns emanates from X (as is the case in Figures 3.9(a) and (b) to follow). In general, a necessary condition for the identifiability of $P(y \mid \hat{x})$ is the absence of a confounding arc between X and any child of X that is an ancestor of Y.

4. Diagrams (a) and (b) in Figure 3.8 contain no back-door paths between X and Y and thus represent experimental designs in which there is no confounding bias between the treatment (X) and the response (Y); hence, $P(y \mid \hat{x}) = P(y \mid x)$. Likewise, diagrams (c) and (d) in Figure 3.8 represent designs in which observed covariates Z block every back-door path between X and Y (i.e., X is "conditionally ignorable" given Z, in the lexicon of Rosenbaum and Rubin 1983); hence, $P(y \mid \hat{x})$ is obtained by standard adjustment for Z (as in (3.19)):

$$P(y \mid \hat{x}) = \sum_z P(y \mid x, z) P(z).$$

5. For each of the diagrams in Figure 3.8, we readily obtain a formula for $P(y \mid \hat{x})$ by using symbolic derivations patterned after those in Section 3.4.3. The derivation

is often guided by the graph topology. For example, diagram (f) in Figure 3.8 dictates the following derivation. Writing

$$P(y \mid \hat{x}) = \sum_{z_1, z_2} P(y \mid z_1, z_2, \hat{x}) \, P(z_1, z_2 \mid \hat{x}),$$

we see that the subgraph containing $\{X, Z_1, Z_2\}$ is identical in structure to that of diagram (e), with (Z_1, Z_2) replacing (Z, Y), respectively. Thus, $P(z_1, z_2 \mid \hat{x})$ can be obtained from (3.43). Likewise, the term $P(y \mid z_1, z_2, \hat{x})$ can be reduced to $P(y \mid z_1, z_2, x)$ by Rule 2, since $(Y \perp\!\!\!\perp X \mid Z_1, Z_2)_{G_{\underline{X}}}$. We therefore have

$$P(y \mid \hat{x}) = \sum_{z_1, z_2} P(y \mid z_1, z_2, x) \, P(z_1 \mid x) \sum_{x'} P(z_2 \mid z_1, x') \, P(x'). \tag{3.48}$$

Applying a similar derivation to diagram (g) of Figure 3.8 yields

$$P(y \mid \hat{x}) = \sum_{z_1} \sum_{z_2} \sum_{x'} P(y \mid z_1, z_2, x') \, P(x' \mid z_2)$$
$$\times \; P(z_1 \mid z_2, x) \, P(z_2). \tag{3.49}$$

Note that the variable Z_3 does not appear in (3.49), which means that Z_3 need not be measured if all one wants to learn is the causal effect of X on Y.

6. In diagrams (e), (f), and (g) of Figure 3.8, the identifiability of $P(y \mid \hat{x})$ is rendered feasible through observed covariates Z that are affected by the treatment X (since members of Z are descendants of X). This stands contrary to the warning – repeated in most of the literature on statistical experimentation – to refrain from adjusting for concomitant observations that are affected by the treatment (Cox 1958; Rosenbaum 1984; Pratt and Schlaifer 1988; Wainer 1989). It is commonly believed that a concomitant Z that is affected by the treatment must be excluded from the analysis of the total effect of the treatment (Pratt and Schlaifer 1988). The reason given for the exclusion is that the calculation of total effects amounts to integrating out Z, which is functionally equivalent to omitting Z to begin with. Diagrams (e), (f), and (g) show cases where the total effects of X are indeed the target of investigation and, even so, the measurement of concomitants that are affected by X (e.g., Z or Z_1) is still necessary. However, the adjustment needed for such concomitants is nonstandard, involving two or more stages of the standard adjustment of (3.19) (see (3.28), (3.48), and (3.49)).

7. In diagrams (b), (c), and (f) of Figure 3.8, Y has a parent whose effect on Y is not identifiable; even so, the effect of X on Y is identifiable. This demonstrates that local identifiability is not a necessary condition for global identifiability. In other words, to identify the effect of X on Y we need not insist on identifying each and every link along the paths from X to Y.

3.5.2 Nonidentifying Models

Figure 3.9 presents typical diagrams in which the total effect of X on Y, $P(y \mid \hat{x})$, is not identifiable. Noteworthy features of these diagrams are as follows.

1. All graphs in Figure 3.9 contain unblockable back-door paths between X and Y, that is, paths ending with arrows pointing to X that cannot be blocked by observed nondescendants of X. The presence of such a path in a graph is, indeed,

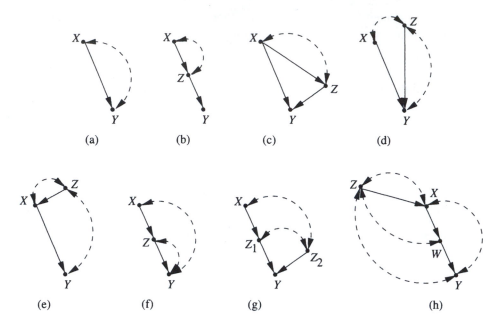

Figure 3.9 Typical models in which $P(y \mid \hat{x})$ is not identifiable.

a necessary test for nonidentifiability (see Theorem 3.3.2). That it is not a sufficient test is demonstrated by Figure 3.8(e), in which the back-door path (dashed) is unblockable and yet $P(y \mid \hat{x})$ is identifiable.

2. A sufficient condition for the nonidentifiability of $P(y \mid \hat{x})$ is the existence of a confounding arc between X and any of its children on a path from X to Y, as shown in Figures 3.9(b) and (c). A stronger sufficient condition is that the graph contain any of the patterns shown in Figure 3.9 as an edge subgraph.

3. Graph (g) in Figure 3.9 (same as Figure 3.7(c)) demonstrates that local identifiability is not sufficient for global identifiability. For example, we can identify $P(z_1 \mid \hat{x})$, $P(z_2 \mid \hat{x})$, $P(y \mid \hat{z}_1)$, and $P(y \mid \hat{z}_2)$ but not $P(y \mid \hat{x})$. This is one of the main differences between nonparametric and linear models; in the latter, all causal effects can be determined from the structural coefficients and each coefficient represents the causal effect of one variable on its immediate successor.

3.6 DISCUSSION

3.6.1 Qualifications and Extensions

The methods developed in this chapter facilitate the drawing of quantitative causal inferences from a combination of qualitative causal assumptions (encoded in the diagram) and nonexperimental observations. The causal assumptions in themselves cannot generally be tested in nonexperimental studies, unless they impose constraints on the observed distributions. The most common type of constraints appears in the form of conditional independencies, as communicated through the d-separation conditions in the diagrams. Another type of constraints takes the form of numerical inequalities. In Chapter 8, for example, we show that the assumptions associated with instrumental variables (Figure 3.7(b)) are subject to falsification tests in the form of inequalities on conditional probabilities (Pearl 1995b). Still, such constraints permit the testing of merely a small fraction of the causal assumptions embodied in the diagrams; the bulk of those assumptions must be substantiated from domain knowledge as obtained from either theoretical considerations (e.g., that falling barometers do not cause rain) or related experimental studies. For example, the experimental study of Moertel et al. (1985), which refuted the hypothesis that vitamin C is effective against cancer, can be used as a substantive assumption in observational studies involving vitamin C and cancer patients; it would be represented as a missing link (between vitamin C and cancer) in the associated diagram. In summary, the primary use of the methods described in this chapter lies not in testing causal assumptions but in providing an effective language for making those assumptions precise and explicit. Assumptions can thereby be isolated for deliberation or experimentation and then (once validated) be integrated with statistical data to yield quantitative estimates of causal effects.

An important issue that will be considered only briefly in this book (see Section 8.5) is sampling variability. The mathematical derivation of causal effect estimands should be considered a first step toward supplementing these estimands with confidence intervals and significance levels, as in traditional analysis of controlled experiments. We should remark, though, that having obtained nonparametric estimands for causal effects does not imply that one should refrain from using parametric forms in the estimation phase of the study. For example, if the assumptions of Gaussian, zero-mean disturbances and additive interactions are deemed reasonable, then the estimand given in (3.28) can be converted to the product $E(Y \mid \hat{x}) = r_{ZX} r_{YZ \cdot X} x$, where $r_{YZ \cdot X}$ is the standardized regression coefficient (Section 5.3.1); the estimation problem then reduces to that of estimating regression coefficients (e.g., by least squares). More sophisticated estimation techniques can be found in Rosenbaum and Rubin (1983), Robins (1989, sec. 17), and Robins et al. (1992, pp. 331–3). For example, the "propensity score" method of Rosenbaum and Rubin (1983) was found useful when the dimensionality of the adjusted covariates is high (Section 11.3.5). Robins (1999) shows that, rather than estimating individual factors in the adjustment formula of (3.19), it is often more advantageous to use $P(y \mid \hat{x}) = \sum_z \frac{P(x,y,z)}{P(x \mid z)}$, where the preintervention distribution remains unfactorized. One can then separately estimate the denominator $P(x \mid z)$, weigh individual samples by the inverse of this estimate, and treat the weighted samples as if they were drawn at random from the postintervention distribution $P(y \mid \hat{x})$. Postintervention parameters, such as $\frac{\partial}{\partial x} E(Y \mid \hat{x})$, can then be estimated by ordinary least squares. This method is especially advantageous in longitudinal studies with time-varying covariates, as in the problems discussed in Sections 3.2.3 (see (3.18)) and 4.4.3.

Several extensions of the methods proposed in this chapter are noteworthy. First, the identification analysis for atomic interventions can be generalized to complex time-varying policies in which a set X of controlled variables is made to respond in a specified way to some set Z of covariates via functional or stochastic strategies, as in Sections 3.2.3 and 4.4.3. In Chapter 4 (Section 4.4.3) it is shown that identifying the effect of such policies requires a sequence of back-door conditions in the associated diagram.

A second extension concerns the use of the intervention calculus (Theorem 3.4.1) in nonrecursive models, that is, in causal diagrams involving directed cycles or feedback loops. The basic definition of causal effects in terms of "wiping out" equations from the model (Definition 3.2.1) still carries over to nonrecursive systems (Strotz and Wold

1960; Sobel 1990), but then two issues must be addressed. First, the analysis of identification must ensure the stability of the remaining submodels (Fisher 1970). Second, the
d-separation criterion for DAGs must be extended to cover cyclic graphs as well. The validity of d-separation has been established for nonrecursive linear models (Spirtes 1995) as
well as for nonlinear systems involving discrete variables (Pearl and Dechter 1996). However, the computation of causal effect estimands will be harder in cyclic nonlinear systems,
because symbolic reduction of $P(y \mid \hat{x})$ to hat-free expressions may require the solution
of nonlinear equations. In Chapter 7 (Section 7.2.1) we demonstrate the evaluation of policies and counterfactuals in nonrecursive linear systems (see also Balke and Pearl 1995a).

A third extension concerns generalizations of intervention calculus (Theorem 3.4.1)
to situations where the data available is not obtained under i.i.d. (independent and identically distributed) sampling. One can imagine, for instance, a physician who prescribes a
certain treatment to patients only when the fraction of survivors among previous patients
drops below some threshold. In such cases, it is required to estimate the causal effect
$P(y \mid \hat{x})$ from nonindependent samples. Vladimir Vovk (1996) gave conditions under
which the rules of Theorem 3.4.1 will be applicable when sampling is not i.i.d., and he
went on to cast the three inference rules as a logical production system.

3.6.2   Diagrams as a Mathematical Language

The benefit of incorporating substantive background knowledge into probabilistic inference was recognized as far back as Thomas Bayes (1763) and Pierre Laplace (1814), and
its crucial role in the analysis and interpretation of complex statistical studies is generally acknowledged by most modern statisticians. However, the mathematical language
available for expressing background knowledge has remained in a rather pitiful state of
development.

Traditionally, statisticians have approved of only one way of combining substantive
knowledge with statistical data: the Bayesian method of assigning subjective priors to distributional parameters. To incorporate causal information within this framework, plain
causal statements such as "Y is not affected by X" must be converted into sentences
or events capable of receiving probability values (e.g., counterfactuals). For instance, to
communicate the innocent assumption that mud does not cause rain, we would have to
use a rather unnatural expression and say that the probability of the counterfactual event
"rain if it were not muddy" is the same as the probability of "rain if it were muddy."
Indeed, this is how the potential-outcome approach of Neyman and Rubin has achieved
statistical legitimacy: causal judgments are expressed as constraints on probability functions involving counterfactual variables (see Section 3.6.3).

Causal diagrams offer an alternative language for combining data with causal information. This language simplifies the Bayesian route by accepting plain causal statements
as its basic primitives. Such statements, which merely indicate whether a causal connection between two variables of interest exists, are commonly used in ordinary discourse and
provide a natural way for scientists to communicate experience and organize knowledge.[7]

---

[7] Remarkably, many readers of this chapter (including two referees of this book) classified the methods presented here as belonging to the "Bayesian camp" and as depending on a "good prior." This

It can be anticipated, therefore, that the language of causal graphs will find applications in problems requiring substantial domain knowledge.

The language is not new. The use of diagrams and structural equation models to convey causal information has been quite popular in the social sciences and econometrics. Statisticians, however, have generally found these models suspect, perhaps because social scientists and econometricians have failed to provide an unambiguous definition of the empirical content of their models – that is, to specify the experimental conditions, however hypothetical, whose outcomes would be constrained by a given structural equation. (Chapter 5 discusses the history of structural equations in the social sciences and economics.) As a result, even such basic notions as "structural coefficients" or "missing links" become the object of serious controversy (Freedman 1987; Goldberger 1992) and misinterpretations (Whittaker 1990, p. 302; Wermuth 1992; Cox and Wermuth 1993).

To a large extent, this history of controversy and miscommunication stems from the absence of an adequate mathematical notation for defining basic notions of causal modeling. For example, standard probabilistic notation cannot express the empirical content of the coefficient b in the structural equation $y = bx + \varepsilon_Y$, even if one is prepared to assume that ε_Y (an unobserved quantity) is uncorrelated with X.[8] Nor can any probabilistic meaning be attached to the analyst's excluding from the equation variables that do not "directly affect" Y.[9]

The notation developed in this chapter gives these (causal) notions a clear empirical interpretation, because it permits one to specify precisely what is being held constant and what is merely measured in a given experiment. (The need for this distinction was recognized by many researchers, most notably Pratt and Schlaifer 1988 and Cox 1992.) The meaning of b is simply $\frac{\partial}{\partial x}E(Y \mid \hat{x})$, that is, the rate of change (in x) of the expectation of Y in an experiment where X is held at x by external control. This interpretation holds regardless of whether ε_Y and X are correlated (e.g., via another equation $x = ay + \varepsilon_x$). Likewise, the analyst's decision as to which variables should be included in a given equation can be based on a hypothetical controlled experiment: A variable Z is excluded from the equation for Y if (for every level of ε_Y) Z has no influence on Y when all other variables (S_{YZ}) are held constant; this implies $P(y \mid \hat{z}, \hat{s}_{YZ}) = P(y \mid \hat{s}_{YZ})$. Specifically, variables that are excluded from the equation $y = bx + \varepsilon_Y$ are not conditionally independent of Y given measurements of X but instead are *causally* irrelevant to Y given settings of X. The operational meaning of the "disturbance term" ε_Y is likewise demystified: ε_Y is defined as the difference $Y - E(Y \mid \hat{s}_Y)$. Two disturbance terms, ε_X and ε_Y, are correlated if $P(y \mid \hat{x}, \hat{s}_{XY}) \neq P(y \mid x, \hat{s}_{XY})$, and so on (see Chapter 5, Section 5.4, for further elaboration).

The distinctions provided by the hat notation clarify the empirical basis of structural equations and should make causal models more acceptable to empirical researchers.

---

classification is misleading. The method does depend on subjective assumptions (e.g., mud does not cause rain), but such assumptions are causal, not statistical, and cannot be expressed as prior probabilities on parameters of joint distributions.

[8] Voluminous literature on the subject of "exogeneity" (e.g., Richard 1980; Engle et al. 1983; Hendry 1995) has emerged from economists' struggle to give statistical interpretation to the causal assertion "X and ε_Y are uncorrelated" (Aldrich 1993; see Section 5.4.3).

[9] The bitter controversy between Goldberger (1992) and Wermuth (1992) revolves around Wermuth's insistence on giving a statistical interpretation to the zero coefficients in structural equations (see Section 5.4.1).

Moreover, since most scientific knowledge is organized around the operation of "holding
X fixed" rather than "conditioning on X," the notation and calculus developed in this
chapter should provide an effective means for scientists to communicate substantive information and to infer its logical consequences.

3.6.3 Translation from Graphs to Potential Outcomes

This chapter uses two representations of causal information: graphs and structural equations, where the former is an abstraction of the latter. Both representations have been
controversial for almost a century. On the one hand, economists and social scientists have
embraced these modeling tools, but they continue to question and debate the causal content of the parameters they estimate (see Sections 5.1 and 5.4 for details); as a result, the
use of structural models in policy-making contexts is often viewed with suspicion. Statisticians, by and large, reject both representations as problematic (Freedman 1987) if not
meaningless (Wermuth 1992; Holland 1995), and they sometimes resort to the
Neyman–Rubin potential-outcome notation when pressed to communicate causal information (Rubin 1990).[10] A detailed formal analysis of the relationships between the structural and potential-outcome approaches is offered in Chapter 7 (Section 7.4.4) and proves
their mathematical equivalence – a theorem in one entails a theorem in the other. In this
section we highlight the key methodological differences.

   The primitive object of analysis in the potential-outcome framework is the unit-based
response variable, denoted $Y(x, u)$ or $Y_x(u)$, read: "the value that Y would obtain in unit u,
had X been x." This counterfactual expression has a formal interpretation in structural
equations models. Consider a structural model M that contains a set of equations

$$x_i = f_i(pa_i, u_i), \quad i = 1, \dots, n, \tag{3.50}$$

as in (3.4). Let U stand for the vector $(U_1, \dots, U_n)$ of background variables, let X and Y
be two disjoint subsets of observed variables, and let M_x be the submodel created by replacing the equations corresponding to variables in X with $X = x$, as in Definition 3.2.1.
The structural interpretation of $Y(x, u)$ is given by

$$Y(x, u) \triangleq Y_{M_x}(u). \tag{3.51}$$

That is, $Y(x, u)$ is the (unique) solution of Y under the realization $U = u$ in the submodel
M_x of M. Although the term *unit* in the potential-outcome literature normally stands for
the identity of a specific individual in a population, a unit may also be thought of as the
set of attributes that characterize that individual, the experimental conditions under study,
the time of day, and so on – all of which are represented as components of the vector u
in structural modeling. In fact, the only requirements on U are (i) that it represent as
many background factors as needed to render the relations among endogenous variables
deterministic and (ii) that the data consist of independent samples drawn from $P(u)$. The

---

[10] A parallel framework was developed in the econometrics literature under the rubric "switching
regression" (Manski 1995, p. 38), which Heckman (1996) attributed to Roy (1951) and Quandt
(1958); but lacking the formal semantics of (3.51), did not progress beyond a "framework."

identity of an individual person in an experiment is often sufficient for this purpose because it represents the anatomical and genetic makings of that individual, which are often sufficient for determining that individual's response to treatments or other programs of interest.

Equation (3.51) defines the key formal connection between the opaque English phrase "the value that Y would obtain in unit u, had X been x" and the physical processes that transfer changes in X into changes in Y. The formation of the submodel M_x explicates precisely how the hypothetical phrase "had X been x" could be realized, as well as what process must give in to make $X = x$ a reality.

Given this interpretation of $Y(x, u)$, it is instructive to contrast the methodologies of causal inference in the potential-outcome versus structural frameworks. If U is treated as a random variable, then the value of the counterfactual $Y(x, u)$ becomes a random variable as well, denoted as $Y(x)$ or Y_x. The potential-outcome analysis proceeds by imagining the observed distribution $P(x_1, \ldots, x_n)$ as the marginal distribution of an augmented probability function P^* defined over both observed and counterfactual variables. Queries about causal effects (written $P(y \mid \hat{x})$ in our structural analysis) are phrased as queries about the marginal distribution of the counterfactual variable of interest, written $P^*(Y(x) = y)$. The new hypothetical entities $Y(x)$ are treated as ordinary random variables; for example, they are assumed to obey the axioms of probability calculus, the laws of conditioning, and the axioms of conditional independence. Moreover, these hypothetical entities are assumed to be connected to observed variables via consistency constraints (Robins 1986) such as[11]

$$X = x \Longrightarrow Y(x) = Y, \tag{3.52}$$

which states that, for every u, if the actual value of X turns out to be x, then the value that Y would take on if X were x is equal to the actual value of Y. Thus, whereas the structural approach views the intervention $do(x)$ as an operation that changes the model (and the distribution) but keeps all variables the same, the potential-outcome approach views the variable Y under $do(x)$ to be a different variable, $Y(x)$, loosely connected to Y through relations such as (3.52). In Chapter 7 we show, using the structural interpretation of $Y(x, u)$, that it is indeed legitimate to treat counterfactuals as random variables in all respects and, moreover, that consistency constraints like (3.52) follow as theorems from the structural interpretation, and no other constraint need ever be considered.

To communicate substantive causal knowledge, the potential-outcome analyst must express causal assumptions as constraints on P^*, usually in the form of conditional independence assertions involving counterfactual variables. For example, to communicate the understanding that – in a randomized clinical trial with imperfect compliance (see Figure 3.7(b)) – the way subjects react (Y) to treatments (X) is statistically independent of the treatment assignment (Z), the potential-outcome analyst would write $Y(x) \perp\!\!\!\perp Z$. Likewise, to convey the understanding that the assignment is randomized and hence independent of how subjects comply with the assignment, the potential-outcome analyst would use the independence constraint $Z \perp\!\!\!\perp X (z)$.

---

[11] Gibbard and Harper (1976, p. 156) expressed this constraint as $A \supset [(A \,\square\!\!\rightarrow S) \equiv S]$.

transform causal effect expressions of the form $P^*(Y(x) = y)$ into expressions involving only measurable variables. When such a transformation is found, the corresponding causal effect is identifiable, since P^* then reduces to P.

The question naturally arises of whether the constraints used by potential-outcome analysts are *complete* – that is, whether they are sufficient for deriving every valid statement about causal processes, interventions, and counterfactuals. To answer this question, the validity of counterfactual statements must be defined relative to more basic mathematical objects, such as possible worlds (Section 1.4.4) or structural equations (equation (3.51)). In the standard potential-outcome framework, however, the question of completeness remains open, because $Y(x, u)$ is taken as a primitive notion and because consistency constraints such as (3.52) – although they appear plausible for the English expression "had X been x" – are not derived from a deeper mathematical object. This question of completeness is settled in Chapter 7, where a necessary and sufficient set of axioms is derived from the structural semantics given to $Y(x, u)$ by (3.51).

In assessing the historical development of structural equations and potential-outcome models, one cannot overemphasize the importance of the conceptual clarity that structural equations offer vis-à-vis the potential-outcome model. The reader may appreciate this importance by attempting to judge whether the condition of (3.61) holds in a given familiar situation. This condition reads: "the value that Z would obtain had X been x is jointly independent of both X and the value that Y would obtain had Z been x." (In the structural representation, the sentence reads: "Z shares no cause with either X or Y, except for X itself, as shown in Figure 3.5.") The thought of having to express, defend, and manage formidable counterfactual relationships of this type may explain why the enterprise of causal inference is currently viewed with such awe and despair among rank-and-file epidemiologists and statisticians – and why most economists and social scientists continue to use structural equations instead of the potential-outcome alternatives advocated in Holland (1988), Angrist et al. (1996), and Sobel (1998). On the other hand, the algebraic machinery offered by the potential-outcome notation, once a problem is properly formalized, can be quite powerful in refining assumptions, deriving probabilities of counterfactuals, and verifying whether conclusions follow from premises – as we demonstrate in Chapter 9. The translation given in (3.51)–(3.56) is the key to unifying the two camps and should help researchers combine the best features of the two approaches.

3.6.4   Relations to Robins's *G*-Estimation

Among the investigations conducted in the potential-outcome framework, the one closest in spirit to the structural analysis described in this chapter is Robins's work on "causally interpreted structured tree graphs" (Robins 1986, 1987). Robins was the first to realize the potential of Neyman's counterfactual notation $Y(x)$ as a general mathematical language for causal inference, and he used it to extend Rubin's (1978) "time-independent treatment" model to studies with direct and indirect effects and time-varying treatments, concomitants, and outcomes.

Robins considered a set $V = \{V_1, \ldots, V_M\}$ of temporally ordered discrete random variables (as in Figure 3.3) and asked under what conditions one can identify the effect of control policy $g : X = x$ on outcomes $Y \subseteq V \backslash X$, where $X = \{X_1, \ldots, X_K\} \subseteq V$ are the temporally ordered and potentially manipulable treatment variables of interest. The causal effect of $X = x$ on Y was expressed as the probability

$$P(y \mid g = x) \triangleq P\{Y(x) = y\},$$

where the counterfactual variable $Y(x)$ stands for the value that outcome variables Y would take had the treatment variables X been x.

Robins showed that $P(y \mid g = x)$ is identified from the distribution $P(v)$ if each component X_k of X is "assigned at random, given the past," a notion explicated as follows. Let L_k be the variables occurring between X_{k-1} and X_k, with L_1 being the variables preceding X_1. Write $\bar{L}_k = (L_1, \ldots, L_k)$, $L = \bar{L}_K$, and $\bar{X}_k = (X_1, \ldots, X_k)$, and define $\bar{X}_0$, $\bar{L}_0$, $\bar{V}_0$ to be identically zero. The treatment $X_k = x_k$ is said to be *assigned at random*, *given the past*, if the following relation holds:

$$(Y(x) \perp\!\!\!\perp X_k \mid \bar{L}_k, \bar{X}_{k-1} = \bar{x}_{k-1}). \tag{3.62}$$

Robins further proved that, if (3.62) holds for every k, then the causal effect is given by

$$P(y \mid g = x) = \sum_{\bar{l}_k} P(y \mid \bar{l}_K, \bar{x}_K) \prod_{k=1}^{K} P(l_k \mid \bar{l}_{k-1}, \bar{x}_{k-1}), \tag{3.63}$$

an expression he called the "G-computation algorithm formula." This expression can be derived by applying condition (3.62) iteratively, as in the derivation of (3.54). If X is univariate, then (3.63) reduces to the standard adjustment formula

$$P(y \mid g = x) = \sum_{l_1} P(y \mid x, l_1) P(l_1),$$

paralleling (3.54). Likewise, in the special structure of Figure 3.3, (3.63) reduces to (3.18).

To place this result in the context of our analysis in this chapter, we need to focus attention on condition (3.62), which facilitated Robins's derivation of (3.63), and ask whether this formal counterfactual independency can be given a meaningful graphical interpretation. The answer will be given in Chapter 4 (Theorem 4.4.1), where we derive a graphical condition for identifying the effect of a plan, i.e., a sequential set of actions. The condition reads as follows: $P(y \mid g = x)$ is identifiable and is given by (3.63) if every action-avoiding back-door path from X_k to Y is blocked by some subset L_k of nondescendants of X_k. (By "action-avoiding" we mean a path containing no arrow entering an X variable later than X_k.) Chapter 11 (Section 11.4.2) shows by examples that this "sequential back-door" criterion is more general than that given in (3.62).

The structural analysis introduced in this chapter supports and generalizes Robins's result from a new theoretical perspective. First, on the technical front, this analysis offers systematic ways of managing models where Robins's starting assumption (3.62) is inapplicable. Examples are Figures 3.8(d)–(g).

Second, on the conceptual front, the structural framework represents a fundamental shift from the vocabulary of counterfactual independencies, to the vocabulary of

(2001), Hernán et al. (2002), Hernán et al. (2004), Greenland and Brumback (2002), Greenland et al. (1999a,b) Kaufman et al. (2005), Petersen et al. (2006), Hernández-Díaz et al. (2006), VanderWeele and Robins (2007) and Glymour and Greenland (2008).

Interesting applications of the front-door criterion (Section 3.3.2) were noted in social science (Morgan and Winship 2007) and economics (Chalak and White 2006).

Some advocates of the "potential outcome" approach have been most resistant to accepting graphs or structural equations as the basis for causal analysis and, lacking these conceptual tools, were unable to address the issue of covariate selection (Rosenbaum 2002, p. 76; Rubin 2007, 2008a) and were led to dismiss important scientific concepts as "ill-defined," "deceptive," "confusing" (Holland 2001; Rubin 2004, 2008b), and worse (Rubin 2009). Lauritzen (2004) and Heckman (2005) have criticized this attitude; Pearl (2009) demonstrates its fallacies.

Equally puzzling are concerns of some philosophers (Cartwright 2007; Woodward 2003) and economists (Heckman 2005) that the *do*-operator is too local to model complex, real-life policy interventions, which sometimes affect several mechanisms at once and often involve conditional decisions, imperfect control, and multiple actions. These concerns emerge from conflating the mathematical definition of a relationship (e.g., causal effect) with the technical feasibility of testing that relationship in the physical world. While the *do*-operator is indeed an ideal mathematical tool (not unlike the *derivative* in differential calculus), it nevertheless permits us to specify and analyze interventional strategies of great complexity. Readers will find examples of such strategies in Chapter 4, and a further discussion of this issue in Chapter 11 (Sections 11.4.3–11.4.6 and Section 11.5.4).

Chapter Road Map to the Main Results

The three key results in this chapter are: 1. The control of confounding, 2. The evaluation of policies, and 3. The evaluation of counterfactuals.

1. The problem of controlling confounding bias is resolved through the back-door condition (Theorem 3.3.2, pp.79–80) – a criterion for selecting a set of covariates that, if adjusted for, would yield an unbiased estimate of causal effects.

2. The policy evaluation problem – to predict the effect of interventions from non-experimental data – is resolved through the *do*-calculus (Theorem 3.4.1, pp. 85–86) and the graphical criteria that it entails (Theorem 3.3.4, p. 83; Theorem 3.6.1, p. 105). The completeness of *do*-calculus implies that any (nonparametric) policy evaluation problem that is not supported by an identifying graph, or an equivalent set of causal assumptions, can be proven "unsolvable."

3. Finally, equation (3.51) provides a formal semantics for counterfactuals, through which joint probabilities of counterfactuals can be defined and evaluated in the framework of scientific theories (see Chapter 7).

CHAPTER FOUR

Actions, Plans, and Direct Effects

He whose actions exceed his wisdom,
his wisdom shall endure.
Rabbi Hanina ben Dosa
(1st century A.D.)

Preface

So far, our analysis of causal effects has focused on primitive interventions of the form $do(x)$, which stood for setting the value of variable X to a fixed constant, x, and asking for the effect of this action on the probabilities of some response variables Y. In this chapter we introduce several extensions of this analysis.

First (Section 4.1), we discuss the status of actions vis-à-vis observations in probability theory, decision analysis, and causal modeling, and we advance the thesis that the main role of causal models is to facilitate the evaluation of the effect of *novel* actions and policies that were unanticipated during the construction of the model.

In Section 4.2 we extend the identification analysis of Chapter 3 to conditional actions of the form "do x if you see z" and stochastic policies of the form "do x with probability p if you see z." We shall see that the evaluation and identification of these more elaborate interventions can be obtained from the analysis of primitive interventions. In Section 4.3, we use the intervention calculus developed in Chapter 3 to give a graphical characterization of a set of semi-Markovian models for which the causal effect of one variable on another can be identified.

We address in Section 4.4 the problem of evaluating the effect of sequential plans – namely, sequences of time-varying actions (some taken concurrently) designed to produce a certain outcome. We provide a graphical method of estimating the effect of such plans from nonexperimental studies in which some of the actions are influenced by observations and former actions, some observations are influenced by the actions, and some confounding variables are unmeasured. We show that there is substantial advantage to analyzing a plan into its constituent actions rather than treating the set of actions as a single entity.

Finally, in Section 4.5 we address the question of distinguishing direct from indirect effects. We show that direct effects can be identified by the graphical method developed in Section 4.4. An example using alleged sex discrimination in college admission will serve to demonstrate the assumptions needed for proper analysis of direct effects.

4.1   INTRODUCTION

4.1.1   Actions, Acts, and Probabilities

Actions admit two interpretations: reactive and deliberative. The reactive interpretation
sees action as a consequence of an agent's beliefs, disposition, and environmental inputs,
as in "Adam ate the apple because Eve handed it to him." The deliberative interpretation
sees action as an option of choice in contemplated decision making, usually involving
comparison of consequences, as in "Adam was wondering what God would do if he ate
the apple." We shall distinguish the two views by calling the first "act" and the second
"action." An act is viewed from the outside, an action from the inside. Therefore, an
act can be predicted and can serve as evidence for the actor's stimuli and motivations
(provided the actor is part of our model). Actions, in contrast, can neither be predicted
nor provide evidence, since (by definition) they are pending deliberation and turn into
acts once executed.

The confusion between actions and acts has led to Newcomb's paradox (Nozick 1969)
and other oddities in the so-called evidential decision theory, which encourages decision
makers to take into consideration the evidence that an action would provide, if enacted.
This bizarre theory seems to have loomed from Jeffrey's influential book *The Logic of
Decision* (Jeffrey 1965), in which actions are treated as ordinary events (rather than interventions) and, accordingly, the effects of actions are obtained through conditionalization
rather than through a mechanism-modifying operation like $do(x)$. (See Stalnaker 1972;
Gibbard and Harper 1976; Skyrms 1980; Meek and Glymour 1994; Hitchcock 1996.)

Commonsensical decision theory[1] instructs rational agents to choose the option x
that maximizes expected utility,[2]

$$U(x) = \sum_y P(y \mid do(x))u(y),$$

where $u(y)$ is the utility of outcome y; in contrast, "evidential decision theory" calls for
maximizing the conditional expectation

$$U_{ev}(x) = \sum_y P(y \mid x)u(y),$$

in which x is (improperly) treated as an observed proposition.

The paradoxes that emerge from this fallacy are obvious: patients should avoid going
to the doctor "to reduce the probability that one is seriously ill" (Skyrms 1980, p. 130);
workers should never hurry to work, to reduce the probability of having overslept; students

---

[1] I purposely avoid the common title "causal decision theory" in order to suppress even the slightest
hint that any alternative, noncausal theory can be used to guide decisions.

[2] Following a suggestion of Stalnaker (1972), Gibbard and Harper (1976) used $P(x \,\square\!\!\rightarrow y)$ in $U(x)$,
rather than $P(y \mid do(x))$, where $x \,\square\!\!\rightarrow y$ stands for the subjunctive conditional "y if it were x."
The semantics of the two operators are closely related (see Section 7.4), but the equation-removal
interpretation of the $do(x)$ operator is less ambiguous and clearly suppresses inference from effect
to cause.

should not prepare for exams, lest this would prove them behind in their studies; and so on. In short, all remedial actions should be banished lest they increase the probability that a remedy is indeed needed.

The oddity in this kind of logic stems from treating actions as acts that are governed by past associations instead of as objects of free choice, as dictated by the semantics of the $do(x)$ operator. This "evidential" decision theory preaches that one should never ignore genuine statistical evidence (in our case, the evidence that an act normally provides regarding whether the act is needed), but decision theory proper reminds us that actions – by their very definition – render such evidence irrelevant to the decision at hand, for actions *change* the probabilities that acts normally obey.[3]

The moral of this story can be summarized in the following mnemonic rhymes:

Whatever evidence an act might provide
On what could have caused the act,
Should never be used to help one decide
On whether to choose that same act.

Evidential decision theory was a passing episode in the philosophical literature, and no philosopher today takes the original version of this theory seriously. Still, some recent attempts have been made to revive interest in Jeffrey's expected utility by replacing $P(y \mid x)$ with $P(y \mid x, K)$, where K stands for various background contexts, chosen to suppress spurious associations (as in (3.13)) (Price 1991; Hitchcock 1996). Such attempts echo an overly restrictive empiricist tradition, according to which rational agents live and die by one source of information – statistical associations – and hence expected utilities should admit no other operation but Bayes's conditionalization. This tradition is rapidly giving way to a more accommodating conception: rational agents should act according to theories of actions; naturally, such theories demand action-specific conditionalization (e.g., $do(x)$) while reserving Bayes's conditionalization for representing passive observations (see Goldszmidt and Pearl 1992; Meek and Glymour 1994; Woodward 1995).

In principle, actions are not part of probability theory, and understandably so: probabilities capture normal relationships in the world, whereas actions represent interventions that perturb those relationships. It is no wonder, then, that actions are treated as foreign entities throughout the literature on probability and statistics; they serve neither as arguments of probability expressions nor as events for conditioning such expressions.

Even in the statistical decision-theoretic literature (e.g., Savage 1954), where actions are the main target of analysis, the symbols given to actions serve merely as indices for distinguishing one probability function from another, not as entities that stand in logical relationships to the variables on which probabilities are defined. Savage (1954, p. 14) defined "act" as a "function attaching a consequence to each state of the world," and he treated a chain of decisions, one leading to another, as a single decision. However, the

---

[3] Such evidence is rendered irrelevant within the actor's own probability space; in multiagent decision situations, however, each agent should definitely be cognizant of how other agents might interpret each of his pending "would-be" acts.

logic that leads us to infer the consequences of actions and strategies from more elementary considerations is left out of the formalism. For example, consider the actions: "raise taxes," "lower taxes," and "raise interest rates." The consequences of all three actions must be specified separately, prior to analysis; none can be inferred from the others. As a result, if we are given two probabilities, P_A and P_B, denoting the probabilities prevailing under actions A and B, respectively, there is no way we can deduce from this input the probability $P_{A \wedge B}$ corresponding to the joint action $A \wedge B$ or indeed any Boolean combination of the propositions A and B. This means that, in principle, the impact of all anticipated joint actions would need to be specified in advance – an insurmountable task.

The peculiar status of actions in probability theory can be seen most clearly in comparison to the status of observations. By specifying a probability function $P(s)$ on the possible states of the world, we automatically specify how probabilities should change with every conceivable observation e, since $P(s)$ permits us to compute (by conditioning on e) the posterior probabilities $P(E \mid e)$ for every pair of events E and e. However, specifying $P(s)$ tells us nothing about how probabilities should change in response to an external action $do(A)$. In general, if an action $do(A)$ is to be described as a function that takes $P(s)$ and transforms it to $P_A(s)$, then $P(s)$ tells us nothing about the nature of $P_A(s)$, even when A is an elementary event for which $P(A)$ is well defined (e.g., "raise the temperature by 1 degree" or "turn the sprinkler on"). With the exception of the trivial requirement that $P_A(s)$ be zero if s implies $\neg A$, a requirement that applies uniformly to every $P(s)$, probability theory does not tell us how $P_A(s)$ should differ from $P'_A(s)$, where $P'(s)$ is some other preaction probability function. Conditioning on A is clearly inadequate for capturing this transformation, as we have seen in many examples in Chapters 1 and 3 (see, e.g., Section 1.3.1), because conditioning represents passive observations in an unchanging world, whereas actions change the world.

Drawing an analogy to visual perception, we may say that the information contained in $P(s)$ is analogous to a precise description of a three-dimensional object; it is sufficient for predicting how that object will be viewed from any angle outside the object, but it is insufficient for predicting how the object will be viewed if manipulated and squeezed by external forces. Additional information about the physical properties of the object must be supplied for making such predictions. By analogy, the additional information required for describing the transformation from $P(s)$ to $P_A(s)$ should identify those elements of the world that remain invariant under the action $do(A)$. This extra information is provided by causal knowledge, and the $do(\cdot)$ operator enables us to capture the invariant elements (thus defining $P_A(s)$) by locally modifying the graph or the structural equations. The next section will compare this device to the way actions are handled in standard decision theory.

4.1.2   Actions in Decision Analysis

Instead of introducing new operators into probability calculus, the traditional approach has been to attribute the differences between seeing and doing to differences in the total evidence available. Consider the statements: "the barometer reading was observed to be x" and "the barometer reading was set to level x." The former helps us predict the weather, the latter does not. While the evidence described in the first statement is limited to the reading of the barometer, the second statement also tells us that the barometer was manipulated by some agent, and conditioning on this additional evidence should render the barometer reading irrelevant to predicting the rain.

The practical aspects of this approach amount to embracing the acting agents as variables in the analysis, constructing an augmented distribution function including the decisions of those agents, and inferring the effect of actions by conditioning those decision variables to particular values. Thus, for example, the agent manipulating the barometer might enter the system as a decision variable "squeezing the barometer"; after incorporating this variable into the probability distribution, we could infer the impact of manipulating the barometer simply by conditioning the augmented distribution on the event "the barometer was squeezed by force y and has reached level x."

For this conditioning method to work properly in evaluating the effect of future actions, the manipulating agent must be treated as an ideal experimenter acting out of free will, and the associated decision variables must be treated as exogenous – causally unaffected by other variables in the system. For example, if the augmented probability function encodes the fact that the current owner of the barometer tends to squeeze the barometer each time she feels arthritis pain, we will be unable to use that function for evaluating the effects of deliberate squeezing of the barometer, even by the same owner. Recalling the difference between acts and actions, whenever we set out to calculate the effect of a pending action, we must ignore all mechanisms that constrained or triggered the execution of that action in the past. Accordingly, the event "The barometer was squeezed" must enter the augmented probability function as independent of all events that occurred prior to the time of manipulation, similar to the way action variable F entered the augmented network in Figure 3.2.

This solution corresponds precisely to the way actions are treated in decision analysis, as depicted in the literature on influence diagrams (IDs) (Howard and Matheson 1981; Shachter 1986; Pearl 1988b, chap. 6; Dawid 2002). Each decision variable is represented as an exogenous variable (a parentless node in the diagram), and its impact on other variables is assessed and encoded in terms of conditional probabilities, similar to the impact of any other parent node in the diagram.[4]

The difficulty with this approach is that we need to anticipate in advance, and represent explicitly, all actions whose effects we might wish to evaluate in the future. This renders the modeling process unduly cumbersome, if not totally unmanageable. In circuit diagnosis, for example, it would be awkward to represent every conceivable act of component replacement (similarly, every conceivable connection to a voltage source, current source, etc.) as a node in the diagram. Instead, the effects of such replacements are implicit in the circuit diagram itself and can be deduced from the diagram, given its causal interpretation. In econometric modeling likewise, it would be awkward to represent every conceivable variant of policy intervention as a new variable in the economic equations. Instead, the effects of such interventions can be deduced from the structural

[4] The ID literature's insistence on divorcing the links in the ID from any causal interpretation (Howard and Matheson 1981; Howard 1990) is at odds with prevailing practice. The causal interpretation is what allows us to treat decision variables as root nodes and construct the proper decision trees for analysis; see Section 11.6 for a demonstration.

interpretation of those equations, if only we can tie the immediate effects of each policy to the corresponding variables and parameters in the equations. The compound action "raise taxes and lower interest rates," for example, need not be introduced as a new variable in the equations, because the effect of that action can be deduced if we have the quantities "taxation level" and "interest rates" already represented as (either exogenous or endogenous) variables in the equations.

The ability to predict the effect of interventions without enumerating those interventions in advance is one of the main advantages we draw from causal modeling and one of the main functions served by the notion of causation. Since the number of actions or action combinations is enormous, they cannot be represented explicitly in the model but rather must be indexed by the propositions that each action enforces directly. Indirect consequences of enforcing those propositions are then inferred from the causal relationships among the variables represented in the model. We will return to this theme in Chapter 7 (Section 7.2.4), where we further explore the invariance assumptions that must be met for this encoding scheme to work.

4.1.3 Actions and Counterfactuals

As an alternative to Bayesian conditioning, philosophers (Lewis 1976; Gardenfors 1988) have studied another probability transformation called "imaging," which was deemed useful in the analysis of subjunctive conditionals and which more adequately represents the transformations associated with actions. Whereas Bayes conditioning of $P(s \mid e)$ transfers the entire probability mass from states excluded by e to the remaining states (in proportion to their current probabilities, $P(s)$), imaging works differently: each excluded state s transfers its mass individually to a select set of states $S^*(s)$ that are considered to be "closest" to s (see Section 7.4.3). Although providing a more adequate and general framework for actions (Gibbard and Harper 1976), imaging leaves the precise specification of the selection function $S^*(s)$ almost unconstrained. Consequently, the problem of enumerating future actions is replaced by the problem of encoding distances among states in a way that would be both economical and respectful of common understanding of the causal laws that operate in the domain. The second requirement is not trivial, considering that indirect ramifications of actions often result in worlds that are quite dissimilar to the one from which we start (Fine 1975).

The difficulties associated with making the closest-world approach conform to causal laws (Section 7.4) are circumvented in the structural approach pursued in this book by basing the notion of interventions directly on causal mechanisms and by capitalizing on the properties of invariance and autonomy that accompany these mechanisms. This mechanism-modification approach can be viewed as a special instance of the closest-world approach, where the closeness measure is crafted so as to respect the causal mechanisms in the domain; the selection function $S^*(s)$ that ensues is represented in (3.11) (see discussion that follows).

The operationality of this mechanism-modification semantics was demonstrated in Chapter 3 and led to the quantitative predictions of the effects of actions, including actions and action combinations that were not contemplated during the model's construction, at which time the modeller can be free to describe how Nature works, unburdened by thoughts of external interventions. In Chapter 7 we further use the mechanism-modification interpretation to provide semantics for counterfactual statements, as outlined in Section 1.4.4. In this chapter, we will extend the applications of the *do* calculus to the analysis of complex policies and decomposition of effects.

4.2   CONDITIONAL ACTIONS AND STOCHASTIC POLICIES

The interventions considered in our analysis of identification (Sections 3.3–3.4) were limited to actions that merely force a variable or a group of variables X to take on some specified value x. In general (see the process control example in Section 3.2.3), interventions may involve complex policies in which a variable X is made to respond in a specified way to some set Z of other variables – say, through a functional relationship $x = g(z)$ or through a stochastic relationship whereby X is set to x with probability $P^*(x \mid z)$. We will show, based on Pearl (1994b), that identifying the effect of such policies is equivalent to identifying the expression $P(y \mid \hat{x}, z)$.

Let $P(y \mid do(X = g(z)))$ stand for the distribution (of Y) prevailing under the policy $do(X = g(z))$. To compute $P(y \mid do(X = g(z)))$, we condition on Z and write

$$P(y \mid do(X = g(z))) = \sum_z P(y \mid do(X = g(z)), z) P(z \mid do(X = g(z)))$$

$$= \sum_z P(y \mid \hat{x}, z)|_{x=g(z)} P(z)$$

$$= E_z[P(y \mid \hat{x}, z)|_{x=g(z)}].$$

The equality

$$P(z \mid do(X = g(z))) = P(z)$$

stems, of course, from the fact that Z cannot be a descendant of X; hence, any control exerted on X can have no effect on the distribution of Z. Thus, we see that the causal effect of a policy $do(X = g(z))$ can be evaluated directly from the expression of $P(y \mid \hat{x}, z)$ simply by substituting $g(z)$ for x and taking the expectation over Z (using the observed distribution $P(z)$).

This identifiability criterion for conditional policy is somewhat stricter than that for unconditional intervention. Clearly, if a policy $do(X = g(z))$ is identifiable, then the simple intervention $do(X = x)$ is identifiable as well, since we can always obtain the latter by setting $g(z) = x$. The converse does not hold, however, because conditioning on Z might create dependencies that will prevent the successful reduction of $P(y \mid \hat{x}, z)$ to a hat-free expression. Kuroki and Miyakawa (1999a, 2003) present graphical criteria.

A stochastic policy, which imposes a new conditional distribution $P^*(x \mid z)$ for x, can be handled in a similar manner. We regard the stochastic intervention as a random process in which the unconditional intervention $do(X = x)$ is enforced with probability $P^*(x \mid z)$. Thus, given $Z = z$, the intervention $do(X = x)$ will occur with probability

$P^*(x \mid z)$ and will produce a causal effect given by $P(y \mid \hat{x}, z)$. Averaging over x and z gives the effect (on Y) of the stochastic policy $P^*(x \mid z)$:

$$P(y)|_{P^*(x \mid z)} = \sum_x \sum_z P(y \mid \hat{x}, z) P^*(x \mid z) P(z).$$

Because $P^*(x \mid z)$ is specified externally, we see again that the identifiability of $P(y \mid \hat{x}, z)$ is a necessary and sufficient condition for the identifiability of any stochastic policy that shapes the distribution of X by the outcome of Z.

Of special importance in planning is a STRIPS-like action (Fikes and Nilsson 1971) whose immediate effects $X = x$ depend on the satisfaction of some enabling precondition $C(w)$ on a set W of variables. To represent such actions, we let $Z = W \cup PA_X$ and set

$$P^*(x \mid z) = \begin{cases} P(x \mid pa_X) & \text{if} \quad C(w) = \text{false}, \\ 1 & \text{if} \quad C(w) = \text{true and } X = x, \\ 0 & \text{if} \quad C(w) = \text{true and } X \neq x. \end{cases}$$

4.3   WHEN IS THE EFFECT OF AN ACTION IDENTIFIABLE?

In Chapter 3 we developed several graphical criteria for recognizing when the effect of one variable on another, $P(y \mid do(x))$, is identifiable in the presence of unmeasured variables. These criteria, like the back-door (Theorem 3.3.2) and front-door (Theorem 3.3.4), are special cases of a more general class of semi-Markovian models for which repeated application of the inference rules of *do*-calculus (Theorem 3.4.1) will reduce $P(y \mid \hat{x})$ to a hat-free expression, thus rendering it identifiable. In this section we characterize a wider class of models in which the causal effect $P(y \mid \hat{x})$ is identifiable. This class is subsumed by the one established by Tian and Pearl (2002a) in Theorem 3.7 and the complete characterization given later in Shpitser and Pearl (2006b). It is brought here for its intuitive appeal.

4.3.1   Graphical Conditions for Identification

Theorem 4.3.1 characterizes a class of models in the form of four graphical conditions, any one of which is sufficient for the identification of $P(y \mid \hat{x})$ when X and Y are singleton nodes in the graph. Theorem 4.3.2 then states that at least one of these four conditions must hold in the model for $P(y \mid \hat{x})$ to be identifiable in *do*-calculus. In view of the completeness of *do*-calculus, we conclude that one of the four conditions is necessary for any method of identification compatible with the semantics of Definition 3.2.4.

Theorem 4.3.1 (Galles and Pearl 1995)
Let X and Y denote two singleton variables in a semi-Markovian model characterized by graph G. A sufficient condition for the identifiability of $P(y \mid \hat{x})$ is that G satisfy one of the following four conditions.

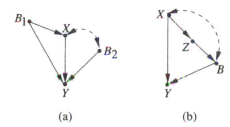

Figure 4.1   Condition 3 of Theorem 4.3.1. In (a), the set $\{B_1, B_2\}$ blocks all back-door paths from X to Y, and $P(b_1, b_2 \mid \hat{x}) = P(b_1, b_2)$. In (b), the node B blocks all back-door paths from X to Y, and $P(b \mid \hat{x})$ is identifiable using Condition 4.

(a)   (b)

1. *There is no back-door path from X to Y in G; that is, $(X \perp\!\!\!\perp Y)_{G_{\underline{X}}}$.*
2. *There is no directed path from X to Y in G.*
3. *There exists a set of nodes B that blocks all back-door paths from X to Y so that $P(b \mid \hat{x})$ is identifiable.* (A special case of this condition occurs when B consists entirely of nondescendants of X, in which case $P(b \mid \hat{x})$ reduces immediately to $P(b)$.)
4. *There exist sets of nodes Z_1 and Z_2 such that:*
   (i)   *Z_1 blocks every directed path from X to Y (i.e., $(Y \perp\!\!\!\perp X \mid Z_1)_{G_{\overline{Z_1}\underline{X}}}$);*
   (ii) *Z_2 blocks all back-door paths between Z_1 and Y (i.e., $(Y \perp\!\!\!\perp Z_1 \mid Z_2)_{G_{\overline{X}\overline{Z_1}}}$);*
   (iii) *Z_2 blocks all back-door paths between X and Z_1 (i.e., $(X \perp\!\!\!\perp Z_1 \mid Z_2)_{G_{\underline{X}}}$; and*

   (iv) *Z_2 does not activate any back-door paths from X to Y (i.e., $(X \perp\!\!\!\perp Y \mid Z_1, Z_2)_{G_{\overline{Z_1}\overline{X(Z_2)}}}$).* (This condition holds if (i)–(iii) are met and no member of Z_2 is a descendant of X.)
   (A special case of condition 4 occurs when $Z_2 = \emptyset$ and there is no back-door path from X to Z_1 or from Z_1 to Y.)

Proof

Condition 1. This condition follows directly from Rule 2 (see Theorem 3.4.1). If $(Y \perp\!\!\!\perp X)_{G_{\underline{X}}}$, then we can immediately change $P(y \mid \hat{x})$ to $P(y \mid x)$, so the query is identifiable.

Condition 2. If there is no directed path from X to Y in G, then $(Y \perp\!\!\!\perp X)_{G_{\overline{X}}}$. Hence, by Rule 3, $P(y \mid \hat{x}) = P(y)$ and so the query is identifiable.

Condition 3. If there is a set of nodes B that blocks all back-door paths from X to Y (i.e., $(Y \perp\!\!\!\perp X \mid B)_{G_{\underline{X}}}$), then we can expand $P(y \mid \hat{x})$ as $\sum_b P(y \mid \hat{x}, b) P(b \mid \hat{x})$ and, by Rule 2, rewrite $P(y \mid \hat{x}, b)$ as $P(y \mid x, b)$. If the query $(b \mid \hat{x})$ is identifiable, then the original query must also be identifiable. See examples in Figure 4.1.

Condition 4. If there is a set of nodes Z_1 that block all directed paths from X to Y and a set of nodes Z_2 that block all back-door paths between Y and Z_1 in $G_{\overline{X}}$, then we expand $P(y \mid \hat{x}) = \sum_{Z_1, Z_2} P(y \mid \hat{x}, z_1, z_2) P(z_1, z_2 \mid \hat{x})$ and rewrite $P(y \mid \hat{x}, z_1, z_2)$ as $P(y \mid \hat{x}, \hat{z}_1, z_2)$ using Rule 2, since all back-door paths between Z_1 and Y are blocked by Z_2 in $G_{\overline{X}}$. We can reduce $P(y \mid \hat{x}, \hat{z}_1, z_2)$ to $P(y \mid \hat{z}_1, z_2)$ using Rule 3, since $(Y \perp\!\!\!\perp X \mid Z_1, Z_2)_{G_{\overline{Z_1}\overline{X(Z_2)}}}$. We can rewrite $P(y \mid \hat{z}_1, z_2)$ as $P(y \mid z_1, z_2)$ if $(Y \perp\!\!\!\perp Z_1 \mid Z_2)_{G_{\underline{Z_1}}}$. The only way that this independence cannot hold is if there is a path from Y to Z_1 through X, since $(Y \perp\!\!\!\perp Z_1 \mid Z_2)_{G_{\overline{X}\overline{Z_1}}}$. However, we can block this path by conditioning and

3. Otherwise, let B = BlockingSet (X, Y) and Pb = ClosedForm $(P(b \mid \hat{x}))$; if $Pb \neq$ FAIL then return $\sum_b P(y \mid b, x) * Pb$.

4. Otherwise, let Z_1 = Children$(X) \cap (Y \cup$ Ancestors$(Y))$, Z_3 = BlockingSet(X, Z_1), Z_4 = BlockingSet(Z_1, Y), and $Z_2 = Z_3 \cup Z_4$; if $Y \notin Z_1$ and $X \notin Z_2$ then return

$$\sum_{z_1, z_2} \sum_{x'} P(y \mid z_1, z_2, x') P(x' \mid z_2) P(z_1 \mid x, z_2) P(z_2).$$

5. Otherwise, return FAIL.

Steps 3 and 4 invoke the function BlockingSet(X, Y), which selects a set of nodes Z that d-separate X from Y. Such sets can be found in polynomial time (Tian et al. 1998). Step 3 contains a recursive call to the algorithm ClosedForm$(b \mid \hat{x})$ itself, in order to obtain an expression for causal effect $P(b \mid \hat{x})$.

4.3.4 Summary

The conditions of Theorem 4.3.1 expand the boundary between the class of identifying models (such as those depicted in Figure 3.8) and nonidentifying models (Figure 3.9). These conditions lead to an effective algorithm for determining the identifiability of control queries of the type $P(y \mid \hat{x})$, where X is a single variable. The algorithm further gives a closed-form expression for the causal effect $P(y \mid \hat{x})$ in terms of estimable probabilities.

Although the completeness results of Shpitser and Pearl (2006a) now offer a precise characterization of the boundary between the identifying and nonidentifying models (see the discussion following Theorem 3.6.1), the conditions of Theorem 4.3.2 may still be useful on account of their simplicity and intuitive appeal.

4.4 THE IDENTIFICATION OF DYNAMIC PLANS

This section, based on Pearl and Robins (1995), concerns the probabilistic evaluation of plans in the presence of unmeasured variables, where each plan consists of several concurrent or sequential actions and each action may be influenced by its predecessors in the plan. We establish a graphical criterion for recognizing when the effects of a given plan can be predicted from passive observations on measured variables only. When the criterion is satisfied, a closed-form expression is provided for the probability that the plan will achieve a specified goal.

4.4.1 Motivation

To motivate the discussion, consider an example discussed in Robins (1993, apx. 2), as depicted in Figure 4.4. The variables X_1 and X_2 stand for treatments that physicians prescribe to a patient at two different times, Z represents observations that the second physician consults to determine X_2, and Y represents the patient's survival. The hidden variables U_1 and U_2 represent, respectively, part of the patient's history and the patient's disposition

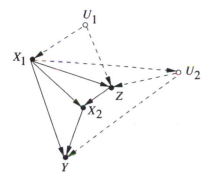

Figure 4.4   The problem of evaluating the effect of the plan $(do(x_1), do(x_2))$ on Y, from nonexperimental data taken on X_1, Z, X_2, and Y.

to recover. A simple realization of such structure could be found among AIDS patients, where Z represents episodes of PCP. This is a common opportunistic infection of AIDS patients that (as the diagram shows) does not have a direct effect on survival Y because it can be treated effectively, but it is an indicator of the patient's underlying immune status (U_2), which can cause death. The terms X_1 and X_2 stand for bactrim, a drug that prevents PCP (Z) and may also prevent death by other mechanisms. Doctors used the patient's earlier PCP history (U_1) to prescribe X_1, but its value was not recorded for data analysis.

The problem we face is as follows. Assume we have collected a large amount of data on the behavior of many patients and physicians, which is summarized in the form of (an estimated) joint distribution P of the observed four variables (X_1, Z, X_2, Y). A new patient comes in, and we wish to determine the impact of the (unconditional) plan $(do(x_1), do(x_2))$ on survival, where x_1 and x_2 are two predetermined dosages of bactrim to be administered at two prespecified times.

In general, our problem amounts to that of evaluating a new plan by watching the performance of other planners whose decision strategies are indiscernible. Physicians do not provide a description of all inputs that prompted them to prescribe a given treatment; all they communicate to us is that U_1 was consulted in determining X_1 and that Z and X_1 were consulted in determining X_2. But U_1, unfortunately, was not recorded. In epidemiology, the plan evaluation problem is known as "*time-varying treatment with time-varying confounders*" (Robins 1993). In artificial intelligence applications, the evaluation of such plans enables one agent to learn to act by observing the performance of another agent, even in cases where the actions of the other agent are predicated on factors that are not visible to the learner. If the learner is permitted to act as well as observe, then the task becomes much easier: the topology of the causal diagram could also be inferred (at least partially), and the effects of some previously unidentifiable actions could be determined.

As in the identification of actions (Chapter 3), the main problem in plan identification is the control of "confounders," that is, unobserved factors that trigger actions and simultaneously affect the response. However, plan identification is further complicated by the fact that some of the confounders (e.g., Z) are affected by control variables. As remarked in Chapter 3, one of the deadliest sins in the design of statistical experiments (Cox 1958, p. 48) is to adjust for such variables, because adjusting for a variable that stands between an action and its consequence interferes with the very quantity we wish to estimate – the total effect of that action. The identification method presented will circumvent such difficulties.

Two other features of Figure 4.4 are worth noting. First, the quantity $P(y \mid \hat{x}_1, \hat{x}_2)$ cannot be computed if we treat the control variables X_1 and X_2 as a single compound variable X. The graph corresponding to such compounding would depict X as connected to Y by both an arrow and a curved arc (through U) and thus would form a bow pattern (see Figure 3.9), which is indicative of nonidentifiability. Second, the causal effect $P(y \mid \hat{x}_1)$ in isolation is not identifiable because U_1 creates a bow pattern around the link $X \to Z$, which lies on a directed path from X to Y (see the discussion in Section 3.5).

The feature that facilitates the identifiability of $P(y \mid \hat{x}_1, \hat{x}_2)$ is the identifiability of $P(y \mid x_1, z, \hat{x}_2)$ – the causal effect of the action $do(X_2 = x_2)$ alone, conditioned on the observations available at the time of this action. This can be verified using the back-door criterion, observing that $\{X_1, Z\}$ blocks all back-door paths between X_2 and Y. Thus, the identifiability of $P(y \mid \hat{x}_1, \hat{x}_2)$ can be readily proven by writing

$$P(y \mid \hat{x}_1, \hat{x}_2) = P(y \mid x_1, \hat{x}_2) \tag{4.1}$$

$$= \sum_z P(y \mid z, x_1, \hat{x}_2) P(z \mid x_1) \tag{4.2}$$

$$= \sum_z P(y \mid z, x_1, x_2) P(z \mid x_1), \tag{4.3}$$

where (4.1) and (4.3) follow from Rule 2, and (4.2) follows from Rule 3. The subgraphs that permit the application of these rules are shown in Figure 4.5 (in Section 4.4.3).

This derivation also highlights how conditional plans can be evaluated. Assume we wish to evaluate the effect of the plan $\{do(X_1 = x_1), do(X_2 = g(x_1, z))\}$. Following the analysis of Section 4.2, we write

$$P(y \mid do(X_1 = x_1), do(X_2 = g(x_1, z))) = P(y \mid x_1, do(X_2 = g(x_1, z)))$$

$$= \sum_z P(y \mid z, x_1, do(X_2 = g(x_1, z))) P(z \mid x_1)$$

$$= \sum_z P(y \mid z, x_1, x_2) P(z \mid x_1)|_{x_2 = g(x_1, z)}.$$

Again, the identifiability of this conditional plan rests on the identifiability of the expression $P(y \mid z, x_1, \hat{x}_2)$, which reduces to $P(y \mid z, x_1, x_2)$ because $\{X_1, Z\}$ blocks all back-door paths between X_2 and Y. (See also Section 11.4.1.)

The criterion developed in the next section will enable us to recognize in general, by graphical means, whether a proposed plan can be evaluated from the joint distribution on the observables and, if so, to identify which covariates should be measured and how they should be adjusted.

4.4.2 Plan Identification: Notation and Assumptions

Our starting point is a knowledge specification scheme in the form of a causal diagram, like the one shown in Figure 4.4, that provides a qualitative summary of the analyst's understanding of the relevant data-generating processes.[5]

---

[5] An alternative specification scheme using counterfactual dependencies was used in Robins (1986, 1987), as described in Section 3.6.4.

Notation

A *control problem* consists of a directed acyclic graph (DAG) G with vertex set V, partitioned into four disjoint sets $V = \{X, Z, U, Y\}$, where

X = the set of control variables (exposures, interventions, treatments, etc.);

Z = the set of observed variables, often called *covariates*;

U = the set of unobserved (latent) variables; and

Y = an outcome variable.

We let the control variables be ordered $X = X_1, X_2, \dots, X_n$ such that every X_k is a non-descendant of X_{k+j} $(j > 0)$ in G, and we let the outcome Y be a descendant of X_n. Let N_k stand for the set of observed nodes that are nondescendants of any element in the set $\{X_k, X_{k+1}, \dots, X_n\}$. A *plan* is an ordered sequence $(\hat{x}_1, \hat{x}_2, \dots, \hat{x}_n)$ of value assignments to the control variables, where $\hat{x}_k$ means "X_k is set to x_k." A *conditional plan* is an ordered sequence $(\hat{g}_1(z_1), \hat{g}_2(z_2), \dots, \hat{g}_n(z_n))$, where each g_k is a function from a set Z_k to X_k and where $\hat{g}_k(z_k)$ stands for the statement "set X_k to $g_k(z_k)$ whenever Z_k attains the value z_k." The support Z_k of each $g_k(z_k)$ function must not contain any variables that are descendants of X_k in G.

Our problem is to *evaluate* an unconditional plan[6] by computing $P(y \mid \hat{x}_1, \hat{x}_2, \dots, \hat{x}_n)$, which represents the impact of the plan $(\hat{x}_1, \dots, \hat{x}_n)$ on the outcome variable Y. The expression $P(y \mid \hat{x}_1, \hat{x}_2, \dots, \hat{x}_n)$ is said to be *identifiable* in G if, for every assignment $(\hat{x}_1, \hat{x}_2, \dots, \hat{x}_n)$, the expression can be determined uniquely from the joint distribution of the observables $\{X, Y, Z\}$. A control problem is identifiable whenever $P(y \mid \hat{x}_1, \hat{x}_2, \dots, \hat{x}_n)$ is identifiable.

Our main identifiability criteria are presented in Theorems 4.4.1 and 4.4.6. These invoke sequential back-door tests on various subgraphs of G, from which arrows that point to future actions are deleted. We denote by $G_{\overline{X}}$ (and $G_{\underline{X}}$, respectively) the graphs obtained by deleting from G all arrows pointing to (emerging from) nodes in X. To represent the deletion of both incoming and outgoing arrows, we use the notation $G_{\overline{X}\underline{Z}}$. Finally, the expression $P(y \mid \hat{x}, z) \triangleq P(y, z \mid \hat{x})/P(z \mid \hat{x})$ stands for the probability of $Y = y$ given that $Z = z$ is observed and X is held constant at x.

4.4.3   Plan Identification: The Sequential Back-Door Criterion

Theorem 4.4.1 (Pearl and Robins 1995)
The probability $P(y \mid \hat{x}_1, \dots, \hat{x}_n)$ is identifiable if, for every $1 \le k \le n$, there exists a set Z_k of covariates satisfying the following (sequential back-door) conditions:

$$Z_k \subseteq N_k, \tag{4.4}$$

(i.e., Z_k consists of nondescendants of $\{X_k, X_{k+1}, \dots, X_n\}$) and

---

[6] Conditional plans are analyzed in Section 11.4.1, using repeated applications of Theorem 4.4.1.

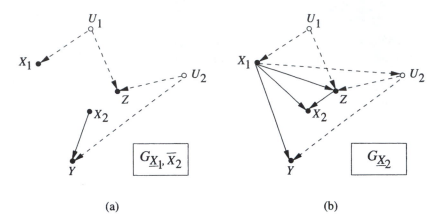

(a)   (b)

Figure 4.5 The two subgraphs of G used in testing the identifiability of the plan $(\hat{x}_1, \hat{x}_2)$ in Figure 4.4.

$$(Y \perp\!\!\!\perp X_k \mid X_1, \ldots, X_{k-1}, Z_1, Z_2, \ldots, Z_k)_{G_{\underline{X}_k, \overline{X}_{k+1}, \ldots \overline{X}_n}}. \tag{4.5}$$

When these conditions are satisfied, the effect of the plan is given by

$$P(y \mid \hat{x}_1, \ldots, \hat{x}_n) = \sum_{z_1, \ldots, z_n} P(y \mid z_1, \ldots, z_n, x_1, \ldots, x_n)$$

$$\times \prod_{k=1}^{n} P(z_k \mid z_1, \ldots, z_{k-1}, x_1, \ldots, x_{k-1}). \tag{4.6}$$

Before presenting its proof, let us demonstrate how Theorem 4.4.1 can be used to test the identifiability of the control problem shown in Figure 4.4. First, we will show that $P(y \mid \hat{x}_1, \hat{x}_2)$ cannot be identified without measuring Z; in other words, that the sequence $Z_1 = \emptyset$, $Z_2 = \emptyset$ would not satisfy conditions (4.4)–(4.5). The two d-separation tests encoded in (4.5) are

$$(Y \perp\!\!\!\perp X_1)_{G_{\underline{X}_1, \overline{X}_2}} \quad \text{and} \quad (Y \perp\!\!\!\perp X_2 \mid X_1)_{G_{\underline{X}_2}}.$$

The two subgraphs associated with these tests are shown in Figure 4.5. We see that $(Y \perp\!\!\!\perp X_1)$ holds in $G_{\underline{X}_1, \overline{X}_2}$ but that $(Y \perp\!\!\!\perp X_2 \mid X_1)$ fails to hold in $G_{\underline{X}_2}$. Thus, in order to pass the test, we must have either $Z_1 = \{Z\}$ or $Z_2 = \{Z\}$; since Z is a descendant of X_1, only the second alternative satisfies (4.4). The tests applicable to the sequence $Z_1 = \emptyset$, $Z_2 = \{Z\}$ are $(Y \perp\!\!\!\perp X_1)_{G_{\underline{X}_1, \overline{X}_2}}$ and $(Y \perp\!\!\!\perp X_2 \mid X_1, Z)_{G_{\underline{X}_2}}$. Figure 4.5 shows that both tests are now satisfied, because $\{X_1, Z\}$ d-separates Y from X_2 in $G_{\underline{X}_2}$. Having satisfied conditions (4.4)–(4.5), equation (4.6) provides a formula for the effect of plan $(\hat{x}_1, \hat{x}_2)$ on Y:

$$P(y \mid \hat{x}_1, \hat{x}_2) = \sum_z P(y \mid z, x_1, x_2) P(z \mid x_1), \tag{4.7}$$

which coincides with (4.3).

The question naturally arises of whether the sequence $Z_1 = \emptyset$, $Z_2 = \{Z\}$ can be identified without exhaustive search. This question will be answered in Corollary 4.4.5 and Theorem 4.4.6.

Proof of Theorem 4.4.1

The proof given here is based on the inference rules of *do*-calculus (Theorem 3.4.1), which facilitate the reduction of causal effect formulas to hat-free expressions. An alternative proof, using latent variable elimination, is given in Pearl and Robins (1995).

Step 1. The condition $Z_k \subseteq N_k$ implies $Z_k \subseteq N_j$ for all $j \geq k$. Therefore, we have

$$P(z_k \mid z_1, \ldots, z_{k-1}, x_1, \ldots, x_{k-1}, \hat{x}_k, \hat{x}_{k+1}, \ldots, \hat{x}_n)$$

$$= P(z_k \mid z_1, \ldots, z_{k-1}, x_1, \ldots, x_{k-1}).$$

This is so because no node in $\{Z_1, \ldots, Z_k, X_1, \ldots, X_{k-1}\}$ can be a descendant of any node in $\{X_k, \ldots, X_n\}$. Hence, Rule 3 allows us to delete the hat variables from the expression.

Step 2. The condition in (4.5) permits us to invoke Rule 2 and write:

$$P(y \mid z_1, \ldots, z_k, x_1, \ldots, x_{k-1}, \hat{x}_k, \hat{x}_{k+1}, \ldots, \hat{x}_n)$$

$$= P(y \mid z_1, \ldots, z_k, x_1, \ldots, x_{k-1}, x_k, \hat{x}_{k+1}, \ldots, \hat{x}_n).$$

Thus, we have

$$P(y \mid \hat{x}_1, \ldots, \hat{x}_n)$$

$$= \sum_{z_1} P(y \mid z_1, \hat{x}_1, \hat{x}_2, \ldots, \hat{x}_n) P(z_1 \mid \hat{x}_1, \ldots, \hat{x}_n)$$

$$= \sum_{z_1} P(y \mid z_1, x_1, \hat{x}_2, \ldots, \hat{x}_n) P(z_1)$$

$$= \sum_{z_2} \sum_{z_1} P(y \mid z_1, z_2, x_1, \hat{x}_2, \ldots, \hat{x}_n) P(z_1) P(z_2) \mid z_1, x_1, \hat{x}_2, \ldots, \hat{x}_n)$$

$$= \sum_{z_2} \sum_{z_1} P(y \mid z_1, z_2, x_1, x_2, \hat{x}_3, \ldots, \hat{x}_n) P(z_1) P(z_2) \mid z_1, x_1)$$

$$\vdots$$

$$= \sum_{z_n} \cdots \sum_{z_2} \sum_{z_1} P(y \mid z_1, \ldots, z_n, x_1, \ldots, x_n)$$

$$\times P(z_1) P(z_2 \mid z_1, x_1) \cdots P(z_n \mid z_1, x_1, z_2, x_2, \ldots, z_{n-1}, x_{n-1})$$

$$= \sum_{z_1, \ldots, z_n} P(y \mid z_1, \ldots, z_n, x_1, \ldots, x_n) \prod_{k=1}^{n} P(z_k \mid z_1, \ldots, z_{k-1}, x_1, \ldots, x_{k-1}). \qquad \square$$

Definition 4.4.2 (Admissible Sequence and *G*-Identifiability)

Any sequence $Z_1, \ldots, Z_n$ of covariates satisfying the conditions in (4.4)–(4.5) will be called admissible, *and any expression $P(y \mid \hat{x}_1, \hat{x}_2, \ldots, \hat{x}_n)$ that is identifiable by the criterion of Theorem 4.4.1 will be called* G-*identifiable.*[7]

---

[7] Note that admissibility (4.5) requires that each subsequence $X_1, \ldots, X_{k-1}, Z_1, \ldots, Z_k$ blocks every "action-avoiding" back-door path from X_k to Y (see page 103).

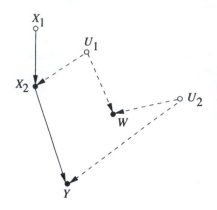

Figure 4.6 An admissible choice $Z_1 = W$ that rules out any admissible choice for Z_2. The choice $Z_1 = \emptyset$ would permit the construction of an admissible sequence $(Z_1 = \emptyset, Z_2 = \emptyset)$.

The following corollary is immediate.

Corollary 4.4.3
A control problem is G-identifiable if and only if it has an admissible sequence.

Note that, despite the completeness of *do*-calculus, the property of *G*-identifiability is sufficient but not necessary for general plan identifiability as defined in Section 4.4.2. The reason is that the kth step in the reduction of (4.6) refrains from conditioning on variables Z_k that are descendants of X_k – namely, variables that may be affected by the action $do(X_k = x_k)$. In certain causal structures, the identifiability of causal effects requires that we condition on such variables, as demonstrated by the front-door criterion (Theorem 3.3.4).

4.4.4 Plan Identification: A Procedure

Theorem 4.4.1 provides a declarative condition for plan identifiability. It can be used to ratify that a proposed formula is valid for a given plan, but it does not provide an effective procedure for deriving such formulas because the choice of each Z_k is not spelled out procedurally. The possibility exists that some unfortunate choice of Z_k satisfying (4.4) and (4.5) might prevent us from continuing the reduction process even though another reduction sequence is feasible.

This is illustrated in Figure 4.6. Here W is an admissible choice for Z_1, but if we make this choice then we will not be able to complete the reduction, since no set Z_2 can be found that satisfies condition (4.5): $(Y \perp\!\!\!\perp X_2 \mid X_1, W, Z_2)_{G_{\overline{X}_2}}$. In this example it would be wiser to choose $Z_1 = Z_2 = \emptyset$, which satisfies both $(Y \perp\!\!\!\perp X_1 \mid \emptyset)_{G_{\underline{X}_1, \overline{X}_2}}$ and $(Y \perp\!\!\!\perp X_2 \mid X_1, \emptyset)_{G_{\overline{X}_2}}$.

The obvious way to avoid bad choices of covariates, like the one illustrated in Figure 4.6, is to insist on always choosing a "minimal" Z_k, namely, a set of covariates satisfying (4.5) that has no proper subset satisfying (4.5). However, since there are usually many such minimal sets (see Figure 4.7), the question remains of whether every choice of a minimal Z_k is "safe": Can we be sure that no choice of a minimal subsequence $Z_1, \ldots, Z_k$ will ever prevent us from finding an admissible Z_{k+1} when some admissible sequence $Z_1^*, \ldots, Z_n^*$ exists?

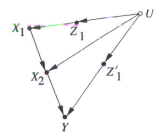

Figure 4.7   Nonuniqueness of minimal admissible sets: Z_1 and Z'_1 are each minimal and admissible, since $(Y \perp\!\!\!\perp \overline{X}_1 \mid Z_1)$ and $(Y \perp\!\!\!\perp X_1 \mid Z'_1)$ both hold in $G_{\underline{X}_1, \overline{X}_2}$.

The next result guarantees the safety of every minimal subsequence $Z_1, \ldots, Z_k$ and hence provides an effective test for G-identifiability.

Theorem 4.4.4

If there exists an admissible sequence $Z_1^*\ldots, Z_n^*$ then, for every minimally admissible subsequence $Z_1, \ldots, Z_{k-1}$ of covariates, there is an admissible set Z_k.

A proof is given in Pearl and Robins (1995).

Theorem 4.4.4 now yields an effective decision procedure for testing G-identifiability as follows.

Corollary 4.4.5

A control problem is G-identifiable if and only if the following algorithm exits with success.

1. *Set $k = 1$.*
2. *Choose any minimal $Z_k \subseteq N_k$ satisfying (4.5).*
3. *If no such Z_k exists then exit with failure; else set $k = k + 1$.*
4. *If $k = n + 1$ then exit with success; else return to step 2.*

A further variant of Theorem 4.4.4 can be stated that avoids the search for minimal sets Z_k. This follows from the realization that, if an admissible sequence exists, we can rewrite Theorem 4.4.1 in terms of an explicit sequence of covariates $W_1, W_2, \ldots, W_n$ that can easily be identified in G.

Theorem 4.4.6

The probability $P(y \mid \hat{x}_1, \ldots, \hat{x}_n)$ is G-identifiable if and only if the following condition holds for every $1 \leq k \leq n$:

$$(Y \perp\!\!\!\perp X_k \mid X_1, \ldots, X_{k-1}, W_1, W_2, \ldots, W_k)_{G_{\underline{X}_k, \overline{X}_{k+1}, \ldots, \overline{X}_n}},$$

where W_k is the set of all covariates in G that are both nondescendants of $\{X_k, X_{k+1}, \ldots, X_n\}$ and have either Y or X_k as descendant in $G_{\underline{X}_k, \overline{X}_{k+1}, \ldots, \overline{X}_n}$. Moreover, if this condition is satisfied then the plan evaluates as

$$P(y \mid \hat{x}_1, \ldots, \hat{x}_n) = \sum_{w_1, \ldots, w_n} P(y \mid w_1, \ldots, w_n, x_1, \ldots, x_n)$$

$$\times \prod_{k=1}^{n} P(w_k \mid w_1, \ldots, w_{k-1}, x_1, \ldots, x_{k-1}). \tag{4.8}$$

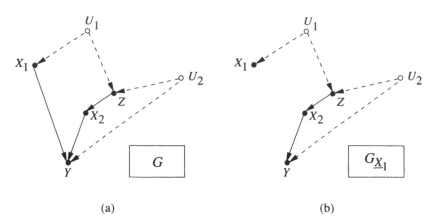

Figure 4.8 Causal diagram G in which proper ordering of the control variables X_1 and X_2 is important.

A proof of Theorem 4.4.6, together with several generalizations, can be found in Pearl and Robins (1995) and Robins (1997). Extensions to G-identifiability are reported in Kuroki (with Miyakawa 1999a,b, 2003; with et al. 2003; with Cai 2004).

The reader should note that, although Corollary 4.4.5 and Theorem 4.4.6 are procedural in the sense of offering systematic tests for plan identifiability, they are still *order-dependent*. It is quite possible that an admissible sequence exists in one ordering of the control variables and not in another when both orderings are consistent with the arrows in G. The graph G in Figure 4.8 illustrates such a case. It is obtained from Figure 4.4 by deleting the arrows $X_1 \to X_2$ and $X_1 \to Z$, so that the two control variables (X_1 and X_2) can be ordered arbitrarily. The ordering (X_1, X_2) would still admit the admissible sequence ($\emptyset, Z$) as before, but no admissible sequence can be found for the ordering (X_2, X_1). This can be seen immediately from the graph $G_{\underline{X}_1}$, in which (according to (4.5) with $k = 1$) we need to find a set Z such that $\{X_2, Z\}$ d-separates Y from X_1. No such set exists.

The implication of this order sensitivity is that, whenever G permits several orderings of the control variables, all orderings need be examined before we can be sure that a plan is not G-identifiable. The graphical criteria of Shpitser and Pearl (2006b) circumvents this search.

4.5 DIRECT AND INDIRECT EFFECTS

4.5.1 Direct versus Total Effects

The causal effect we have analyzed so far, $P(y \mid \hat{x})$, measures the *total* effect of a variable (or a set of variables) X on a response variable Y. In many cases, this quantity does not adequately represent the target of investigation, and attention is focused instead on the direct effect of X on Y. The term "direct effect" is meant to quantify an effect that is not mediated by other variables in the model or, more accurately, the sensitivity of Y to changes in X while all other factors in the analysis are held fixed. Naturally, holding those factors fixed would sever all causal paths from X to Y with the exception of the direct link $X \to Y$, which is not intercepted by any intermediaries.

A classical example of the ubiquity of direct effects (see Hesslow 1976; Cartwright 1989) tells the story of a birth-control pill that is suspected of producing thrombosis in women and, at the same time, has a negative indirect effect on thrombosis by reducing the rate of pregnancies (pregnancy is known to encourage thrombosis). In this example, interest is focused on the direct effect of the pill because it represents a stable biological relationship that, unlike the total effect, is invariant to marital status and other social factors that may affect women's chances of getting pregnant or of sustaining pregnancy.

Another class of examples involves legal disputes over race or sex discrimination in hiring. Here, neither the effect of sex or race on applicants' qualification nor the effect of qualification on hiring are targets of litigation. Rather, defendants must prove that sex and race do not *directly* influence hiring decisions, whatever indirect effects they might have on hiring by way of applicant qualification.

In all these examples, the requirement of holding the mediating variables fixed must be interpreted as (hypothetically) setting these variables to constants by physical intervention, not by conditioning, or adjustment (a misguided habit that dates back to Fisher 1935). For example, it will not be sufficient to measure the association between the birth-control pill and thrombosis separately among pregnant and nonpregnant women and then aggregate the results. Instead, we must perform the study among women who became pregnant before the use of the pill and among women who prevented pregnancy by means other than the drug. The reason is that, by conditioning on an intermediate variable (pregnancy in the example), we may create spurious associations between X and Y even when there is no direct effect of X on Y. This can easily be illustrated in the model $X \rightarrow Z \leftarrow U \rightarrow Y$, where X has no direct effect on Y. Physically holding Z constant would permit no association between X and Y, as can be seen by deleting all arrows entering Z. But if we were to condition on Z, a spurious association would be created through U (unobserved) that might be construed as a direct effect of X on Y.

4.5.2   Direct Effects, Definition, and Identification

Controlling all variables in a problem is obviously a major undertaking, if not an impossibility. The analysis of identification tells us under what conditions direct effects can be estimated from nonexperimental data even without such control. Using our $do(x)$ notation (or $\hat{x}$ for short), we can express the direct effect as follows.

Definition 4.5.1 (Direct Effect)
The direct effect of X on Y is given by $P(y \mid \hat{x}, \hat{s}_{XY})$, where S_{XY} is the set of all endogenous variables except X and Y in the system.

We see that the measurement of direct effects is ascribed to an ideal laboratory; the scientist controls for all possible conditions S_{XY} and need not be aware of the structure of the diagram or of which variables are truly intermediaries between X and Y. Much of the experimental control can be eliminated, however, if we know the structure of the diagram. For one thing, there is no need to actually hold *all* other variables constant; holding constant the direct parents of Y (excluding X) should suffice. Thus, we obtain the following equivalent definition of a direct effect.

Corollary 4.5.2

The direct effect of X on Y is given by $P(y \mid \hat{x}, \widehat{pa}_{Y \setminus X})$ where $pa_{Y \setminus X}$ stands for any realization of the parents of Y, excluding X.

Clearly, if X does not appear in the equation for Y (equivalently, if X is not a parent of Y), then $P(y \mid \hat{x}, \widehat{pa}_{Y \setminus X})$ defines a constant distribution on Y that is independent of x, thus matching our understanding of "having no direct effect." In general, assuming that X is a parent of Y, Corollary 4.5.2 implies that the direct effect of X on Y is identifiable whenever $P(y \mid \widehat{pa}_Y)$ is identifiable. Moreover, since the conditioning part of this expression corresponds to a plan in which the parents of Y are the control variables, we conclude that a direct effect is identifiable whenever the effect of the corresponding parents' plan is identifiable. We can now use the analysis of Section 4.4 and apply the graphical criteria of Theorems 4.4.1 and 4.4.6 to the analysis of direct effects. In particular, we can state our next theorem.

Theorem 4.5.3

Let $PA_Y = \{X_1, \ldots, X_k, \ldots, X_m\}$. The direct effect of any X_k on Y is identifiable whenever the conditions of Corollary 4.4.5 hold for the plan $(\hat{x}_1, \hat{x}_2, \ldots, \hat{x}_m)$ in some admissible ordering of the variables. The direct effect is then given by (4.8).

Theorem 4.5.3 implies that if the effect of one parent of Y is identifiable, then the effect of every parent of Y is identifiable as well. Of course, the magnitude of the effect would differ from parent to parent, as seen in (4.8).

The following corollary is immediate.

Corollary 4.5.4

Let X_j be a parent of Y. The direct effect of X_j on Y is, in general, nonidentifiable if there exists a confounding arc that embraces any link $X_k \rightarrow Y$.

4.5.3   Example: Sex Discrimination in College Admission

To illustrate the use of this result, consider the study of Berkeley's alleged sex bias in graduate admission (Bickel et al. 1975), where data showed a higher rate of admission for male applicants overall but, when broken down by departments, a slight bias toward female applicants. The explanation was that female applicants tend to apply to the more competitive departments, where rejection rates are high; based on this finding, Berkeley was exonerated from charges of discrimination. The philosophical aspects of such reversals, known as Simpson's paradox, will be discussed more fully in Chapter 6. Here we focus on the question of whether adjustment for department is appropriate for assessing sex discrimination in college admission. Conventional wisdom has it that such adjustment is appropriate because "We know that applying to a popular department (one with considerably more applicants than positions) is just the kind of thing that causes rejection" (Cartwright 1983, p. 38), but we will soon see that additional factors should be considered.

Let us assume that the relevant factors in the Berkeley example are configured as in Figure 4.9, with the following interpretation of the variables:

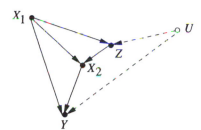

Figure 4.9 Causal relationships relevant to Berkeley's sex discrimination study. Adjusting for department choice (X_2) or career objective (Z) (or both) would be inappropriate in estimating the direct effect of gender on admission. The appropriate adjustment is given in (4.10).

X_1 = applicant's gender;

X_2 = applicant's choice of department;

Z = applicant's (pre-enrollment) career objectives;

Y = admission outcome (accept/reject);

U = applicant's aptitude (unrecorded).

Note that U affects applicant's career objective and also the admission outcome Y (say, through verbal skills (unrecorded)).

Adjusting for department choice amounts to computing the following expression:

$$E_{x_2} P(y \mid \hat{x}_1, x_2) = \sum_{x_2} P(y \mid x_1, x_2) P(x_2). \qquad (4.9)$$

In contrast, the direct effect of X_1 on Y, as given by (4.7), reads

$$P(y \mid \hat{x}_1, \hat{x}_2) = \sum_{z} P(y \mid z, x_1, x_2) P(z \mid x_1). \qquad (4.10)$$

It is clear that the two expressions may differ substantially. The first measures the (average) effect of sex on admission among applicants to a given department, a quantity that is sensitive to the fact that some gender–department combinations may be associated with high admission rates merely because such combinations are indicative of a certain aptitude (U) that was unrecorded. The second expression eliminates such spurious associations by separately adjusting for career objectives (Z) in each of the two genders.

To verify that (4.9) does not properly measure the direct effect of X_1 on Y, we note that the expression depends on the value of X_1 even in cases where the arrow between X_1 and Y is absent. Equation (4.10), on the other hand, becomes insensitive to x_1 in such cases – an exercise that we leave for the reader to verify.[8]

To cast this analysis in a concrete numerical setting, let us imagine a college consisting of two departments, A and B, both admitting students on the basis of qualification, Q, alone. Let us further assume (i) that the applicant pool consists of 100 males and 100 females and (ii) that 50 applicants of each gender have high qualifications (hence are admitted) and 50 have low qualifications (hence are rejected). Clearly, this college cannot be accused of sex discrimination.

---

[8] *Hint*: Factorize $P(y, u, z \mid \hat{x}_1, \hat{x}_2)$ using the independencies in the graph and eliminate u as in the derivation of (3.27). Cole and Hernán (2002) present examples in epidemiology.

Table 4.1. *Admission Rate among Males and Females in Each Department*

	Males		Females		Total	
	Admitted	Applied	Admitted	Applied	Admitted	Applied
Dept. A	50	50	0	0	50	50
Dept. B	0	50	50	100	50	150
Unadjusted	50%		50%		50%	
Adjusted	25%		37.5%			

A different result would surface, however, if we adjust for departments while ignoring qualifications, which amounts to using (4.9) to estimate the effect of gender on admission. Assume that the nature of the departments is such that *all and only* qualified male applicants apply to department *A*, while all females apply to department *B* (see Table 4.1).

We see from the table that adjusting for department would falsely indicate a bias of $37.5:25$ ($= 3:2$) in favor of female applicants. An unadjusted (sometimes called "crude") analysis happens to give the correct result in this example – a 50% admission rate for males and females alike – thus exonerating the school from charges of sex discrimination.

Our analysis is not meant to imply that the Berkeley study of Bickel et al. (1975) is defective, or that adjustment for department was not justified in that study. The purpose is to emphasize that no adjustment is guaranteed to give an unbiased estimate of causal effects, direct or indirect, absent a careful examination of the causal assumptions that ensure identification. Theorem 4.5.3 provides us with the understanding of those assumptions and with a mathematical means of expressing them. We note that if applicants' qualifications were not recorded in the data, then the direct effect of gender on admission will not be identifiable unless we can measure some proxy variable that stands in the same relation to Q as Z stands to U in Figure 4.9.

4.5.4 Natural Direct Effects

Readers versed in structural equation models (SEMs) will note that, in linear systems, the direct effect $E(Y \mid \hat{x}, \widehat{pa}_{Y \setminus X})$ is fully specified by the path coefficient attached to the link from X to Y; therefore, the direct effect is independent of the values $pa_{Y \setminus X}$ at which we hold the other parents of Y. In nonlinear systems, those values would, in general, modify the effect of X on Y and thus should be chosen carefully to represent the target policy under analysis. For example, the direct effect of a pill on thrombosis would most likely be different for pregnant and nonpregnant women. Epidemiologists call such differences "effect modification" and insist on separately reporting the effect in each subpopulation.

Although the direct effect is sensitive to the levels at which we hold the parents of the outcome variable, it is sometimes meaningful to average the direct effect over those levels. For example, if we wish to assess the degree of discrimination in a given school without reference to specific departments, we should replace the controlled difference

$$P(\text{admission} \mid \widehat{\text{male}}, \widehat{\text{dept}}) - P(\text{admission} \mid \widehat{\text{female}}, \widehat{\text{dept}})$$

with some average of this difference over all departments. This average should measure the increase in admission rate in a hypothetical experiment in which we instruct all female candidates to retain their department preferences but change their gender identification (on the application form) from female to male.

Conceptually, we can define the average direct effect $DE_{x,x'}(Y)$ as the expected change in Y induced by changing X from x to x' while keeping all mediating factors constant at whatever value they would have obtained under $do(x)$. This hypothetical change, which Robins and Greenland (1991) called "pure" and Pearl (2001c) called "natural," is precisely what lawmakers instruct us to consider in race or sex discrimination cases: "The central question in any employment-discrimination case is whether the employer would have taken the same action had the employee been of a different race (age, sex, religion, national origin etc.) and everything else had been the same." (In *Carson versus Bethlehem Steel Corp.*, 70 FEP Cases 921, 7th Cir. (1996)).

Using the parenthetical notation of equation 3.51, Pearl (2001c) gave the following definition for the "natural direct effect":

$$DE_{x,x'}(Y) = E[(Y(x', Z(x))) - E(Y(x)]. \tag{4.11}$$

Here, Z represents all parents of Y excluding X, and the expression $Y(x', Z(x))$ represents the value that Y would attain under the operation of setting X to x' and, simultaneously, setting Z to whatever value it would have obtained under the setting $X = x$. We see that $DE_{x,x'}(Y)$, the natural direct effect of the transition from x to x', involves probabilities of *nested counterfactuals* and cannot be written in terms of the $do(x)$ operator. Therefore, the natural direct effect cannot in general be identified, even with the help of ideal, controlled experiments (see Robins and Greenland 1992 and Section 7.1 for intuitive explanation). Pearl (2001c) has nevertheless shown that, if certain assumptions of "no confounding" are deemed valid,[9] the natural direct effect can be reduced to

$$DE_{x,x'}(Y) = \sum_z [E(Y \mid do(x', z)) - E(Y \mid do(x, z))] P(z \mid do(x)). \tag{4.12}$$

The intuition is simple; the natural direct effect is the weighted average of controlled direct effects, using the causal effect $P(z \mid do(x))$ as a weighing function. Under such assumptions, the sequential back-door criteria developed in Section 4.4 for identifying control-specific plans, $P(y \mid \hat{x}_1, \hat{x}_2, \ldots, \hat{x}_n)$, become applicable.

In particular, expression (4.12) is both valid and identifiable in Markovian models, where all *do*-operators can be eliminated using Corollary 3.2.6; for example,

$$P(z \mid do(x)) = \sum_t P(z \mid x, pa_X = t) P(pa_X = t) \tag{4.13}$$

---

[9] One sufficient condition is that $Z(x) \perp\!\!\!\perp Y(x', z) \mid W$ holds for some set W of measured covariates. See details and graphical criteria in Pearl (2001c, 2005a) and in Petersen et al. (2006).

4.5.5 Indirect Effects

Remarkably, the definition of the natural direct effect (4.11) can easily be turned around and provide an operational definition for the *indirect effect* – a concept shrouded in mystery and controversy, because it is impossible, using the $do(x)$ operator, to disable the direct link from X to Y so as to let X influence Y solely via indirect paths.

The natural indirect effect, IE, of the transition from x to x' is defined as the expected change in Y affected by holding X constant, at $X = x$, and changing Z to whatever value it would have attained had X been set to $X = x'$. Formally, this reads (Pearl 2001c):

$$IE_{x,x'}(Y) \triangleq E[(Y(x, Z(x'))) - E(Y(x))], \tag{4.14}$$

which is almost identical to the direct effect (equation (4.11)) save for exchanging x and x'.

Indeed, it can be shown that, in general, the total effect TE of a transition is equal to the *difference* between the direct effect of that transition and the indirect effect of the reverse transition. Formally,

$$TE_{x,x'}(Y) \triangleq E(Y(x) - Y(x')) = DE_{x,x'}(Y) - IE_{x',x}(Y). \tag{4.15}$$

In linear systems, where reversal of transitions amounts to negating the signs of their effects, we have the standard additive formula

$$TE_{x,x'}(Y) = DE_{x,x'}(Y) + IE_{x,x'}(Y). \tag{4.16}$$

Since each term above is based on an independent operational definition, this quality constitutes a formal justification for the additive formula.

Note that the indirect effect has clear policy-making implications. For example: in a hiring discrimination environment, a policy maker may be interested in predicting the gender mix in the work force if gender bias is eliminated and all applicants are treated equally – say, the same way that males are currently treated. This quantity will be given by the indirect effect of gender on hiring, mediated by factors such as education and aptitude, which may be gender-dependent.

More generally, a policy maker may be interested in the effect of issuing a directive to a select set of subordinate employees, or in carefully controlling the routing of messages in a network of interacting agents. Such applications motivate the analysis of *path-specific effects*, that is, the effect of X on Y through a selected set of paths (Avin et al. 2005).

Note that in all these cases, the policy intervention invokes the selection of signals to be sensed, rather than variables to be fixed. Pearl (2001c) has suggested therefore that signal sensing is more fundamental to the notion of causation than manipulation; the latter being but a crude way of stimulating the former in experimental setup. (See Section 11.4.5.)

It is remarkable that counterfactual quantities like DE and IE that could not be expressed in terms of $do(x)$ operators, and appear therefore void of empirical content, can, under certain conditions, be estimated from empirical studies. A general analysis of those conditions is given in Shpitser and Pearl (2007).

We shall see additional examples of this "marvel of formal analysis" in Chapters 7, 9, and 11. It constitutes an unassailable argument in defense of counterfactual analysis, as expressed in Pearl (2000) against the stance of Dawid (2000).

CHAPTER FIVE

Causality and Structural Models in Social Science and Economics

Do two men travel together
unless they have agreed?
                    Amos 3:3

Preface

Structural equation modeling (SEM) has dominated causal analysis in economics and the social sciences since the 1950s, yet the prevailing interpretation of SEM differs substantially from the one intended by its originators and also from the one expounded in this book. Instead of carriers of substantive causal information, structural equations are often interpreted as carriers of probabilistic information; economists view them as convenient representations of density functions, and social scientists see them as summaries of covariance matrices. The result has been that many SEM researchers have difficulty articulating the causal content of SEM, and the most distinctive capabilities of SEM are currently ill understood and underutilized.

This chapter is written with the ambitious goal of reinstating the causal interpretation of SEM. We shall demonstrate how developments in the areas of graphical models and the logic of intervention can alleviate the current difficulties and thus revitalize structural equations as the primary language of causal modeling. Toward this end, we recast several of the results of Chapters 3 and 4 in parametric form (the form most familiar to SEM researchers) and demonstrate how practical and conceptual issues of model testing and parameter identification can be illuminated through graphical methods. We then move back to nonparametric analysis, from which an operational semantics will evolve that offers a coherent interpretation of what structural equations are all about (Section 5.4). In particular, we will provide answers to the following fundamental questions: What do structural equations claim about the world? What portion of those claims is testable? Under what conditions can we estimate structural parameters through regression analysis?

In Section 5.1 we survey the history of SEM and suggest an explanation for the current erosion of its causal interpretation. The testable implications of structural models are explicated in Section 5.2. For recursive models (herein termed *Markovian*), we find that the statistical content of a structural model can be fully characterized by a set of zero partial correlations that are entailed by the model. These zero partial correlations can be read off the graph using the *d-separation* criterion, which in linear models applies to graphs with cycles and correlated errors as well (Section 5.2). The application of this criterion to model testing is discussed in Section 5.2.2, which advocates local over global testing strategies. Section 5.2.3 provides simple graphical tests of model equivalence and thus clarifies the *nontestable* part of structural models.

In Section 5.3 we deal with the issue of determining the identifiability of structural parameters prior to gathering any data. In Section 5.3.1, simple graphical tests of identifiability are developed for linear Markovian and semi-Markovian models (i.e., acyclic diagrams with correlated errors). These tests result in a simple procedure for determining when a path coefficient can be equated to a regression coefficient and, more generally, when structural parameters can be estimated through regression analysis. Section 5.3.2 discusses the connection between parameter identification in linear models and causal effect identification in nonparametric models, and Section 5.3.3 offers the latter as a semantical basis for the former.

Finally, in Section 5.4 we discuss the logical foundations of SEM and resolve a number of difficulties that were kept dormant in the past. These include operational definitions for structural equations, structural parameters, error terms, and total and direct effects, as well as a formal definition of exogeneity in econometrics.

5.1 INTRODUCTION

5.1.1 Causality in Search of a Language

The word *cause* is not in the vocabulary of standard probability theory. It is an embarrassing yet inescapable fact that probability theory, the official mathematical language of many empirical sciences, does not permit us to express sentences such as "Mud does not cause rain"; all we can say is that the two events are mutually correlated, or dependent – meaning that if we find one, we can expect to encounter the other. Scientists seeking causal explanations for complex phenomena or rationales for policy decisions must therefore supplement the language of probability with a vocabulary for causality, one in which the symbolic representation for the causal relationship "Mud does not cause rain" is distinct from the symbolic representation for "Mud is independent of rain." Oddly, such distinctions have yet to be incorporated into standard scientific analysis.[1]

Two languages for causality have been proposed: path analysis or structural equation modeling (SEM) (Wright 1921; Haavelmo 1943) and the Neyman–Rubin potential-outcome model (Neyman 1923; Rubin 1974). The former has been adopted by economists and social scientists (Goldberger 1972; Duncan 1975), while a group of statisticians champion the latter (Rubin 1974; Holland 1988; Rosenbaum 2002). These two languages are mathematically equivalent (see Chapter 7, Section 7.4.4), yet neither has become standard in causal modeling – the structural equation framework because it has been greatly misused and inadequately formalized (Freedman 1987) and the potential-outcome framework because it has been only partially formalized and (more significantly) because it rests on an esoteric and seemingly metaphysical vocabulary of randomized experiments and counterfactual variables that bears no apparent relation to ordinary understanding of cause–effect processes in nonexperimental settings (see Section 3.6.3).

---

[1] A summary of attempts by philosophers to reduce causality to probabilities is given in Chapter 7 (Section 7.5).

Currently, potential-outcome models are understood by few and used by even fewer. Structural equation models are used by many, but their causal interpretation is generally questioned or avoided, even by their leading practitioners. In Chapters 3 and 4 we described how structural equation models, in nonparametric form, can provide the semantic basis for theories of interventions. In Sections 1.4 and 3.6.3 we outlined how these models provide the semantical basis for a theory of counterfactuals as well. It is somewhat embarrassing that these distinctive features are hardly recognized and rarely utilized in the modern SEM literature. The current dominating philosophy treats SEM as just a convenient way to encode density functions (in economics) or covariance information (in social science). Ironically, we are witnessing one of the most bizarre circles in the history of science: causality in search of a language and, simultaneously, speakers of that language in search of its meaning.

The purpose of this chapter is to formulate the causal interpretation and outline the proper use of structural equation models, thereby reinstating confidence in SEM as the primary formal language for causal analysis in the social and behavioral sciences. First, however, we present a brief analysis of the current crisis in SEM research in light of its historical development.

5.1.2   SEM: How Its Meaning Became Obscured

Structural equation modeling was developed by geneticists (Wright 1921) and economists (Haavelmo 1943; Koopmans 1950, 1953) so that qualitative cause–effect information could be combined with statistical data to provide quantitative assessment of cause–effect relationships among variables of interest. Thus, to the often asked question, "Under what conditions can we give causal interpretation to structural coefficients?" Wright and Haavelmo would have answered, "Always!" According to the founding fathers of SEM, the conditions that make the equation $y = \beta x + \varepsilon$ *structural* is precisely the claim that the causal connection between X and Y is β and nothing about the statistical relationship between x and ε can ever change this interpretation of β. Amazingly, this basic understanding of SEM has all but disappeared from the literature, leaving modern econometricians and social scientists in a quandary over β.

Most SEM researchers today are of the opinion that extra ingredients are necessary for structural equations to qualify as carriers of causal claims. Among social scientists, James, Mulaik, and Brett (1982, p. 45), for example, stated that a condition called *self-containment* is necessary for consecrating the equation $y = \beta x + \varepsilon$ with causal status, where self-containment stands for $\text{cov}(x, \varepsilon) = 0$. According to James et al. (1982), if self-containment does not hold, then "neither the equation nor the functional relation represents a causal relation." Bollen (1989, p. 44) reiterated the necessity of self-containment (under the rubric *isolation* or *pseudo-isolation*) – contrary to the understanding that structural equations attain their causal interpretation prior to, and independently of, any statistical relationships among their constituents. Since the early 1980s, it has become exceedingly rare to find an open endorsement of the original SEM logic: that β defines the sensitivity of $E(Y)$ to experimental manipulations (or counterfactual variations) of X; that ε is defined in terms of β, not the other way around; and that the orthogonality condition $\text{cov}(x, \varepsilon) = 0$ is neither necessary nor sufficient for the causal interpretation of β (see Sections 3.6.2 and 5.4.1).[2] It is therefore not surprising that many SEM text-books have given up on causal interpretation altogether: "We often see the terms cause, effect, and causal modeling used in the research literature. We do not endorse this practice and therefore do not use these terms here" (Schumaker and Lomax 1996, p. 90).

Econometricians have just as much difficulty with the causal reading of structural parameters. Leamer (1985, p. 258) observed, "It is my surprising conclusion that econo-mists know very well what they mean when they use the words 'exogenous,' 'structural,' and 'causal,' yet no textbook author has written adequate definitions." There has been little change since Leamer made these observations. Econometric textbooks invariably devote most of their analysis to estimating structural parameters, but they rarely discuss the role of these parameters in policy evaluation. The few books that deal with policy analysis (e.g., Goldberger 1991; Intriligator et al. 1996, p. 28) assume that policy vari-ables satisfy the orthogonality condition by their very nature, thus rendering structural information superfluous. Hendry (1995, p. 62), for instance, explicitly tied the interpre-tation of β to the orthogonality condition, stating as follows:

> the status of β may be unclear until the conditions needed to estimate the postulated model are specified. For example, in the model:
>
> $$y_t = z_t\beta + u_t \quad \text{where} \quad u_t \sim \text{IN}[0, \sigma_u^2],$$
>
> until the relationship between z_t and u_t is specified the meaning of β is uncertain since $E[z_t u_t]$ could be either zero or nonzero on the information provided.

LeRoy (1995, p. 211) goes even further: "It is a commonplace of elementary instruction in economics that endogenous variables are not generally causally ordered, implying that the question 'What is the effect of y_1 on y_2' where y_1 and y_2 are endogenous variables is generally meaningless." According to LeRoy, causal relationships cannot be attributed to any variable whose causes have separate influence on the effect variable, a position that denies any causal reading to most of the structural parameters that economists and social scientists labor to estimate.

Cartwright (1995b, p. 49), a renowned philosopher of science, addresses these diffi-culties by initiating a renewed attack on the tormenting question, "*Why* can we assume that we can read off causes, including causal order, from the parameters in equations whose exogenous variables are uncorrelated?" Cartwright, like SEM's founders, rec-ognizes that causes cannot be derived from statistical or functional relationships alone and that causal assumptions are prerequisite for validating any causal conclusion. Unlike Wright and Haavelmo, however, she launches an all-out search for the assumptions that would endow the parameter β in the regression equation $y = \beta x + \varepsilon$ with a legitimate causal meaning and endeavors to prove that the assumptions she proposes are indeed sufficient. What is revealing in Cartwright's analysis is that she does not consider the an-swer Haavelmo would have provided – namely, that the assumptions needed for drawing

---

[2] In fact, this condition is not necessary even for the *identification* of β, once β is interpreted (see the identification of α in Figures 5.7 and 5.9).

causal conclusions from parameters are communicated to us by the scientist who declared the equation "structural"; they are already encoded in the *syntax* of the equations and can be read off the associated graph as easily as a shopping list;[3] they need not be searched for elsewhere, nor do they require new proofs of sufficiency. Again, Haavelmo's answer applies to models of any size and shape, including models with correlated exogenous variables.

These examples bespeak an alarming tendency among economists and social scientists to view a structural equation as an algebraic object that carries functional and statistical assumptions but is void of causal content. This statement from one leading social scientist is typical: "It would be very healthy if more researchers abandoned thinking of and using terms such as cause and effect" (Muthen 1987, p. 180). Perhaps the boldest expression of this tendency was voiced by Holland (1995, p. 54): "I am speaking, of course, about the equation: $\{y = a + bx + \varepsilon\}$. What does it mean? The only meaning I have ever determined for such an equation is that it is a shorthand way of describing the conditional distribution of $\{y\}$ given $\{x\}$."[4]

The founders of SEM had an entirely different conception of structures and models. Wright (1923, p. 240) declared that "prior knowledge of the causal relations is assumed as prerequisite" in the theory of path coefficients, and Haavelmo (1943) explicitly interpreted each structural equation as a statement about a hypothetical controlled experiment. Likewise, Marschak (1950), Koopmans (1953), and Simon (1953) stated that the purpose of postulating a structure behind the probability distribution is to cope with the hypothetical changes that can be brought about by policy. One wonders, therefore, what has happened to SEM over the past 50 years, and why the basic (and still valid) teachings of Wright, Haavelmo, Marschak, Koopmans, and Simon have been forgotten.

Some economists attribute the decline in the understanding of structural equations to Lucas's (1976) critique, according to which economic agents anticipating policy interventions would tend to act contrary to SEM's predictions, which often ignore such anticipations. However, since this critique merely shifts the model's invariants and the burden of structural modeling – from the behavioral level to a deeper level that involves agents' motivations and expectations – it does not exonerate economists from defining and representing the causal content of structural equations at some level of discourse.

I believe that the causal content of SEM has gradually escaped the consciousness of SEM practitioners mainly for the following reasons.

---

[3] These assumptions are explicated and operationalized in Section 5.4. Briefly, if G is the graph associated with a causal model that renders a certain parameter identifiable, then two assumptions are sufficient for authenticating the causal reading of that parameter: (1) every missing arrow, say between X and Y, represents the assumption that X has no effect on Y once we intervene and hold the parents of Y fixed; and (2) every missing bidirected arc $X \blacktriangleleft - - \blacktriangleright Y$ represents the assumption that all omitted factors that affect Y are uncorrected with those that affect X. Each of these assumptions is *testable* in experimental settings, where interventions are feasible (Section 5.4.1).

[4] All but forgotten, the structural interpretation of the equation (Haavelmo 1943) says nothing whatsoever about the conditional distribution of $\{y\}$ given $\{x\}$. Paraphrased in our vocabulary, it reads: "In an ideal experiment where we control X to x and any other set Z of variables (not containing X or Y) to z, Y will attain a value y given by $a + bx + \varepsilon$, where ε is a random variable that is (pointwise) independent of the settings x and z" (see Section 5.4.1). This statement implies that $E[Y \mid do(x), do(z)] = a + bx + c$ but says nothing about $E(Y \mid X = x)$.

1. SEM practitioners have sought to gain respectability for SEM by keeping causal assumptions implicit, since statisticians, the arbiters of respectability, abhor assumptions that are not directly testable.

2. The algebraic language that has dominated SEM lacks the notational facility needed to make causal assumptions, as distinct from statistical assumptions, explicit. By failing to equip causal relations with precise mathematical notation, the founding fathers in fact committed the causal foundations of SEM to oblivion. Their disciples today are seeking foundational answers elsewhere.

Let me elaborate on the latter point. The founders of SEM understood quite well that, in structural models, the equality sign conveys the asymmetrical relation "is determined by" and hence behaves more like an assignment symbol (:=) in programming languages than like an algebraic equality. However, perhaps for reasons of mathematical purity, they refrained from introducing a symbol to represent the asymmetry. According to Epstein (1987), in the 1940s Wright gave a seminar on path diagrams to the Cowles Commission (the breeding ground for SEM), but neither side saw particular merit in the other's methods. Why? After all, a diagram is nothing but a set of nonparametric structural equations in which, to avoid confusion, the equality signs are replaced with arrows.

My explanation is that the early econometricians were extremely careful mathematicians who thought they could keep the mathematics in purely equational–statistical form and just reason about structure in their heads. Indeed, they managed to do so surprisingly well, because they were truly remarkable individuals who *could* do it in their heads. The consequences surfaced in the early 1980s, when their disciples began to mistake the equality sign for an algebraic equality. The upshot was that suddenly the "so-called disturbance terms" did not make any sense at all (Richard 1980, p. 3). We are living with the sad end to this tale. By failing to express their insights in mathematical notation, the founders of SEM brought about the current difficulties surrounding the interpretation of structural equations, as summarized by Holland's "What does it mean?"

5.1.3 Graphs as a Mathematical Language

Recent developments in graphical methods promise to bring causality back into the mainstream of scientific modeling and analysis. These developments involve an improved understanding of the relationships between graphs and probabilities, on the one hand, and graphs and causality, on the other. But the crucial change has been the emergence of graphs as a mathematical language. This mathematical language is not simply a heuristic mnemonic device for displaying algebraic relationships, as in the writings of Blalock (1962) and Duncan (1975). Rather, graphs provide a fundamental notational system for concepts and relationships that are not easily expressed in the standard mathematical languages of algebraic equations and probability calculus. Moreover, graphical methods now provide a powerful symbolic machinery for deriving the consequences of causal assumptions when such assumptions are combined with statistical data.

A concrete example that illustrates the power of the graphical language – and that will set the stage for the discussions in Sections 5.2 and 5.3 – is Simpson's paradox, discussed in Section 3.3 and further analyzed in Section 6.1. This paradox concerns the reversal of an association between two variables (e.g., gender and admission to school) that occurs when we partition a population into finer groups, (e.g., departments). Simpson's reversal has been the topic of much statistical research since its discovery in 1899. This research has focused on conditions for escaping the reversal instead of addressing the practical questions posed by the reversal: "Which association is more valid, before or after partitioning?" In linear analysis, the problem surfaces through the choice of regressors – for example, determining whether a variate Z can be added to a regression equation without biasing the result. Such an addition may easily reverse the sign of the coefficients of the other regressors, a phenomenon known as "suppressor effect" (Darlington 1990).

Despite a century of analysis, questions of regressor selection or adjustment for co-variates continue to be decided informally, case by case, with the decision resting on folklore and intuition rather than on hard mathematics. The standard statistical literature is remarkably silent on this issue. Aside from noting that one should not adjust for a covariate that is affected by the putative cause (X),[5] the literature provides no guidelines as to what covariates might be admissible for adjustment and what assumptions would be needed for making such a determination formally. The reason for this silence is clear: the solution to Simpson's paradox and the covariate selection problem (as we have seen in Sections 3.3.1 and 4.5.3) rests on causal assumptions, and such assumptions cannot be expressed formally in the standard language of statistics.[6]

In contrast, formulating the covariate selection problem in the language of graphs immediately yields a general solution that is both natural and formal. The investigator expresses causal knowledge (or assumptions) in the familiar qualitative terminology of path diagrams, and once the diagram is completed, a simple procedure decides whether a proposed adjustment (or regression) is appropriate relative to the quantity under evaluation. This procedure, which we called the *back-door criterion* in Definition 3.3.1, was applicable when the quantity of interest is the total effect of X on Y. If instead the direct effect is to be evaluated, then the graphical criterion of Theorem 4.5.3 is applicable. A modified criterion for identifying direct effects (i.e., a path coefficient) in linear models will be given in Theorem 5.3.1.

This example is not an isolated instance of graphical methods affording clarity and understanding. In fact, the conceptual basis for SEM achieves a new level of precision through graphs. What makes a set of equations "structural," what assumptions are expressed by the authors of such equations, what the testable implications of those assumptions are, and what policy claims a given set of structural equations advertises are some of the questions that receive simple and mathematically precise answers via graphical methods. These issues, shunned even by modern SEM writers (Heckman and Vytlacil 2007; see Section 11.5.4), will be discussed in the following sections.

---

[5] This advice, which rests on the causal relationship "not affected by," is (to the best of my knowledge) the *only* causal notion that has found a place in statistics textbooks. The advice is neither necessary nor sufficient, as readers can verify from the discussion in Chapter 3.

[6] Simpson's reversal, as well as the supressor effect, are paradoxical only when we attach a causal reading to the associations involved; see Section 6.1.

5.2   GRAPHS AND MODEL TESTING

In 1919, Wright developed his "method of path coefficients," which allows researchers to compute the magnitudes of cause–effect coefficients from correlation measurements provided the path diagram represents correctly the causal processes underlying the data. Wright's method consists of writing a set of equations, one for each pair of variables (X_i, X_j), and equating the (standardized) correlation coefficient ρ_{ij} with a sum of products of path coefficients and residual correlations along the various paths connecting X_i and X_j. One can then attempt to solve these equations for the path coefficients in terms of the observed correlations. Whenever the resulting equations give a unique solution to some path coefficient p_{mn} that is independent of the (unobserved) residual correlations, that coefficient is said to be *identifiable*. If every set of correlation coefficients ρ_{ij} is compatible with some choice of path coefficients, then the model is said to be *untestable* or *unfalsifiable* (also called *saturated, just identified,* etc.), because it is capable of perfectly fitting any data whatsoever.

Whereas Wright's method is partly graphical and partly algebraic, the theory of directed graphs permits us to analyze questions of testability and identifiability in purely graphical terms, prior to data collection, and it also enables us to extend these analyses from linear to nonlinear or nonparametric models. This section deals with issues of testability in linear and nonparametric models.

5.2.1   The Testable Implications of Structural Models

When we hypothesize a model of the data-generating process, that model often imposes restrictions on the statistics of the data collected. In observational studies, these restrictions provide the only view under which the hypothesized model can be tested or falsified. In many cases, such restrictions can be expressed in the form of zero partial correlations; more significantly, the restrictions are implied by the structure of the path diagram alone, independent of the numerical values of the parameters, as revealed by the *d*-separation criterion.

Preliminary Notation

Before addressing the testable implication of structural models, let us first review some definitions from Section 1.4 and relate them to the standard notation used in the SEM literature.

The graphs we discuss in this chapter represent sets of structural equations of the form

$$x_i = f_i(pa_i, \varepsilon_i), \quad i = 1, \dots, n, \tag{5.1}$$

where pa_i (connoting *parents*) stands for (values of) the set of variables judged to be immediate causes of X_i and where the ε_i represent errors due to omitted factors. Equation (5.1) is a nonlinear, nonparametric generalization of the standard linear equations

$$x_i = \sum_{k \neq i} \alpha_{ik} x_k + \varepsilon_i, \quad i = 1, \dots, n, \tag{5.2}$$

in which pa_i correspond to those variables on the r.h.s. of (5.2) that have nonzero co-efficients. A set of equations in the form of (5.1) will be called a *causal model* if each equation represents the process by which the value (not merely the probability) of variable X_i is selected. The graph G obtained by drawing an arrow from every member of pa_i to X_i will be called a *causal diagram*. In addition to full arrows, a causal diagram should contain a bidirected (i.e., double-arrowed) arc between any pair of variables whose corresponding errors are dependent.

It is important to emphasize that causal diagrams (as well as traditional path diagrams) should be distinguished from the wide variety of graphical models in the statistical litera-ture whose construction and interpretation rest solely on properties of the joint distribution (Kiiveri et al. 1984; Edwards 2000; Cowell et al. 1999; Whittaker 1990; Cox and Wermuth 1996; Lauritzen 1996; Andersson et al. 1998). The missing links in those sta-tistical models represent conditional independencies, whereas the missing links in causal diagrams represent absence of causal connections (see note 3 and Section 5.4), which may or may not imply conditional independencies in the distribution.

A causal model will be called *Markovian* if its graph contains no directed cycles and if its ε_i are mutually independent (i.e., if there are no bidirected arcs). A model is *semi-Markovian* if its graph is acyclic and if it contains dependent errors.

If the ε_i are multivariate normal (a common assumption in the SEM literature), then the X_i in (5.2) will also be multivariate normal and will be fully characterized by the cor-relation coefficients ρ_{ij}. A useful property of multivariate normal distributions is that the conditional variance $\sigma^2_{X|z}$, conditional covariance $\sigma_{XY|z}$, and conditional correlation co-efficient $\rho_{XY|z}$ are all independent of the value z. These are known as *partial* variance, covariance, and correlation coefficient and are denoted by $\sigma_{X \cdot Z}$, $\sigma_{XY \cdot Z}$, and $\rho_{XY \cdot Z}$ (respec-tively), where X and Y are single variables and Z is a set of variables. Moreover, the partial correlation coefficient $\rho_{XY \cdot Z}$ is zero if and only if $(X \perp\!\!\!\perp Y \mid Z)$ holds in the distribution.

The *partial regression coefficient* is given by

$$r_{YX \cdot Z} = \rho_{YX \cdot Z} \frac{\sigma_{Y \cdot Z}}{\sigma_{X \cdot Z}};$$

it is equal to the coefficient of X in the linear regression of Y on X and Z (the order of the subscripts is essential). In other words, the coefficient of x in the regression equation

$$y = ax + b_1 z_1 + \cdots + b_k z_k$$

is given by

$$a = r_{YX \cdot Z_1 Z_2 \cdots Z_k}.$$

These coefficients can therefore be estimated by the method of least squares (Crámer 1946).

d-Separation and Partial Correlations

Markovian models (the parallel term in the SEM literature is *recursive models*;[7] Bollen 1989) satisfy the Markov property of Theorem 1.2.7; as a result, the statistical parameters

---

[7] The term *recursive* is ambiguous; some authors exclude correlated errors, but others do not.

of Markovian models can be estimated by ordinary regression analysis. In particular, the
d-separation criterion is valid in such models (here we restate Theorem 1.2.4).

Theorem 5.2.1 (Verma and Pearl 1988; Geiger et al. 1990)
*If sets X and Y are d-separated by Z in a DAG G, then X is independent of Y conditional
on Z in every Markovian model structured according to G. Conversely, if X and Y are
not d-separated by Z in a DAG G, then X and Y are dependent conditional on Z in almost
all Markovian models structured according to G.*

Because conditional independence implies zero partial correlation, Theorem 5.2.1 translates into a graphical test for identifying those partial correlations that must vanish in the
model.

Corollary 5.2.2
*In any Markovian model structured according to a DAG G, the partial correlation $\rho_{XY \cdot Z}$
vanishes whenever the nodes corresponding to the variables in Z d-separate node X from
node Y in G, regardless of the model's parameters. Moreover, no other partial correlation would vanish for all the model's parameters.*

Unrestricted semi-Markovian models can always be emulated by Markovian models that
include latent variables, with the latter accounting for all dependencies among error terms.
Consequently, the *d*-separation criterion remains valid in such models if we interpret bidirected arcs as emanating from latent common parents. This may not be possible in
some linear semi-Markovian models where each latent variable is restricted to influence
at most two observed variables (Spirtes et al. 1996). However, it has been shown that the
d-separation criterion remains valid in such restricted systems (Spirtes et al. 1996) and,
moreover, that the validity is preserved when the network contains cycles (Spirtes et al.
1998; Koster 1999). These results are summarized in the next theorem.

Theorem 5.2.3 (*d*-Separation in General Linear Models)
*For any linear model structured according to a diagram D, which may include cycles
and bidirected arcs, the partial correlation $\rho_{XY \cdot Z}$ vanishes if the nodes corresponding to
the set of variables Z d-separate node X from node Y in D. (Each bidirected arc $i \blacktriangleleft -- \blacktriangleright j$
is interpreted as a latent common parent $i \leftarrow L \rightarrow j$.)*

For linear structural equation models (see (5.2)), Theorem 5.2.3 implies that those (and
only those) partial correlations identified by the *d*-separation test are guaranteed to vanish independent of the model parameters α_{ik} and independent of the error variances. This
suggests a simple and direct method for testing models: rather than going through the
standard exercise of finding a maximum likelihood estimate for the model's parameters
and scoring those estimates for fit to the data, we can directly test for each zero partial
correlation implied by the free model. The advantages of using such tests were noted by
Shipley (1997), who also devised implementations of these tests.
     However, the question arises of whether it is feasible to test for the vast number of
zero partial correlations entailed by a given model. Fortunately, these partial correlations

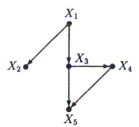

Figure 5.1 Model testable with two regressors for each missing link (equation (5.3)).

are not independent of each other; they can be derived from a relatively small number of partial correlations that constitutes a *basis* for the entire set (Pearl and Verma 1987).

Definition 5.2.4 (Basis)

Let S be a set of partial correlations. A basis *B for S is a set of zero partial correlations where* (i) *B implies* (*using the laws of probability*) *the zero of every element of S and* (ii) *no proper subset of B sustains such implication.*

An obvious choice of a basis for the zero partial correlations entailed by a DAG D is the set of equalities $B = \{\rho_{ij \cdot pa_i} = 0 \mid i > j\}$, where i ranges over all nodes in D and j ranges over all predecessors of i in any order that agrees with the arrows of D. In fact, this set of equalities reflects the "parent screening" property of Markovian models (Theorem 1.2.7), which is the source of all the probabilistic information encoded in a DAG. Testing for these equalities is therefore sufficient for testing all the statistical claims of a linear Markovian model. Moreover, when the parent sets PA_i are large, it may be possible to select a more economical basis, as shown in the next theorem.[8]

Theorem 5.2.5 (Graphical Basis)

Let (i, j) *be a pair of nonadjacent nodes in a DAG D, and let* Z_{ij} *be any set of nodes that are closer to i than j is to i and such that* Z_{ij} *d-separates i from j. The set of zero partial correlations* $B = \{\rho_{ij \cdot Z_{ij}} = 0 \mid i > j\}$, *consisting of one element per nonadjacent pair, constitutes a basis for the set of all zero partial correlations entailed by D.*

Theorem 5.2.5 states that the set of zero partial correlations corresponding to *any* separation between nonadjacent nodes in the diagram encapsulates all the statistical information conveyed by a linear Markovian model. A proof of Theorem 5.2.5 is given in Pearl and Meshkat (1999).

Examining Figure 5.1, we see that each of following two sets forms a basis for the model in the figure:

$$B_1 = \{\rho_{32 \cdot 1} = 0, \;\; \rho_{41 \cdot 3} = 0, \;\; \rho_{42 \cdot 3} = 0, \;\; \rho_{51 \cdot 43} = 0, \;\; \rho_{52 \cdot 43} = 0\},$$

$$B_2 = \{\rho_{32 \cdot 1} = 0, \;\; \rho_{41 \cdot 3} = 0, \;\; \rho_{42 \cdot 1} = 0, \;\; \rho_{51 \cdot 3} = 0, \;\; \rho_{52 \cdot 1} = 0\}. \tag{5.3}$$

The basis B_1 employs the parent set PA_i for separating i from j ($i > j$). Basis B_2, on the other hand, employs smaller separating sets and thus leads to tests that involve fewer

---

[8] The possibility that linear models may possess more economical bases came to my awareness during a conversation with Rod McDonald.

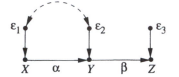

Figure 5.2 A testable model containing unidentified parameter (α).

regressors. Note that each member of a basis corresponds to a missing arrow in the DAG; therefore, the number of tests required to validate a DAG is equal to the number of missing arrows it contains. The sparser the graph, the more it constrains the covariance matrix and more tests are required to verify those constraints.

5.2.2 Testing the Testable

In linear structural equation models, the hypothesized causal relationships between variables can be expressed in the form of a directed graph annotated with coefficients, some fixed a priori (usually to zero) and some free to vary. The conventional method for testing such a model against the data involves two stages. First, the free parameters are estimated by iteratively maximizing a fitness measure such as the likelihood function. Second, the covariance matrix implied by the estimated parameters is compared to the sample covariances and a statistical test is applied to decide whether the latter could originate from the former (Bollen 1989; Chou and Bentler 1995).

There are two major weaknesses to this approach:

1. if some parameters are not identifiable, then the first phase may fail to reach stable estimates for the parameters and the investigator must simply abandon the test;

2. if the model fails to pass the data fitness test, the investigator receives very little guidance about which modeling assumptions are wrong.

For example, Figure 5.2 shows a path model in which the parameter α is not identifiable if $\mathrm{cov}(\varepsilon_1, \varepsilon_2)$ is assumed to be unknown, which means that the maximum likelihood method may fail to find a suitable estimate for α, thus precluding the second phase of the test. Still, this model is no less testable than the one in which $\mathrm{cov}(\varepsilon_1, \varepsilon_2) = 0$, α is identifiable, and the test can proceed. These models impose the same restrictions on the covariance matrix – namely, that the partial correlation $\rho_{XZ \cdot Y}$ should vanish (i.e., $\rho_{XZ} = \rho_{XY}\rho_{YZ}$) – yet the model with free $\mathrm{cov}(\varepsilon_1, \varepsilon_2)$, by virtue of α being nonidentifiable, cannot be tested for this restriction.

Figure 5.3 illustrates the weakness associated with model diagnosis. Suppose the true data-generating model has a direct causal connection between X and W, as shown in Figure 5.3(a), while the hypothesized model (Figure 5.3(b)) has no such connection. Statistically, the two models differ in the term $\rho_{XW \cdot Z}$, which should vanish according to Figure 5.3(b) and is left free according to Figure 5.3(a). Once the nature of the discrepancy is clear, the investigator must decide whether substantive knowledge justifies alteration of the model by adding either a link or a curved arc between X and W. However, because the effect of the discrepancy will be spread over several covariance terms, global fitness tests will not be able to isolate the discrepancy easily. Even multiple fitness tests

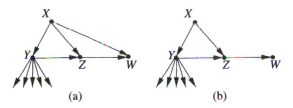

Figure 5.3 Models differing in one local test, $\rho_{XW \cdot Z} = 0$.

(a)   (b)

on various local modifications of the model (such tests are provided by LISREL) may not help much, because the results may be skewed by other discrepancies in different parts of the model, such as the subgraph rooted at Y. Thus, testing for global fitness is often of only minor use in model debugging.

   An attractive alternative to global fitness testing is local fitness testing, which involves listing the restrictions implied by the model and testing them one by one. A restriction such as $\rho_{XW \cdot Z} = 0$, for example, can be tested locally without measuring Y or any of its descendants, thus keeping errors associated with those measurements from interfering with the test for $\rho_{XW \cdot Z} = 0$, which is the real source of the lack of fit. More generally, typical SEM models are often close to being "saturated," claiming but a few restrictions in the form of a few edges missing from large, otherwise unrestrictive diagrams. Local and direct tests for those restrictions are more reliable than global tests, since they involve fewer degrees of freedom and are not contaminated with irrelevant measurement errors. The missing edges approach described in Section 5.2.1 provides a systematic way of detecting and enumerating the local tests needed for testing a given model.

5.2.3   Model Equivalence

In Section 2.3 (Definition 2.3.3) we defined two structural equation models to be observationally equivalent if every probability distribution that is generated by one of the models can also be generated by the other. In standard SEM, models are assumed to be linear and data are characterized by covariance matrices. Thus, two such models are observationally indistinguishable if they are *covariance equivalent*, that is, if every covariance matrix generated by one model (through some choice of parameters) can also be generated by the other. It can be easily verified that the equivalence criterion of Theorem 1.2.8 extends to covariance equivalence.

Theorem 5.2.6

Two Markovian linear-normal models are covariance equivalent if and only if they entail the same sets of zero partial correlations. Moreover, two such models are covariance equivalent if and only if their corresponding graphs have the same sets of edges and the same sets of v-structures.

The first part of Theorem 5.2.6 defines the testable implications of Markovian models. It states that, in nonmanipulative studies, Markovian structural equation models cannot be tested for any feature other than those zero partial correlations that the d-separation test reveals. It also provides a simple test for equivalence that requires, instead of checking all the d-separation conditions, merely a comparison of corresponding edges and their directionalities.

In semi-Markovian models (DAGs with correlated errors), the d-separation criterion is still valid for testing independencies (see Theorem 5.2.3), but independence equivalence no longer implies observational equivalence.[9] Two models that entail the same set of zero partial correlations among the observed variables may yet impose different inequality constraints on the covariance matrix. Nevertheless, Theorems 5.2.3 and 5.2.6 still provide necessary conditions for testing equivalence.

Generating Equivalent Models

By permitting arrows to be reversed as long as no v-structures are destroyed or created, we can use Theorem 5.2.6 to generate equivalent alternatives to any Markovian model. Meek (1995) and Chickering (1995) showed that $X \rightarrow Y$ can be replaced by $X \leftarrow Y$ if and only if all parents of X are also parents of Y. They also showed that, for any two equivalent models, there is always some sequence of such edge reversals that takes one model into the other. This simple rule for edge reversal coincides with those proposed by Stelzl (1986) and Lee and Hershberger (1990).

In semi-Markovian models, the rules for generating equivalent models are more complicated. Nevertheless, Theorem 5.2.6 yields convenient graphical principles for testing the correctness of edge-replacement rules. The basic principle is that if we regard each bidirected arc $X \twoheadleftarrow\!\!-\!\!\twoheadrightarrow Y$ as representing a latent common cause $X \leftarrow L \rightarrow Y$, then the "if" part of Theorem 5.2.6 remains valid; that is, any edge-replacement transformation that does not destroy or create a v-structure is allowed. Thus, for example, an edge $X \rightarrow Y$ can be replaced by a bidirected arc $X \twoheadleftarrow\!\!-\!\!\twoheadrightarrow Y$ whenever X and Y have no other parents, latent or observed. Likewise, an edge $X \rightarrow Y$ can be replaced by a bidirected arc $X \twoheadleftarrow\!\!-\!\!\twoheadrightarrow Y$ whenever (1) X and Y have no latent parents and (2) every parent of X or Y is a parent of both. Such replacements do not introduce new v-structures. However, since v-structures may now involve latent variables, we can tolerate the creation or destruction of some v-structures as long as this does not affect partial correlations among the observed variables. Figure 5.4(a) demonstrates that the creation of certain v-structures can be tolerated. By reversing the arrow $X \rightarrow Y$ we create two converging arrows $Z \rightarrow X \leftarrow Y$ whose tails are connected, not directly, but through a latent common cause. This is tolerated because, although the new convergence at X blocks the path (Z, X, Y), the connection between Z and Y (through the arc $Z \twoheadleftarrow\!\!-\!\!\twoheadrightarrow Y$) remains unblocked and, in fact, cannot be blocked by any set of observed variables.

We can carry this principle further by generalizing the concept of v-structure. Whereas in Markovian models a v-structure is defined as two converging arrows whose tails are not connected by a link, we now define v-structure as any two converging arrowheads whose tails are "separable." By *separable* we mean that there exists a conditioning set S capable of d-separating the two tails. Clearly, the two tails will not be separable if they are connected by an arrow or by a bidirected arc. But a pair of nodes in a semi-Markovian model can be inseparable even when not connected by an edge (Verma and Pearl 1990). With this generalization in mind, we can state necessary conditions for edge replacement as follows.

---

[9] Verma and Pearl (1990) presented an example using a nonparametric model, and Richardson devised an example using linear models with correlated errors (Spirtes and Richardson 1996).

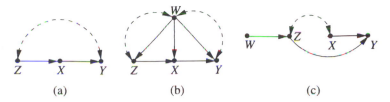

Figure 5.4 Models permitting ((a) and (b)) and forbidding (c) the reversal of $X \rightarrow Y$.

Rule 1: An arrow $X \rightarrow Y$ is interchangeable with $X \leftarrow\!-\!\rightarrow Y$ only if every neighbor or parent of X is inseparable from Y. (By *neighbor* we mean a node connected (to X) through a bidirected arc.)

Rule 2: An arrow $X \rightarrow Y$ can be reversed into $X \leftarrow Y$ only if, before reversal, (i) every neighbor or parent of Y (excluding X) is inseparable from X and (ii) every neighbor or parent of X is inseparable from Y.

For example, consider the model $Z \leftarrow\!-\!\rightarrow X \rightarrow Y$. The arrow $X \rightarrow Y$ cannot be replaced with a bidirected arc $X \leftarrow\!-\!\rightarrow Y$ because Z (a neighbor of X) is separable from Y by the set $S = \{X\}$. Indeed, the new v-structure created at X would render Z and Y marginally independent, contrary to the original model.

As another example, consider the graph in Figure 5.4(a). Here, it is legitimate to replace $X \rightarrow Y$ with $X \leftarrow\!-\!\rightarrow Y$ or with a reversed arrow $X \leftarrow Y$ because X has no neighbors and Z, the only parent of X, is inseparable from Y. The same considerations apply to Figure 5.4(b); variables Z and Y, though nonadjacent, are inseparable, because the paths going from Z to Y through W cannot be blocked.

A more complicated example, one that demonstrates that rules 1 and 2 are not sufficient to ensure the legitimacy of a transformation, is shown in Figure 5.4(c). Here, it appears that replacing $X \rightarrow Y$ with $X \leftarrow\!-\!\rightarrow Y$ would be legitimate because the (latent) v-structure at X is shunted by the arrow $Z \rightarrow Y$. However, the original model shows the path from W to Y to be d-connected given Z, whereas the postreplacement model shows the same path d-separated given Z. Consequently, the partial correlation $\rho_{WY \cdot Z}$ vanishes in the postreplacement model but not in the prereplacement model. A similar disparity also occurs relative to the partial correlation $\rho_{WY \cdot ZX}$. The original model shows that the path from W to Y is blocked, given $\{Z, X\}$, but the postreplacement model shows that path to be d-connected, given $\{Z, X\}$. Consequently, the partial correlation $\rho_{WY \cdot ZX}$ vanishes in the prereplacement model but is unconstrained in the postreplacement model.[10] Evidently, it is not enough to impose rules on the parents and neighbors of X; remote ancestors (e.g., W) should be considered, too.

These rules are just a few of the implications of the d-separation criterion when applied to semi-Markovian models. A necessary and sufficient criterion for testing the d-separation equivalence of two semi-Markovian models was devised by Spirtes and Verma (1992). Spirtes and Richardson (1996) extended that criterion to include models with feedback cycles. However, we should keep in mind that, because two semi-Markovian

---

[10] This example was brought to my attention by Jin Tian, and a similar one by two anonymous reviewers.

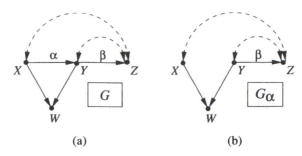

(a)   (b)

Figure 5.6 Test of whether structural parameter α can be equated with regression coefficient r_{YX}.

we know that $I_{YX} = 0$ and $\alpha = r_{YX}$; hence, α is identified. Such entailment can be established graphically by testing whether X is d-separated from Y (by the empty set $Z = \{\emptyset\}$) in the subgraph. Figure 5.6 illustrates this simple test for identification: all paths between X and Y in the subgraph G_α are blocked by converging arrows, and α can immediately be equated with r_{YX}.

We can extend this basic idea to cases where I_{YX} is not zero but can be made zero by adjusting for a set of variables $Z = \{Z_1, Z_2, ..., Z_k\}$ that lie on various d-connected paths between X and Y. Consider the partial regression coefficient $r_{YX \cdot Z} = \rho_{YX \cdot Z} \, \sigma_{Y \cdot Z} / \sigma_{X \cdot Z}$, which represents the residual correlation between Y and X after Z is "partialled out." If Z contains no descendant of Y, then again we can write[12]

$$r_{YX \cdot Z} = \alpha + I_{YX \cdot Z},$$

where $I_{YX \cdot Z}$ represents the partial correlation between X and Y resulting from setting α to zero, that is, the partial correlation in a model whose graph G_α lacks the edge $X \rightarrow Y$ but is otherwise identical to G. If Z d-separates X from Y in G_α, then $I_{YX \cdot Z}$ would indeed be zero in such a model, and so we can conclude that, in our original model, α is identified and is equal to $r_{YX \cdot Z}$. Moreover, since $r_{YX \cdot Z}$ is given by the coefficient of x in the regression of Y on X and Z, α can be estimated using the regression

$$y = \alpha x + \beta_1 z_1 + \cdots + \beta_k z_k + \varepsilon.$$

This result provides a simple graphical answer to the questions, alluded to in Section 5.1.3, of (i) what constitutes an adequate set of regressors and (ii) when a regression coefficient provides a consistent estimate of a path coefficient. The answers are summarized in the following theorem.[13]

Theorem 5.3.1 (Single-Door Criterion for Direct Effects)
Let G be any path diagram in which α is the path coefficient associated with link $X \rightarrow Y$, and let G_α denote the diagram that results when $X \rightarrow Y$ is deleted from G. The coefficient α is identifiable if there exists a set of variables Z such that (i) Z contains no

---

[12] This can be seen when the relation between Y and its parents, $Y = \alpha x + \Sigma_i \beta_i w_i + \varepsilon$, is substituted into the expression for $r_{YX \cdot Z}$, which yields α plus an expression $I_{YX \cdot Z}$ involving partial correlations among the variables $\{X, W_1, ..., W_k, Z, \varepsilon\}$. Because Y is assumed not to be an ancestor of any of these variables, their joint density is unaffected by the equation for Y; hence, $I_{YX \cdot Z}$ is independent of α.

[13] This result is presented in Pearl (1998a) and Spirtes et al. (1998).

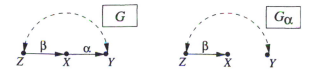

Figure 5.7 The identification of α with $r_{YX \cdot Z}$ (Theorem 5.3.1) is confirmed by G_α.

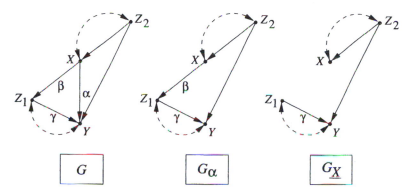

Figure 5.8 Graphical identification of the total effect of X on Y, yielding $\alpha + \beta\gamma = r_{YX \cdot Z_2}$.

descendant of Y and (ii) *Z d-separates X from Y in G_α. If Z satisfies these two conditions, then α is equal to the regression coefficient $r_{YX \cdot Z}$. Conversely, if Z does not satisfy these conditions, then $r_{YX \cdot Z}$ is not a consistent estimand of α (except in rare instances of measure zero).*

The use of Theorem 5.3.1 can be illustrated as follows. Consider the graphs G and G_α in Figure 5.7. The only path connecting X and Y in G_α is the one traversing Z, and since that path is *d*-separated (blocked) by Z, α is identifiable and is given by $\alpha = r_{YX \cdot Z}$. The coefficient β is identifiable, of course, since Z is *d*-separated from X in G_β (by the empty set $\emptyset$) and thus $\beta = r_{XZ}$. Note that this "single-door" test differs slightly from the back-door criterion for total effects (Definition 3.3.1); the set Z here must block *all* indirect paths from X to Y, not only back-door paths. Condition (i) is identical to both cases, because if X is a parent of Y then every descendant of Y must also be a descendant of X.

We now extend the identification of structural parameters through the identification of total effects (rather than direct effects). Consider the graph G in Figure 5.8. If we form the graph G_α by removing the link $X \rightarrow Y$, we observe that there is no set Z of nodes that *d*-separates all paths from X to Y. If Z contains Z_1, then the path $X \rightarrow Z_1 \leftarrow\!\!-\!\!\rightarrow Y$ will be unblocked through the converging arrows at Z_1. If Z does not contain Z_1, the path $X \rightarrow Z_1 \rightarrow Y$ is unblocked. Thus we conclude that α cannot be identified using our previous method. However, suppose we are interested in the total effect of X on Y, which is given by $\alpha + \beta\gamma$. For this sum to be identified by r_{YX}, there should be no contribution to r_{YX} from paths other than those leading from X to Y. However, we see that two such paths, called *confounding* or *back-door* paths, exist in the graph – namely, $X \leftarrow Z_2 \rightarrow Y$ and $X \leftarrow\!\!-\!\!\rightarrow Z_2 \rightarrow Y$. Fortunately, these paths are blocked by Z_2 and so we may conclude that adjusting for Z_2 would render $\alpha + \beta\gamma$ identifiable; thus we have

$$\alpha + \beta\gamma = r_{YX \cdot Z_2}.$$

This line of reasoning is captured by the back-door criterion of Definition 3.3.1, which we restate here for completeness.

Theorem 5.3.2 (Back-Door Criterion)

For any two variables X and Y in a causal diagram G, the total effect of X on Y is identifiable if there exists a set of measurements Z such that

1. *no member of Z is a descendant of X; and*
2. *Z d-separates X from Y in the subgraph $G_{\underline{X}}$ formed by deleting from G all arrows emanating from X.*

Moreover, if the two conditions are satisfied, then the total effect of X on Y is given by $r_{YX \cdot Z}$.

The two conditions of Theorem 5.3.2, as we have seen in Section 3.3.1, are also valid in nonlinear non-Gaussian models as well as in models with discrete variables. The test ensures that, after adjustment for Z, the variables X and Y are not associated through confounding paths, which means that the regression coefficient $r_{YX \cdot Z}$ is equal to the total effect. In fact, we can view Theorems 5.3.1 and 5.3.2 as special cases of a more general scheme: In order to identify any *partial effect*, as defined by a select bundle of causal paths from X to Y, we ought to find a set Z of measured variables that block all nonselected paths between X and Y. The partial effect will then equal the regression coefficient $r_{YX \cdot Z}$.

Figure 5.8 demonstrates that some total effects can be determined directly from the graphs without having to identify their individual components. Standard SEM methods (Bollen 1989; Chou and Bentler 1995) that focus on the identification and estimation of individual parameters may miss the identification and estimation of effects such as the one in Figure 5.8, which can be estimated reliably even though some of the constituents remain unidentified.

Some total effects cannot be determined directly as a unit but instead require the determination of each component separately. In Figure 5.7, for example, the effect of Z on Y $(= \alpha\beta)$ does not meet the back-door criterion, yet this effect can be determined from its constituents α and β, which meet the back-door criterion individually and evaluate to

$$\beta = r_{XZ}, \qquad \alpha = r_{YX \cdot Z}.$$

There is yet a third kind of causal parameter: one that cannot be determined either directly or through its constituents but rather requires the evaluation of a broader causal effect of which it is a part. The structure shown in Figure 5.9 represents an example of this case. The parameter α cannot be identified either directly or from its constituents (it has none), yet it can be determined from $\alpha\beta$ and β, which represent the effect of Z on Y and of Z on X, respectively. These two effects can be identified directly, since there are no back-door paths from Z to either Y or X; therefore, $\alpha\beta = r_{YZ}$ and $\beta = r_{XZ}$. It follows that

$$\alpha = r_{YZ}/r_{XZ},$$

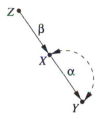

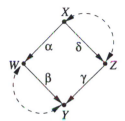

Figure 5.9 Graphical identification of α using instrumental variable Z.

Figure 5.10 Graphical identification of α, β, and γ.

which is familiar to us as the *instrumental variable* formula (Bowden and Turkington 1984; see also Section 3.5, equation (3.46)).

The example shown in Figure 5.10 combines all three methods considered thus far. The total effect of X on Y is given by $\alpha\beta + \gamma\delta$, which is not identifiable because it does not meet the back-door criterion and is not part of another identifiable structure. However, suppose we wish to estimate β. By conditioning on Z, we block all paths going through Z and obtain $\alpha\beta = r_{YX \cdot Z}$, which is the effect of X on Y mediated by W. Because there are no back-door paths from X to W, α itself evaluates directly to $\alpha = r_{WX}$. We therefore obtain

$$\beta = r_{YX \cdot Z} / r_{WX}.$$

On the other hand, γ can be evaluated directly by conditioning on X (thus blocking all back-door paths from Z to Y through X), which gives

$$\gamma = r_{YZ \cdot X}.$$

The methods that we have been using suggest the following systematic procedure for recognizing identifiable coefficients in a graph.

1. Start by searching for identifiable causal effects among pairs of variables in the graph, using the back-door criterion and Theorem 5.3.1. These can be either direct effects, total effects, or partial effects (i.e., effects mediated by specific sets of variables).

2. For any such identified effect, collect the path coefficients involved and put them in a bucket.

3. Begin labeling the coefficients in the buckets according to the following procedure:

   (a) if a bucket is a singleton, label its coefficient I (denoting *identifiable*);

   (b) if a bucket is not a singleton but contains only a single unlabeled element, label that element I.

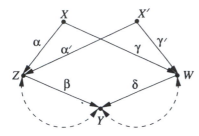

Figure 5.11 Identifying β and δ using two instrumental variables.

4. Repeat this process until no new labeling is possible.

5. List all labeled coefficients; these are identifiable.

The process just described is not complete, because our insistence on labeling coefficients one at a time may cause us to miss certain opportunities. This is shown in Figure 5.11. Starting with the pairs (X, Z), (X, W), (X', Z), and (X', W), we discover that α, γ, α', and γ' are identifiable. Going to (X, Y), we find that $\alpha\beta + \delta\gamma$ is identifiable; likewise, from (X', Y) we see that $\alpha'\beta + \gamma'\delta$ is identifiable. This does not yet enable us to label β or δ, but we can solve two equations for the unknowns β and δ as long as the determinant $\begin{vmatrix} \alpha & \gamma \\ \alpha' & \gamma' \end{vmatrix}$ is nonzero. Since we are not interested in identifiability at a point but rather in identifiability "almost everywhere" (Koopmans et al. 1950; Simon 1953), we need not compute this determinant. We merely inspect the symbolic form of the determinant's rows to make sure that the equations are nonredundant; each imposes a new constraint on the unlabeled coefficients for at least one value of the labeled coefficients.

With a facility to detect redundancies, we can increase the power of our procedure by adding the following rule:

3*. If there are k nonredundant buckets that contain at most k unlabeled coefficients, label these coefficients and continue.

Another way to increase the power of our procedure is to list not only identifiable effects but also expressions involving correlations due to bidirected arcs, in accordance with Wright's rules. Finally, one can endeavor to list effects of several variables jointly as is done in Section 4.4. However, such enrichments tend to make the procedure more complex and might compromise our main objective of providing investigators with a way to immediately recognize the identified coefficients in a given model and immediately understand those features in the model that influence the identifiability of the target quantity. We now relate these results to the identification in nonparametric models, such as those treated in Section 3.3.

5.3.2   Comparison to Nonparametric Identification

The identification results of the previous section are significantly more powerful than those obtained in Chapters 3 and 4 for nonparametric models. Nonparametric models should nevertheless be studied by parametric modelers for both practical and conceptual

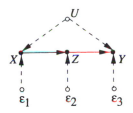

Figure 5.12 Path diagram corresponding to equations (5.4)–(5.6), where $\{X, Z, Y\}$ are observed and $\{U, \varepsilon_1, \varepsilon_2, \varepsilon_3\}$ are unobserved.

reasons. On the practical side, investigators often find it hard to defend the assumptions of linearity and normality (or other functional-distributional assumptions), especially when categorical variables are involved. Because nonparametric results are valid for nonlinear functions and for any distribution of errors, having such results allows us to gauge how sensitive standard techniques are to assumptions of linearity and normality. On the conceptual side, nonparametric models illuminate the distinctions between structural and algebraic equations. The search for nonparametric quantities analogous to path coefficients forces explication of what path coefficients really mean, why one should labor at their identification, and why structural models are not merely a convenient way of encoding covariance information.

In this section we cast the problem of nonparametric causal effect identification (Chapter 3) in the context of parameter identification in linear models.

Parametric versus Nonparametric Models: An Example

Consider the set of structural equations

$$x = f_1(u, \varepsilon_1), \tag{5.4}$$

$$z = f_2(x, \varepsilon_2), \tag{5.5}$$

$$y = f_3(z, u, \varepsilon_3), \tag{5.6}$$

where X, Z, Y are observed variables, f_1, f_2, f_3 are unknown arbitrary functions, and $U, \varepsilon_1, \varepsilon_2, \varepsilon_3$ are unobservables that we can regard either as latent variables or as disturbances. For the sake of this discussion, we will assume that $U, \varepsilon_1, \varepsilon_2, \varepsilon_3$ are mutually independent and arbitrarily distributed. Graphically, these influences can be represented by the path diagram of Figure 5.12.

The problem is as follows. We have drawn a long stream of independent samples of the process defined by (5.4) – (5.6) and have recorded the values of the observed variables X, Z, and Y; we now wish to estimate the unspecified quantities of the model to the greatest extent possible.

To clarify the scope of the problem, we consider its linear version, which is given by

$$x = u + \varepsilon_1, \tag{5.7}$$

$$z = \alpha x + \varepsilon_2, \tag{5.8}$$

$$y = \beta z + \gamma u + \varepsilon_3, \tag{5.9}$$

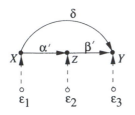

Figure 5.13 Diagram representing model M' of (5.12)–(5.14).

where U, ε_1, ε_2, ε_3 are uncorrelated, zero-mean disturbances.[14] It is not hard to show that parameters α, β, and γ can be determined uniquely from the correlations among the observed quantities X, Z, and Y. This identification was demonstrated already in the example of Figure 5.7, where the back-door criterion yielded

$$\beta = r_{YZ \cdot X}, \quad \alpha = r_{ZX}, \tag{5.10}$$

and hence

$$\gamma = r_{YX} - \alpha\beta. \tag{5.11}$$

Thus, returning to the nonparametric version of the model, it is tempting to generalize that, for the model to be identifiable, the functions $\{f_1, f_2, f_3\}$ must be determined uniquely from the data. However, the prospect of this happening is unlikely, because the mapping between functions and distributions is known to be many-to-one. In other words, given any nonparametric model M, if there exists one set of functions $\{f_1, f_2, f_3\}$ compatible with a given distribution $P(x, y, z)$, then there are infinitely many such functions (see Figure 1.6). Thus, it seems that nothing useful can be inferred from loosely specified models such as the one given by (5.4)–(5.6).

Identification is not an end in itself, however, even in linear models. Rather, it serves to answer practical questions of prediction and control. At issue is not whether the data permit us to identify the form of the equations but, instead, whether the data permit us to provide unambiguous answers to questions of the kind traditionally answered by parametric models.

When the model given by (5.4)–(5.6) is used strictly for prediction (i.e., to determine the probabilities of some variables given a set of observations on other variables), the question of identification loses much (if not all) of its importance; all predictions can be estimated directly from either the covariance matrices or the sample estimates of those covariances. If dimensionality reduction is needed (e.g., to improve estimation accuracy) then the covariance matrix can be encoded in a variety of simultaneous equation models, all of the same dimensionality. For example, the correlations among X, Y, and Z in the linear model M of (5.7)–(5.9) might well be represented by the model M' (Figure 5.13):

$$x = \varepsilon_1, \tag{5.12}$$

$$z = \alpha'x + \varepsilon_2, \tag{5.13}$$

$$y = \beta'z + \delta x + \varepsilon_3. \tag{5.14}$$

---

[14] An equivalent version of this model is obtained by eliminating U from the equations and allowing ε_1 and ε_3 to be correlated, as in Figure 5.7.

This model is as compact as (5.7)–(5.9) and is covariance equivalent to M with respect to the observed variables X, Y, Z. Upon setting $\alpha' = \alpha, \beta' = \beta$, and $\delta = \gamma$, model M' will yield the same probabilistic predictions as those of the model of (5.7)–(5.9). Still, when viewed as data-generating mechanisms, the two models are not equivalent. Each tells a different story about the processes generating X, Y, and Z, so naturally their predictions differ concerning the changes that would result from subjecting these processes to external interventions.

5.3.3   Causal Effects: The Interventional Interpretation of Structural Equation Models

The differences between models M and M' illustrate precisely where the structural reading of simultaneous equation models comes into play, and why even causally shy researchers consider structural parameters more "meaningful" than covariances and other statistical parameters. Model M', defined by (5.12)–(5.14), regards X as a direct participant in the process that determines the value of Y, whereas model M, defined by (5.7)–(5.9), views X as an indirect factor whose effect on Y is mediated by Z. This difference is not manifested in the data itself but rather in the way the data would change in response to outside interventions. For example, suppose we wish to predict the expectation of Y after we intervene and fix the value of X to some constant x; this is denoted $E(Y \mid do(X = x))$. After $X = x$ is substituted into (5.13) and (5.14), model M' yields

$$E[Y \mid do(X = x)] = E[\beta'\alpha'x + \beta'\varepsilon_2 + \delta x + \varepsilon_3] \tag{5.15}$$
$$= (\beta'\alpha' + \delta)x; \tag{5.16}$$

model M yields

$$E[Y \mid do(X = x)] = E[\beta\alpha x + \beta\varepsilon_2 + \gamma u + \varepsilon_3] \tag{5.17}$$
$$= \beta\alpha x. \tag{5.18}$$

Upon setting $\alpha' = \alpha, \beta' = \beta$, and $\delta = \gamma$ (as required for covariance equivalence; see (5.10) and (5.11)), we see clearly that the two models assign different magnitudes to the (total) causal effect of X on Y: model M predicts that a unit change in x will change $E(Y)$ by the amount $\beta\alpha$, whereas model M' puts this amount at $\beta\alpha + \gamma$.

At this point, it is tempting to ask whether we should substitute $x - \varepsilon_1$ for u in (5.9) prior to taking expectations in (5.17). If we permit the substitution of (5.8) into (5.9), as we did in deriving (5.17), why not permit the substitution of (5.7) into (5.9) as well? After all (the argument runs), there is no harm in upholding a mathematical equality, $u = x - \varepsilon_1$, that the modeler deems valid. This argument is fallacious, however.[15] Structural equations are not meant to be treated as immutable mathematical equalities. Rather, they are meant to define a state of equilibrium – one that is *violated* when the equilibrium is perturbed by outside interventions. In fact, the power of structural equation models is

---

[15] Such arguments have led to Newcomb's paradox in the so-called evidential decision theory (see Section 4.1.1).

that they encode not only the initial equilibrium state but also the information necessary for determining which equations must be violated in order to account for a new state of equilibrium. For example, if the intervention consists merely of holding X constant at x, then the equation $x = u + \varepsilon_1$, which represents the preintervention process determining X, should be overruled and replaced with the equation $X = x$. The solution to the new set of equations then represents the new equilibrium. Thus, the essential characteristic of structural equations that sets them apart from ordinary mathematical equations is that the former stand not for one but for many sets of equations, each corresponding to a subset of equations taken from the original model. Every such subset represents some hypothetical physical reality that would prevail under a given intervention.

If we take the stand that the value of structural equations lies not in summarizing distribution functions but in encoding causal information for predicting the effects of policies (Haavelmo 1943; Marschak 1950; Simon 1953), it is natural to view such predictions as the proper generalization of structural coefficients. For example, the proper generalization of the coefficient β in the linear model M would be the answer to the control query, "What would be the change in the expected value of Y if we were to intervene and change the value of Z from z to $z + 1$?", which is different, of course, from the observational query, "What would be the difference in the expected value of Y if we were to *find Z* at level $z + 1$ instead of level z?" Observational queries, as we discussed in Chapter 1, can be answered directly from the joint distribution $P(x, y, z)$, while control queries require causal information as well. Structural equations encode this causal information in their syntax by treating the variable on the left-hand side of the equality sign as the effect and treating those on the right as causes. In Chapter 3 we distinguished between the two types of queries through the symbol $do(\cdot)$. For example, we wrote

$$E(Y \mid do(x)) \triangleq E\,[Y \mid do(X = x)] \tag{5.19}$$

for the controlled expectation and

$$E(Y \mid x) \triangleq E(Y \mid X = x) \tag{5.20}$$

for the standard conditional or observational expectation. That $E(Y \mid do(x))$ does not equal $E(Y \mid x)$ can easily be seen in the model of (5.7)–(5.9), where $E(Y \mid do(x)) = \alpha\beta x$ but $E(Y \mid x) = r_{YX}x = (\alpha\beta + y)\,x$. Indeed, the passive observation $X = x$ should not violate any of the equations, and this is the justification for substituting both (5.7) and (5.8) into (5.9) before taking the expectation.

In linear models, the answers to questions of direct control are encoded in the path (or structural) coefficients, which can be used to derive the total effect of any variable on another. For example, the value of $E(Y \mid do(x))$ in the model defined by (5.7)–(5.9) is $\alpha\beta x$, that is, x times the product of the path coefficients along the path $X \rightarrow Z \rightarrow Y$. Computation of $E(Y \mid do(x))$ would be more complicated in the nonparametric case, even if we knew the functions f_1, f_2, and f_3. Nevertheless, this computation is well defined; it requires the solution (for the expectation of Y) of a modified set of equations in which f_1 is "wiped out" and X is replaced by the constant x:

$$z = f_2(x, \varepsilon_2), \tag{5.21}$$

$$y = f_3(z, u, \varepsilon_3). \tag{5.22}$$

Thus, computation of $E(Y \mid do(x))$ requires evaluation of

$$E(Y \mid do(x)) = E\{f_3 \ [f_2\,(x, \varepsilon_2), u, \varepsilon_3]\},$$

where the expectation is taken over U, ε_2, and ε_3. Remarkably, graphical methods perform this computation without knowledge of f_2, f_3, and $P(\epsilon_2, \epsilon_3, u)$ (Section 3.3.2).

This is indeed the essence of identifiability in nonparametric models. The ability to answer interventional queries *uniquely*, from the data and the graph, is precisely how Definition 3.2.3 interprets the identification of the causal effect $P(y \mid do(x))$. As we have seen in Chapters 3 and 4, that ability can be discerned graphically, almost by inspection, from the diagrams that accompany the equations.

5.4   SOME CONCEPTUAL UNDERPINNINGS

5.4.1   What Do Structural Parameters Really Mean?

Every student of SEM has stumbled on the following paradox at some point in his or her career. If we interpret the coefficient β in the equation

$$y = \beta x + \varepsilon$$

as the change in $E(Y)$ per unit change of X, then, after rewriting the equation as

$$x = (y - \varepsilon)/\beta,$$

we ought to interpret $1/\beta$ as the change in $E(X)$ per unit change of Y. But this conflicts both with intuition and with the prediction of the model: the change in $E(X)$ per unit change of Y ought to be *zero* if Y does not appear as an independent variable in the original, structural equation for X.

Teachers of SEM generally evade this dilemma via one of two escape routes. One route involves denying that β has any causal reading and settling for a purely statistical interpretation, in which β measures the reduction in the variance of Y explained by X (see, e.g., Muthen 1987). The other route permits causal reading of only those coefficients that meet the "isolation" restriction (Bollen 1989; James et al. 1982): the explanatory variable must be uncorrelated with the error in the equation. Because ε cannot be uncorrelated with both X and Y (or so the argument goes), β and $1/\beta$ cannot both have causal meaning, and the paradox dissolves.

The first route is self-consistent, but it compromises the founders' intent that SEM function as an aid to policy making and clashes with the intuition of most SEM users. The second is vulnerable to attack logically. It is well known that every pair of bivariate normal variables, X and Y, can be expressed in two equivalent ways,

$$y = \beta x + \varepsilon_1 \quad \text{and} \quad x = \alpha y + \varepsilon_2,$$

where $\text{cov}(X, \varepsilon_1) = \text{cov}(Y, \varepsilon_2) = 0$ and $\alpha = r_{XY} = \beta\sigma_X^2/\sigma_Y^2$. Thus, if the condition $\text{cov}(X, \varepsilon_1) = 0$ endows β with causal meaning, then $\text{cov}(Y, \varepsilon_2) = 0$ ought to endow α with causal meaning as well. But this too conflicts with both intuition and the intentions behind SEM; the change in $E(X)$ per unit change of Y ought to be zero, not r_{XY}, if there is no causal path from Y to X.

What then *is* the meaning of a structural coefficient? Or a structural equation? Or an error term? The interventional interpretation of causal effects, when coupled with the $do(x)$ notation, provides simple answers to these questions. The answers explicate the operational meaning of structural equations and thus should end, I hope, an era of controversy and confusion regarding these entities.

Structural Equations: Operational Definition

Definition 5.4.1 (Structural Equations)

An equation $y = \beta x + \varepsilon$ is said to be structural *if it is to be interpreted as follows: In an ideal experiment where we control X to x and any other set Z of variables (not containing X or Y) to z, the value y of Y is given by $\beta x + \varepsilon$, where ε is not a function of the settings x and z.*

This definition is operational because all quantities are observable, albeit under conditions of controlled manipulation. That manipulations cannot be performed in most observational studies does not negate the operationality of the definition, much as our inability to observe bacteria with the naked eye does not negate their observability under a microscope. The challenge of SEM is to extract the maximum information concerning what we wish to observe from the little we actually can observe.

Note that the operational reading just given makes no claim about how X (or any other variable) will behave when we control Y. This asymmetry makes the equality signs in structural equations different from algebraic equality signs; the former act symmetrically in relating observations on X and Y (e.g., observing $Y = 0$ implies $\beta x = -\varepsilon$), but they act asymmetrically when it comes to interventions (e.g., setting Y to zero tells us nothing about the relation between x and ε). The arrows in path diagrams make this dual role explicit, and this may account for the insight and inferential power gained through the use of diagrams.

The strongest empirical claim of the equation $y = \beta x + \varepsilon$ is made by excluding other variables from the r.h.s. of the equation, thus proclaiming X the *only* immediate cause of Y. This translates into a testable claim of *invariance*: the statistics of Y under condition $do(x)$ should remain invariant to the manipulation of any other variable in the model (see Section 1.3.2).[16] This claim can be written symbolically as

$$P(y \mid do(x), do(z)) = P(y \mid do(x)) \tag{5.23}$$

for all Z disjoint of $\{X \cup Y\}$.[17] In contrast, regression equations make no empirical claims whatsoever.

---

[16] The basic notion that structural equations remain invariant to certain changes in the system goes back to Marschak (1950) and Simon (1953), and it has received mathematical formulation at various levels of abstraction in Hurwicz (1962), Mesarovic (1969), Sims (1977), Cartwright (1989), Hoover (1990), and Woodward (1995). The simplicity, precision, and clarity of (5.23) is unsurpassed, however.

[17] This claim is, in fact, only part of the message conveyed by the equation; the other part consists of a dynamic or counterfactual claim: If we were to control X to x' instead of x, then Y would attain

Note that this invariance holds relative to manipulations, not observations, of Z. The statistics of Y under condition $do(x)$ given the measurement $Z = z$, written $P(y \mid do(x), z)$, would certainly depend on z if the measurement were taken on a consequence (i.e., descendant) of Y. Note also that the ordinary conditional probability $P(y \mid x)$ does not enjoy such a strong property of invariance, since $P(y \mid x)$ is generally sensitive to manipulations of variables other than X in the model (unless X and ε are independent). Equation (5.23), in contrast, remains valid regardless of the statistical relationship between ε and X.

Generalized to a set of several structural equations, (5.23) explicates the assumptions underlying a given causal diagram. If G is the graph associated with a set of structural equations, then the assumptions are embodied in G as follows: (1) every missing arrow – say, between X and Y – represents the assumption that X has no causal effect on Y once we intervene and hold the parents of Y fixed; and (2) every missing bidirected link between X and Y represents the assumption that the omitted factors that (directly) influence X are uncorrected with those that (directly) influence Y. We shall define the operational meaning of the latter assumption in (5.25)–(5.27).

The Structural Parameters: Operational Definition

The interpretation of a structural equation as a statement about the behavior of Y under a hypothetical intervention yields a simple definition for the structural parameters. The meaning of β in the equation $y = \beta x + \varepsilon$ is simply

$$\beta = \frac{\partial}{\partial x} E[Y \mid do(x)], \tag{5.24}$$

that is, the rate of change (relative to x) of the expectation of Y in an experiment where X is held at x by external control. This interpretation holds regardless of whether ε and X are correlated in nonexperimental studies (e.g., via another equation $x = \alpha y + \delta$).

We hardly need to add at this point that β has nothing to do with the regression coefficient r_{YX} or, equivalently, with the conditional expectation $E(Y \mid x)$, as suggested in many textbooks. The conditions under which β coincides with the regression coefficient are spelled out in Theorem 5.3.1.

It is important nevertheless to compare the definition of (5.24) with theories that acknowledge the invariant character of β but have difficulties explicating which changes β is invariant to. Cartwright (1989, p. 194), for example, characterizes β as an invariant of nature that she calls "capacity." She states correctly that β remains constant under change but explains that, as the statistics of X changes, "it is the ratio $[\beta = E(YX)/E(X^2)]$ which remains fixed no matter how the variances shift." This characterization is imprecise on two accounts. First, β may in general not be equal to the stated ratio nor to any other combination of statistical parameters. Second – and this is the main point of Definition 5.4.1 – structural parameters are invariant to local interventions (i.e., changes in

---

the value $\beta x' + \varepsilon$. In other words, plotting the value of Y under various hypothetical controls of X, and under the same external conditions (ε), should result in a straight line with slope β. Such deterministic dynamic claims concerning system behavior under successive control conditions can be tested only under the assumption that ε, representing external conditions or properties of experimental units, remains unaltered as we switch from x to x'. Such counterfactual claims constitute the empirical content of every scientific law (see Section 7.2.2).

specific equations in the system) and not to general changes in the statistics of the variables. If we start with $\text{cov}(X, \varepsilon) = 0$ and the variance of X changes because we (or Nature) locally modify the *process* that generates X, then Cartwright is correct; the ratio $\beta = E(YX)/E(X^2)$ will remain constant. However, if the variance of X changes for any other reason – say, because we observed some evidence $Z = z$ that depends on both X and Y or because the process generating X becomes dependent on a wider set of variables – then that ratio will not remain constant.

The Mystical Error Term: Operational Definition

The interpretations given in Definition 5.4.1 and (5.24) provide an operational definition for that mystical error term

$$\varepsilon = y - E[Y \mid do(x)], \tag{5.25}$$

which, despite being unobserved in nonmanipulative studies, is far from being metaphysical or definitional as suggested by some researchers (e.g. Richard 1980; Holland 1988, p. 460; Hendry 1995, p. 62). Unlike errors in regression equations, ε measures the deviation of Y from its controlled expectation $E[Y \mid do(x)]$ and not from its conditional expectation $E[Y \mid x]$. The statistics of ε can therefore be measured from observations on Y once X is controlled. Alternatively, because β remains the same regardless of whether X is manipulated or observed, the statistics of $\varepsilon = y - \beta x$ can be measured in observational studies if we know β.

Likewise, correlations among errors can be estimated empirically. For any two nonadjacent variables X and Y, (5.25) yields

$$E[\varepsilon_Y \varepsilon_X] = E[YX \mid do(pa_Y, pa_X)] - E[Y \mid do(pa_Y)]E[X \mid do(pa_X)]. \tag{5.26}$$

Once we have determined the structural coefficients, the controlled expectations $E[Y \mid do(pa_Y)]$, $E[X \mid do(pa_X)]$, and $E[YX \mid do(pa_Y, pa_X)]$ become known linear functions of the observed variables pa_Y and pa_X; hence, the expectations on the r.h.s. of (5.26) can be estimated in observational studies. Alternatively, if the coefficients are not determined, then the expression can be assessed directly in interventional studies by holding pa_X and pa_Y fixed (assuming X and Y are not in parent–child relationship) and estimating the covariance of X and Y from data obtained under such conditions.

Finally, we are often interested not in assessing the numerical value of $E[\varepsilon_Y \varepsilon_X]$ but rather in determining whether ε_Y and ε_X can be assumed to be uncorrected. For this determination, it suffices to test whether the equality

$$E[Y \mid x, do(s_{XY})] = E[Y \mid do(x), do(s_{XY})] \tag{5.27}$$

holds true, where s_{XY} stands for (any setting of) all variables in the model excluding X and Y. This test can be applied to any two variables in the model *except* when Y is a parent of X, in which case the symmetrical equation (with X and Y interchanged) is applicable.

The Mystical Error Term: Conceptual Interpretation

The authors of SEM textbooks usually interpret error terms as representing the influence of omitted factors. Many SEM researchers are reluctant to accept this interpretation,

however, partly because unspecified omitted factors open the door to metaphysical speculations and partly because arguments based on such factors were improperly used as a
generic, substance-free license to omit bidirected arcs from path diagrams (McDonald
1997). Such concerns are answered by the operational interpretation of error terms, (5.25),
since it prescribes how errors are measured, not how they originate.

It is important to note, though, that this operational definition is no substitute for the
omitted-factors conception when it comes to deciding whether pairs of error terms can
be assumed to be uncorrected. Because such decisions are needed at a stage when the
model's parameters are still "free," they cannot be made on the basis of numerical assessments of correlations but must rest instead on qualitative structural knowledge about
how mechanisms are tied together and how variables affect each other. Such judgmental decisions are hardly aided by the operational criterion of (5.26), which instructs the
investigator to assess whether two deviations – taken on two different variables under
complex experimental conditions – would be correlated or uncorrected. Such assessments are cognitively unfeasible.

In contrast, the omitted-factors conception instructs the investigator to judge whether
there could be factors that simultaneously influence several observed variables. Such
judgments are cognitively manageable because they are qualitative and rest on purely
structural knowledge – the only knowledge available during this phase of modeling.

Another source of error correlation that should be considered by investigators is *selection bias*. If two uncorrected unobserved factors have a common effect that is omitted
from the analysis but influences the selection of samples for the study, then the corresponding error terms will be correlated in the sampled population; hence, the expectation
in (5.26) will not vanish when taken over the sampled population (see discussion of Berkson's paradox in Section 1.2.3).

We should emphasize, however, that the arcs *missing* from the diagram, not those *in*
the diagram, demand the most attention and careful substantive justification. Adding an
extra bidirected arc can at worst compromise the identifiability of parameters, but deleting an existing bidirected arc may produce erroneous conclusions as well as a false sense
of model testability. Thus, bidirected arcs should be assumed to exist, by default, between any two nodes in the diagram. They should be deleted only by well-motivated
justifications, such as the unlikely existence of a common cause for the two variables and
the unlikely existence of selection bias. Although we can never be cognizant of all the
factors that may affect our variables, substantive knowledge sometimes permits us to state
that the influence of a possible common factor is not likely to be significant.

Thus, as often happens in the sciences, the way we measure physical entities does
not offer the best way of thinking about them. The omitted-factor conception of errors,
because it rests on structural knowledge, is a more useful guide than the operational definition when building, evaluating, and thinking about causal models.

5.4.2   Interpretation of Effect Decomposition

Structural equation modeling prides itself, and rightly so, on providing principled methodology for distinguishing direct from indirect effects. We have seen in Section 4.5 that such
distinction is important in many applications, ranging from process control to legal disputes, and that SEM indeed provides a coherent methodology of defining, identifying, and

estimating direct and indirect effects. However, the reluctance of most SEM researchers to admit the causal reading of structural parameters – coupled with their preoccupation with algebraic manipulations – has resulted in inadequate definitions of direct and indirect effects, as pointed out by Freedman (1987) and Sobel (1990). In this section we hope to correct this confusion by adhering to the operational meaning of the structural coefficients.

We start with the general notion of a causal effect $P(y \mid do(x))$ as in Definition 3.2.1. We then specialize it to define direct effect, as in Section 4.5, and finally express the definitions in terms of structural coefficients.

Definition 5.4.2 (Total Effect)

The total effect *of X on Y is given by $P(y \mid do(x))$, namely, the distribution of Y while X is held constant at x and all other variables are permitted to run their natural course.*

Definition 5.4.3 (Direct Effect)

The direct effect *of X on Y is given by $P(y \mid do(x), do(s_{XY}))$, where S_{XY} is the set of all observed variables in the system except X and Y.*

In linear analysis, Definitions 5.4.2 and 5.4.3 yield, after differentiation with respect to x, the familiar path coefficients in terms of which direct and indirect effects are usually defined. Yet they differ from conventional definitions in several important aspects. First, direct effects are defined in terms of hypothetical experiments in which intermediate variables are held constant by *physical intenvention*, not by statistical adjustment (which is often disguised under the misleading phrase "control for"). Figure 5.10 depicts a simple example where adjusting for the intermediate variables (Z and W) would not give the correct value of zero for the direct effect of X on Y, whereas $\frac{\partial}{\partial x} E(Y \mid do(x, z, w))$ does yield the correct value: $\frac{\partial}{\partial x}(\beta w + \gamma z) = 0$. Section 4.5.3 (Table 4.1) provides another such example, one that involves dichotomous variables.

Second, there is no need to limit control only to intermediate variables; *all* variables in the system may be held constant (except for X and Y). Hypothetically, the scientist controls for all possible conditions S_{XY}, and measurements may commence without knowing the structure of the diagram. Finally, our definitions differ from convention by interpreting total and direct effects independently of each other, as outcomes of two different experiments. Textbook definitions (e.g., Bollen 1989, p. 376; Mueller 1996, p. 141; Kline 1998, p. 175) usually equate the total effect with a power series of path coefficient matrices. This algebraic definition coincides with the operational definition (Definition 5.4.2) in recursive (semi-Markovian) systems, but it yields erroneous expressions in models with feedback. For instance, given the pair of equations $\{y = \beta x + \varepsilon, x = \alpha y + \delta\}$, the total effect of X on Y is simply β, not $\beta(1 - \alpha\beta)^{-1}$ as stated in Bollen (1989, p. 379). The latter has no operational significance worthy of the phrase "effect of X."[18]

We end this section of effect decomposition with a few remarks that should be of interest to researchers dealing with dichotomous variables. The relations among such

---

[18] This error was noted by Sobel (1990) but, perhaps because constancy of path coefficients was presented as a new and extraneous assumption, Sobel's correction has not brought about a shift in practice or philosophy.

variables are usually nonlinear, so the results of Section 4.5 should be applicable. In particular, the direct effect of X on Y will depend on the levels at which we hold the other parents of Y. If we wish to average over these values, we obtain the expression given in (4.11), which invokes nested counterfactuals, and may be reduced to (4.12).

The manipulative account that we have invoked in defining the empirical content of structural equations (Definition 5.4.1) is adequate in linear systems, where most causal quantities of interest can be inferred from experimental studies at the population level (see Section 11.7.1). In nonlinear and nonparametric models, we occasionally need to go down to the individual unit level and invoke the (more fundamental) counterfactual reading of structural equations, as articulated in equation (3.51) and footnote 17, page 160. The analysis of indirect effects is a case in point; its definition (4.14) rests on nested counterfactuals and cannot be expressed in terms of population averages. Such analysis is necessary to give indirect effects operational meaning, independent of total and direct effects (see Section 11.4.2). With the help of counterfactual language, however, we can give indirect effects a simple operational definition: The indirect effect of X on Y is the increase we would see in Y while holding X constant and increasing the mediating variables Z to whatever value Z would have attained under a unit increase of X (see Section 4.5.5 for a formal definition). In linear systems, this definition coincides, indeed, with the difference between the total and direct effects. See Chapter 11 for further discussion of the role of indirect effects in social science and policy analysis (Pearl 2005a).

5.4.3   Exogeneity, Superexogeneity, and Other Frills

Economics textbooks invariably warn readers that the distinction between exogenous and endogenous variables is, on the one hand, "most important for model building" (Darnell 1994, p. 127) and, on the other hand, "a subtle and sometimes controversial complication" (Greene 1997, p. 712). Economics students would naturally expect the concepts and tools developed in this chapter to shed some light on the subject, and rightly so. We next offer a simple definition of exogeneity that captures the important nuances appearing in the literature and that is both palatable and precise.

It is fashionable today to distinguish three types of exogeneity: weak, strong, and super (Engle et al. 1983); the former two are statistical and the latter causal. However, the importance of exogeneity – and the reason for its controversial status – lies in its implications for policy interventions. Some economists believe, therefore, that only the causal aspect (i.e., superexogeneity) deserves the "exogenous" title and that the statistical versions are unwarranted intruders that tend to confuse issues of identification and interpretability with those of estimation efficiency (Ed Leamer, personal communication).[19] I will serve both camps by starting with a simple definition of causal exogeneity and then offering a more general definition, from which both the causal and the statistical aspects would follow as special cases. Thus, what we call "exogeneity" corresponds to what Engle et al. called "superexogeneity," a notion that captures the structural invariance of certain relationships under policy intervention.

---

[19] Similiar opinions have also been communicated by John Aldrich and James Heckman. See also Aldrich (1993).

Suppose that we consider intervening on a set of variables X and that we wish to characterize the statistical behavior of a set Y of outcome variables under the intervention $do(X = x)$. Denote the postintervention distribution of Y by the usual expression $P(y \mid do(x))$. If we are interested in a set λ of parameters of that distribution, then our task is to estimate $\lambda [P(y \mid do(x))]$ from the available data. However, the data available is typically generated under a different set of conditions: X was not held constant but instead was allowed to vary with whatever economic pressures and expectations prompted decision makers to set X in the past. Denoting the process that generated data in the past by M and the probability distribution associated with M by $P_M(v)$, we ask whether $\lambda [P_M(y \mid do(x))]$ can be estimated consistently from samples drawn from $P_M(v)$, given our background knowledge T (connoting "theory") about M. This is essentially the problem of identification that we have analyzed in this and previous chapters, with one important difference; we now ask whether $\lambda[P(y \mid do(x))]$ can be identified from the *conditional* distribution $P(y \mid x)$ alone, instead of from the entire joint distribution $P(v)$. When identification holds under this restricted condition, X is said to be *exogenous* relative to (Y, λ, T).

We may state this formally as follows.

Definition 5.4.4 (Exogeneity)
Let X and Y be two sets of variables, and let λ be any set of parameters of the postintervention probability $P(y \mid do(x))$. We say that X is exogenous *relative to (Y, λ, T) if λ is identifiable from the conditional distribution $P(y \mid x)$, that is, if*

$$P_{M_1}(y \mid x) = P_{M_2}(y \mid x) \Longrightarrow \lambda[P_{M_1}(y \mid do(x))] = \lambda[P_{M_2}(y \mid do(x))] \tag{5.28}$$

for any two models, M_1 and M_2, satisfying theory T.

In the special case where λ constitutes a complete specification of the postintervention probabilities, (5.28) reduces to the implication

$$P_{M_1}(y \mid x) = P_{M_2}(y \mid x) \Longrightarrow P_{M_1}(y \mid do(x)) = P_{M_2}(y \mid do(x)). \tag{5.29}$$

If we further assume that, for every $P(y \mid x)$, our theory T does not a priori exclude some model M_2 satisfying $P_{M_2}(Y \mid do(x)) = P_{M_2}(y \mid x)$,[20] then (5.29) reduces to the equality

$$P(y \mid do(x)) = P(y \mid x), \tag{5.30}$$

a condition we recognize as "no confounding" (see Sections 3.3 and 6.2). Equation (5.30) follows (from (5.29)) because (5.29) must hold for all M_1 in T. Note that, since the theory T is not mentioned explicitly, (5.30) can be applied to any individual model M and can be taken as yet another definition of exogeneity – albeit a stronger one than (5.28).

The motivation for insisting that λ be identifiable from the conditional distribution $P(y \mid x)$ alone, even though the marginal distribution $P(x)$ is available, lies in its ramification for the process of estimation. As stated in (5.30), discovering that X is exogenous

---

[20] For example, if T stands for all models possessing the same graph structure, then such M_2 is not a priori excluded.

permits us to predict the effect of interventions (in X) directly from passive observations, without even adjusting for confounding factors. Our analyses in Sections 3.3 and 5.3 further provide a graphical test of exogeneity: X is exogenous for Y if there is no unblocked back-door path from X to Y (Theorem 5.3.2). This test supplements the declarative definition of (5.30) with a procedural definition and thus completes the formalization of exogeneity. That the invariance properties usually attributable to superexogeneity are discernible from the topology of the causal diagram should come as no surprise, considering that each causal diagram represents a structural model and that each structural model already embodies the invariance assumptions necessary for policy predictions (see Definition 5.4.1).

Leamer (1985) defined X to be exogenous if $P(y \mid x)$ remains invariant to changes in the "process that generates" X. This definition coincides[21] with (5.30) because $P(y \mid do(x))$ is governed by a structural model in which the equations determining X are wiped out; thus, $P(y \mid x)$ must be insensitive to the nature of those equations. In contrast, Engle et al. (1983) defined exogeneity (i.e., their superexogeneity) in terms of changes in the "marginal density" of X; as usual, the transition from process language to statistical terminology leads to ambiguities. According to Engle et al. (1983, p. 284), exogeneity requires that all the parameters of the conditional distribution $P(y \mid x)$ be "invariant for any change in the distribution of the conditioning variables"[22] (i.e., $P(x)$). This requirement of constancy under *any* change in $P(x)$ is too strong – changing conditions or new observations can easily alter both $P(x)$ and $P(y \mid x)$ even when X is perfectly exogenous. (To illustrate, consider a change that turns a randomized experiment, where X is indisputably exogenous, into a nonrandomized experiment; we should not insist on $P(y \mid x)$ remaining invariant under such a change.) The class of changes considered must be restricted to local modification of the mechanisms (or equations) that determine X, as stated by Leamer, and this restriction must be incorporated into any definition of exogeneity. In order to make this restriction precise, however, the vocabulary of SEMs must be invoked as in the definition of $P(y \mid do(x))$; the vocabulary of marginal and conditional densities is far too coarse to properly define the changes against which $P(y \mid x)$ ought to remain invariant.

We are now ready to define a more general notion of exogeneity, one that includes "weak" and "super" exogeneities under the same umbrella.[23] Toward that end, we remove from Definition 5.4.4 the restriction that λ must represent features of the postintervention distribution. Instead, we allow λ to represent *any* feature of the underlying model M, including structural features such as path coefficients, causal effects, and counterfactuals, and including statistical features (which could, of course, be ascertained from the joint distribution alone). With this generalization, we also obtain a simpler definition of exogeneity.

---

[21]   Provided that changes are confined to modification of functions without changing the set of arguments (i.e., parents) in each function.

[22]   This requirement is repeated verbatim in Darnell (1994, p. 131) and Maddala (1992, p. 192).

[23]   We leave out discussion of "strong" exogeneity, which is a slightly more involved version of weak exogeneity applicable to time-series analysis.

Definition 5.4.5 (General Exogeneity)
Let X and Y be two sets of variables, and let λ be any set of parameters defined on a structural model M in a theory T. We say that X is exogenous relative to (Y, λ, T) if λ is identifiable from the conditional distribution $P(y \mid x)$, that is, if

$$P_{M_1}(y \mid x) = P_{M_2}(y \mid x) \implies \lambda(M_1) = \lambda(M_2) \tag{5.31}$$

for any two models, M_1 and M_2, satisfying theory T.

When λ consists of structural parameters, such as path coefficients or causal effects, (5.31) expresses invariance to a variety of interventions, not merely $do(X = x)$. Although the interventions themselves are not mentioned explicitly in (5.31), the equality $\lambda(M_1) = \lambda(M_2)$ reflects such interventions through the structural character of λ. In particular, if λ stands for the values of the causal effect function $P(y \mid do(x))$ at selected points of x and y, then (5.31) reduces to the implication

$$P_{M_1}(y \mid x) = P_{M_2}(y \mid x) \implies P_{M_1}(y \mid do(x)) = P_{M_2}(y \mid do(x)), \tag{5.32}$$

which is identical to (5.29). Hence the causal properties of exogeneity follow.

When λ consists of strictly statistical parameters – such as means, modes, regression coefficients, or other distributional features – the structural features of M do not enter into consideration; we have $\lambda(M) = \lambda(P_M)$, and so (5.31) reduces to

$$P_1(y \mid x) = P_2(y \mid x) \implies \lambda(P_1) = \lambda(P_2) \tag{5.33}$$

for any two probability distributions $P_1(x, y)$ and $P_2(x, y)$ that are consistent with T. We have thus obtained a statistical notion of exogeneity that permits us to ignore the marginal $P(x)$ in the estimation of λ and that we may call "weak exogeneity."[24]

Finally, if λ consists of causal effects among variables in Y (excluding X), we obtain a generalized definition of *instrumental variables*. For example, if our interest lies in the causal effect $\lambda = P(w \mid do(z))$, where W and Z are two sets of variables in Y, then the exogeneity of X relative to this parameter ensures the identification of $P(w \mid do(z))$ from the conditional probability $P(z, w \mid x)$. This is indeed the role of an instrumental variable – to assist in the identification of causal effects not involving the instrument. (See Figure 5.9, with Z, X, Y representing X, Z, W, respectively.)

A word of caution regarding the language used in most textbooks: exogeneity is frequently defined by asking whether parameters "enter" into the expressions of the conditional or the marginal density. For example, Maddala (1992, p. 392) defined weak exogeneity as the requirement that the marginal distribution $P(x)$ "does not involve" λ. Such definitions are not unambiguous, because the question of whether a parameter "enters" a density or whether a density "involves" a parameter are syntax-dependent; different algebraic representations may make certain parameters explicit or obscure. For example,

---

[24] Engle et al. (1983) further imposed a requirement called "variation-free," which is satisfied by default when dealing with genuinely structural models M in which mechanisms do not constrain one another.

if X and Y are dichotomous, then the marginal probability $P(x)$ certainly "involves" parameters such as

$$\lambda_1 = P(x_0, y_0) + P(x_0, y_1) \quad \text{and} \quad \lambda_2 = P(x_0, y_0),$$

as well as their ratio:

$$\lambda = \lambda_2 / \lambda_1.$$

Therefore, writing $P(x_0) = \lambda_2/\lambda$ shows that both λ and λ_2 are involved in the marginal probability $P(x_0)$, and one may be tempted to conclude that X is not exogenous relative to λ. Yet X *is* in fact exogenous relative to λ, because the ratio $\lambda = \lambda_2/\lambda_1$ is none other than $P(y_0 \mid x_0)$; hence it is determined uniquely by $P(y_0 \mid x_0)$ as required by (5.33).[25]

The advantage of the definition given in (5.31) is that it depends not on the syntactic representation of the density function but rather on its semantical content alone. Parameters are treated as quantities *computed from* a model, and not as mathematical symbols that *describe* a model. Consequently, the definition applies to both statistical and structural parameters and, in fact, to any quantity λ that can be computed from a structural model M, regardless of whether it serves (or may serve) in the description of the marginal or conditional densities.

The Mystical Error Term Revisited

Historically, the definition of exogeneity that has evoked most controversy is the one expressed in terms of correlation between variables and errors. It reads as follows.

Definition 5.4.6 (Error-Based Exogeneity)
A variable X is exogenous (relative to $\lambda = P(y \mid do(x))$) if X is independent of all errors that influence Y, except those mediated by X.

This definition, which Hendry and Morgan (1995) trace to Orcutt (1952), became standard in the econometric literature between 1950 and 1970 (e.g., Christ 1966, p. 156; Dhrymes 1970, p. 169) and still serves to guide the thoughts of most econometricians (as in the selection of instrumental variables; Bowden and Turkington 1984). However, it came under criticism in the early 1980s when the distinction between structural errors (equation (5.25)) and regression errors became obscured (Richard 1980). (Regression errors, by definition, are orthogonal to the regressors.) The Cowles Commission logic of structural equations (see Section 5.1) has not reached full mathematical maturity and – by denying notational distinction between structural and regressional parameters – has left all notions based on error terms suspect of ambiguity. The prospect of establishing an entirely new foundation of exogeneity – seemingly free of theoretical terms such as "errors" and "structure" (Engle et al. 1983) – has further dissuaded economists from tidying up the Cowles Commission logic, and criticism of the error-based definition of exogeneity has become increasingly fashionable. For example, Hendry and Morgan (1995) wrote that

---

[25] Engle et al. (1983, p. 281) and Hendry (1995, pp. 162–3) attempted to overcome this ambiguity by using "reparameterization" – an unnecessary complication.

"the concept of exogeneity rapidly evolved into a loose notion as a property of an observable variable being uncorrelated with an unobserved error," and Imbens (1997) readily agreed that this notion "is inadequate."[26]

These critics are hardly justified if we consider the precision and clarity with which structural errors can be defined when using the proper notation (e.g., (5.25)). When applied to structural errors, the standard error-based criterion of exogeneity coincides formally with that of (5.30), as can be verified using the back-door test of Theorem 5.3.2 (with $Z = \emptyset$). Consequently, the standard definition conveys the same information as that embodied in more complicated and less communicable definitions of exogeneity. I am therefore convinced that the standard definition will eventually regain the acceptance and respectability that it has always deserved.

Relationships between graphical and counterfactual definitions of exogeneity and instrumental variables will be discussed in Chapter 7 (Section 7.4.5).

5.5 CONCLUSION

Today the enterprise known as structural equation modeling is increasingly under fire. The founding fathers have retired, their teachings are forgotten, and practitioners, teachers, and researchers currently find the methodology they inherited difficult to either defend or supplant. Modern SEM textbooks are preoccupied with parameter estimation and rarely explicate the role that those parameters play in causal explanations or in policy analysis; examples dealing with the effects of interventions are conspicuously absent, for instance. Research in SEM now focuses almost exclusively on model fitting, while issues pertaining to the meaning and usage of SEM's models are subjects of confusion and controversy. Some of these confusions are reflected in the many questions that I have received from readers (Section 11.5), to whom I dedicated an "SEM Survival Kit" (Section 11.5.3) – a set of arguments for defending the causal reading of SEM and its scientific rationale.

I am thoroughly convinced that the contemporary crisis in SEM originates in the lack of a mathematical language for handling the causal information embedded in structural equations. Graphical models have provided such a language. They have thus helped us answer many of the unsettled questions that drive the current crisis:

1. Under what conditions can we give causal interpretation to structural coefficients?
2. What are the causal assumptions underlying a given structural equation model?
3. What are the statistical implications of any given structural equation model?
4. What is the operational meaning of a given structural coefficient?
5. What are the policy-making claims of any given structural equation model?
6. When is an equation not structural?

---

[26] Imbens prefers definitions in terms of experimental metaphors such as "random assignment assumption," fearing, perhaps, that "[t]ypically the researcher does not have a firm idea what these disturbances really represent" (Angrist et al. 1996, p. 446). I disagree; "random assignment" is a misleading metaphor, while "omitted factors" shines in clarity.

This chapter has described the conceptual developments that now resolve such foundational questions. (Sections 11.5.2 and 11.5.3 provide further elaboration.) In addition, we have presented several tools to be used in answering questions of practical importance:

1. When are two structural equation models observationally indistinguishable?

2. When do regression coefficients represent path coefficients?

3. When would the addition of a regressor introduce bias?

4. How can we tell, prior to collecting any data, which path coefficients can be identified?

5. When can we dispose of the linearity–normality assumption and still extract causal information from the data?

I remain hopeful that researchers will recognize the benefits of these concepts and tools and use them to revitalize causal analysis in the social and behavioral sciences.

5.6 Postscript for the Second Edition

5.6.1 An Econometric Awakening?

After decades of neglect of causal analysis in economics, a surge of interest seems to be in progress. In a recent series of papers, Jim Heckman (2000, 2003, 2005, 2007 (with Vytlacil)) has made great efforts to resurrect and reassert the Cowles Commission interpretation of structural equation models, and to convince economists that recent advances in causal analysis are rooted in the ideas of Haavelmo (1943), Marschak (1950), Roy (1951), and Hurwicz (1962). Unfortunately, Heckman still does not offer econometricians clear answers to the questions posed in this chapter (pp. 133, 170, 171, 215–217). In particular, unduly concerned with implementational issues, Heckman rejects Haavelmo's "equation wipe-out" as a basis for defining counterfactuals and fails to provide econometricians with an alternative definition, namely, a procedure, like that of equation (3.51), for computing the counterfactual $Y(x, u)$ in a well-posed economic model, with X and Y two arbitrary variables in the models. (See Sections 11.5.4–5.) Such a definition is essential for endowing the "potential outcome" approach with a formal semantics, based on SEM, and thus unifying the two econometric camps currently working in isolation.

Another sign of positive awakening comes from the social sciences, through the publication of Morgan and Winship's book *Counterfactual and Causal Inference* (2007), in which the causal reading of SEM is clearly reinstated.[27]

5.6.2 Identification in Linear Models

In a series of papers, Brito and Pearl (2002a,b, 2006) have established graphical criteria that significantly expand the class of identifiable semi-Markovian linear models beyond those discussed in this chapter. They first proved that identification is ensured in all

---

[27] Though the SEM basis of counterfactuals is unfortunately not articulated.

graphs that do not contain bow-arcs, that is, no error correlation is allowed between a cause and its *direct* effect, while no restrictions are imposed on errors associated with indirect causes (Brito and Pearl 2002b). Subsequently, generalizing the concept of instrumental variables beyond the classical patterns of Figures 5.9 and 5.11, they establish a general identification condition that is testable in polynomial time and subsumes all conditions known in the literature. See also McDonald (2002a).

5.6.3 Robustness of Causal Claims

Causal claims in SEM are established through a combination of data and the set of causal assumptions embodied in the model. For example, the claim that the causal effect $E(Y \mid do(x))$ in Figure 5.9 is given by $\alpha x = r_{YZ}/r_{XZ}\, x$ is based on the assumptions: $cov(e_Z, e_Y) = 0$ and $E(Y \mid do(x, z)) = E(Y \mid do(z))$; both are shown in the graph. A claim is *robust* when it is insensitive to violations of some of the assumptions in the model. For example, the claim above is insensitive to the assumption $cov(e_Z, e_X) = 0$, which is shown in the model.

When several distinct sets of assumptions give rise to k distinct estimands for a parameter α, that parameter is called k-identified; the higher the k, the more robust are claims based on α, because equality among these estimands imposes $k - 1$ constraints on the covariance matrix which, if satisfied in the data, indicate an agreement among k distinct sets of assumptions, thus supporting their validity. A typical example emerges when several (independent) instrumental variables are available $Z_1, Z_2, ..., Z_k$ for a single link $X \rightarrow Y$, which yield the equalities $\alpha = r_{YZ_1}/r_{XZ_1} = r_{YZ_2}/r_{XZ_2} = \cdots = r_{YZ_k}/r_{XZ_k}$.

Pearl (2004) gives a formal definition for this notion of robustness, and established graphical conditions for quantifying the degree of robustness of a given causal claim. k-identification generalizes the notion of *degree of freedom* in standard SEM analysis; the latter characterizes the entire model, while the former applies to individual parameters and, more generally, to individual causal claims.

Acknowledgments

This chapter owes its inspiration to the generations of statisticians who have asked, with humor and disbelief, how SEM's methodology could make sense to any rational being – and to the social scientists who (perhaps unwittingly) have saved the SEM tradition from drowning in statistical interpretations. The comments of Herman Ader, Peter Bentler, Kenneth Bollen, Jacques Hagenaars, Rod McDonald, Les Hayduk, and Stan Mulaik have helped me gain a greater understanding of SEM practice and vocabulary. John Aldrich, Nancy Cartwright, Arthur Goldberger, James Heckman, Kevin Hoover, Ed Leamer, and Herbert Simon helped me penetrate the mazes of structural equations and exogeneity in econometrics. Jin Tian was instrumental in revising Sections 5.2.3 and 5.3.1.

CHAPTER SIX

Simpson's Paradox, Confounding, and Collapsibility

He who confronts the paradoxical
exposes himself to reality.
    Friedrick Durrenmatt (1962)

Preface

Confounding represents one of the most fundamental impediments to the elucidation of causal inferences from empirical data. As a result, the consideration of confounding underlies much of what has been written or said in areas that critically rely on causal inferences; this includes epidemiology, econometrics, biostatistics, and the social sciences. Yet, apart from the standard analysis of randomized experiments, the topic is given little or no discussion in most statistics texts. The reason for this is simple: confounding is a causal concept and hence cannot be expressed in standard statistical models. When formal statistical analysis is attempted, it often leads to confusions or complexities that make the topic extremely hard for the nonexpert to comprehend, let alone master.

One of my main objectives in writing this book is to see these confusions resolved – to see problems involving the control of confounding reduced to simple mathematical routines. The mathematical techniques introduced in Chapter 3 have indeed culminated in simple graphical routines for detecting the presence of confounding and for identifying variables that should be controlled in order to obtain unconfounded effect estimates. In this chapter, we address the difficulties encountered when we attempt to define and control confounding by using statistical criteria.

We start by analyzing the interesting history of Simpson's paradox (Section 6.1) and use it as a magnifying glass to examine the difficulties that generations of statisticians have had in their attempts to capture causal concepts in the language of statistics. In Sections 6.2 and 6.3, we examine the feasibility of replacing the causal definition of confounding with statistical criteria that are based solely on frequency data and measurable statistical associations. We will show that, although such replacement is generally not feasible (Section 6.3), a certain kind of nonconfounding conditions, called *stable*, can be given statistical or semistatistical characterization (Section 6.4). This characterization leads to operational tests, similar to collapsibility tests, that can alert investigators to the existence of either instability or bias in a given effect estimate (Section 6.4.3). Finally, Section 6.5 clarifies distinctions between collapsibility and no-confounding, confounders and confounding, and between the structural and exchangeability approaches to representing problems of confounding.

6.1 SIMPSON'S PARADOX: AN ANATOMY

The reversal effect known as Simpson's paradox has been briefly discussed twice in this book: first in connection with the covariate selection problem (Section 3.3) and then in connection with the definition of direct effects (Section 4.5.3). In this section we analyze the reasons why the reversal effect has been (and still is) considered paradoxical and why its resolution has been so late in coming.

6.1.1 A Tale of a Non-Paradox

Simpson's paradox (Simpson 1951; Blyth 1972), first encountered by Pearson in 1899 (Aldrich 1995), refers to the phenomenon whereby an event C increases the probability of E in a given population p and, at the same time, decreases the probability of E in every subpopulation of p. In other words, if F and $\neg F$ are two complementary properties describing two subpopulations, we might well encounter the inequalities

$$P(E \mid C) > P(E \mid \neg C), \tag{6.1}$$

$$P(E \mid C, F) < P(E \mid \neg C, F), \tag{6.2}$$

$$P(E \mid C, \neg F) < P(E \mid \neg C, \neg F). \tag{6.3}$$

Although such order reversal might not surprise students of probability, it is paradoxical when given causal interpretation. For example, if we associate C (connoting *cause*) with taking a certain drug, E (connoting *effect*) with recovery, and F with being a female, then – under the causal interpretation of (6.2)–(6.3) – the drug seems to be harmful to both males and females yet beneficial to the population as a whole (equation (6.1)). Intuition deems such a result impossible, and correctly so.

   The tables in Figure 6.1 represent Simpson's reversal numerically. We see that, overall, the recovery rate for patients receiving the drug (C) at 50% exceeds that of the control ($\neg C$) at 40%, and so the drug treatment is apparently to be preferred. However, when we inspect the separate tables for males and females, the recovery rate for the untreated patients is 10% higher than that for the treated ones, for males and females both.

   The explanation for Simpson's paradox should be clear to readers of this book, since we have taken great care in distinguishing *seeing* from *doing*. The conditioning operator in probability calculus stands for the evidential conditional "given that we see," whereas the $do(\cdot)$ operator was devised to represent the causal conditional "given that we do." Accordingly, the inequality

$$P(E \mid C) > P(E \mid \neg C)$$

is not a statement about C having a positive effect on E, properly written

$$P(E \mid do(C)) > P(E \mid do(\neg C)),$$

but rather about C being positive *evidence* for E, which may be due to spurious confounding factors that cause both C and E. In our example, the drug appears beneficial

Combined	E	$\neg E$		Recovery Rate
(a) Drug (C)	20	20	40	50%
No drug $(\neg C)$	16	24	40	40%
	36	44	80	

Males	E	$\neg E$		Recovery Rate
(b) Drug (C)	18	12	30	60%
No drug $(\neg C)$	7	3	10	70%
	25	15	40	

Females	E	$\neg E$		Recovery Rate
(c) Drug (C)	2	8	10	20%
No drug $(\neg C)$	9	21	30	30%
	11	29	40	

Figure 6.1   Recovery rates under treatment (C) and control $(\neg C)$ for males, females, and combined.

overall because the males, who recover (regardless of the drug) more often than the females, are also more likely than the females to use the drug. Indeed, finding a drug-using patient (C) of unknown gender, we would do well inferring that the patient is more likely to be a male and hence more likely to recover, in perfect harmony with (6.1)–(6.3).

The standard method for dealing with potential confounders of this kind is to "hold them fixed,"[1] namely, to condition the probabilities on any factor that might cause both C and E. In our example, if being a male $(\neg F)$ is perceived to be a cause for both recovery (E) and drug usage (C), then the effect of the drug needs to be evaluated separately for men and women (as in (6.2)–(6.3)) and then averaged accordingly. Thus, assuming F is the only confounding factor, (6.2)–(6.3) properly represent the efficacy of the drug in the respective populations, while (6.1) represents merely its evidential weight in the absence of gender information, and the paradox dissolves.

6.1.2   A Tale of Statistical Agony

Thus far, we have described the paradox as it is understood, or should be understood, by modern students of causality (see, e.g., Cartwright 1983;[2] Holland and Rubin 1983; Greenland and Robins 1986; Pearl 1993b; Spirtes et al. 1993; Meek and Glymour 1994). Most

---

[1] The phrases "hold F fixed" and "control for F," used by both philosophers (e.g., Eells 1991) and statisticians (e.g., Pratt and Schlaifer 1988), connote external interventions and may therefore be misleading. In statistical analysis, all one can do is *simulate* "holding F fixed" by considering cases with equal values of F – that is, "conditioning" on F and $\neg F$ – an operation that I will call "adjusting for F."

[2] Cartwright states, though, that the third factor F should be "held fixed" if and only if F is causally relevant to E (p. 37); the correct (back-door) criterion is somewhat more involved (see Definition 3.3.1).

statisticians, however, are reluctant to entertain the idea that Simpson's paradox emerges from causal considerations. The general attitude is as follows: The reversal is real and disturbing, because it actually shows up in the numbers and may actually mislead statisticians into incorrect conclusions. If something is real, then it cannot be causal, because causality is a mental construct that is not well defined. Thus, the paradox must be a statistical phenomenon that can be detected, understood, and avoided using the tools of statistical analysis. *The Encyclopedia of Statistical Sciences*, for example, warns us sternly of the dangers lurking from Simpson's paradox with no mention of the words "cause" or "causality" (Agresti 1983). *The Encyclopedia of Biostatistics* (Dong 1998) and *The Cambridge Dictionary of Statistics in Medical Sciences* (Everitt 1995) uphold the same conception.

I know of only two articles in the statistical literature that explicitly attribute the peculiarity of Simpson's reversal to causal interpretations. The first is Pearson et al. (1899), where the discovery of the phenomenon[3] is enunciated in these terms:

> To those who persist on looking upon all correlation as cause and effect, the fact that correlation can be produced between two quite uncorrelated characters A and B by taking an artificial mixture of the two closely allied races, must come as rather a shock. (p. 278)

Influenced by Pearson's life-long campaign, statisticians have refrained from causal talk whenever possible and, for over half a century, the reversal phenomenon has been treated as a curious mathematical property of 2×2 tables, stripped of its causal origin. Finally, Lindley and Novick (1981) analyzed the problem from a new angle, and made the second published connection to causality:

> In the last paragraph the concept of a "cause" has been introduced. One possibility would be to use the language of causation, rather than that of exchangeability or identification of populations. We have not chosen to do this; nor to discuss causation, because the concept, although widely used, does not seem to be well-defined. (p. 51)

What is amazing about the history of Simpson's reversal is that, from Pearson et al. to Lindley and Novick, none of the many authors who wrote on the subject dared ask why the phenomenon should warrant our attention and why it evokes surprise. After all, seeing probabilities change magnitude upon conditionalization is commonplace, and seeing such changes turn into sign reversal (by taking differences and mixtures of those probabilities) is not uncommon either. Thus, if it were not for some misguided yet persistent illusion, what is so shocking about inequalities reversing direction?

Pearson understood that the shock originates with distorted causal interpretations, which he set out to correct through the prisms of statistical correlations and contingency tables (see the Epilogue following Chapter 10). His disciples took him rather seriously, and some even asserted that causation is none but a species of correlation (Niles 1922). In so denying any attention to causal intuition, researchers often had no choice but to attribute Simpson's reversal to some evil feature of the data, one that ought to be avoided

---

[3] Pearson et al. (1899) and Yule (1903) reported a weaker version of the paradox in which (6.2)–(6.3) are satisfied with equality. The reversal was discovered later by Cohen and Nagel (1934, p. 449).

by scrupulous researchers. Dozens of papers have been written since the 1950s on the statistical aspects of Simpson's reversal; some dealt with the magnitude of the effect (Blyth 1972; Zidek 1984), some established conditions for its disappearance (Bishop et al. 1975; Whittemore 1978; Good and Mittal 1987; Wermuth 1987), and some even proposed remedies as drastic as replacing $P(E \mid C)$ with $P(C \mid E)$ as a measure of treatment efficacy (Barigelli and Scozzafava 1984) – the reversal had to be avoided at all cost.

A typical treatment of the topic can be found in the influential book of Bishop, Fienberg, and Holland (1975). Bishop et al. (1975, pp. 41–2) presented an example whereby an apparent association between amount of prenatal care and infant survival disappears when the data are considered separately for each clinic participating in the study. They concluded: "If we were to look only at this [the combined] table we would erroneously conclude that survival *was related* [my italics] to the amount of care received." Ironically, survival *was* in fact *related* to the amount of care received in the study considered. What Bishop et al. meant to say is that, looking uncritically at the combined table, we would erroneously conclude that survival was *causally* related to the amount of care received. However, since causal vocabulary had to be avoided in the 1970s, researchers like Bishop et al. were forced to use statistical surrogates such as "related" or "associated" and so naturally fell victim to the limitations of the language; statistical surrogates could not express the causal relationships that researchers meant to convey.

Simpson's paradox helps us to appreciate both the agony and the achievement of this generation of statisticians. Driven by healthy causal intuition, yet culturally forbidden from admitting it and mathematically disabled from expressing it, they managed nevertheless to extract meaning from dry tables and to make statistical methods the standard in the empirical sciences. But the spice of Simpson's paradox turned out to be nonstatistical after all.

6.1.3 Causality versus Exchangeability

Lindley and Novick (1981) were the first to demonstrate the nonstatistical character of Simpson's paradox – that there is no statistical criterion that would warn the investigator against drawing the wrong conclusions or would indicate which table represents the correct answer.

In the tradition of Bayesian decision theory, they first shifted attention to the practical side of the phenomenon and boldly asked: A new patient comes in; do we use the drug or do we not? Equivalently: Which table do we consult, the combined or the gender-specific? "The apparent answer is," confesses Novick (1983, p. 45), "that when we know that the gender of the patient is male or when we know that it is female we do not use the treatment, but if the gender is unknown we should use the treatment! Obviously that conclusion is ridiculous." Lindley and Novick then go through lengthy informal discussion, concluding (as we did in Section 6.1.1) that we should consult the gender-specific tables and not use the drug.

The next step was to ask whether some additional statistical information could in general point us to the right table. This question Lindley and Novick answered in the negative by showing that, with the very same data, we sometimes should decide the opposite and

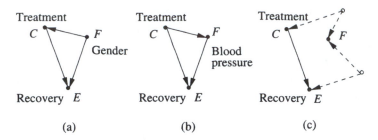

Figure 6.2   Three causal models capable of generating the data in Figure 6.1. Model (a) dictates use of the gender-specific tables, whereas (b) and (c) dictate use of the combined table.

consult the combined table. They asked: Suppose we keep the same numbers and merely change the story behind the data, imagining that F stands for some property that is affected by C – say, low blood pressure, as shown in Figure 6.2(b).[4] By inspecting the diagram in Figure 6.2(b), the reader should immediately conclude that the combined table represents the answer we want; we should not condition on F because it resides on the very causal pathway that we wish to evaluate. (Equivalently, by comparing patients with the same posttreatment blood pressure, we mask the effect of one of the two pathways through which the drug operates to bring about recovery.)

When two causal models generate the same statistical data (Figures 6.2(a) and (b) are observationally equivalent), and in one we decide to use the drug yet in the other not to use it, it is obvious that our decision is driven by causal and not by statistical considerations. Some readers might suspect that temporal information is involved in the decision, noting that gender is established before the treatment and blood pressure afterwards. But this is not the case; Figure 6.2(c) shows that F may occur before *or* after C and still the correct decision should remain to consult the combined table (i.e., not to condition on F, as can be seen from the back-door criterion).

We have just demonstrated by example what we already knew in Section 6.1.1 – namely, that every question related to the effect of actions must be decided by causal considerations; statistical information alone is insufficient. Moreover, the question of choosing the correct table on which to base our decision is a special case of the covariate selection problem that was given a general solution in Section 3.3 using causal calculus. Lindley and Novick, on the other hand, stopped short of this realization and attributed the difference between the two examples to a meta-statistical[5] concept called *exchangeability*, first proposed by DeFinetti (1974).

Exchangeability concerns the question of choosing an appropriate reference class, or subpopulation, for making predictions about an individual unit. Insurance companies, for example, would like to estimate the life expectancy of a new customer using mortality records of a class of persons most closely resembling the characteristics of the

---

[4] The example used in Lindley and Novick (1981) was taken from agriculture, and the causal relationship between C and F was not mentioned, but the structure was the same as in Figure 6.2(b).

[5] By "meta-statistical" I mean a criterion – not itself discernible from statistical data – for judging the adequacy of a certain statistical method.

new customer. De Finetti gave this question a formal twist by translating judgment about resemblance into judgment of probabilities. According to this criterion, an $(n + 1)$th unit is *exchangeable* in property X, relative to a group of n other units, if the joint probability distribution $P(X_1, \ldots, X_n, X_{n+1})$ is invariant under permutation. To De Finetti, the question of how such invariance can be established was a psychological question of secondary importance; the main point was to cast the target of this psychological exercise in the form of mathematical expression so that it could be communicated and discussed in scientific terms. It is this concept that Lindley and Novick tried to introduce into Simpson's reversal phenomenon and with which they hoped to show that the appropriate subpopulations in the $F =$ gender example are the male and female whereas, in the $F =$ blood pressure example, the whole population of patients should be considered.

Readers of Lindley and Novick's article would quickly realize that, although these authors decorate their discussion with talks of *exchangeability* and *subpopulations*, what they actually do is present informal cause–effect arguments for their intuitive conclusions. Meek and Glymour (1994) keenly observed that the only comprehensible part of Lindley and Novick's discussion of exchangeability is the one based on causal considerations, which suggests that "an explicit account of the interaction of causal beliefs and probabilities is necessary to understand when exchangeability should and should not be assumed" (Meek and Glymour 1994, p. 1013).

This is indeed the case; exchangeability in experimental studies depends on causal understanding of the mechanisms that generate the data. The determination of whether the response of a new unit should be judged by previous response of a group of units is predicated upon the question of whether the experimental conditions to which we contemplate subjecting the new unit are equal to those prevailing when the group was observed. The reason we cannot use the combined table (Figure 6.1(a)) for determining the response of a new patient (of unknown gender) is that the experimental conditions have changed; whereas the group was studied with patients selecting treatment by choice, the new patient will be given treatment by decree, perhaps against his or her natural inclination. A mechanism will therefore be altered in the new experiment, and no judgment of exchangeability is feasible without first making causal assumptions regarding whether the probabilities involved would or would not remain invariant to such alteration. The reason we could use the combined table in the blood pressure example of Figure 6.2(b) is that the altered treatment selection mechanism in that setup is assumed to have no effect on the conditional probability $P(E \mid C)$; that is, C is assumed to be exogenous. (This can clearly be seen in the absence of any back-door path in the graph.)

Note that the same consideration holds if the next patient is a member of the group under study (assuming hypothetically that treatment and effect can be replicated and that the next patient is of unknown gender and identity); a randomly selected sample from a population is not "exchangeable" with that population if we subject the sample to new experimental conditions. Alteration of causal mechanisms must be considered in order to determine whether exchangability holds under the new circumstances. And once causal mechanisms are considered, separate judgment of exchangeability is not needed.

But why did Lindley and Novick choose to speak so elliptically (via exchangeability) when they could have articulated their ideas directly by talking openly about causal relations? They partially answered this question as follows: "[causality], although widely used, does not seem to be well-defined." One may naturally wonder how exchangeability can be more "well-defined" than the very considerations by which it is judged! The answer can be understood only when we consider the mathematical tools available to statisticians in 1981. When Lindley and Novick wrote that causality is not well defined, what they really meant is that causality cannot be written down in any mathematical form to which they were accustomed. The potentials of path diagrams, structural equations, and Neyman–Rubin notation as mathematical languages were generally unrecognized in 1981, for reasons described in Sections 5.1 and 7.4.3. Indeed, had Lindley and Novick wished to convey their ideas in causal terms, they would have been unable to express mathematically even the simple yet crucial fact that gender is not affected by the drug and a fortiori to derive less obvious truths from that fact.[6] The only formal language with which they were familiar was probability calculus, but as we have seen on several occasions already, this calculus cannot adequately handle causal relationships without the proper extensions.

Fortunately, the mathematical tools that have been developed in the past ten years permit a more systematic and friendly resolution of Simpson's paradox.

6.1.4   A Paradox Resolved (Or: What Kind of Machine Is Man?)

Paradoxes, like optical illusions, are often used by psychologists to reveal the inner workings of the mind, for paradoxes stem from (and amplify) dormant clashes among implicit sets of assumptions. In the case of Simpson's paradox, we have a clash between (i) the assumption that causal relationships are governed by the laws of probability calculus and (ii) the set of implicit assumptions that drive our causal intuitions. The first assumption tells us that the three inequalities in (6.1)–(6.3) are consistent, and it even presents us with a probability model to substantiate the claim (Figure 6.1). The second tells us that no miracle drug can ever exist that is harmful to both males and females and is simultaneously beneficial to the population at large.

To resolve the paradox we must either (a) show that our causal intuition is misleading or incoherent or (b) deny the premise that causal relationships are governed by the laws of standard probability calculus. As the reader surely suspects by now, we will choose the second option; our stance here, as well as in the rest of the book, is that causality is governed by its own logic and that this logic requires a major extension of probability calculus. It still behooves us to explicate the logic that governs our causal intuition and to show, formally, that this logic precludes the existence of such a miracle drug.

The logic of the *do* (·) operator is perfectly suitable for this purpose. Let us first translate the statement that our miracle drug C has a harmful effect on both males and females into formal statements in causal calculus:

---

[6] Lindley and Novick (1981, p. 50) did try to express this fact in probabilistic notation. But not having the *do* (·) operator at their disposal, they improperly wrote $P(F \mid C)$ instead of $P(F \mid do\ (C))$ and argued unconvincingly that we should equate $P(F \mid C)$ and $P(F)$: "Instead you might judge that the decision to use the treatment or the control is not affected by the unknown sex, so that F and C are independent." Oddly, this decision is also not affected by the unknown blood pressure, and yet, if we write $P(F \mid C) = P(F)$ in the example of Figure 6.2(b), we obtain the wrong result.

$$P(E \mid do(C), F) < P(E \mid do(\neg C), F), \tag{6.4}$$

$$P(E \mid do(C), \neg F) < P(E \mid do(\neg C), \neg F). \tag{6.5}$$

We need to demonstrate that C must be harmful to the population at large; that is, the inequality

$$P(E \mid do(C)) > P(E \mid do(\neg C)) \tag{6.6}$$

must be shown to be inconsistent with what we know about drugs and gender.

Theorem 6.1.1 (Sure-Thing Principle)[7]
An action C that increases the probability of an event E in each subpopulation must also increase the probability of E in the population as a whole, provided that the action does not change the distribution of the subpopulations.

Proof
We will prove Theorem 6.1.1 in the context of our example, where the population is partitioned into males and females; generalization to multiple partitions is straightforward. In this context, we need to prove that the reversal in the inequalities of (6.4)–(6.6) is inconsistent with the assumption that drugs have no effect on gender:

$$P(F \mid do(C)) = P(F \mid do(\neg C)) = P(F). \tag{6.7}$$

Expanding $P(E \mid do(C))$ and using (6.7) yields

$$\begin{aligned}
P(E \mid do(C)) &= P(E \mid do(C), F)\,P\,(F \mid do(C)) \\
&\quad + P(E \mid do(C), \neg F)\,P\,(\neg F \mid do(C)) \\
&= P(E \mid do(C), F)\,P(F) + P(E \mid do(C), \neg F)\,P(\neg F).
\end{aligned} \tag{6.8}$$

Similarly, for $do(\neg C)$ we obtain

$$\begin{aligned}
P(E \mid do(\neg C)) &= P(E \mid do(\neg C), F)\,P\,(F) \\
&\quad + P(E \mid do(\neg C), \neg F)\,P\,(\neg F).
\end{aligned} \tag{6.9}$$

Since every term on the right-hand side of (6.8) is smaller than the corresponding term in (6.9), we conclude that

---

[7] Savage (1954, p. 21) proposed the sure-thing principle as a basic postulate of preferences (on actions), tacitly assuming the no-change provision in the theorem. Blyth (1972) used this omission to devise an apparent counterexample. Theorem 6.1.1 shows that the sure-thing principle need not be stated as a separate postulate – it follows logically from the semantics of actions as modifiers of structural equations (or mechanisms). See Gibbard and Harper (1976) for a counterfactual analysis. Note that the no-change provision is probabilistic; it permits the action to change the classification of individual units as long as the relative sizes of the subpopulations remain unaltered.

$$P(E \mid do\,(C)) < P\,(E \mid do\,(\neg C)),$$

proving Theorem 6.1.1.   □

We thus see where our causal intuition comes from: an obvious but crucial assumption in our intuitive logic has been that drugs do not influence gender. This explains why our intuition changes so drastically when F is interpreted as an intermediate event affected by the drug, as in Figure 6.2(b). In this case, our intuitive logic tells us that it is perfectly consistent to find a drug satisfying the three inequalities of (6.4)–(6.6) and, moreover, that it would be inappropriate to adjust for F. If F is affected by the C, then (6.8) cannot be derived, and the difference $P(E \mid do\,(C)) - P\,(E \mid do\,(\neg C))$ may be positive or negative, depending on the relative magnitudes of $P\,(F \mid do\,(C))$ and $P(F \mid do\,(\neg C))$. Provided C and E have no common cause, we should then assess the efficacy of C directly from the combined table (equation (6.1)) and not from the F-specific tables (equations (6.2)–(6.3)).

Note that nowhere in our analysis have we assumed either that the data originate from a randomized study (i.e., $P(E \mid do(C)) = P(E \mid C)$) or from a balanced study (i.e., $P\,(C \mid F) = P\,(C \mid \neg F)$). On the contrary, given the tables of Figure 6.1, our causal logic accepts gracefully that we are dealing with an unbalanced study but nevertheless refuses to accept the consistency of (6.4)–(6.6). People, likewise, can see clearly from the tables that the males were more likely to take the drug than the females; still, when presented with the reversal phenomenon, people are "shocked" to discover that differences in recovery rates can be reversed by combining tables.

The conclusions we may draw from these observations are that humans are generally oblivious to rates and proportions (which are transitory) and that they constantly search for causal relations (which are invariant). Once people interpret proportions as causal relations, they continue to process those relations by causal calculus and not by the calculus of proportions. Were our minds governed by the calculus of proportions, Figure 6.1 would have evoked no surprise at all, and Simpson's paradox would never have generated the attention that it did.

6.2 WHY THERE IS NO STATISTICAL TEST FOR CONFOUNDING, WHY MANY THINK THERE IS, AND WHY THEY ARE ALMOST RIGHT

6.2.1   Introduction

Confounding is a simple concept. If we undertake to estimate the effect[8] of one variable (X) on another (Y) by examining the statistical association between the two, we ought to ensure that the association is not produced by factors other than the effect under study. The presence of spurious association – due, for example, to the influence of extraneous variables – is called *confounding* because it tends to confound our reading and to

---

[8] We will confine the use of the terms "effect," "influence," and "affect" to their causal interpretations; the term "association" will be set aside for statistical dependencies.

bias our estimate of the effect studied. Conceptually, therefore, we can say that X and Y are confounded when there is a third variable Z that influences both X and Y; such a variable is then called a *confounder* of X and Y.

As simple as this concept is, it has resisted formal treatment for decades, and for good reason: The very notions of "effect" and "influence" – relative to which "spurious association" must be defined – have resisted mathematical formulation. The empirical definition of effect as an association that *would* prevail in a controlled randomized experiment cannot easily be expressed in the standard language of probability theory, because that theory deals with static conditions and does not permit us to predict, even from a full specification of a population density function, what relationships would prevail if conditions were to change – say, from observational to controlled studies. Such predictions require extra information in the form of causal or counterfactual assumptions, which are not discernible from density functions (see Sections 1.3 and 1.4). The $do(\cdot)$ operator used in this book was devised specifically for distinguishing and managing this extra information.

These difficulties notwithstanding, epidemiologists, biostatisticians, social scientists, and economists[9] have made numerous attempts to define confounding in statistical terms, partly because statistical definitions – free of theoretical terms of "effect" or "influence" – can be expressed in conventional mathematical form and partly because such definitions may lead to practical tests of confounding and thereby alert investigators to possible bias and need for adjustment. These attempts have converged in the following basic criterion.

Associational Criterion
Two variables X and Y are not confounded if and only if every variable Z that is not affected by X is either

(U_1) *unassociated with X or*
(U_2) *unassociated with Y, conditional on X.*

This criterion, with some variations and derivatives (often avoiding the "only if" part), can be found in almost every epidemiology textbook (Schlesselman 1982; Rothman 1986; Rothman and Greenland 1998) and in almost every article dealing with confounding. In fact, the criterion has become so deeply entrenched in the literature that authors (e.g., Gail 1986; Hauck et al. 1991; Becher 1992; Steyer et al. 1996) often take it to be the *definition* of no-confounding, forgetting that ultimately confounding is useful only so far as it tells us about effect bias.[10]

The purpose of this and the next section is to highlight several basic limitations of the associational criterion and its derivatives. We will show that the associational criterion

---

[9] In econometrics, the difficulties have focused on the notion of "exogeneity" (Engle et al. 1983; Leamer 1985; Aldrich 1993), which stands essentially for "no confounding" (see Section 5.4.3).
[10] Hauck et al. (1991) dismiss the effect-based definition of confounding as "philosophic" and consider a difference between two measures of association to be a "bias." Grayson (1987) even goes so far as to state that the change-in-parameter method, a derivative of the associational criterion, is the only fundamental definition of confounding (see Greenland et al. 1989 for critiques of Grayson's position).

neither ensures unbiased effect estimates nor follows from the requirement of unbiased-ness. After demonstrating, by examples, the absence of logical connections between the statistical and the causal notions of confounding, we will define a stronger notion of un-biasedness, called "stable" unbiasedness, relative to which a modified statistical criterion will be shown necessary and sufficient. The necessary part will then yield a practical test for stable unbiasedness that, remarkably, does not require knowledge of all potential confounders in a problem. Finally, we will argue that the prevailing practice of sub-stituting statistical criteria for the effect-based definition of confounding is not entirely misguided, because stable unbiasedness is in fact (i) what investigators have been (and perhaps should be) aiming to achieve and (ii) what statistical criteria can test.

6.2.2 Causal and Associational Definitions

In order to facilitate the discussion, we shall first cast the causal and statistical definitions of no-confounding in mathematical forms.[11]

Definition 6.2.1 (No-Confounding; Causal Definition)
Let M be a causal model of the data-generating process – that is, a formal description of how the value of each observed variable is determined. Denote by $P(y \mid do(x))$ the probability of the response event $Y = y$ under the hypothetical intervention $X = x$, cal-culated according to M. We say that X and Y are not confounded in M if and only if

$$P(y \mid do(x)) = P(y \mid x) \tag{6.10}$$

for all x and y in their respective domains, where $P(y \mid x)$ is the conditional probability generated by M. If (6.10) holds, we say that $P(y \mid x)$ is unbiased.

For the purpose of our discussion here, we take this causal definition as the meaning of the expression "no confounding." The probability $P(y \mid do(x))$ was defined in Chapter 3 (Definition 3.2.1, also abbreviated $P(y \mid \hat{x})$); it may be interpreted as the conditional proba-bility $P^*(Y = y \mid X = x)$ corresponding to a controlled experiment in which X is ran-domized. We recall that this probability can be calculated from a causal model M either directly, by simulating the intervention $do(X = x)$, or (if $P(x, s) > 0$) via the adjustment formula (equation (3.19))

$$P(y \mid do(x)) = \sum_{s} P(y \mid x, s) P(s),$$

where S stands for any set of variables, observed as well as unobserved, that satisfy the back-door criterion (Definition 3.3.1). Equivalently, $P(y \mid do(x))$ can be written $P(Y(x) = y)$, where $Y(x)$ is the potential-outcome variable as defined in (3.51) or in

---

[11] For simplicity, we will limit our discussion to unadjusted confounding; extensions involving mea-surement of auxiliary variables are straightforward and can be obtained from Section 3.3. We also use the abbreviated expression "X and Y are not confounded," though "the effect of X on Y is not confounded" is more exact.

Rubin (1974). We bear in mind that the operator do $(\cdot)$, and hence also effect estimates and confounding, must be defined relative to a specific causal or data-generating model M because these notions are not statistical in character and cannot be defined in terms of joint distributions.

Definition 6.2.2 (No-Confounding; Associational Criterion)
Let T be the set of variables in a problem that are not affected by X. We say that X and Y are not confounded in the presence of T if each member Z of T satisfies at least one of the following conditions:

> (U_1) *Z is not associated with X (i.e., $P(x \mid z) = P(x)$);*
> (U_2) *Z is not associated with Y, conditional on X (i.e., $P(y \mid z, x) = P(y \mid x)$).*

Conversely, X and Y are said to be confounded if any member Z of T violates both (U_1) and (U_2).

Note that the associational criterion in Definition 6.2.2 is not purely statistical in that it invokes the predicate "affected by," which is not discernible from probabilities but rests instead on causal information. This exclusion of variables that are affected by treatments (or exposures) is unavoidable and has long been recognized as a necessary judgmental input to every analysis of treatment effect in observational and experimental studies alike (Cox 1958, p. 48; Greenland and Neutra 1980). We shall assume throughout that investigators possess the knowledge required for distinguishing variables that are affected by the treatment X from those that are not. We shall then explore what additional causal knowledge is needed, if any, for establishing a test of confounding.

6.3   HOW THE ASSOCIATIONAL CRITERION FAILS

We will say that a criterion for no-confounding is *sufficient* if it never errs when it classifies a case as no-confounding and *necessary* if it never errs when it classifies a case as confounding. There are several ways that the associational criterion of Definition 6.2.2 fails to match the causal criterion of Definition 6.2.1. Failures with respect to sufficiency and necessity will be addressed in turn.

6.3.1   Failing Sufficiency via Marginality

The criterion in Definition 6.2.2 is based on testing each element of T individually. A situation may well be present where two factors, Z_1 and Z_2, jointly confound X and Y (in the sense of Definition 6.2.2) and yet each factor separately satisfies (U_1) or (U_2). This may occur because statistical independence between X and individual members of T does not guarantee the independence of X and groups of variables taken from T. For example, let Z_1 and Z_2 be the outcomes of two independent fair coins, each affecting both X and Y. Assume that X occurs when Z_1 and Z_2 are equal and that Y occurs whenever Z_1 and Z_2 are unequal. Clearly, X and Y are highly confounded by the pair $T = (Z_1, Z_2)$; they are, in fact, perfectly correlated (negatively) without causally affecting

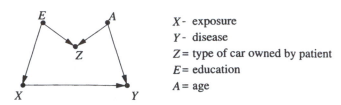

X - exposure
Y - disease
Z = type of car owned by patient
E = education
A = age

Figure 6.3 X and Y are not confounded, though Z is associated with both.

each other. Yet neither Z_1 nor Z_2 is associated with either X or Y; discovering the outcome of any one coin does not change the probability of X (or of Y) from its initial value of $\frac{1}{2}$.

An attempt to remedy Definition 6.2.2 by replacing Z with arbitrary subsets of T in (U_1) and (U_2) would be much too restrictive, because the set of *all* causes of X and Y, when treated as a group, would almost surely fail the tests of (U_1) and (U_2). In Section 6.5.2 we identify the subsets that should replace Z in (U_1) and (U_2) if sufficiency is to be restored.

6.3.2   Failing Sufficiency via Closed-World Assumptions

By "closed-world" assumption I mean the assumption that our model accounts for all relevant variables and, specifically to Definition 6.2.2, that the set T of variables consists of *all* potential confounders in a problem. In order to correctly classify every case of no-confounding, the associational criterion requires that condition (U_1) or (U_2) be satisfied for every potential confounder Z in a problem. In practice, since investigators can never be sure whether a given set T of potential confounders is complete, the associational criterion will falsely classify certain confounded cases as unconfounded.

This limitation actually implies that any statistical test whatsoever is destined to be insufficient. Since practical tests always involve proper subsets of T, the most we can hope to achieve by statistical means is *necessity* – that is, a test that would correctly label cases as confounding when criteria such as (U_1) and (U_2) are violated by an arbitrary subset of T. This prospect too is not fulfilled by Definition 6.2.2, as we now demonstrate.

6.3.3   Failing Necessity via Barren Proxies

Example 6.3.1 Imagine a situation where exposure (X) is influenced by a person's education (E), disease (Y) is influenced by both exposure and age (A), and car type (Z) is influenced by both age (A) and education (E). These relationships are shown schematically in Figure 6.3.

The car-type variable (Z) violates the two conditions in Definition 6.2.2 because: (1) car type is indicative of education and hence is associated with the exposure variable; and (2) car type is indicative of age and hence is associated with the disease among the exposed and the nonexposed. However, in this example the effect of X on Y is not confounded; the type of car owned by a person has no effect on either exposure or disease and is merely one among many irrelevant properties that are associated with both via intermediaries. The analysis of Chapter 3 establishes that, indeed, (6.10) is satisfied in this model[12] and that, moreover, adjustment for Z would generally yield a biased result:

$$\sum_z P(Y = y \mid X = x, Z = z)\, P(Z = z) \neq P(Y = y \mid do(x)).$$

Thus we see that the traditional criterion based on statistical association fails to identify an unconfounded effect and would tempt one to adjust for the wrong variable. This failure occurs whenever we apply (U_1) and (U_2) to a variable Z that is a *barren proxy* – that is, a variable that has no influence on X or Y but is a proxy for factors that do have such influence.

Readers may not consider this failure to be too serious, because experienced epidemiologists would rarely regard a variable as a confounder unless it is suspect of having some influence on either X or Y. Nevertheless, adjustment for proxies is a prevailing practice in epidemiology and should be done with great caution (Greenland and Neutra 1980; Weinberg 1993). To regiment this caution, the associational criterion must be modified to exclude barren proxies from the test set T. This yields the following modified criterion in which T consists only of variables that (causally) influence Y (possibly through X).

Definition 6.3.2 (No-Confounding; Modified Associational Criterion)
Let T be the set of variables in a problem that are not affected by X but may potentially affect Y. We say that X and Y are unconfounded by the presence of T if and only if every member Z of T satisfies either (U_1) or (U_2) of Definition 6.2.2.

Stone (1993) and Robins (1997) proposed alternative modifications of Definition 6.2.2 that avoid the problems created by barren proxies without requiring one to judge whether a variable has an effect on Y. Instead of restricting the set T to potential causes of Y, we let T remain the set of *all* variables unaffected by X,[13] requiring instead that T be composed of two disjoint subsets, T_1 and T_2, such that

(U_1^*) T_1 is unassociated with X *and*
(U_2^*) T_2 is unassociated with Y given X and T_1.

In the model of Figure 6.3, for instance, conditions $((U_1^*)$ and (U_2^*) are satisfied by the choice $T_1 = A$ and $T_2 = \{Z, E\}$, because (using the *d*-separation test) A is independent of X and $\{E, Z\}$ is independent of Y, given $\{X, A\}$.

This modification of the associational criterion further rectifies the problem associated with marginality (see Section 6.3.1) because (U_1^*) and (U_2^*) treat T_1 and T_2 as compound

---

[12] Because the (back-door) path $X \leftarrow E \rightarrow Z \leftarrow A \rightarrow Y$ is blocked by the colliding arrows at Z (see Definition 3.3.1).
[13] Alternatively, T can be confined to any set S of variables sufficient for control of confounding:

$$P(y \mid do(x)) = \sum_s P(y \mid x, s)P(s).$$

Again, however, we can never be sure if the measured variables in the model contain such a set, or which of T's subsets possess this property.

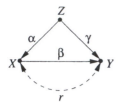

Figure 6.4 Z is associated with both X and Y, yet the effect of X on Y is not confounded (when $r = -\alpha\gamma$).

variables. However, the modification falls short of restoring necessity. Because the set $T = (T_1, T_2)$ must include *all* variables unaffected by X (see note 13) and because practical tests are limited to proper subsets of T, we cannot conclude that confounding is present solely upon the failure of (U_1^*) and (U_2^*), as specified in Section 6.3.2. This criterion too is thus inadequate as a basis for practical detection of confounding.

We now discuss another fundamental limitation on our ability to detect confounding by statistical means.

6.3.4 Failing Necessity via Incidental Cancellations

Here we present a case that is devoid of barren proxies and in which the effect of X on Y (i) is not confounded in the sense of (6.10) but (ii) is confounded according to the modified associational criterion of Definition 6.3.2.

Example 6.3.3 Consider a causal model defined by the linear equations

$$x = \alpha z + \varepsilon_1, \tag{6.11}$$

$$y = \beta x + \gamma z + \varepsilon_2, \tag{6.12}$$

where ε_1 and ε_2 are correlated unmeasured variables with $\text{cov}(\varepsilon_1, \varepsilon_2) = r$ and where Z is an exogenous variable that is uncorrelated with ε_1 or ε_2. The diagram associated with this model is depicted in Figure 6.4. The effect of X on Y is quantified by the path coefficient β, which gives the rate of change of $E(Y \mid do(x))$ per unit change in x.[14]

It is not hard to show (assuming standardized variables) that the regression of Y on X gives

$$y = (\beta + r + \alpha\gamma)x + \varepsilon,$$

where $\text{cov}(x, \varepsilon) = 0$. Thus, whenever the equality $r = -\alpha\gamma$ holds, the regression coefficient of $r_{YX} = \beta + r + \alpha\gamma$ is an unbiased estimate of β, meaning that the effect of X on Y is unconfounded (no adjustment is necessary). Yet the associational conditions (U_1) and (U_2) are both violated by the variable Z; Z is associated with X (if $\alpha \neq 0$) and conditionally associated with Y, given X (except for special values of γ for which $\rho_{yz \cdot x} = 0$).

---

[14] See Sections 3.5–3.6 or (5.24) in Section 5.4.1.

This example demonstrates that the condition of unbiasedness (Definition 6.2.1) does not imply the modified criterion of Definition 6.3.2. The associational criterion might falsely classify some unconfounded situations as confounded and, worse yet, adjusting for the false confounder (Z in our example) will introduce bias into the effect estimate.[15]

6.4   STABLE VERSUS INCIDENTAL UNBIASEDNESS

6.4.1   Motivation

The failure of the associational criterion in the previous example calls for a reexamination of the notion of confounding and unbiasedness as defined in (6.10). The reason that X and Y were classified as unconfounded in Example 6.3.3 was that, by setting $r = -\alpha\gamma$, we were able to make the spurious association represented by r *cancel* the one mediated by Z. In practice, such perfect cancellation would be an incidental event specific to a peculiar combination of study conditions, and it would not persist when the parameters of the problem (i.e., α, γ, and r) undergo slight changes – say, when the study is repeated in a different location or at a different time. In contrast, the condition of no-confounding found in Example 6.3.1 does not exhibit such volatility. In this example, the unbiasedness expressed in (6.10) would continue to hold regardless of the strength of connection between education and exposure and regardless on how education and age influence the type of car that a patient owns. We call this type of unbiasedness *stable*, since it is robust to change in parameters and remains intact as long as the configuration of causal connections in the model remains the same.

In light of this distinction between stable and incidental unbiasedness, we need to reexamine whether we should regard a criterion as inadequate if it misclassifies (as confounded) cases that are rendered unconfounded by mere incidental cancellation and, more fundamentally, whether we should insist on including such peculiar cases in the definition of unbiasedness (given the precarious conditions under which (6.10) would be satisfied in these cases). Although answers to these questions are partly a matter of choice, there is ample evidence that our intuition regarding confounding is driven by considerations of stable unbiasedness, not merely incidental ones. How else can we explain why generations of epidemiologists and biostatisticians would advocate confounding criteria that fail in cases involving incidental cancellation? On the pragmatic side, failing to detect situations of incidental unbiasedness should not introduce appreciable error in observational studies because those situations are short-lived and are likely to be refuted by subsequent studies, under slightly different conditions.[16]

Assuming that we are prepared to classify as unbiased only cases in which unbiasedness remains robust to changes in parameters, two questions remain: (1) How can we give this new notion of "stable unbiasedness" a formal, nonparametric formulation? (2) Are practical statistical criteria available for testing stable unbiasedness? Both questions can be answered using structural models.

---

[15] Note that the Stone–Robins modifications of Definition 6.3.2 would also fail in this example, unless we can measure the factors responsible for the correlation between ε_1 and ε_2.

[16] As we have seen in Example 6.3.3, any statistical test capable of recognizing such cases would require measurement of *all* variables in T.

Chapter 3 describes a graphical criterion, called the "back-door criterion," for identifying conditions of unbiasedness in a causal diagram.[17] In the simple case of no adjustment (for measured covariates), the criterion states that X and Y are unconfounded if every path between X and Y that contains an arrow pointing into X must also contain a pair of arrows pointing head-to-head (as in Figure 6.3); this criterion is valid whenever the missing links in the diagram represent absence of causal connections among the corresponding variables. Because the causal assumptions embedded in the missing links are so explicit, the back-door criterion has two remarkable features. First, no statistical information is needed; the topology of the diagram suffices for reliably determining whether an effect is unconfounded (in the sense of Definition 6.2.1) and whether an adjustment for a set of variables is sufficient for removing confounding when one exists. Second, any model that meets the back-door criterion would in fact satisfy (6.10) for an infinite class of models (or situations), each generated by assigning different parameters to the causal connections in the diagram.

To illustrate, consider the diagram depicted in Figure 6.3. The back-door criterion will identify the pair (X, Y) as unconfounded, because the only path ending with an arrow into X is the one traversing (X, E, Z, A, Y), and this path contains two arrows pointing head-to-head at Z. Moreover, since the criterion is based only on graphical relationships, it is clear that (X, Y) will continue to be classified as unconfounded regardless of the strength or type of causal relationships that are represented by the arrows in the diagram. In contrast, consider Figure 6.4 in Example 6.3.3, where two paths end with arrows into X. Since none of these paths contains head-to-head arrows, the back-door criterion will fail to classify the effect of X on Y as unconfounded, acknowledging that an equality $r = -\alpha\gamma$ (if it prevails) would not represent a stable case of unbiasedness.

The vulnerability of the back-door criterion to causal assumptions can be demonstrated in the context of Figure 6.3. Assume the investigator suspects that variable Z (car type) has some influence on the outcome variable Y. This would amount to adding an arrow from Z to Y in the diagram, classifying the situation as confounded, and suggesting an adjustment for E (or $\{A, Z\}$). Yet no adjustment is necessary if, owing to the specific experimental conditions in the study, Z has in fact no influence on Y. It is true that the adjustment suggested by the back-door criterion would introduce no bias, but such adjustment could be costly if it calls for superfluous measurements in a no-confounding situation.[18] The added cost is justified in light of (i) the causal information at hand (i.e., that Z may potentially influence Y) and (ii) our insistence on ensuring stable unbiasedness – that is, avoiding bias in all situations compatible with the information at hand.

---

[17] A gentle introduction to applications of the back-door criterion in epidemiology can be found in Greenland et al. (1999a).

[18] On the surface, it appears as though the Stone–Robins criterion would correctly recognize the absence of confounding in this situation, since it is based on associations that prevail in the probability distribution that actually generates the data (according to which $\{E, Z\}$ should be independent of Y, given $\{A, X\}$). However, these associations are of no help in deciding whether certain measurements can be *avoided*; such decisions must be made prior to gathering the data and must rely therefore on subjective assumptions about the disappearance of conditional associations. Such assumptions are normally supported by causal, not associational, knowledge (see Section 1.3).

6.4.2 Formal Definitions

To formally distinguish between *stable* and *incidental* unbiasedness, we use the following general definition.

Definition 6.4.1 (Stable Unbiasedness)
Let A be a set of assumptions (or restrictions) on the data-generating process, and let C_A be a class of causal models satisfying A. The effect estimate of X on Y is said to be stably unbiased given A if $P(y \mid do(x)) = P(y \mid x)$ holds in every model M in C_A. Correspondingly, we say that the pair (X, Y) is stably unconfounded given A.

The assumptions commonly used to specify causal models can be either parametric or topological. For example, the structural equation models used in the social sciences and economics are usually restricted by the assumptions of linearity and normality. In this case, C_A would consist of all models created by assigning different values to the unspecified parameters in the equations and in the covariance matrix of the error terms. Weaker, nonparametric assumptions emerge when we specify merely the topological structure of the causal diagram but let the error distributions and the functional form of the equations remain undetermined. We now explore the statistical ramifications of these nonparametric assumptions.

Definition 6.4.2 (Structurally Stable No-Confounding)
Let A_D be the set of assumptions embedded in a causal diagram D. We say that X and Y are stably unconfounded given A_D if $P(y \mid do(x)) = P(y \mid x)$ holds in every parameterization of D. By "parameterization" we mean an assignment of functions to the links of the diagram and prior probabilities to the background variables in the diagram.

Explicit interpretation of the assumptions embedded in a causal diagram are given in Chapters 3 and 5. Put succinctly, if D is the diagram associated with the causal model, then:

1. every missing arrow (between, say, X and Y) represents the assumption that X has no effect on Y once we intervene and hold the parents of Y fixed;

2. every missing bidirected link between X and Y represents the assumption that there are no common causes for X and Y, except those shown in D.

Whenever the diagram D is acyclic, the back-door criterion provides a necessary and sufficient test for stable no-confounding, given A_D. In the simple case of no adjustment for covariates, the criterion reduces to the nonexistence of a common ancestor, observed or latent, of X and Y.[19] Thus, we have our next theorem.

---

[19] The colloquial term "common ancestors" should exclude nodes that have no other connection to Y except through X (e.g., node E in Figure 6.3) and include latent nodes for correlated errors. In the diagram of Figure 6.4, for example, X and Y are understood to have two common ancestors; the first is Z and the second is the (implicit) latent variable responsible for the double-arrowed arc between X and Y (i.e., the correlation between ε_1 and ε_2).

Theorem 6.4.3 (Common-Cause Principle)

Let A_D be the set of assumptions embedded in an acyclic causal diagram D. Variables X and Y are stably unconfounded given A_D if and only if X and Y have no common ancestor in D.

Proof

The "if" part follows from the validity of the back-door criterion (Theorem 3.3.2). The "only if" part requires the construction of a specific model in which (6.10) is violated whenever X and Y have a common ancestor in D. This is easily done using linear models and Wright's rules for path coefficients.   □

Theorem 6.4.3 provides a necessary and sufficient condition for stable no-confounding without invoking statistical data, since it relies entirely on the information embedded in the diagram. Of course, the diagram itself has statistical implications that can be tested (Sections 1.2.3 and 5.2.1), but those tests do not specify the diagram uniquely (see Chapter 2 and Section 5.2.3).

Suppose, however, that we do not possess all the information required for constructing a causal diagram and instead know merely for each variable Z whether it is safe to assume that Z has no effect on Y and whether X has no effect on Z. The question now is whether this more modest information, together with statistical data, is sufficient to qualify or disqualify a pair (X, Y) as stably unconfounded. The answer is positive.

6.4.3   Operational Test for Stable No-Confounding

Theorem 6.4.4 (Criterion for Stable No-Confounding)

Let A_Z denote the assumptions that (i) the data are generated by some (unspecified) acyclic model M and (ii) Z is a variable in M that is unaffected by X but may possibly affect Y.[20] If both of the associational criteria (U_1) and (U_2) of Definition 6.2.2 are violated, then (X, Y) are not stably unconfounded given A_Z.

Proof

Whenever X and Y are stably unconfounded, Theorem 6.4.3 rules out the existence of a common ancestor of X and Y in the diagram associated with the underlying model. The absence of a common ancestor, in turn, implies the satisfaction of either (U_1) or (U_2) whenever Z satisfies A_Z. This is a consequence of the d-separation rule (Section 1.2.3) for reading the conditional independence relationships entailed by a diagram.[21]   □

Theorem 6.4.4 implies that the traditional associational criteria (U_1) and (U_2) could be used in a simple operational test for stable no-confounding, a test that does not require us to know the causal structure of the variables in the domain or even to enumerate the set of relevant variables. Finding just *any* variable Z that satisfies A_Z and violates (U_1)

---

[20] By "possibly affecting Y" we mean: A_Z does not contain the assumption that Z does not affect Y. In other words, the diagram associated with M must contain a directed path from Z to Y.
[21] It also follows from Theorem 7(a) in Robins (1997).

and (U_2) permits us to disqualify (X, Y) as stably unconfounded (though (X, Y) may be incidentally unconfounded in the particular experimental conditions prevailing in the study).

Theorem 6.4.4 communicates a formal connection between statistical associations and confounding that is not based on the closed-world assumption.[22] It is remarkable that the connection can be formed under such a weak set of added assumptions: the qualitative assumption that a variable may have influence on Y and is not affected by X suffices to produce a necessary statistical test for stable no-confounding.

6.5   CONFOUNDING, COLLAPSIBILITY, AND EXCHANGEABILITY

6.5.1   Confounding and Collapsibility

Theorem 6.4.4 also establishes a formal connection between confounding and "collapsibility" – a criterion under which a measure of association remains invariant to the omission of certain variables.

Definition 6.5.1 (Collapsibility)
Let g [P (x, y)] be any functional[23] that measures the association between Y and X in the joint distribution P(x, y). We say that g is collapsible *on a variable Z if*

$$E_z g[P(x, y \mid z)] = g[P(x, y)].$$

It is not hard to show that if g stands for any linear functional of $P(y \mid x)$ – for example, the risk difference $P(y \mid x_1) - P(y \mid x_2)$ – then collapsibility holds whenever Z is either unassociated with X or unassociated with Y given X. Thus, any violation of collapsibility implies violation of the two statistical criteria of Definition 6.2.2, and that is probably why many believed noncollapsibility to be intimately connected with confounding. However, the examples in this chapter demonstrate that violation of these two conditions is neither sufficient nor necessary for confounding. Thus, noncollapsibility and confounding are in general two distinct notions; neither implies the other.

Some authors tend to believe that this distinction is a peculiar property of nonlinear effect measures g, such as the odds or likelihood ratios, and that "when the effect measure is an expectation over population units, confounding and noncollapsibility are algebraically equivalent" (Greenland 1998, p. 906). This chapter shows that confounding and noncollapsibility need not correspond even in linear functionals. For example, the effect measure $P(y \mid x_1) - P(y \mid x_2)$ (the risk difference) is not collapsible over Z in Figure 6.3 (for almost every parameterization of the graph) and yet the effect measure is unconfounded (for every parameterization).

---

[22] I am not aware of another such connection in the literature.

[23] A *functional* is an assignment of a real number to any function from a given set of functions. For example, the mean $E(X) = \Sigma_x x P(x)$ is a functional, since it assigns a real number $E(X)$ to each probability function $P(x)$.

The logical connection between confounding and collapsibility is formed through the notion of *stable no-confounding*, as formulated in Definition 6.4.2 and Theorem 6.4.4. Because any violation of collapsibility means violation of (U_1) and (U_2) in Definition 6.2.2, it also implies (by Theorem 6.4.4) violation of stable unbiasedness (or stable no-confounding). Thus we can state the following corollary.

Corollary 6.5.2 (Stable No-Confounding Implies Collapsibility)

Let Z be any variable that is not affected by X and that may possibly affect Y. Let g [P (x, y)] be any linear functional that measures the association between X and Y. If g is not collapsible on Z, then X and Y are not stably unconfounded.

This corollary provides a rationale for the widespread practice of testing confoundedness by the change-in-parameter method, that is, labeling a variable Z a confounder whenever the "crude" measure of association, $g [P (x, y)]$, is not equal to the Z-specific measures of association averaged over the levels of Z (Breslow and Day 1980; Kleinbaum et al. 1982; Yanagawa 1984; Grayson 1987). Theorem 6.4.4 suggests that the intuitions responsible for this practice were shaped by a quest for a stable condition of no-confounding, not merely an incidental one. Moreover, condition A_Z in Theorem 6.4.4 justifies a requirement made by some authors that a confounder must be a causal determinant of, and not merely associated with, the outcome variable Y.

6.5.2 Confounding versus Confounders

The focus of our discussion in this chapter has been the phenomenon of confounding, which we equated with that of effect bias (Definition 6.2.1). Much of the literature on this topic has been concerned with the presence or absence of *confounders*, presuming that some variables possess the capacity to confound and some do not. This notion may be misleading if interpreted literally, and caution should be exercised before we label a variable as a confounder.

Rothman and Greenland (1998, p. 120), for example, offer this definition: "The extraneous factors responsible for difference in disease frequency between the exposed and unexposed are called *confounders*"; they go on to state that: "In general, a confounder must be associated with both the exposure under study and the disease under study to be confounding" (p. 121). Rothman and Greenland qualify their statement with "In general," and for good reason: We have seen (in the two-coin example of Section 6.3.1) that each individual variable in a problem can be *unassociated* with both the exposure (X) and the disease (Y) under study and still the effect of X on Y remains confounded. A similar situation can also be seen in the linear model depicted in Figure 6.5. Although Z is clearly a confounder for the effect of X on Y and must therefore be controlled, the association between Z and Y may actually vanish (at each level of X); similarly, the association between Z and X may vanish. This can occur if the indirect association mediated by the path $Z \leftarrow A \rightarrow Y$ happens to cancel the direct association carried by the arrow $Z \rightarrow Y$. This cancellation does not imply the absence of confounding, because the path $X \leftarrow E \rightarrow Z \rightarrow Y$ is unblocked while $X \leftarrow E \rightarrow Z \leftarrow A \rightarrow Y$ is blocked. Thus, Z is a confounder that is not associated with the disease (Y).

The intuition behind Rothman and Greenland's statement just quoted can be explicated formally through the notion of stability: a variable that is *stably* unassociated with

Figure 6.5 Z may be unassociated with both X and Y and still be a confounder (i.e., a member of every sufficient set).

either X or Y can safely be excluded from adjustment. Alternatively, Rothman and Greenland's statement can be supported (without invoking stability) by using the notion of a *nontrivial sufficient set* (Section 3.3) – a set of variables for which adjustment will remove confounding bias. It can be shown (see the end of this section) that each such set S, taken as a unit, must indeed be associated with X and be conditionally associated with Y, given X. Thus, Rothman and Greenland's condition is valid for nontrivial sufficient (i.e., admissible) sets but not for the individual variables in the set.

The practical ramifications of this condition are as follows. If we are given a set S of variables that is claimed to be sufficient (for removing bias by adjustment), then that claim can be given a necessary statistical test: S as a compound variable must be associated both with X and with Y (given X). In Figure 6.5, for example, $S_1 = \{A, Z\}$ and $S_2 = \{E, Z\}$ are sufficient and nontrivial; both must satisfy the condition stated.

Note, however, that although this test can screen out some obviously bad sets S claimed to be sufficient, it has nothing to do with sufficiency or confounding; it merely tests for nontriviality, i.e., that adjusting for S would change the association between X and Y. When we find a nontrivial set S, we still cannot be sure whether the association was unbiased to start with (as in Figure 6.3) or that it turned unbiased after the adjustment.

Proof of Necessity

To prove that (U_1) and (U_2) must be violated whenever Z stands for a nontrivial sufficient set S, consider the case where X has no effect on Y. In this case, confounding amounts to a nonvanishing association between X and Y. A well-known property of conditional independence, called *contraction* (Section 1.1.5), states that (U_1), $X \perp\!\!\!\perp S$, together with sufficiency, $X \perp\!\!\!\perp Y \mid S$, implies violation of nontriviality, $X \perp\!\!\!\perp Y$:

$$X \perp\!\!\!\perp S \ \ \& \ \ X \perp\!\!\!\perp Y \mid S \ \Longrightarrow \ X \perp\!\!\!\perp Y.$$

Likewise, another property of conditional independence, called *intersection,* states that (U_2), $S \perp\!\!\!\perp Y \mid X$, together with sufficiency, $X \perp\!\!\!\perp Y \mid S$, also implies violation of nontriviality, $X \perp\!\!\!\perp Y$.

$$S \perp\!\!\!\perp Y \mid X \ \ \& \ \ X \perp\!\!\!\perp Y \mid S \ \Longrightarrow \ X \perp\!\!\!\perp Y.$$

Thus, both (U_1) and (U_2) must be violated by any nontrivial sufficient set S.

Note, however, that intersection holds only for strictly positive probability distributions, which means that the Rothman–Greenland condition may be violated if deterministic relationships hold among some variables in a problem. This can be seen from a simple example in which both X and Y stand in a one-to-one functional relationship to a third variable, Z. Clearly, Z is a nontrivial sufficient set yet is not associated with Y given X; once we know the value of X, the probability of Y is determined, and would no longer change with learning the value of Z.

6.5.3 Exchangeability versus Structural Analysis of Confounding

Students of epidemiology complain bitterly about the confusing way in which the fundamental concept of confounding has been treated in the literature. A few authors have acknowledged the confusion (e.g., Greenland and Robins 1986; Wickramaratne and Holford 1987; Weinberg 1993) and have suggested new ways of looking at the problem that might lead to more systematic analysis. Greenland and Robins (GR), in particular, have recognized the same basic principles and results that we have expounded here in Sections 6.2 and 6.3. Their analysis represents one of the few bright spots in the vast literature on confounding in that it treats confounding as an unknown causal quantity that is not directly measurable from observed data. They further acknowledge (as do Miettinen and Cook 1981) that the presence or absence of confounding should not be equated with absence or presence of collapsibility and that confounding should not be regarded as a parameter-dependent phenomenon.

However, the structural analysis presented in this chapter differs in a fundamental way from that of GR, who have pursued an approach based on judgment of "exchangeability." In Section 6.1 we encountered a related notion of exchangeability, one with which Lindley and Novick (1981) attempted to view Simpson's paradox; GR's idea of exchangeability is more concrete and more clearly applicable. Conceptually, the connection between confounding and exchangeability is as follows. If we undertake to assess the effect of some treatment, we ought to make sure that any response differences between the treated and the untreated group is due to the treatment itself and not to some intrinsic differences between the groups that are unrelated to the treatment. In other words, the two groups must resemble each other in all characteristics that have bearing on the response variable. In principle, we could have ended the definition of confounding at this point, declaring simply that the effect of treatment is unconfounded if the treated and untreated groups resemble each other in all relevant features. This definition, however, is too verbal in the sense that it is highly sensitive to interpretation of the terms "resemblance" and "relevance." To make it less informal, GR used De Finetti's twist of hypothetical permutation; instead of judging whether two groups are similar, the investigator is instructed to imagine a hypothetical *exchange* of the two groups (the treated group becomes untreated, and vice versa) and then to judge whether the observed data under the swap would be distinguishable from the actual data.

One can justifiably ask what has been gained by this mental exercise, relative to judging directly if the two groups are effectively identical. The gain is twofold. First, people are quite good at envisioning dynamic processes and can simulate the outcome of this swapping scenario from basic understanding of the processes that govern the response to treatment and the factors that affect the choice of treatment. Second, moving from judgment about resemblance to judgment about probabilities permits us to cast those judgments in probabilistic notation and hence to invite the power and respectability of probability calculus.

Greenland and Robins made an important first step toward this formalization by bringing notation closer to where judgment originates – the human understanding of causal processes. The structural approach pursued in this book takes the next, natural step: formalizing the causal processes themselves.

Let A and B stand (respectively) for the treated and untreated groups, and let $P_{A1}(y)$ and $P_{A0}(y)$ stand (respectively) for the response distribution of group A under two hypothetical conditions, treatment and no treatment.[24] If our interest lies in some parameter μ of the response distribution, we designate by μ_{A1} and μ_{A0} the values of that parameter in the corresponding distribution $P_{A1}(y)$ and $P_{A0}(y)$, with μ_{B1} and μ_{B0} defined similarly for group B. In actuality, we measure the pair (μ_{A1}, μ_{B0}); after the hypothetical swap, we would measure (μ_{B1}, μ_{A0}). We define the groups to be *exchangeable* relative to parameter μ if the two pairs are indistinguishable, that is, if

$$(\mu_{A1}, \mu_{B0}) = (\mu_{B1}, \mu_{A0}).$$

In particular, if we define the causal effect by the difference $\text{CE} = \mu_{A1} - \mu_{A0}$, then exchangeability permits us to replace μ_{A0} with μ_{B0} and so obtain $\text{CE} = \mu_{A1} - \mu_{B0}$, which is measurable because both quantities are observed. Greenland and Robins thus declare the causal effect CE to be *unconfounded* if $\mu_{A0} = \mu_{B0}$.

If we compare this definition to that of (6.10), $P(y \mid do(x)) = P(y \mid x)$, we find that the two coincide if we rewrite the latter as $\mu[P(y \mid do(x))] = \mu[P(y \mid x)]$, where μ is the parameter of interest in the response distribution. However, the major difference between the structural and the GR approaches lies in the level of analysis. Structural modeling extends the formalization of confounding in two important directions. First, (6.10) is not submitted to direct human judgment but is derived mathematically from more elementary judgments concerning causal processes.[25] Second, the input judgments needed for the structural model are both qualitative and stable.

A simple example will illustrate the benefits of these features. Consider the following statement (Greenland 1998):

(Q^*)   "if the effect measure is the difference or ratio of response proportions, then the above phenomenon – noncollapsibility without confounding – cannot occur, nor can confounding occur without noncollapsibility." (pp. 905–6)

We have seen in this chapter that statement (Q^*) should be qualified in several ways and that, in general, noncollapsibility and confounding are two distinct notions – neither implying the other, regardless of the effect measure (Section 6.5.1). However, the

---

[24] In $do(\cdot)$ notation, we would write $P_{A1}(y) = P_A(y \mid do(X = 1))$.

[25] Recall that the do ($\cdot$) operator is defined mathematically in terms of equation deletion in structural equation models; consequently, the verification of the nonconfounding condition $P(y \mid do(x)) = P(y \mid x)$ in a given model is not a matter of judgment but a subject of mathematical analysis.

question we wish to discuss here is methodological: What formalism would be appropriate for validating, refuting, or qualifying statements of this sort? Clearly, since (Q^*) makes a general claim about all instances, one counterexample would suffice to refute its general validity. But how do we construct such a counterexample? More generally, how do we construct examples that embody properties of confounding, effect bias, causal effects, experimental versus nonexperimental data, counterfactuals, and other causality-based concepts?

In probability theory, if we wish to refute a general statement about parameters and their relationship we need only present one density function f for which that relationship fails to hold. In propositional logic, in order to show that a sentence is false, we need only present one truth table T that satisfies the premises and violates the conclusions. What, then, is the mathematical object that should replace f or T when we wish to refute causal claims like statement (Q^*)? The corresponding object used in the exchangeability framework of Greenland and Robins is a counterfactual contingency table (see, e.g., Greenland et al. 1999b, p. 905, or Figure 1.7 in Section 1.4.4). For instance, to illustrate confounding, we need two such tables: one describing the hypothetical response of the treated group A to both treatment and nontreatment, and one describing the hypothetical response of the untreated group B to both treatment and nontreatment. If the tables show that the parameter μ_{A0}, computed from the hypothetical response of the treated group to no treatment, differs from μ_{B0}, computed from the actual response of the untreated group, then we have confounding on our hands.

Tables of this type can be constructed for simple problems involving one treatment and one response variable, but they become a nightmare when several covariates are involved or when we wish to impose certain constraints on those covariates. For example, we may wish to incorporate the standard assumption that a covariate Z does not lie on the causal pathway between treatment and response, or that Z has causal influence on Y, but such assumptions cannot conveniently be expressed in counterfactual contingency tables. As a result, the author of the claim to be refuted could always argue that the tables used in the counterexample may be inconsistent with the agreed assumptions.[26]

Such difficulties do not plague the structural representation of confounding. In this formalism, the appropriate object for exemplifying or refuting causal statements is a causal model, as defined in Chapter 3 and used throughout this book. Here, hypothetical responses (μ_{A0} and μ_{B0}) and contingency tables are not primitive quantities but rather are derivable from a set of equations that already embody the assumptions we wish to respect. Every parameterization of a structural model implies (using (3.51) or the $do(\cdot)$ operator) a specific set of counterfactual contingency tables that satisfies the input assumptions and exhibits the statistical properties displayed in the graph. For example, any parameterization of the graph in Figure 6.3 generates a set of counterfactual contingency tables that already embodies the assumptions that Z is not on the causal pathway between X and Y and that Z has no causal effect on Y, and almost every such parameterization will generate a counterexample to claim (Q^*). Moreover, we can also disprove (Q^*) by a casual inspection of the diagram and without generating numerical counterexamples.

---

[26] Readers who attempt to construct a counterexample to statement (Q^*) using counterfactual contingency tables will certainly appreciate this difficulty.

Figure 6.3, for example, shows vividly that the risk difference $P(y \mid x_1) - P(y \mid x_2)$ is not collapsible on Z and, simultaneously, that X and Y are (stably) unconfounded.

The difference between the two formulations is even more pronounced when we come to substantiate, not refute, generic claims about confounding. Here it is not enough to present a single contingency table; instead, we must demonstrate the validity of the claim for all tables that can possibly be constructed in compliance with the input assumptions. This task, as the reader surely realizes, is a hopeless exercise within the framework of contingency tables; it calls for a formalism in which assumptions can be stated succinctly and in which conclusions can be deduced by mathematical derivations. The structural semantics offers such formalism, as demonstrated by the many generic claims proven in this book (examples include Theorem 6.4.4 and Corollary 6.5.2).

As much as I admire the rigor introduced by Greenland and Robins's analysis through the framework of exchangeability, I am thoroughly convinced that the opacity and inflexibility of counterfactual contingency tables are largely responsible for the slow acceptance of the GR framework among epidemiologists and, as a by-product, for the lingering confusion that surrounds confounding in the statistical literature at large. I am likewise convinced that formulating claims and assumptions in the language of structural models will make the mathematical analysis of causation accessible to rank-and-file researchers and thus lead eventually to a total and natural disconfounding of confounding.

6.6 CONCLUSIONS

Past efforts to establish a theoretical connection between statistical associations (or collapsibility) and confounding have been unsuccessful for three reasons. First, the lack of mathematical language for expressing claims about causal relationships and effect bias has made it difficult to assess the disparity between the requirement of effect unbiasedness (Definition 6.2.1) and statistical criteria purporting to capture unbiasedness.[27] Second, the need to exclude barren proxies (Figure 6.3) from consideration has somehow escaped the attention of researchers. Finally, the distinction between stable and incidental unbiasedness has not received the attention it deserves and, as we observed in Example 6.3.3, no connection can be formed between associational criteria (or collapsibility) and confounding without a commitment to the notion of stability. Such commitment rests critically on the conception of a causal model as an assembly of autonomous mechanisms that may vary independently of one another (Aldrich 1989). It is only in anticipation of such independent variations that we are not content with incidental unbiasedness but rather seek conditions of stable unbiasedness. The mathematical formalization of this conception has led to related notions of *DAG-isomorph* (Pearl 1988b, p. 128), *stability*

---

[27] The majority of papers on collapsibility (e.g., Bishop 1971; Whittemore 1978; Wermuth 1987; Becher 1992; Geng 1992) motivate the topic by citing Simpson's paradox and the dangers of obtaining confounded effect estimates. Of these, only a handful pursue the study of confounding or effect estimates; most prefer to analyze the more manageable phenomenon of collapsibility as a stand-alone target. Some go as far as naming collapsibility "nonconfoundedness" (Grayson 1987; Steyer et al. 1997).

(Pearl and Verma 1991), and *faithfulness* (Spirtes et al. 1993), which assist in the elucidation of causal diagrams from sparse statistical associations (see Chapter 2). The same
conception has evidently been shared by authors who aspired to connect associational
criteria with confounding.

The advent of structural model analysis, assisted by graphical methods, offers a mathematical framework in which considerations of confounding can be formulated and managed more effectively. Using this framework, this chapter explicates the criterion of stable
unbiasedness and shows that this criterion (i) has implicitly been the target of many investigations in epidemiology and biostatistics, and (ii) can be given operational statistical
tests similar to those invoked in testing collapsibility. We further show (Section 6.5.3)
that the structural framework overcomes basic cognitive and methodological barriers that
have made confounding one of the most confused topics in the literature. It is therefore
natural to predict that this framework will become the primary mathematical basis for
future studies of confounding.

Acknowledgments

Discussions with James Robins and Sander Greenland were extremely valuable. Sander,
in particular, gave many constructive comments on two early drafts and helped to keep
them comprehensible to epidemiologists. Jan Koster called my attention to the connection between Stone's and Robins's criteria of no-confounding and caught several over-
sights in an earlier draft. Other helpful discussants were Michelle Pearl, Bill Shipley,
Rolf Steyer, Stephen Stigler, and David Trichler.

Postscript for the Second Edition

Readers would be amused to learn that the first printing of this chapter did not stop statisticians' fascination with Simpson's nonparadox. Textbooks continue to marvel at the
phenomenon (Moore and McCabe 2005), and researchers continue to chase its mathematical intricacies (Cox and Wermuth 2003) and visualizations (Rücker and
Schumacher 2008) with the passion of the 1970–90s, without once mentioning the word
"cause" and without once stopping to ask: "What's the point?" A notable exception is
Larry Wasserman's *All Of Statistics* (Wasserman 2004), the first statistics textbook to
treat Simpson's reversal in its correct causal context. My confidence in the eventual triumph of the causal understanding of this nonparadox has also been reinforced by a discussion published in the epidemiological literature, concluding in no ambiguous terms:
"The explanations and solutions lie in causal reasoning which relies on background
knowledge, not statistical criteria" (Arah 2008). It appears that, as we enter the age of
causation, we should look to epidemiologists for guidance and wisdom.

CHAPTER SEVEN

The Logic of Structure-Based Counterfactuals

And the Lord said,
"If I find in the city of Sodom fifty good men,
I will pardon the whole place for their sake."
Genesis 18:26

Preface

This chapter provides a formal analysis of structure-based *counterfactuals*, a concept introduced briefly in Chapters 1 and 3 that will occupy the rest of our discussion in this book. Through this analysis, we will obtain sharper mathematical definitions of other concepts that were introduced in earlier chapters, including causal models, action, causal effects, causal relevance, error terms, and exogeneity.

After casting the concepts of causal model and counterfactuals in formal mathematical terms, we will demonstrate by examples how counterfactual questions can be answered from both deterministic and probabilistic causal models (Section 7.1). In Section 7.2.1, we will argue that policy analysis is an exercise in counterfactual reasoning and demonstrate this thesis in a simple example taken from econometrics. This will set the stage for our discussion in Section 7.2.2, where we explicate the empirical content of counterfactuals in terms of policy predictions. Section 7.2.3 discusses the role of counterfactuals in the interpretation and generation of causal explanations. Section 7.2 concludes with discussions of how causal relationships emerge from actions and mechanisms (Section 7.2.4) and how causal directionality can be induced from a set of symmetric equations (Section 7.2.5).

In Section 7.3 we develop an axiomatic characterization of counterfactual and causal relevance relationships as they emerge from the structural model semantics. Section 7.3.1 will identify a set of properties, or axioms, that allow us to derive new counterfactual relations from assumptions, and Section 7.3.2 demonstrates the use of these axioms in algebraic derivation of causal effects. Section 7.3.3 introduces axioms for the relationship of causal relevance and, using their similarity to the axioms of graphs, describes the use of graphs for verifying relevance relationships.

The axiomatic characterization developed in Section 7.3 enables us to compare structural models with other approaches to causality and counterfactuals, most notably those based on Lewis's closest-world semantics (Sections 7.4.1–7.4.4). The formal equivalence of the structural approach and the Neyman–Rubin potential-outcome framework is discussed in Section 7.4.4. Finally, we revisit the topic of exogeneity and extend our discussion of Section 5.4.3 with counterfactual definitions of exogenous and instrumental variables in Section 7.4.5.

---

any sentence of the form $P(A \mid B) < p$, where A and B are Boolean expressions representing events. A *causal model*, naturally, should encode the truth values of sentences that deal with causal relationships; these include action sentences (e.g., "A will be true if we do B"), counterfactuals (e.g., "A would have been different were it not for B"), and plain causal utterances (e.g., "A may cause B" or "B occurred because of A"). Such sentences cannot be interpreted in standard propositional logic or probability calculus because they deal with changes that occur in the external world rather than with changes in our beliefs about a static world. Causal models encode and distinguish information about external changes through an explicit representation of the mechanisms that are altered in such changes.

Definition 7.1.1 (Causal Model)
A causal model *is a triple*

$$M = \langle U, V, F \rangle,$$

where:

(i) *U is a set of* background *variables, (also called* exogenous),[2] *that are determined by factors outside the model;*

(ii) *V is a set $\{V_1, V_2, ..., V_n\}$ of variables, called* endogenous, *that are determined by variables in the model – that is, variables in $U \cup V$; and*

(iii) *F is a set of functions $\{f_1, f_2, ..., f_n\}$ such that each f_i is a mapping from (the respective domains of) $U_i \cup PA_i$ to V_i, where $U_i \subseteq U$ and $PA_i \subseteq V \backslash V_i$ and the entire set F forms a mapping from U to V. In other words, each f_i in*

$$v_i = f_i(pa_i, u_i), \quad i = 1, ..., n,$$

assigns a value to V_i that depends on (the values of) a select set of variables in $V \cup U$, and the entire set F has a unique solution $V(u)$.[3,4]

Every causal model M can be associated with a directed graph, $G(M)$, in which each node corresponds to a variable and the directed edges point from members of PA_i and U_i toward V_i. We call such a graph the *causal diagram* associated with M. This graph merely identifies the endogenous and background variables that have direct influence on each V_i; it does not specify the functional form of f_i. The convention of confining the parent set PA_i to variables in V stems from the fact that the background variables are often unobservable. In general, however, we can extend the parent sets to include observed variables in U.

---

[2] We will try to refrain from using the term "exogenous" in referring to background conditions, because this term has acquired more refined technical connotations (see Sections 5.4.3 and 7.4). The term "predetermined" is used in the econometric literature.
[3] The choice of PA_i (connoting *parents*) is not arbitrary, but expresses the modeller's understanding of which variables Nature must consult before deciding the value of V_i.
[4] Uniqueness is ensured in recursive (i.e., acyclic) systems. Halpern (1998) allows multiple solutions in nonrecursive systems.

The final part of this chapter (Section 7.5) compares the structural account of causality with that based on probabilistic relationships. We elaborate our preference toward the structural account and highlight the difficulties that the probabilistic account is currently facing.

7.1 STRUCTURAL MODEL SEMANTICS

How do scientists predict the outcome of one experiment from the results of other experiments run under totally different conditions? Such predictions require us to envision what the world would be like under various hypothetical changes and so invoke *counterfactual* inference. Though basic to scientific thought, counterfactual inference cannot easily be formalized in the standard languages of logic, algebraic equations, or probability. The formalization of counterfactual inference requires a language within which the invariant relationships in the world are distinguished from transitory relationships that represent one's beliefs about the world, and such distinction is not supported by standard algebras, including the algebra of equations, Boolean algebra, and probability calculus. Structural models offer such distinction, and this section presents a structural model semantics of counterfactuals as defined in Balke and Pearl (1994a,b), Galles and Pearl (1997, 1998), and Halpern (1998),[1] which stands in sharp contrast to the experimental perspective of Rubin (1974). Related approaches have been proposed in Simon and Rescher (1966) and Ortiz (1999).

We start with a deterministic definition of a causal model, which consists (as we have discussed in earlier chapters) of functional relationships among variables of interest, each relationship representing an autonomous mechanism. Causal and counterfactual relationships are defined in this model in terms of response to local modifications of those mechanisms. Probabilistic relationships emerge naturally by assigning probabilities to background conditions. After demonstrating, by examples, how this model facilitates the computation of counterfactuals in both deterministic and probabilistic contexts (Section 7.1.2), we then present a general method of computing probabilities of counterfactual expressions using causal diagrams (Section 7.1.3).

7.1.1 Definitions: Causal Models, Actions, and Counterfactuals

A "model," in the common use of the word, is an idealized representation of reality that highlights some aspects and ignores others. In logical systems, however, a model is a mathematical object that assigns truth values to sentences in a given language, where each sentence represents some aspect of reality. Truth tables, for example, are models in propositional logic; they assign a truth value to any Boolean expression, which may represent an event or a set of conditions in the domain of interest. A joint probability function, as another example, is a model in probability logic; it assigns a truth value to

---

[1] Similar models, called "neuron diagrams" (Lewis 1986, p. 200; Hall 2004), are used informally by philosophers to illustrate chains of causal processes.

Definition 7.1.2 (Submodel)

Let M be a causal model, X a set of variables in V, and x a particular realization of X. A submodel M_x of M is the causal model

$$M_x = \langle U, V, F_x \rangle,$$

where

$$F_x = \{f_i : V_i \notin X\} \cup \{X = x\}. \tag{7.1}$$

In words, F_x is formed by deleting from F all functions f_i corresponding to members of set X and replacing them with the set of constant functions $X = x$.

Submodels are useful for representing the effect of local actions and hypothetical changes, including those implied by counterfactual antecedents. If we interpret each function f_i in F as an independent physical mechanism and define the action $do(X = x)$ as the minimal change in M required to make $X = x$ hold true under any u, then M_x represents the model that results from such a minimal change, since it differs from M by only those mechanisms that directly determine the variables in X. The transformation from M to M_x modifies the algebraic content of F, which is the reason for the name "modifiable structural equations" used in Galles and Pearl (1998).[5]

Definition 7.1.3 (Effect of Action)

Let M be a causal model, X a set of variables in V, and x a particular realization of X. The effect of action $do(X = x)$ on M is given by the submodel M_x.

Definition 7.1.4 (Potential Response)

Let X and Y be two subsets of variables in V. The potential response of Y to action $do(X = x)$, denoted $Y_x(u)$, is the solution for Y of the set of equations F_x,[6] that is, $Y_x(u) = Y_{M_x}(u)$.

We will confine our attention to actions in the form of $do(X = x)$. Conditional actions of the form "$do(X = x)$ if $Z = z$" can be formalized using the replacement of equations by functions of Z, rather than by constants (Section 4.2). We will not consider disjunctive actions of the form "$do(X = x$ or $Z = z)$," since these complicate the probabilistic treatment of counterfactuals.

Definition 7.1.5 (Counterfactual)

Let X and Y be two subsets of variables in V. The counterfactual sentence "Y would be y (in situation u), had X been x" is interpreted as the equality $Y_x(u) = y$, with $Y_x(u)$ being the potential response of Y to $X = x$.

---

[5] Structural modifications date back to Haavelmo (1943), Marschak (1950), and Simon (1953). An explicit translation of interventions into "wiping out" equations from the model was first proposed by Strotz and Wold (1960) and later used in Fisher (1970), Sobel (1990), Spirtes et al. (1993), and Pearl (1995a). A similar notion of submodel was introduced by Fine (1985), though not specifically for representing actions and counterfactuals.

[6] If Y is a set of variables $Y = (Y_1, Y_2, \dots)$, then $Y_x(u)$ stands for a vector of functions $(Y_{1_x}(u), Y_{2_x}(u), \dots)$.

Definition 7.1.5 thus interprets the counterfactual phrase "had X been x" in terms of a hypothetical modification of the equations in the model; it simulates an external action (or spontaneous change) that modifies the actual course of history and enforces the condition "$X = x$" with minimal change of mechanisms. This is a crucial step in the semantics of counterfactuals (Balke and Pearl 1994b), as it permits x to differ from the current value of $X(u)$ without creating logical contradiction; it also suppresses abductive inferences (or backtracking) from the counterfactual antecedent $X = x$.[7] In Chapter 3 (Section 3.6.3) we used the notation $Y(x, u)$ to denote the unit-based conditional "the value that Y would obtain in unit u, had X been x," as used in the Neyman–Rubin potential-outcome model. Throughout the rest of this book we will use the notation $Y_x(u)$ to denote counterfactuals tied specifically to the structural model interpretation of Definition 7.1.5 (paralleling (3.51)); $Y(x, u)$ will be reserved for generic subjunctive conditionals, uncommitted to any specific semantics.

Definition 7.1.5 endows the atomic mechanisms $\{f_i\}$ themselves with interventional–counterfactual interpretation, because $v_i = f_i(pa_i, u_i)$ is the value of V_i in the submodel $M_{v\setminus v_i}$. In other words, $f_i(pa_i, u_i)$ stands for the potential response of V_i when we intervene on *all* other variables in V.

This formulation generalizes naturally to probabilistic systems as follows.

Definition 7.1.6 (Probabilistic Causal Model)
A probabilistic causal model is a pair

$$\langle M, P(u)\rangle,$$

where M is a causal model and P(u) is a probability function defined over the domain of U.

The function $P(u)$, together with the fact that each endogenous variable is a function of U, defines a probability distribution over the endogenous variables. That is, for every set of variables $Y \subseteq V$, we have

$$P(y) \triangleq P(Y = y) = \sum_{\{u|Y(u)=y\}} P(u). \tag{7.2}$$

The probability of counterfactual statements is defined in the same manner, through the function $Y_x(u)$ induced by the submodel M_x:

$$P(Y_x = y) = \sum_{\{u|Y_x(u)=y\}} P(u). \tag{7.3}$$

Likewise, a causal model defines a joint distribution on counterfactual statements. That is, $P(Y_x = y, Z_w = z)$ is defined for any (not necessarily disjoint) sets of variables Y, X, Z, and W. In particular, $P(Y_x = y, X = x')$ and $P(Y_x = y, Y_{x'} = y')$ are well defined for $x \neq x'$ and are given by

---

[7] Simon and Rescher (1966, p. 339) did not include this step in their account of counterfactuals and noted that backward inferences triggered by the antecedents can lead to ambiguous interpretations.

$$P(Y_x = y, X = x') = \sum_{\{u \mid Y_x(u) = y \ \& \ X(u) = x'\}} P(u) \tag{7.4}$$

and

$$P(Y_x = y, Y_{x'} = y') = \sum_{\{u \mid Y_x(u) = y \ \& Y_{x'}(u) = y'\}} P(u). \tag{7.5}$$

If x and x' are incompatible then Y_x and $Y_{x'}$ cannot be measured simultaneously, and it may seem meaningless to attribute probability to the joint statement "Y would be y if $X = x$ *and* Y would be y' if $X = x'$." Such concerns have been a source of recent objections to treating counterfactuals as jointly distributed random variables (Dawid 2000). The definition of Y_x and Y_x' in terms of the solution for Y in two distinct submodels, governed by a standard probability space over U, neutralizes these objections by interpreting the contradictory joint statement as an ordinary event in U-space.

Of particular interest to us would be probabilities of counterfactuals that are conditional on actual observations. For example, the probability that event $X = x$ "was the cause" of event $Y = y$ may be interpreted as the probability that Y would not be equal to y had X not been x, given that $X = x$ and $Y = y$ have in fact occurred (see Chapter 9 for an in-depth discussion of the probabilities of causation). Such probabilities are well defined in the model just described; they require the evaluation of expressions of the form $P(Y_{x'} = y' \mid X = x, Y = y)$, with x' and y' incompatible with x and y, respectively. Equation (7.4) allows the evaluation of this quantity as follows:

$$P(Y_{x'} = y' \mid X = x, Y = y) = \frac{P(Y_{x'} = y', X = x, Y = y)}{P(X = x, Y = y)}$$

$$= \sum_u P(Y_{x'}(u) = y')P(u \mid x, y). \tag{7.6}$$

In other words, we first update $P(u)$ to obtain $P(u \mid x, y)$ and then use the updated distribution $P(u \mid x, y)$ to compute the expectation of the equality $Y_{x'}(u) = y'$.

This substantiates the three-step procedure introduced in Section 1.4, which we now summarize in a theorem.

Theorem 7.1.7
Given model $\langle M, P(u) \rangle$, the conditional probability $P(B_A \mid e)$ of a counterfactual sentence "If it were A then B," given evidence e, can be evaluated using the following three steps.

1. **Abduction** – *Update $P(u)$ by the evidence e to obtain $P(u \mid e)$.*
2. **Action** – *Modify M by the action $do(A)$, where A is the antecedent of the counterfactual, to obtain the submodel M_A.*
3. **Prediction** – *Use the modified model $\langle M_A, P(u \mid e) \rangle$ to compute the probability of B, the consequence of the counterfactual.*

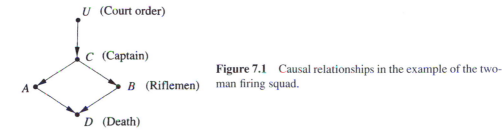

Figure 7.1   Causal relationships in the example of the two-man firing squad.

To complete this section, we introduce two additional objects that will prove useful in subsequent discussions: *worlds*[8] and *theories*.

Definition 7.1.8 (Worlds and Theories)
A causal world w is a pair $\langle M, u \rangle$, where M is a causal model and u is a particular realization of the background variables U. A causal theory *is a set of causal worlds.*

A world w can be viewed as a degenerate probabilistic model for which $P(u) = 1$. Causal theories will be used to characterize partial specifications of causal models, for example, models sharing the same causal diagram or models in which the functions f_i are linear with undetermined coefficients.

7.1.2   Evaluating Counterfactuals: Deterministic Analysis

In Section 1.4.1 we presented several examples demonstrating the interpretation of actions and counterfactuals in structural models. We now apply the definitions of Section 7.1.1 to demonstrate how counterfactual queries, both deterministic and probabilistic, can be answered formally using structural model semantics.

Example 1: The Firing Squad
Consider a two-man firing squad as depicted in Figure 7.1, where A, B, C, D, and U stand for the following propositions:

U = court orders the execution;

C = captain gives a signal;

A = rifleman A shoots;

B = rifleman B shoots;

D = prisoner dies.

Assume that the court's decision is unknown, that both riflemen are accurate, alert, and law-abiding, and that the prisoner is not likely to die from fright or other extraneous causes. We wish to construct a formal representation of the story, so that the following sentences can be evaluated mechanically.

---

[8] Adnan Darwiche called my attention to the importance of this object.

S1   *Prediction* – If rifleman A did not shoot, then the prisoner is alive:

$$\neg A \Longrightarrow \neg D.$$

S2   *Abduction* – If the prisoner is alive, then the captain did not signal:

$$\neg D \Longrightarrow \neg C.$$

S3   *Transduction* – If rifleman A shot, then B shot as well:

$$A \Longrightarrow B.$$

S4   *Action* – If the captain gave no signal and rifleman A decides to shoot, then the prisoner will die and B will not shoot.

$$\neg C \Longrightarrow D_A \;\&\; \neg B_A.$$

S5   *Counterfactual* – If the prisoner is dead, then the prisoner would be dead even if rifleman A had not shot:

$$D \Longrightarrow D_{\neg A}.$$

Evaluating Standard Sentences

To prove the first three sentences we need not invoke causal models; these sentences involve standard logical connectives and thus can be handled using standard logical deduction. The story can be captured in any convenient logical theory (a set of propositional sentences), for example,

$$T_1 \colon U \Longleftrightarrow C, \; C \Longleftrightarrow A, \; C \Longleftrightarrow B, \; A \vee B \Longleftrightarrow D$$

or

$$T_2 \colon U \Longleftrightarrow C \Longleftrightarrow A \Longleftrightarrow B \Longleftrightarrow D,$$

where each theory admits the two logical models

$$m_1 \colon \{U, C, A, B, D\} \quad \text{and} \quad m_2 \colon \{\neg U, \neg C, \neg A, \neg B, \neg D\}.$$

In words, any theory T that represents our story should imply that either all five propositions are true or all are false; models m_1 and m_2 present these two possibilities explicitly. The validity of S1–S3 can easily be verified, either by derivation from T or by noting that the antecedent and consequent in each sentence are both part of the same model.

Two remarks are worth making before we go on to analyze sentences S4 and S5. First, the two-way implications in T_1 and T_2 are necessary for supporting abduction; if we were to use one-way implications (e.g., $C \Longrightarrow A$), then we would not be able to conclude C from A. In standard logic, this symmetry removes all distinctions between the tasks of prediction (reasoning forward in time), abduction (reasoning from evidence to explanation), and transduction (reasoning from evidence to explanation and then from explanation to predictions). Using two-way implication, these three modes of reasoning differ only in the interpretations they attach to antecedents and consequents of conditional sentences – not in their methods of inference. In nonstandard logics (e.g., logic programming), where the implication sign dictates the direction of inference and even contraposition is not licensed, metalogical inference machinery must be invoked to perform abduction (Eshghi and Kowalski 1989).

Second, the feature that renders S1–S3 manageable in standard logic is that they all deal with *epistemic* inference – that is, inference from beliefs to beliefs about a static world. Sentence S2, for example, can be explicated to state: If we find that the prisoner is alive, then we have the license to believe that the captain did not give the signal. The material implication sign ($\Longrightarrow$) in logic does not extend beyond this narrow meaning, to be contrasted next with the counterfactual implication.

Evaluating Action Sentences

Sentence S4 invokes a deliberate action, "rifleman A decides to shoot." From our discussion of actions (see, e.g., Chapter 4 or Definition 7.1.3), any such action must violate some premises, or mechanisms, in the initial theory of the story. To formally identify what remains invariant under the action, we must incorporate causal relationships into the theory; logical relationships alone are not sufficient. The causal model corresponding to our story is as follows.

Model M

$$
\begin{array}{ll}
 & (U) \\
C = U & (C) \\
A = C & (A) \\
B = C & (B) \\
D = A \vee B & (D)
\end{array}
$$

Here we use equality rather than implication in order to (i) permit two-way inference and (ii) stress that, unlike logical sentences, each equation represents an autonomous mechanism (an "integrity constraint" in the language of databases) – it remains invariant unless specifically violated. We further use parenthetical symbols next to each equation in order to identify explicitly the dependent variable (on the l.h.s.) in the equation, thus representing the causal asymmetry associated with the arrows in Figure 7.1.

To evaluate S4, we follow Definition 7.1.3 and form the submodel M_A, in which the equation $A = C$ is replaced by A (simulating the decision of rifleman A to shoot regardless of signals).

Model M_A

$$
\begin{array}{ll}
 & (U) \\
C = U & (C) \\
A & (A) \\
B = C & (B) \\
D = A \vee B & (D)
\end{array}
$$

Facts: $\neg C$

Conclusions: $A, D, \neg B, \neg U, \neg C$

We see that, given $\neg C$, we can easily deduce D and $\neg B$ and thus confirm the validity of S4.

It is important to note that "problematic" sentences like S4, whose antecedent violates one of the basic premises in the story (i.e., that both riflemen are law-abiding) are handled naturally in the same deterministic setting in which the story is told. Traditional logicians and probabilists tend to reject sentences like S4 as contradictory and insist on reformulating the problem probabilistically so as to tolerate exceptions to the law $A = C$.[9] Such reformulations are unnecessary; the structural approach permits us to process commonplace causal statements in their natural deterministic habitat without first immersing them in nondeterministic decor. In this framework, all laws are understood to represent "defeasible" default expressions – subject to breakdown by deliberate intervention. The basic laws of physics remain immutable, of course, but their applicability to any given scenario is subject to modification by agents' actions or external intervention.

Evaluating Counterfactuals

We are now ready to evaluate the counterfactual sentence S5. Following Definition 7.1.5, the counterfactual $D_{\neg A}$ stands for the value of D in submodel $M_{\neg A}$. This value is ambiguous because it depends on the value of U, which is not specified in $M_{\neg A}$. The observation D removes this ambiguity; upon finding the prisoner dead we can infer that the court has given the order (U) and, consequently, if rifleman A had refrained from shooting then rifleman B would have shot and killed the prisoner, thus confirming $D_{\neg A}$.

Formally, we can derive $D_{\neg A}$ by using the steps of Theorem 7.1.7 (though no probabilities are involved). We first add the fact D to the original model M and evaluate U; then we form the submodel $M_{\neg A}$ and reevaluate the truth of D in $M_{\neg A}$, using the value of U found in the first step. These steps are explicated as follows.

Step 1

Model M

$$
\begin{array}{ll}
 & (U) \\
C = U & (C) \\
A = C & (A) \\
B = C & (B) \\
D = A \vee B & (D)
\end{array}
$$

Facts: D

Conclusions: U, A, B, C, D

Step 2

Model $M_{\neg A}$

$$
\begin{array}{ll}
 & (U) \\
C = U & (C) \\
\neg A & (A) \\
B = C & (B) \\
D = A \vee B & (D)
\end{array}
$$

Facts: U

Conclusions: $U, \neg A, C, B, D$

---

[9] This problem, I speculate, was one of the primary forces for the emergence of probabilistic causality in the 1960s (see Section 7.5 for review).

Note that it is only the value of U, the background variable, that is carried over from
step 1 to step 2; all other propositions must be reevaluated subject to the new modification
of the model. This reflects the understanding that background factors U are not affected
by either the variables or the mechanisms in the model $\{f_i\}$; hence, the counterfactual consequent (in our case, D) must be evaluated under the same background conditions as those
prevailing in the actual world. In fact, the background variables are the main carriers of
information from the actual world to the hypothetical world; they serve as the "guardians
of invariance" (or persistence) in the dynamic process that transforms the former into the
latter (an observation by David Heckerman, personal communication).

Note also that this two-step procedure for evaluating counterfactuals can be combined into one. If we use an asterisk to distinguish postmodification from premodifica-
tion variables, then we can combine M and M_x into one logical theory and prove the
validity of S5 by purely logical deduction in the combined theory. To illustrate, we
write S5 as $D \Longrightarrow D^*_{\neg A*}$ (read: If D is true in the actual world, then D would also be true
in the hypothetical world created by the modification $\neg A^*$) and prove the validity of D^*
in the combined theory as follows.

Combined Theory

		(U)
$C^* = U$	$C = U$	(C)
$\neg A^*$	$A = C$	(A)
$B^* = C^*$	$B = C$	(B)
$D^* = A^* \vee B^*$	$D = A \vee B$	(D)

Facts: D

Conclusions: $U, A, B, C, D, \neg A^*, C^*, B^*, D^*$

Note that U need not be "starred," reflecting the assumption that background conditions
remain unaltered.

It is worth reflecting at this point on the difference between S4 and S5. The two appear to be syntactically identical, as both involve a fact implying a counterfactual, and yet
we labeled S4 an "action" sentence and S5 a "counterfactual" sentence. The difference
lies in the relationship between the given fact and the antecedent of the counterfactual
(i.e., the "action" part). In S4, the fact given ($\neg C$) is not affected by the antecedent (A);
in S5, the fact given (D) is potentially affected by the antecedent ($\neg A$). The difference
between these two situations is fundamental, as can be seen from their methods of evaluation. In evaluating S4, we knew in advance that C would not be affected by the model
modification $do(A)$; therefore, we were able to add C directly to the modified model M_A.
In evaluating S5, on the other hand, we were contemplating a possible reversal, from D
to $\neg D$, attributable to the modification $do(\neg A)$. As a result, we first had to add fact D to
the preaction model M, summarize its impact via U, and reevaluate D once the modification $do(\neg A)$ takes place. Thus, although the causal effect of actions can be expressed
syntactically as a counterfactual sentence, this need to route the impact of known facts
through U makes counterfactuals a different species than actions (see Section 1.4).

We should also emphasize that most counterfactual utterances in natural language
presume, often implicitly, knowledge of facts that are affected by the antecedent. For

example, when we say that "B would be different were it not for A," we imply knowledge of what the actual value of B is and that B is susceptible to A. It is this sort of relationship that gives counterfactuals their unique character – distinct from action sentences – and, as we saw in Section 1.4, it is this sort of sentence that would require a more detailed specification for its evaluation: some knowledge of the functional mechanisms f_i (pa_i, u_i) would be necessary.

7.1.3 Evaluating Counterfactuals: Probabilistic Analysis

To demonstrate the probabilistic evaluation of counterfactuals (equations (7.3)–(7.5)), let us modify the firing-squad story slightly, assuming that:

1. there is a probability $P(U) = p$ that the court has ordered the execution;

2. rifleman A has a probability q of pulling the trigger out of nervousness; and

3. rifleman A's nervousness is independent of U.

With these assumptions, we wish to compute the quantity $P(\neg D_{\neg A} | D)$ – namely, the probability that the prisoner would be alive if A had not shot, given that the prisoner is in fact dead.

Intuitively, we can figure out the answer by noting that $\neg D_{\neg A}$ is true if and only if the court has not issued an order. Thus, our task amounts to that of computing $P(\neg U | D)$, which evaluates to $q(1 - p)/[1 - (1 - q)(1 - p)]$. However, our aim is to demonstrate a general and formal method of deriving such probabilities, based on (7.4), that makes little use of intuition.

The probabilistic causal model (Definition 7.1.6) associated with the new story contains two background variables, U and W, where W stands for rifleman A's nervousness. This model is given as follows.

Model $\langle M, P(u, w) \rangle$

$$
\begin{array}{ll}
 & (U, W) \sim P(u, w) \\
C = U & (C) \\
A = C \vee W & (A) \\
B = C & (B) \\
D = A \vee B & (D)
\end{array}
$$

In this model, the background variables are distributed as

$$
P(u, w) = \begin{cases}
pq & \text{if } u = 1, w = 1, \\
p(1 - q) & \text{if } u = 1, w = 0, \\
(1 - p)q & \text{if } u = 0, w = 1, \\
(1 - p)(1 - q) & \text{if } u = 0, w = 0.
\end{cases} \tag{7.7}
$$

Following Theorem 7.1.7, our first step (abduction) is to compute the posterior probability $P(u, w | D)$, accounting for the fact that the prisoner is found dead. This is easily evaluated to:

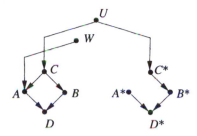

Figure 7.2 Twin network representation of the firing squad.

$$P(u, w \mid D) = \begin{cases} \frac{p(u, w)}{1-(1-p)(1-q)} & \text{if } u = 1 \text{ or } w = 1, \\ 0 & \text{if } u = 0 \text{ and } w = 0. \end{cases} \qquad (7.8)$$

The second step (action) is to form the submodel $M_{\neg A}$ while retaining the posterior probability of (7.8).

Model $\langle M_{\neg A}, P(u, w \mid D) \rangle$

$$
\begin{array}{ll}
 & (U, W){\sim}P(u, w \mid D) \\
C = U & (C) \\
\neg A & (A) \\
B = C & (B) \\
D = A \vee B & (D)
\end{array}
$$

The last step (prediction) is to compute $P(\neg D)$ in this probabilistic model. Noting that $\neg D \Longrightarrow \neg U$, the result (as expected) is

$$P(\neg D_{\neg A} \mid D) = P(\neg U \mid D) = \frac{q(1-p)}{1-(1-q)(1-p)}.$$

7.1.4 The Twin Network Method

A major practical difficulty in the procedure just described is the need to compute, store, and use the posterior distribution $P(u \mid e)$, where u stands for the set of all background variables in the model. As illustrated in the preceding example, even when we start with a Markovian model in which the background variables are mutually independent, conditioning on e normally destroys this independence and so makes it necessary to carry over a full description of the joint distribution of U, conditional on e. Such description may be prohibitively large if encoded in the form of a table, as we have done in (7.8).

A graphical method of overcoming this difficulty is described in Balke and Pearl (1994b); it uses two networks, one to represent the actual world and one to represent the hypothetical world. Figure 7.2 illustrates this construction for the firing-squad story analyzed.

The two networks are identical in structure, save for the arrows entering A^*, which have been deleted to mirror the equation deleted from $M_{\neg A}$. Like Siamese twins, the two networks share the background variables (in our case, U and W), since those remain invariant under modification. The endogenous variables are replicated and labeled distinctly, because they may obtain different values in the hypothetical versus the actual

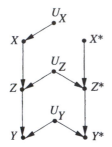

Figure 7.3 Twin network representation of the counterfactual Y_x in the model $X \to Z \to Y$.

world. The task of computing $P(\neg D)$ in the model $\langle M_{\neg A}, P(u, v \mid z)\rangle$ thus reduces to that of computing $P(\neg D^* \mid D)$ in the twin network shown, setting A^* to false.

In general, if we wish to compute the counterfactual probability $P(Y_x = y \mid z)$, where X, Y, and Z are arbitrary sets of variables (not necessarily disjoint), Theorem 7.1.7 instructs us to compute $P(y)$ in the submodel $\langle M_x, P(u \mid z)\rangle$, which reduces to computing an ordinary conditional probability $P(y^* \mid z)$ in an augmented Bayesian network. Such computation can be performed by standard evidence propagation techniques. The advantages of delegating this computation to inference in a Bayesian network are that the distribution $P(u \mid z)$ need not be explicated, conditional independencies can be exploited, and local computation methods can be employed (such as those summarized in Section 1.2.4).

The twin network representation also offers a useful way of testing independencies among counterfactual quantities. To illustrate, suppose that we have a chainlike causal diagram, $X \to Z \to Y$, and that we wish to test whether Y_x is independent of X given Z (i.e., $Y_x \perp\!\!\!\perp X \mid Z$). The twin network associated with this chain is shown in Figure 7.3. To test whether $Y_x \perp\!\!\!\perp X \mid Z$ holds in the original model, we test whether Z d-separates X from Y^* in the twin network. As can be easily seen (via Definition 1.2.3), conditioning on Z renders the path between X and Y^* d-connected through the collider at Z, and hence $Y_x \perp\!\!\!\perp X \mid Z$ does not hold in the model. This conclusion is not easily discernible from the chain model itself or from the equations in that model. In the same fashion, we can see that whenever we condition either on Y or on $\{Y, Z\}$, we form a connection between Y^* and X; hence, Y_x and X are not independent conditional on those variables. The connection is disrupted, however, if we do not condition on either Y or Z, in which case $Y_x \perp\!\!\!\perp X$.

The twin network reveals an interesting interpretation of counterfactuals of the form Z_{pa_Z}, where Z is any variable and PA_Z stands for the set of Z's parents. Consider the question of whether Z_x is independent of some given set of variables in the model of Figure 7.3. The answer to this question depends on whether Z^* is d-separated from that set of variables. However, any variable that is d-separated from Z^* would also be d-separated from U_Z, so the node representing U_Z can serve as a one-way proxy for the counterfactual variable Z_x. This is not a coincidence, considering that Z is governed by the equation $z = f_Z(x, u_Z)$. By definition, the probability of Z_x is equal to the probability of Z under the condition where X is held fixed at x. Under such condition, Z may vary only if U_Z varies. Therefore, if U_Z obeys a certain independence relationship, then Z_x (more generally,

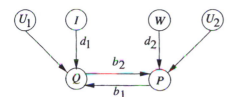

Figure 7.4   Causal diagram illustrating the relationship between price (P) and demand (Q).

Z_{pa_Z}) must obey that relationship as well. We thus obtain a simple graphical representation for any counterfactual variable of the form Z_{pa_Z}, in terms of the so called "error-term" U_z. Using this representation, we can easily verify from Figure 7.3 that $(U_Y \perp\!\!\!\perp X \mid \{Y^*, Z^*\})_G$ and $(U_Y \perp\!\!\!\perp U_Z \mid \{Y, Z\})_G$ both hold in the twin-network and, therefore,

$$Y_z \perp\!\!\!\perp X \mid \{Y_x, Z_x\} \quad \text{and} \quad Y_z \perp\!\!\!\perp Z_x \mid \{Y, Z\}$$

must hold in the model. Additional considerations involving twin networks, including generalizations to multi-networks (representing counterfactuals under different antecedants) are reported in Shpitser and Pearl (2007). See Sections 11.3.2 and 11.7.3.

7.2   APPLICATIONS AND INTERPRETATION OF STRUCTURAL MODELS

7.2.1   Policy Analysis in Linear Econometric Models: An Example

In Section 1.4 we illustrated the nature of structural equation modeling using the canonical economic problem of demand and price equilibrium (see Figure 7.4). In this chapter, we use this problem to answer policy-related questions.

To recall, this example consists of the two equations

$$q = b_1 p + d_1 i + u_1, \tag{7.9}$$

$$p = b_2 q + d_2 w + u_2, \tag{7.10}$$

where q is the quantity of household demand for a product A, p is the unit price of product A, i is household income, w is the wage rate for producing product A, and u_1 and u_2 represent error terms – omitted factors that affect quantity and price, respectively (Goldberger 1992).

This system of equations constitutes a causal model (Definition 7.1.1) if we define $V = \{Q, P\}$ and $U = \{U_1, U_2, I, W\}$ and assume that each equation represents an autonomous process in the sense of Definition 7.1.3. It is normally assumed that I and W are observed, while U_1 and U_2 are unobservable and independent of I and W. Since the error terms U_1 and U_2 are unobserved, a complete specification of the model must include the distribution of these errors, which is usually taken to be Gaussian with the covariance matrix $\Sigma_{ij} = \text{cov}(u_i, u_j)$. It is well known in economics (dating back to Wright 1928) that the assumptions of linearity, normality, and the independence of $\{I, W\}$ and $\{U_1, U_2\}$ permit consistent estimation of all model parameters, including the covariance matrix Σ_{ij}. However, the focus of this book is not the estimation of parameters but rather their utilization in policy predictions. Accordingly, we will demonstrate how to evaluate the following three queries.

1. What is the expected value of the demand Q if the price is *controlled at* $P = p_0$?
2. What is the expected value of the demand Q if the price is *reported to be* $P = p_0$?
3. Given that the current price is $P = p_0$, what would be the expected value of the demand Q if we *were to control* the price at $P = p_1$?

The reader should recognize these queries as representing (respectively) actions, predictions, and counterfactuals – our three-level hierarchy. The second query, representing prediction, is standard in the literature and can be answered directly from the covariance matrix without reference to causality, structure, or invariance. The first and third queries rest on the structural properties of the equations and, as expected, are not treated in the standard literature of structural equations.[10]

In order to answer the first query, we replace (7.10) with $p = p_0$, leaving

$$p = b_1 p + d_1 i + u_1, \tag{7.11}$$

$$p = p_0, \tag{7.12}$$

with the statistics of U_1 and I unaltered. The controlled demand is then $q = b_1 p_0 + d_1 i + u_1$, and its expected value (conditional on $I = i$) is given by

$$E[Q \mid do(P = p_0), i] = b_1 p_0 + d_1 i + E(U_1 \mid i). \tag{7.13}$$

Since U_1 is independent of I, the last term evaluates to

$$E(U_1 \mid i) = E(U_1) = E(Q) - b_1 E(P) - d_1 E(I)$$

and, substituted into (7.13), yields

$$E[Q \mid do(P = p_0), i] = E(Q) + b_1(p_0 - E(P)) + d_i(i - E(I)).$$

The answer to the second query is obtained by conditioning (7.9) on the current observation $\{P = p_0, I = i, W = w\}$ and taking the expectation,

$$E(Q \mid p_0, i, w) = b_1 p_0 + d_1 i + E(U_1 \mid p_0, i, w). \tag{7.14}$$

The computation of $E[U_1 \mid p_0, i, w]$ is a standard procedure once Σ_{ij} is given (Whittaker 1990, p. 163). Note that, although U_1 was assumed to be independent of I and W, this independence no longer holds once $P = p_0$ is observed. Note also that (7.9) and (7.10)

---

[10] I have presented this example to well over a hundred econometrics students and faculty across the United States. Respondents had no problem answering question 2, one person was able to solve question 1, and none managed to answer question 3. Chapter 5 (Section 5.1) suggests an explanation, and Section 11.5.4 a more recent assessment based on Heckman and Vytlacil (2007).

both participate in the solution and that the observed value p_0 will affect the expected demand Q (through $E(U_1 \mid p_0, i, w)$) even when $b_1 = 0$, which is not the case in query 1.

The third query requires the expectation of the counterfactual quantity $Q_{p=p_1}$, conditional on the current observations $\{P = p_0, I = i, W = w\}$ (see Section 11.7.1). According to Definition 7.1.5, $Q_{p=p_1}$ is governed by the submodel

$$q = b_1 p + d_1 i + u_1, \tag{7.15}$$

$$p = p_1; \tag{7.16}$$

the density of u_1 should be conditioned on the observations $\{P = p_0, I = i, W = \omega\}$. We therefore obtain

$$E(Q_{p=p_1} \mid p_0, i, w) = b_1 p_1 + d_1 i + E(U_1 \mid p_0, i, w). \tag{7.17}$$

The expected value $E(U_1 \mid p_0, i, w)$ is the same as in the solution to the second query; the latter differs only in the term $b_1 p_1$. A general matrix method for evaluating counterfactual queries in linear Gaussian models is described in Balke and Pearl (1995a).

At this point, it is worth emphasizing that the problem of computing counterfactual expectations is not an academic exercise; it represents in fact the typical case in almost every decision-making situation. Whenever we undertake to predict the effect of policy, two considerations apply. First, the policy variables (e.g., price and interest rates in economics, pressure and temperature in process control) are rarely exogenous. Policy variables are endogenous when we observe a system under operation; they become exogenous in the planning phase, when we contemplate actions and changes. Second, policies are rarely evaluated in the abstract; rather, they are brought into focus by certain eventualities that demand remedial correction. In troubleshooting, for example, we observe undesirable effects e that are influenced by other conditions $X = x$ and wish to predict whether an action that brings about a change in X would remedy the situation. The information provided by e is extremely valuable, and it must be processed (using abduction) before we can predict the effect of any action. This step of abduction endows practical queries about actions with a counterfactual character, as we have seen in the evaluation of the third query (7.17).

The current price p_0 reflects economic conditions (e.g., Q) that prevail at the time of decision, and these conditions are presumed to be changeable by the policies considered. Thus, the price P represents an endogenous decision variable (as shown in Figure 7.4) that becomes exogenous in deliberation, as dictated by the submodel $M_{p=p_1}$. The hypothetical mood of query 3 translates into a practical problem of policy analysis: "Given that the current price is $P = p_0$, find the expected value of the demand (Q) if we change the price *today* to $P = p_1$." The reasons for using hypothetical phrases in practical decision-making situations are discussed in the next section, as well as 11.7.2.

7.2.2   The Empirical Content of Counterfactuals

The word "counterfactual" is a misnomer, since it connotes a statement that stands contrary to facts or, at the very least, a statement that escapes empirical verification. Counterfactuals are in neither category; they are fundamental to scientific thought and carry as clear an empirical message as any scientific law.

Consider Ohm's law, $V = IR$. The empirical content of this law can be encoded in two alternative forms.

1. *Predictive form*: If at time t_0 we measure current I_0 and voltage V_0 then, ceteris paribus, at any future times $t > t_0$, if the current flow is $I(t)$ then the voltage will be

$$V(t) = \frac{V_0}{I_0} I(t).$$

2. *Counterfactual form*: If at time t_0 we measure current I_0 and voltage V_0 then, had the current flow at time t_0 been I' instead of I_0, the voltage would have been

$$V' = \frac{V_0 I'}{I_0}.$$

On the surface, it seems that the predictive form makes meaningful and testable empirical claims, whereas the counterfactual form merely speculates about events that have not (and could not have) occurred, since it is impossible to apply two different currents into the same resistor at the same time. However, if we interpret the counterfactual form to be neither more nor less than a conversational shorthand of the predictive form, the empirical content of the former shines through clearly. Both enable us to make an infinite number of predictions from just one measurement (I_0, V_0), and both derive their validity from a scientific law that ascribes a time-invariant property (the ratio V/I) to any object that conducts electricity.

But if counterfactual statements are merely a roundabout way of stating sets of predictions, why do we resort to such convoluted modes of expression instead of using the predictive mode directly? One obvious answer is that we often use counterfactuals to convey not the predictions themselves but rather the logical ramifications of those predictions. For example, the intent of saying: "if A were not to have shot, then the prisoner would still be alive" may be merely to convey the factual information that B did not shoot. The counterfactual mood, in this case, serves to supplement the fact conveyed with logical justification based on a general law. The less obvious answer rests with the ceteris paribus (all else held equal) qualification that accompanies the predictive claim, which is not entirely free of ambiguities. What should be held constant when we change the current in a resistor – the temperature? the laboratory equipment? the time of day? Certainly not the reading on the voltmeter!

Such matters must be carefully specified when we pronounce predictive claims and take them seriously. Many of these specifications are implicit (and hence superfluous) when we use counterfactual expressions, especially when we agree on the underlying causal model. For example, we do not need to specify under what temperature and pressure the predictions should hold true; these are implied by the statement "had the current flow at time t_0 been I' instead of I_0." In other words, we are referring to precisely those conditions that prevailed in our laboratory at time t_0. The statement also implies that we do not really mean for anyone to hold the reading on the voltmeter constant; variables should run their natural course, and the only change we should envision is in the mechanism that (according to our causal model) is currently determining the current.

To summarize, a counterfactual statement might well be interpreted as conveying a set of predictions under a well-defined set of conditions – those prevailing in the factual part of the statement. For these predictions to be valid, two components must remain invariant: the laws (or mechanisms) and the boundary conditions. Cast in the language of structural models, the laws correspond to the equations $\{f_i\}$, and the boundary conditions correspond to the state of the background variables U. Thus, a precondition for the validity of the predictive interpretation of a counterfactual statement is the assumption that U will not change when our predictive claim is to be applied or tested.

This is best illustrated by using a betting example. We must bet heads or tails on the outcome of a fair coin toss; we win a dollar if we guess correctly and lose one if we don't. Suppose we bet heads and win a dollar, without glancing at the outcome of the coin. Consider the counterfactual "Had I bet differently I would have lost a dollar." The predictive interpretation of this sentence translates into the implausible claim: "If my next bet is tails, I will lose a dollar." For this claim to be valid, two invariants must be assumed: the payoff policy and the outcome of the coin. Whereas the former is a plausible assumption in a betting context, the latter would be realized only in rare circumstances. It is for this reason that the predictive utility of the statement "Had I bet differently I would have lost a dollar" is rather low, and some would even regard it as hindsighted nonsense. It is the persistence across time of U and $f(x, u)$ that endows counterfactual expressions with predictive power; absent this persistence, the counterfactual loses its obvious predictive utility.

However, there is an element of utility in counterfactuals that does not translate immediately to predictive payoff and thus may serve to explain the ubiquity of counterfactuals in human discourse. I am thinking of explanatory value. Suppose, in the betting story, coins were tossed afresh for every bet. Is there no value whatsoever to the statement "Had I bet differently I would have lost a dollar?" I believe there is; it tells us that we are not dealing here with a whimsical bookie but instead with one who at least glances at the bet, compares it to some standard, and decides a win or a loss using a consistent policy. This information may not be very useful to us as players, but it may be useful to, say, state inspectors who come every so often to calibrate the gambling machines and so ensure the state's take of the profit. More significantly, it may be useful to us players, too, if we venture to cheat slightly – say, by manipulating the trajectory of the coin, or by installing a tiny transmitter to tell us which way the coin landed. For such cheating to work, we should know the payoff policy $y = f(x, u)$, and the statement "Had I bet differently I would have lost a dollar" reveals important aspects of that policy.

Is it far-fetched to argue for the merit of counterfactuals by hypothesizing unlikely situations where players cheat and rules are broken? I suggest that such unlikely operations are precisely the norm for gauging the explanatory value of sentences. It is the nature of any causal explanation that its utility be proven not over standard situations but rather over novel settings that require innovative manipulations of the standards. The utility of understanding how television works comes not from turning the knobs correctly but from the ability to repair a TV set when it breaks down. Recall that every causal model advertises not one but rather a host of submodels, each created by violating some laws. The autonomy of the mechanisms in a causal model thus stands for an open invitation to

The Logic of Structure-Based Counterfactuals

remove or replace those mechanisms, and it is only natural that the explanatory value of sentences be judged by how well they predict the ramifications of such replacements.

Counterfactuals with Intrinsic Nondeterminism

Recapping our discussion, we see that counterfactuals may earn predictive value under two conditions: (1) when the unobserved uncertainty-producing variables (U) remain constant (until our next prediction or action); or (2) when the uncertainty-producing variables offer the potential of being observed sometime in the future (before our next prediction or action). In both cases, we also need to ensure that the outcome-producing mechanism $f(x, u)$ persists unaltered.

These conclusions raise interesting questions regarding the use of counterfactuals in microscopic phenomena, as none of these conditions holds for the type of uncertainty that we encounter in quantum theory. Heisenberg's die is rolled afresh billions of times each second, and our measurement of U will never be fine enough to remove all uncertainty from the response equation $y = f(x, u)$. Thus, when we include quantum-level processes in our analysis we face a dilemma: either dismiss all talk of counterfactuals (a strategy recommended by some researchers, including Dawid 2000) or continue to use counterfactuals but limit their usage to situations where they assume empirical meaning. This amounts to keeping in our analysis only those U that satisfy conditions (1) and (2) of the previous paragraph. Instead of hypothesizing U that completely remove all uncertainties, we admit only those U that are either (1) persistent or (2) potentially observable.

Naturally, coarsening the granularity of the background variables has its price: the mechanism equations $u_i, = f_i(pa_i, u_i)$ lose their deterministic character and hence should be made stochastic. Instead of constructing causal models from a set of deterministic equations $\{f_i\}$, we should consider models made up of stochastic functions $\{f_i^*\}$, where each f_i^* is a mapping from $V \cup U$ to some intrinsic probability distribution $P^*(v_i)$ over the states of V_i. This option leads to a causal Bayesian network (Section 1.3) in which the conditional probabilities $P^*(v_i \mid pa_i, u_i)$ represent intrinsic nondeterminism (sometimes called "objective chance"; Skyrms 1980) and in which the root nodes represent background variables U that are either persistent or potentially observable. In this representation, counterfactual probabilities $P(Y_x = y \mid e)$ can still be evaluated using the three steps (abduction, action, and prediction) of Theorem 7.1.7. In the abduction phase, we condition the prior probability $P(u)$ of the root nodes on the evidence available, e, and so obtain $P(u \mid e)$. In the action phase, we delete the arrows entering variables in set X and instantiate their values to $X = x$. Finally, in the prediction phase, we compute the probability of $Y = y$ resulting from the updated manipulated network.

This evaluation can, of course, be implemented in ordinary causal Bayesian networks (i.e., not only in ones that represent intrinsic nondeterminism), but in that case the results computed would not represent the probability of the counterfactual $Y_x = y$. Such evaluation amounts to assuming that units are homogeneous, with each possessing the stochastic properties of the population – namely, $P(v_i \mid pa_i, u) = P(v_i \mid pa_i)$. Such an assumption may be adequate in quantum-level phenomena, where units stands for specific experimental conditions, but it will not be adequate in macroscopic phenomena, where units may differ appreciably from each other. In the example of Chapter 1 (Section 1.4.4, Figure 1.6), the stochastic attribution amounts to assuming that no individual is affected by the drug (as dictated by model 1) while ignoring the possibility that some individuals may, in fact, be more sensitive to the drug than others (as in model 2).

7.2.3   Causal Explanations, Utterances, and Their Interpretation

It is a commonplace wisdom that explanation improves understanding and that he who understands more can reason and learn more effectively. It is also generally accepted that the notion of explanation cannot be divorced from that of causation; for example, a symptom may explain our *belief* in a disease, but it does not explain the disease itself. However, the precise relationship between causes and explanations is still a topic of much discussion (Cartwright 1989; Woodward 1997). Having a formal theory of causality and counterfactuals in both deterministic and probabilistic settings casts new light on the question of what constitutes an adequate explanation, and it opens new possibilities for automatic generation of explanations by machine (Halpern and Pearl 2005a,b).

A natural starting point for generating explanations would be to use a causal Bayesian network (Section 1.3) in which the events to be explained (explanadum) consist of some combination e of instantiated nodes in the network, and where the task is to find an instantiation c of a subset of e's ancestors (i.e., causes) that maximizes some measure of "explanatory power," namely, the degree to which c explains e. However, the proper choice of this measure is unsettled. Many philosophers and statisticians argue for the likelihood ratio $L = \frac{P(e \mid c)}{P(e \mid c')}$ as the proper measure of the degree to which c is a better explanation of e than c'. In Pearl (1988b, chap. 5) and Peng and Reggia (1986), the best explanation is found by maximizing the posterior probability $P(c \mid e)$. Both measures have their faults and have been criticized by several researchers, including Pearl (1988b), Shimony (1991, 1993), Suermondt and Cooper (1993), and Chajewska and Halpern (1997). To remedy these faults, more intricate combinations of the probabilistic parameters $[P(e \mid c), P(e \mid c'), P(c), P(c')]$ have been suggested, none of which seems to capture well the meaning people attach to the word "explanation."

The problem with probabilistic measures is that they cannot capture the strength of a *causal* connection between c and e; any proposition h whatsoever can, with a small stretch of imagination, be thought of as having some influence on e, however feeble. This would then qualify h as an ancestor of e in the causal network and would permit h to compete and win against genuine explanations by virtue of h having strong spurious association with e.

To rid ourselves of this difficulty, we must go beyond probabilistic measures and concentrate instead on causal parameters, such as causal effects $P(y \mid do(x))$ and counterfactual probabilities $P(Y_{x'} = y' \mid x, y)$, as the basis for defining explanatory power. Here x and x' range over the set of alternative explanations, and Y is the set of response variables observed to take on the value y. The expression $P(Y_{x'} = y' \mid x, y)$ is read as: the probability that Y would take on a different value, y', had X been x' (instead of the actual values x). (Note that $P(y \mid do(x)) \triangleq P(Y_x = y)$.) The developments of computational models for evaluating causal effects and counterfactual probabilities now make it possible to combine these parameters with standard probabilistic parameters and so synthesize a more faithful measure of explanatory power that may guide the selection and generation of adequate explanations.

These possibilities trigger an important basic question: Is "explanation" a concept based on *general* causes (e.g., "Drinking hemlock causes death") or *singular* causes (e.g., "Socrates' drinking hemlock caused his death")? Causal effect expressions $P(y \mid do(x))$ belong to the first category, whereas counterfactual expressions $P(Y_{x'} = y' \mid x, y)$ belong to the second, since conditioning on x and y narrows down world scenarios to those compatible with the most specific information at hand: $X = x$ and $Y = y$.

The classification of causal statements into general and singular categories has been the subject of intensive research in philosophy (see, e.g., Good 1961; Kvart 1986; Cartwright 1989; Eells 1991; see also discussions in Sections 7.5.4 and 10.1.1). This research has attracted little attention in cognitive science and artificial intelligence, partly because it has not entailed practical inferential procedures and partly because it is based on problematic probabilistic semantics (see Section 7.5 for discussion of probabilistic causality). In the context of machine-generated explanations, this classification assumes both cognitive and computational significance. We discussed in Chapter 1 (Section 1.4) the sharp demarcation line between two types of causal queries, those that are answerable from the pair $\langle P(M), G(M) \rangle$ (the probability and diagram, respectively, associated with model M) and those that require additional information in the form of functional specification. Generic causal statements (e.g., $P(y \mid do(x))$) often fall into the first category (as in Chapter 3), whereas counterfactual expressions (e.g., $P(Y_{x'} = y \mid x, y)$) fall into the second, thus demanding more detailed specifications and higher computational resources.

The proper classification of explanation into a general or singular category depends on whether the cause c attains its explanatory power relative to its effect e by virtue of c's general *tendency* to produce e (as compared with the weaker tendencies of c's alternatives) or by virtue of c being *necessary* for triggering a specific chain of events leading to e in the specific situation at hand (as characterized by e and perhaps other facts and observations). Formally, the difference hinges on whether, in evaluating explanatory powers of various hypotheses, we should condition our beliefs on the events c and e that actually occurred.

Formal analysis of these alternatives is given in Chapters 9 and 10, where we discuss the necessary and sufficient aspects of causation as well as the notion of single-event causation. In the balance of this section we will be concerned with the interpretation and generation of explanatory utterances, taking the necessary aspect as a norm.

The following list, taken largely from Galles and Pearl (1997), provides examples of utterances used in explanatory discourse and their associated semantics within the modifiable structural model approach described in Section 7.1.1.

- "X is a cause of Y" if there exist two values x and x' of X and a value u of U such that $Y_x(u) \neq Y_{x'}(u)$.
- "X is a cause of Y in the context $Z = z$" if there exist two values x and x' of X and a value u of U such that $Y_{xz}(u) \neq Y_{x'z}(u)$.
- "X is a direct cause of Y" if there exist two values x and x' of X and a value u of U such that $Y_{xr}(u) \neq Y_{x'r}(u)$, where r is some realization of $V \setminus \{X, Y\}$.
- "X is an indirect cause of Y" if X is a cause of Y and X is not a direct cause of Y.

- "Event $X = x$ always causes $Y = y$" if:
  - (i) $Y_x(u) = y$ for all u; and
  - (ii) there exists a value u' of U such that $Y_x(u') \neq y$ for some $x' \neq x$.
- "Event $X = x$ may have caused $Y = y$" if:
  - (i) $X = x$ and $Y = y$ are true; and
  - (ii) there exists a value u of U such that $X(u) = x$, $Y(u) = y$, and $Y_{x'}(u) \neq y$ for some $x' \neq x$.
- "The unobserved event $X = x$ is a likely cause of $Y = y$" if:
  - (i) $Y = y$ is true; and
  - (ii) $P(Y_x = y, Y_{x'} \neq y \mid Y = y)$ is high for all $x' \neq x$.
- "Event $Y = y$ occurred despite $X = x$" if:
  - (i) $X = x$ and $Y = y$ are true; and
  - (ii) $P(Y_x = y)$ is low.

The preceding list demonstrates the flexibility of modifiable structural models in formalizing nuances of causal expressions. Additional nuances (invoking such notions as enabling, preventing, sustaining, producing, etc.) will be analyzed in Chapters 9 and 10. Related expressions include: "Event A explains the occurrence of event B"; "A would explain B if C were the case"; "B occurred despite A because C was true." The ability to interpret and generate such explanatory sentences, or to select the expression most appropriate for the context, is one of the most intriguing challenges of research in man–machine conversation.

7.2.4   From Mechanisms to Actions to Causation

The structural model semantics described in Section 7.1.1 suggests solutions to two problems in cognitive science and artificial intelligence: the representation of actions and the role of causal ordering. We will discuss these problems in turn, since the second builds on the first.

Action, Mechanisms, and Surgeries

Whether we take the probabilistic paradigm that actions are transformations from probability distributions to probability distributions or the deterministic paradigm that actions are transformations from states to states, such transformations could in principle be infinitely complex. Yet in practice, people teach each other rather quickly the normal results of actions in the world, and people predict the consequences of most actions without much trouble. How?

Structural models answer this question by assuming that the actions we normally invoke in common reasoning can be represented as *local surgeries*. The world consists of a huge number of autonomous and invariant linkages or mechanisms, each corresponding to a physical process that constrains the behavior of a relatively small group of variables. If we understand how the linkages interact with each other (usually, they simply share variables), then we should also be able to understand what the effect of any given action would be: simply respecify those few mechanisms that are perturbed by the action; then let the mechanisms in the modified assembly interact with one another and see what state

between events or variables. We say, for example: "If tile i is tipped to the right, it causes tile $i + 1$ to tip to the right as well"; we do not communicate such knowledge in terms of the tendencies of each domino tile to maintain its physical shape, to respond to gravitational pull, and to obey Newtonian mechanics.

7.2.5 Simon's Causal Ordering

Our ability to talk directly in terms of one event causing another (rather than an action altering a mechanism and the alteration, in turn, producing the effect) is computationally very useful, but at the same time it requires that the assembly of mechanisms in our domain satisfy certain conditions that accommodate causal directionality. Indeed, the formal definition of causal models given in Section 7.1.1 assumes that each equation is designated a distinct privileged variable, situated on its left-hand side, that is considered "dependent" or "output." In general, however, a mechanism may be specified as a functional constraint

$$G_k(x_1, \ldots, x_l; u_1, \ldots, u_m) = 0$$

without identifying any "dependent" variable.

Simon (1953) devised a procedure for deciding whether a collection of such symmetric G functions dictates a unique way of selecting an endogenous dependent variable for each mechanisms (excluding the background variables, since they are determined outside the system). Simon asked: When can we order the variables $(V_1, V_2, \ldots, V_n)$ in such a way that we can solve for each V_i without solving for any of V_i's successors? Such an ordering, if it exists, dictates the direction we attribute to causation. This criterion might at first sound artificial, since the order of solving equations is a matter of computational convenience, whereas causal directionality is an objective attribute of physical reality. (For discussion of this issue see De Kleer and Brown 1986; Iwasaki and Simon 1986; Druzdzel and Simon 1993.) To justify the criterion, let us rephrase Simon's question in terms of actions and mechanisms. Assume that each mechanism (i.e., equation) can be modified independently of the others, and let A_k be the set of actions capable of modifying equation G_k (while leaving other equations unaltered). Imagine that we have chosen an action a_k from A_k and that we have modified G_k in such a way that the set of solutions $(V_1(u), V_2(u), \ldots, V_n(u))$ to the entire system of equations differs from what it was prior to the action. If X is the set of variables directly constrained by G_k, we can ask whether there is one member of X, say X_k, that accounts for the changes in all the other solutions. If the identity of that predictive member remains the same for all choices of a_k and u, then we designate X_k as the *dependent* variable in G_k.

Formally, this property means that changes in a_k induce a *functional mapping* from the domain of X_k to the domain of $\{V \setminus X_k\}$; all changes in the system (generated by a_k) can be attributed to changes in X_k. It would make sense, in such a case, to designate X_k as a "representative" of the mechanism G_k, and we would be justified in replacing the sentence "action a_k caused event $Y = y$" with "event $X_k = x_k$ caused $Y = y$" (where Y is any variable in the system). The invariance of X_k to the choice of a_k is the basis for treating an action as a modality $do(X_k = x_k)$ (Definition 7.1.3). It provides a license for characterizing an action by its immediate consequence(s), independent of the instrument that actually brought about those consequences, and it defines in fact the notion of "local action" or "local surgery."

It can be shown (Nayak 1994) that the uniqueness of X_k can be determined by a simple criterion that involves purely topological properties of the equation set (i.e., how variables are grouped into equations). The criterion is that one should be able to form a one-to-one correspondence between equations and variables and that the correspondence be unique. This can be decided by solving the "matching problem" (Serrano and Gossard 1987) between equations and variables. If the matching is unique, then the choice of dependent variable in each equation is unique and the directionality induced by that choice defines a directed acyclic graph (DAG). In Figure 7.1, for example, the directionality of the arrows need not be specified externally; it can be determined mechanically from the set of symmetrical constraints (i.e., logical propositions)

$$S = \{G_1(C, U), G_2(A, C), G_3(B, C), G_4(A, B, D)\} \tag{7.18}$$

that characterizes the problem. The reader can easily verify that the selection of a privileged variable from each equation is unique and hence that the causal directionality of the arrows shown in Figure 7.1 is inevitable.

Thus, we see that causal directionality, according to Simon, emerges from two assumptions: (1) the partition of variables into background (U) and endogenous (V) sets; and (2) the overall configuration of mechanisms in the model. Accordingly, a variable designated as "dependent" in a given mechanism may well be labeled "independent" when that same mechanism is embedded in a different model. Indeed, the engine causes the wheels to turn when the train goes uphill but changes role in going downhill.

Of course, if we have no way of determining the background variables, then several causal orderings may ensue. In (7.18), for example, if we were not given the information that U is a background variable, then either one of $\{U, A, B, C\}$ could be chosen as background, and each such choice would induce a different ordering on the remaining variables. (Some would conflict with commonsense knowledge, e.g., that the captain's signal influences the court's decision.) However, the directionality of $A \rightarrow D \leftarrow B$ would be maintained in all those orderings. The question of whether there exists a partition $\{U, V\}$ of the variables that would yield a causal ordering in a system of symmetric constraints can also be solved (in polynomial time) by topological means (Dechter and Pearl 1991).

Simon's ordering criterion fails when we are unable to solve the equations one at a time and so must solve a block of k equations simultaneously. In such a case, all the k variables determined by the block would be mutually unordered, though their relationships with other blocks may still be ordered. This occurs, for example, in the economic model of Figure 7.4, where (7.9) and (7.10) need to be solved simultaneously for P and Q and hence the correspondence between equations and variables is not unique; either Q or P could be designated as "independent" in either of the two equations. Indeed, the information needed for classifying (7.9) as the "demand" equation (and, respectively, (7.10) as the "price" equation) comes not from the way variables are assigned to equations but rather from subject-matter considerations. Our understanding that household income directly affects household demand (and not prices) plays a major role in this classification.

In cases where we tend to assert categorically that the flow of causation in a feedback loop goes clockwise, this assertion is normally based on the relative magnitudes of forces. For example, turning the faucet would lower the water level in the water tank, but there is practically nothing we can do to the water in the tank that would turn the faucet. When such information is available, causal directionality is determined by appealing, again, to the notion of hypothetical intervention and asking whether an external control over one variable in the mechanism necessarily affects the others. This consideration then constitutes the operational semantics for identifying the dependent variables V_i in nonrecursive causal models (Definition 7.1.1).

The asymmetry that characterizes causal relationships in no way conflicts with the symmetry of physical equations. By saying that "X causes Y and Y does not cause X," we mean to say that changing a mechanism in which X is normally the dependent variable has a different effect on the world than changing a mechanism in which Y is normally the dependent variable. Because two separate mechanisms are involved, the statement stands in perfect harmony with the symmetry we find in the equations of physics.

Simon's theory of causal ordering has profound repercussions on Hume's problem of causal induction, that is, how causal knowledge is acquired from experience (see Chapter 2). The ability to deduce causal directionality from an assembly of symmetrical mechanisms (together with a selection of a set of endogenous variables) means that the acquisition of causal relationships is no different than the acquisition (e.g., by experiments) of ordinary physical laws, such as Hooke's law of suspended springs or Newton's law of acceleration. This does not imply that acquiring physical laws is a trivial task, free of methodological and philosophical subtleties. It does imply that the problem of causal induction – one of the toughest in the history of philosophy – can be reduced to the more familiar problem of scientific induction.

7.3   AXIOMATIC CHARACTERIZATION

Axioms play important roles in the characterization of formal systems. They provide a parsimonious description of the essential properties of the system, thus allowing comparisons among alternative formulations and easy tests of equivalence or subsumption among such alternatives. Additionally, axioms can often be used as rules of inference for deriving (or verifying) new relationships from a given set of premises. In the next subsection, we will establish a set of axioms that characterize the relationships among counterfactual sentences of the form $Y_x(u) = y$ in both recursive and nonrecursive systems. Using these axioms, we will then demonstrate (in Section 7.3.2) how the identification of causal effects can be verified by symbolic means, paralleling the derivations of Chapter 3 (Section 3.4). Finally, Section 7.3.3 establishes axioms for the notion of *causal relevance*, contrasting those that capture informational relevance.

7.3.1   The Axioms of Structural Counterfactuals

We present three properties of counterfactuals – composition, effectiveness, and reversibility – that hold in all causal models.

Property 1 (Composition)
For any three sets of endogenous variables X, Y, and W in a causal model, we have

$$W_x(u) = w \implies Y_{xw}(u) = Y_x(u). \tag{7.19}$$

Composition states that, if we force a variable (W) to a value w that it would have had without our intervention, then the intervention will have no effect on other variables in the system. That invariance holds in all fixed conditions $do(X = x)$.

Since composition allows for the removal of a subscript (i.e., reducing $Y_{xw}(u)$ to $Y_x(u)$), we need an interpretation for a variable with an empty set of subscripts, which (naturally) we identify with the variable under no interventions.

Definition 7.3.1 (Null Action)

$Y_{\emptyset}(u) \triangleq Y(u)$.

Corollary 7.3.2 (Consistency)
For any set of variables Y and X in a causal model, we have

$$X(u) = x \implies Y(u) = Y_x(u). \tag{7.20}$$

Proof
Substituting X for W and $\emptyset$ for X in (7.19), we obtain $Y_{\emptyset}(u) = x \implies Y_{\emptyset}(u) = Y_k(u)$. Null action (Definition 7.3.1) allows us to drop the $\emptyset$, leaving $X(u) = x \implies Y(u) = Y_x(u)$. $\square$

The implication in (7.20) was called "consistency" by Robins (1987).[13]

Property 2 (Effectiveness)
For all sets of variables X and W, $X_{xw}(u) = x$.

Effectiveness specifies the effect of an intervention on the manipulated variable itself – namely, that if we force a variable X to have the value x, then X will indeed take on the value x.

Property 3 (Reversibility)
For any two variables Y and W and any set of variables X,

$$(Y_{xw}(u) = y) \, \& \, (W_{xy}(u) = w) \implies Y_x(u) = y. \tag{7.21}$$

Reversibility precludes multiple solutions due to feedback loops. If setting W to a value w results in a value y for Y, and if setting Y to the value y results in W achieving the

---

[13] Consistency and composition are used routinely in economics (Manski 1990; Heckman 1996) and statistics (Rosenbaum 1995) within the potential-outcome framework (Section 3.6.3). Consistency was stated formally by Gibbard and Harper (1976, p. 156) and Robins (1987) (see equation (3.52)). Composition is stated in Holland (1986, p. 968) and was brought to my attention by J. Robins.

value w, then W and Y will naturally obtain the values w and y (respectively), without any external setting. In recursive systems, reversibility follows directly from composition. This can easily be seen by noting that, in a recursive system, either $Y_{xw}(u) = Y_x(u)$ or $W_{xy}(u) = W_x(u)$. Thus, reversibility reduces to $(Y_{xw}(u) = y)$ & $(W_x(u) = w) \Longrightarrow Y_x(u) = y$ (another form of composition) or to $(Y_x(u) = y)$ & $(W_{xy}(u) = w) \Longrightarrow Y_x(u) = y$ (which is trivially true).

Reversibility reflects "memoryless" behavior: the state of the system, V, tracks the state of U regardless of U's history. A typical example of irreversibility is a system of two agents who adhere to a "tit-for-tat" strategy (e.g., the prisoners' dilemma). Such a system has two stable solutions – cooperation and defection – under the same external conditions U, and thus it does not satisfy the reversibility condition; forcing either one of the agents to cooperate results in the other agent's cooperation ($Y_w(u) = y$, $W_y(u) = w$), yet this does not guarantee cooperation from the start ($Y(u) = y$, $W(u) = w$). In such systems, irreversibility is a product of using a state description that is too coarse, one where not all of the factors that determine the ultimate state of the system are included in U. In a tit-for-tat system, a complete state description should include factors such as the previous actions of the players, and reversibility is restored once the missing factors are included.

In general, the properties of composition, effectiveness, and reversibility are independent – none is a consequence of the other two. This can be shown (Galles and Pearl 1997) by constructing specific models in which two of the properties hold and the third does not. In recursive systems, composition and effectiveness are independent while reversibility holds trivially, as just shown.

The next theorem asserts the *soundness*[14] of properties 1–3, that is, their validity.

Theorem 7.3.3 (Soundness)
Composition, effectiveness, and reversibility are sound in structural model semantics; that is, they hold in all causal models.

A proof of Theorem 7.3.3 is given in Galles and Pearl (1997).

Our next theorem establishes the *completeness* of the three properties when treated as axioms or rules of inference. Completeness amounts to sufficiency; *all* other properties of counterfactual statements follow from these three. Another interpretation of completeness is as follows: Given any set S of counterfactual statements that is consistent with properties 1–3, there exists a causal model M in which S holds true.

A formal proof of completeness requires the explication of two technical properties – existence and uniqueness – that are implicit in the definition of causal models (Definition 7.1.1).

Property 4 (Existence)
For any variable X and set of variables Y,

$$\exists x \in X \quad \text{s.t.} \quad X_y(u) = x. \tag{7.22}$$

---

[14] The terms *soundness* and *completeness* are sometimes referred to as *necessity* and *sufficiency*, respectively.

Property 5 (Uniqueness)
For every variable X and set of variables Y,

$$X_y(u) = x \ \ \& \ \ X_y(u) = x' \Longrightarrow x = x'. \tag{7.23}$$

Definition 7.3.4 (Recursiveness)
Let X and Y be singleton variables in a model, and let $X \rightarrow Y$ stand for the inequality $Y_{xw}(u) \neq Y_w(u)$ for some values of x, w, and u. A model M is recursive if, for any sequence $X_1, X_2, \ldots, X_k$, we have

$$X_1 \rightarrow X_2, X_2 \rightarrow X_3, \ldots, X_{k-1} \rightarrow X_k \Longrightarrow X_k \nrightarrow X_1. \tag{7.24}$$

Clearly, any model M for which the causal diagram $G(M)$ is acyclic must be recursive.

Theorem 7.3.5 (Recursive Completeness)
Composition, effectiveness, and recursiveness are complete (Galles and Pearl 1998; Halpern 1998).[15]

Theorem 7.3.6 (Completeness)
Composition, effectiveness, and reversibility are complete for all causal models (Halpern 1998).

The practical importance of soundness and completeness surfaces when we attempt to test whether a certain set of conditions is sufficient for the identifiability of some counterfactual quantity Q. Soundness, in this context, guarantees that if we symbolically manipulate Q using the three axioms and manage to reduce it to an expression that involves ordinary probabilities (free of counterfactual terms), then Q is identifiable (in the sense of Definition 3.2.3). Completeness guarantees the converse: if we do not succeed in reducing Q to a probabilistic expression, then Q is nonidentifiable – our three axioms are as powerful as can be.

The next section demonstrates a proof of identifiability that uses effectiveness and decomposition as inference rules.

7.3.2   Causal Effects from Counterfactual Logic: An Example

We revisit the smoking–cancer example analyzed in Section 3.4.3. The model associated with this example is assumed to have the following structure (see Figure 7.5):

$$V = \{X \text{ (smoking)}, Y \text{ (lung cancer)}, Z \text{ (tar in lungs)}\},$$

$$U = \{U_1, U_2\}, U_1 \perp\!\!\!\perp U_2,$$

---

[15] Galles and Pearl (1997) proved recursive completeness assuming that, for any two variables, one knows which of the two (if any) is an ancestor of the other. Halpern (1998) proved recursive completeness without this assumption, provided only that (7.24) is known to hold for any two variables in the model. Halpern further provided a set of axioms for cases where the solution of $Y_x(u)$ is not unique or does not exist.

Task 3

Compute $P(Y_x = y)$ (i.e., the causal effect of smoking on cancer).

For any variable Z, by composition we have

$$Y_x(u) = Y_{xz}(u) \quad \text{if} \quad Z_x(u) = z.$$

Since $Y_{xz}(u) = Y_z(u)$ (from (7.29)),

$$Y_X(u) = Y_{xz_x}(u) = Y_z(u), \quad \text{where } z_x = Z_x(u). \tag{7.35}$$

Thus,

$$
\begin{aligned}
P(Y_x = y) &= P(Y_{zx} = y) & \text{from (7.35)} \\
&= \sum_z P(Y_{zx} = y \mid Z_x = z)\, P(Z_x = z) & \\
&= \sum_z P(Y_z = y \mid Z_x = z)\, P(Z_x = z) & \text{by composition} \\
&= \sum_z P(Y_z = y)\, P(Z_x = z). & \text{from (7.30)} \quad (7.36)
\end{aligned}
$$

The probabilities $P(Y_z = y)$ and $P(Z_x = z)$ were computed in (7.34) and (7.31), respectively. Substituting gives us

$$P(Y_x = y) = \sum_z P(z \mid x) \sum_{x'} P(y \mid z, x')\ P(x'). \tag{7.37}$$

The right-hand side of (7.37) can be computed from $P(x, y, z)$ and coincides with the front-door formula derived in Section 3.4.3 (equation (3.42)).

Thus, $P(Y_x = y)$ can be reduced to expressions involving probabilities of observed variables and is therefore identifiable. More generally, our completeness result (Theorem 7.3.5) implies that *any* identifiable counterfactual quantity can be reduced to the correct expression by repeated application of composition and effectiveness (assuming recursiveness).

7.3.3 Axioms of Causal Relevance

In Section 1.2 we presented a set of axioms for a class of relations called *graphoids* (Pearl and Paz 1987; Geiger et al. 1990) that characterize informational relevance.[16] We now develop a parallel set of axioms for *causal relevance*, that is, the tendency of certain events to affect the occurrence of other events in the physical world, independent of the observer–reasoner. Informational relevance is concerned with questions of the form: "Given that we know Z, would gaining information about X give us new information

---

[16] "Relevance" will be used primarily as a generic name for the relationship of being relevant or irrelevant. It will be clear from the context when "relevance" is intended to negate "irrelevance."

about Y?" Causal relevance is concerned with questions of the form: "Given that Z is fixed, would changing X alter Y?" We show that causal relevance complies with all the axioms of path interception in directed graphs except transitivity.

The notion of causal relevance has its roots in the philosophical works of Suppes (1970) and Salmon (1984), who attempted to give probabilistic interpretations to cause–effect relationships and recognized the need to distinguish causal from statistical relevance (see Section 7.5). Although these attempts did not produce a probabilistic definition of causal relevance, they led to methods for testing the consistency of relevance statements against a given probability distribution and a given temporal ordering among the variables (see Section 7.5.2). Here we aim at axiomatizing relevance statements in themselves – with no reference to underlying probabilities or temporal orderings.

The axiomization of causal relevance may be useful to experimental researchers in domains where exact causal models do not exist. If we know, through experimentation, that some variables have no causal influence on others in a system, then we may wish to determine whether other variables will exert causal influence (perhaps under different experimental conditions) or may ask what additional experiments could provide such information. For example, suppose we find that a rat's diet has no effect on tumor growth while the amount of exercise is kept constant and, conversely, that exercise has no effect on tumor growth while diet is kept constant. We would like to be able to infer that controlling only diet (while paying no attention to exercise) would still have no influence on tumor growth. A more subtle inference problem is deciding whether changing the ambient temperature in the cage would have an effect on the rat's physical activity, given that we have established that temperature has no effect on activity when diet is kept constant and that temperature has no effect on (the rat's choice of) diet when activity is kept constant.

Galles and Pearl (1997) analyzed both probabilistic and deterministic interpretations of causal irrelevance. The probabilistic interpretation, which equates causal irrelevance with inability to change the probability of the effect variable, has intuitive appeal but is inferentially very weak; it does not support a very expressive set of axioms unless further assumptions are made about the underlying causal model. If we add the stability assumption (i.e., that no irrelevance can be destroyed by changing the nature of the individual processes in the system), then we obtain the same set of axioms for probabilistic causal irrelevance as the set governing path interception in directed graphs.

In this section we analyze a deterministic interpretation that equates causal irrelevance with inability to change the effect variable in any state u of the world. This interpretation is governed by a rich set of axioms without our making any assumptions about the causal model: many of the path interception properties in directed graphs hold for deterministic causal irrelevance.

Definition 7.3.7 (Causal Irrelevance)

A variable X is causally irrelevant to Y, given Z (written $X \not\rightarrow Y \mid Z$) if, for every set W disjoint of $X \cup Y \cup Z$, we have

$$\forall\, (u, z, x, x', w), \quad Y_{xzw}(u) = Y_{x'zw}(u), \tag{7.38}$$

where x and x' are two distinct values of X.

$$V = \{X, W, Y\} \ \text{binary}$$
$$U = \{U_1, U_2\} \ \text{binary}$$

$$x = u_1$$
$$y = \begin{cases} u_2 & \text{if } x = w \\ x & \text{otherwise} \end{cases}$$
$$w = x$$

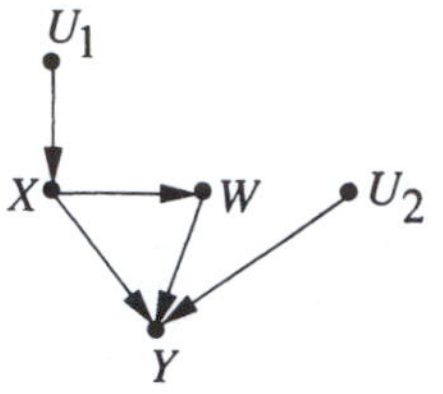

Figure 7.6 Example of a causal model that requires the examination of submodels to determine causal relevance.

This definition captures the intuition "If X is causally irrelevant to Y, then X cannot affect Y under any circumstance u or under any modification of the model that includes $do(Z = z)$."

To see why we require the equality $Y_{xzw}(u) = Y_{x'zw}(u)$ to hold in every context $W = w$, consider the causal model of Figure 7.6. In this example, $Z = \emptyset$, W follows X, and hence Y follows X; that is, $Y_{x=0}(u) = Y_{x=1}(u) = u_2$. However, since $y(x, w, u_2)$ is a nontrivial function of x, X is perceived to be causally relevant to Y. Only holding W constant would reveal the causal influence of X on Y. To capture this intuition, we must consider all contexts $W = w$ in Definition 7.3.7.

With this definition of causal irrelevance, we have the following theorem.

Theorem 7.3.8

For any causal model, the following sentences must hold.

Weak Right Decomposition:[17]

$$(X \nrightarrow YW \mid Z) \ \& \ (X \nrightarrow Y \mid ZW) \implies (X \nrightarrow Y \mid Z).$$

Left Decomposition:

$$(XW \nrightarrow Y \mid Z) \implies (X \nrightarrow Y \mid Z) \ \& \ (W \nrightarrow Y \mid Z).$$

Strong Union:

$$(X \nrightarrow Y \mid Z) \implies (X \nrightarrow Y \mid ZW) \ \forall W.$$

Right Intersection:

$$(X \nrightarrow Y \mid ZW) \ \& \ (X \nrightarrow W \mid ZY) \implies (X \nrightarrow YW \mid Z).$$

Left Intersection:

$$(X \nrightarrow Y \mid ZW) \ \& \ (W \nrightarrow Y \mid ZX) \implies (XW \nrightarrow Y \mid Z).$$

This set of axioms bears a striking resemblance to the properties of path interception in a directed graph. Paz and Pearl (1994) showed that the axioms of Theorem 7.3.8, together with transitivity and right decomposition, constitute a complete characterization of the

---

[17] Galles and Pearl (1997) used a stronger version of right decomposition: $(X \nrightarrow YW \mid Z) \implies (X \nrightarrow Y \mid Z)$. But Bonet (2001) showed that it must be weakened to render the axiom system sound.

relation $(X \nrightarrow Y \mid Z)_G$ when interpreted to mean that every directed path from X to Y in a directed graph G contains at least one node in Z (see also Paz et al. 1996).

Galles and Pearl (1997) showed that, despite the absence of transitivity, Theorem 7.3.8 permits one to infer certain properties of causal irrelevance from properties of directed graphs. For example, suppose we wish to validate a generic statement such as: "If X has an effect on Y, but ceases to have an effect when we fix Z, then Z must have an effect on Y." That statement can be proven from the fact that, in any directed graph, if all paths from X to Y are intercepted by Z and there are no paths from Z to Y, then there is no path from X to Y.

Remark on the Transitivity of Causal Dependence

That causal dependence is not transitive is clear from Figure 7.6. In any state of (U_1, U_2), X is capable of changing the state of W and W is capable of changing Y, yet X is incapable of changing Y. Galles and Pearl (1997) gave examples where causal relevance in the weak sense of Definition 7.3.7 is also nontransitive, even for binary variables. The question naturally arises as to why transitivity is so often conceived of as an inherent property of causal dependence or, more formally, what assumptions we tacitly make when we classify causal dependence as transitive.

One plausible answer is that we normally interpret transitivity to mean the following: "If (1) X causes Y and (2) Y causes Z regardless of X, then (3) X causes Z." The suggestion is that questions about transitivity bring to mind chainlike processes, where X influences Y and Y influences Z but where X does not have a *direct* influence over Z. With this qualification, transitivity for binary variables can be proven immediately from composition (equation (7.19)) as follows.

Let the sentence "$X = x$ causes $Y = y$," denoted $x \to y$, be interpreted as the joint condition $\{X(u) = x, Y(u) = y, Y_{x'}(u) = y' \neq y\}$ (in words, x and y hold, but changing x to x' would change y to y'). We can now prove that if X has no direct effect on Z, that is, if

$$Z_{y'x'} = Z_{y'}, \tag{7.39}$$

then

$$x \to y \ \& \ y \to z \ \implies \ x \to z. \tag{7.40}$$

Proof

The l.h.s. of (7.40) reads

$$X(u) = x, \quad Y(u) = y, \quad Z(u) = z, \quad Y_{x'}(u) = y', \quad Z_{y'}(u) = z'.$$

From (7.39) we can rewrite the last term as $Z_{y'x'}(u) = z'$. Composition further permits us to write

$$Y_{x'}(u) = y' \ \& \ Z_{y'x'}(u) = z' \ \implies \ Z_{x'}(u) = z',$$

which, together with $X(u) = x$ and $Z(u) = z$, implies $x \to z$. □

Weaker forms of causal transitivity are discussed in Chapter 9 (Lemmas 9.2.7 and 9.2.8).

7.4 STRUCTURAL AND SIMILARITY-BASED COUNTERFACTUALS

7.4.1 Relations to Lewis's Counterfactuals

Causality from Counterfactuals

In one of his most quoted sentences, David Hume tied together two aspects of causation, regularity of succession and counterfactual dependency:

> we may define a cause to be an object followed by another, and where all the objects, similar to the first, are followed by objects similar to the second, Or, in other words, where, if the first object had not been, the second never had existed. (Hume 1748/1958, sec. VII).

This two-faceted definition is puzzling on several accounts. First, regularity of succession, or "correlation" in modern terminology, is not sufficient for causation, as even nonstatisticians know by now. Second, the expression "in other words" is too strong, considering that regularity rests on observations, whereas counterfactuals rest on mental exercise. Third, Hume had introduced the regularity criterion nine years earlier,[18] and one wonders what jolted him into supplementing it with a counterfactual companion. Evidently, Hume was not completely happy with the regularity account, and must have felt that the counterfactual criterion is less problematic and more illuminating. But how can convoluted expressions of the type "if the first object had not been, the second never had existed" illuminate simple commonplace expressions like "*A* caused *B*"?

The idea of basing causality on counterfactuals is further echoed by John Stuart Mill (1843), and it reached fruition in the works of David Lewis (1973b, 1986). Lewis called for abandoning the regularity account altogether and for interpreting "*A* has caused *B*" as "*B* would not have occurred if it were not for *A*." Lewis (1986, p. 161) asked: "Why not take counterfactuals at face value: as statements about possible alternatives to the actual situation . . . ?"

Implicit in this proposal lies a claim that counterfactual expressions are less ambiguous to our mind than causal expressions. Why else would the expression "*B* would be false if it were not for *A*" be considered an *explication* of "*A* caused *B*," and not the other way around, unless we could discern the truth of the former with greater certitude than that of the latter? Taken literally, discerning the truth of counterfactuals requires generating and examining possible alternatives to the actual situation as well as testing whether certain propositions hold in those alternatives – a mental task of nonnegligible proportions. Nonetheless, Hume, Mill, and Lewis apparently believed that going through this mental exercise is simpler than intuiting directly on whether it was *A* that caused *B*. How can this be done? What mental representation allows humans to process counterfactuals so swiftly and reliably, and what logic governs that process so as to maintain uniform standards of coherence and plausibility?

---

[18] In *Treatise of Human Nature*, Hume wrote: "We remember to have had frequent instances of the existence of one species of objects; and also remember, that the individuals of another species of objects have always attended them, and have existed in a regular order of contiguity and succession with regard to them" (Hume 1739, p. 156).

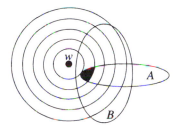

Figure 7.7 Graphical representation of Lewis's closest-world semantics. Each circular region corresponds to a set of worlds that are equally similar to w. The shaded region represents the set of closest A-worlds; since all these worlds satisfy B, the counterfactual sentence $A \,\square\!\!\rightarrow B$ is declared true in w.

Structure versus Similarity

According to Lewis's account (1973b), the evaluation of counterfactuals involves the notion of *similarity*: one orders possible worlds by some measure of similarity, and the counterfactual $A \,\square\!\!\rightarrow B$ (read: "B if it were A") is declared true in a world w just in case B is true in all the closest A-worlds to w (see Figure 7.7).[19]

This semantics still leaves questions of representation unsettled. What choice of similarity measure would make counterfactual reasoning compatible with ordinary conceptions of cause and effect? What mental representation of worlds ordering would render the computation of counterfactuals manageable and practical (in both man and machine)?

In his initial proposal, Lewis was careful to keep the formalism as general as possible; save for the requirement that every world be closest to itself, he did not impose any structure on the similarity measure. However, simple observations tell us that similarity measures cannot be arbitrary. The very fact that people communicate with counterfactuals already suggests that they share a similarity measure, that this measure is encoded parsimoniously in the mind, and hence that it must be highly structured. Kit Fine (1975) further demonstrated that similarity of appearance is inadequate. Fine considers the counterfactual "Had Nixon pressed the button, a nuclear war would have started," which is generally accepted as true. Clearly, a world in which the button happened to be disconnected is many times more similar to our world, as we know it, than the one yielding a nuclear blast. Thus we see not only that similarity measures could not be arbitrary but also that they must respect our conception of causal laws.[20] Lewis (1979) subsequently set up an intricate system of weights and priorities among various aspects of similarity – size of "miracles" (violations of laws), matching of facts, temporal precedence, and so forth – in attempting to bring similarity closer to causal intuition. But these priorities are rather post hoc and still yield counterintuitive inferences (J. Woodward, personal communication).

Such difficulties do not enter the structural account. In contrast with Lewis's theory, counterfactuals are not based on an abstract notion of similarity among hypothetical worlds; instead, they rest directly on the mechanisms (or "laws," to be fancy) that produce those worlds and on the invariant properties of those mechanisms. Lewis's elusive "miracles" are replaced by principled minisurgeries, $do(X = x)$, which represent the minimal change (to a model) necessary for establishing the antecedent $X = x$ (for all u).

---

[19] Related possible-world semantics were introduced in artificial intelligence to represent actions and database updates (Ginsberg 1986; Ginsberg and Smith 1987; Winslett 1988; Katsuno and Mendelzon 1991).

[20] In this respect, Lewis's reduction of causes to counterfactuals is somewhat circular.

In sum, for recursive models, the causal model framework does not add any restrictions to counterfactual statements beyond those imposed by Lewis's framework; the very general concept of closest worlds is sufficient. Put another way, the assumption of recursiveness is so strong that it already embodies all other restrictions imposed by structural semantics. When we consider nonrecursive systems, however, we see that reversibility is not enforced by Lewis's framework. Lewis's axiom (3) is similar to but not as strong as reversibility; that is, even though $Y = y$ may hold in all closest w-worlds and $W = w$ in all closest y-worlds, $Y = y$ still may not hold in the actual world. Nonetheless, we can safely conclude that, in adopting the causal interpretation of counterfactuals (together with the representational and algorithmic machinery of modifiable structural equation models), we are not introducing any restrictions on the set of counterfactual statements that are valid relative to recursive systems.

7.4.3 Imaging versus Conditioning

If action is a transformation from one probability function to another, one may ask whether every such transformation corresponds to an action, or if there are some constraints that are peculiar to those transformations that originate from actions. Lewis's (1976) formulation of counterfactuals indeed identifies such constraints: the transformation must be an *imaging* operator.

Whereas Bayes conditioning $P(s \mid e)$ transfers the entire probability mass from states excluded by e to the remaining states (in proportion to their current $P(s)$), imaging works differently; each excluded state s transfers its mass individually to a select set of states $S^*(s)$ that are considered "closest" to s. Indeed, we saw in (3.11) that the transformation defined by the action $do(X_i = x'_i)$ can be interpreted in terms of such a mass-transfer process; each excluded state (i.e., one in which $X_i \neq x'_i$) transferred its mass to a select set of nonexcluded states that shared the same value of pa_i. This simple characterization of the set $S^*(s)$ of closest states is valid for Markovian models, but imaging generally permits the selection of any such set.

The reason why imaging is a more adequate representation of transformations associated with actions can be seen through a representation theorem due to Gardenfors (1988, thm. 5.2, p. 113; strangely, the connection to actions never appears in Gardenfors's analysis). Gardenfors's theorem states that a probability update operator $P(s) \rightarrow P_A(s)$ is an imaging operator if and only if it preserves mixtures; that is,

$$[\alpha P(s) + (1 - \alpha) P'(s)]_A = \alpha P_A(s) + (1 - \alpha)P'_A(s) \tag{7.43}$$

for all constants $1 > \alpha > 0$, all propositions A, and all probability functions P and P'. In other words, the update of any mixture is the mixture of the updates.[22]

This property, called *homomorphy*, is what permits us to specify actions in terms of transition probabilities, as is usually done in stochastic control and Markov decision processes. Denoting by $P_A(s \mid s')$ the probability resulting from acting A on a known state s', the homomorphism (7.43) dictates that

---

[22] Property (7.43) is reflected in the (U8) postulate of Katsuno and Mendelzon (1991): $(K_1 \vee K_2)o\mu = (K_{1o\mu}) \vee (K_{2o\mu})$, where o is an update operator, similar to our $do(\cdot)$ operator.

$$P_A(s) = \sum_{s'} P_A(s \mid s') \ P(s');$$ (7.44)

this means that, whenever s' is not known with certainty, $P_A(s)$ is given by a weighted sum of $P_A(s \mid s')$ over s', with the weight being the current probability function $P(s')$.

This characterization, however, is too permissive; although it requires any action-based transformation to be describable in terms of transition probabilities, it also accepts any transition probability specification, howsoever whimsical, as a descriptor of some action. The valuable information that actions are defined as *local* surgeries is ignored in this characterization. For example, the transition probability associated with the atomic action $A_i = do(X_i = x_i)$ originates from the deletion of just one mechanism in the assembly. Hence, the transition probabilities associated with the set of atomic actions would normally constrain one another. Such constraints emerge from the axioms of effectiveness, composition, and reversibility when probabilities are assigned to the states of U (Galles and Pearl 1997).

7.4.4 Relations to the Neyman–Rubin Framework

A Language in Search of a Model

The notation $Y_x(u)$ that we used for denoting counterfactual quantities is borrowed from the potential-outcome framework of Neyman (1923) and Rubin (1974), briefly introduced in Section 3.6.3, which was devised for statistical analysis of treatment effects.[23] In that framework, $Y_x(u)$ (often written $Y(x, u)$) stands for the outcome of experimental unit u (e.g., an individual or an agricultural lot) under a hypothetical experimental condition $X = x$. In contrast to the structural modeling, however, this variable is not derived from a causal model or from any formal representation of scientific knowledge, but is taken as a primitive – that is, an unobserved variable that reveals its value only when x coincides with the treatment actually received, as dictated by the consistency rule $X = x \Longrightarrow Y_x = Y$ (equation (7.20)). Consequently, the potential-outcome framework does not provide a mathematical model from which such rules could be *derived* or on the basis of which an axiomatic characterization could be attempted in order to decide, for example, whether additional rules should be deployed, or whether a given collection of potential-outcome expressions is redundant or contradictory.

The structural equation model formulated in Section 7.1 provides in fact the formal semantics lacking in the potential-outcome framework, since each such model assigns coherent truth values to the counterfactual quantities used in potential-outcome studies. From the structural perspective, the quantity $Y_x(u)$ is not a primitive but rather is derived mathematically from a set of equations F that represents, transparently, one's knowledge about the subject matter. This knowledge is expressed qualitatively through the variables participating in those equations, without committing to their precise functional form. The variable U represents any set of background factors relevant to the analysis, not necessarily the identity of a specific individual in the population.

---

[23] A related (if not identical) framework that has been used in economics is the *switching regression*. For a brief review of such models, see Heckman (1996; see also Heckman and Honoré 1990 and Manski 1995). Winship and Morgan (1999) provided an excellent overview of the two schools.

Using this semantics, in Section 7.3 we established an axiomatic characterization of the potential-response function $Y_x(u)$ and its relationships with the observed variables $X(u)$ and $Y(u)$. These basic axioms include or imply restrictions such as the consistency rule (equation (7.20)), which were taken as given by potential-outcome researchers.

The completeness result further assures us that derivations involving counterfactual relationships in recursive models may safely be managed with two axioms only, effectiveness and composition. All truths implied by structural equation semantics are also derivable using these two axioms. Likewise – in constructing hypothetical contingency tables for recursive models (see Section 6.5.3) – we are guaranteed that, once a table satisfies effectiveness and composition, there exists at least one causal model that would generate that table. In essence, this establishes the formal equivalence of structural equation modeling, which is popular in economics and the social sciences (Goldberger 1991), and the potential-outcome framework as used in statistics (Rubin 1974; Holland 1986; Robins 1986).[24] In nonrecursive models, however, this is not the case. Attempts to evaluate counterfactual statements using only composition and effectiveness may fail to certify some valid conclusions (i.e., true in all causal models) whose validity can only be recognized through the use of reversibility.

Graphical versus Counterfactual Analysis

This formal equivalence between the structural and potential-outcome frameworks covers issues of semantics and expressiveness but does not imply equivalence in conceptualization or practical usefulness. Structural equations and their associated graphs are particularly useful as means of expressing assumptions about cause-effect relationships. Such assumptions rest on prior experiential knowledge, which – as suggested by ample evidence – is encoded in the human mind in terms of interconnected assemblies of autonomous mechanisms. These mechanisms are thus the building blocks from which judgments about counterfactuals are derived. Structural equations $\{f_i\}$ and their graphical abstraction $G(M)$ provide direct mappings for these mechanisms and therefore constitute a natural language for articulating or verifying causal knowledge or assumptions. The major weakness of the potential-outcome framework lies in the requirement that assumptions be articulated as conditional independence relationships involving counterfactual variables. For example, an assumption such as the one expressed in (7.30) is not easily comprehended even by skilled investigators, yet its structural image $U_1 \perp\!\!\!\perp U_2$ evokes an immediate process-based interpretation.[25]

---

[24] This equivalence was anticipated in Holland (1988), Pratt and Schlaifer (1988), Pearl (1995a), and Robins (1995) and became a mathematical fact through the explicit translation of equation (3.51) followed by the completeness result of Theorem 7.3.6.

[25] These views are diametrically opposite to those expressed by Angrist et al. (1996), who stated: "Typically the researcher does not have a firm idea what these disturbances really represent." Researchers who are knowledgeable in their respective subjects have a very clear idea what these disturbances represent, and those who don't would certainly not be able to make realistic judgments about counterfactual dependencies. Indeed, researchers who shun structural equations or graphs tend to avoid subject matter knowledge in their analyses (e.g., Rubin 2009).

A happy symbiosis between graphs and counterfactual notation was demonstrated in Section 7.3.2. In that example, assumptions were expressed in graphical form, then translated into counterfactual notation (using the rules of (7.25) and (7.26)), and finally submitted to algebraic derivation. Such symbiosis offers a more effective method of analysis than methods that insist on expressing assumptions directly as counterfactuals. Additional examples will be demonstrated in Chapter 9, where we analyze probability of causation. Note that, in the derivation of Section 7.3.2, the graph continued to assist the procedure by displaying independence relationships that are not easily derived by algebraic means alone. For example, it is hardly straightforward to show that the assumptions of (7.27)–(7.30) imply the conditional independence $(Y_z \perp\!\!\!\perp Z_x \,|\, \{Z, X\})$ but do not imply the conditional independence $(Y_z \perp\!\!\!\perp Z_x \,|\, Z)$. Such implications can, however, easily be tested in the graph of Figure 7.5 or in the twin network construction of Section 7.1.3 (see Figure 7.3).

The most compelling reason for molding causal assumptions in the language of graphs is that such assumptions are needed before the data are gathered, at a stage when the model's parameters are still "free" (i.e., still to be determined from the data). The usual temptation is to mold those assumptions in the language of statistical independence, which carries an aura of testability and hence of scientific legitimacy. (Chapter 6 exemplifies the futility of such temptations.) However, conditions of statistical independence – regardless of whether they relate to V variables, U variables, or counterfactuals – are generally sensitive to the values of the model's parameters, which are not available at the model construction phase. The substantive knowledge available at the modeling phase cannot support such assumptions unless they are *stable*, that is, insensitive to the values of the parameters involved. The implications of graphical models, which rest solely on the interconnections among mechanisms, satisfy this stability requirement and can therefore be ascertained from generic substantive knowledge *before* data are collected. For example, the assertion $(X \perp\!\!\!\perp Y \,|\, Z, U_1)$, which is implied by the graph of Figure 7.5, remains valid for any substitution of functions in $\{f_i\}$ and for any assignment of prior probabilities to U_1 and U_2.

These considerations apply not only to the formulation of causal assumptions but also to the language in which causal concepts are defined and communicated. Many concepts in the social and medical sciences are defined in terms of relationships among unobserved U variables, also known as "errors" or "disturbance terms." We have seen in Chapter 5 (Section 5.4.3) that key econometric notions such as exogeneity and instrumental variables have traditionally been defined in terms of absence of correlation between certain observed variables and certain error terms. Naturally, such definitions attract criticism from strict empiricists, who regard unobservables as metaphysical or definitional (Richard 1980; Engle et al. 1983; Holland 1988), and also (more recently) from potential-outcome analysts, who mistakenly regard the use of structural models as an unwarranted commitment to a particular functional form (Angrist et al. 1996). This criticism will be considered in the following section.

7.4.5   Exogeneity and Instruments: Counterfactual and Graphical Definitions

The analysis of this chapter provides a counterfactual interpretation of the error terms in structural equation models, supplementing the operational definition of (5.25). We have seen that the meaning of the error term u_Y in the equation $Y = f_Y(pa_Y, u_Y)$ is captured by the counterfactual variable Y_{pa_Y}. In other words, the variable U_Y can be interpreted as a modifier of the functional mapping from PA_Y to Y. The statistics of such modifications is observable when pa_Y is held fixed. This translation into counterfactual notation may facilitate algebraic manipulations of U_Y without committing to the functional form of f_Y. However, from the viewpoint of model specification, the error terms should still be viewed as (summaries of) omitted factors.

Armed with this interpretation, we can obtain graphical and counterfactual definitions of causal concepts that were originally given error-based definitions. Examples of such concepts are causal influence, exogeneity, and instrumental variables (Section 5.4.3). In clarifying the relationships among error-based, counterfactual, and graphical definitions of these concepts, we should first note that these three modes of description can be organized in a simple hierarchy. Since graph separation implies independence, but independence does not imply graph separation (Theorem 1.2.4), definitions based on graph separation should imply those based on error-term independence. Likewise, since for any two variables X and Y the independence relation $U_X \perp\!\!\!\perp U_y$ implies the counterfactual independence $X_{pa_X} \perp\!\!\!\perp Y_{pa_Y}$ (but not the other way around), it follows that definitions based on error independence should imply those based on counterfactual independence. Overall, we have the following hierarchy:

$$\text{graphical criteria} \implies \text{error-based criteria} \implies \text{counterfactual criteria.}$$

The concept of exogeneity may serve to illustrate this hierarchy. The pragmatic definition of exogeneity is best formulated in counterfactual or interventional terms as follows.

Exogeneity (Counterfactual Criterion)
A variable X is exogenous relative to Y if and only if the effect of X on Y is identical to the conditional probability of Y given X – that is, if

$$P(Y_x = y) = P(y \mid x) \tag{7.45}$$

or, equivalently,

$$P(Y = y \mid do\,(x)) = P(y \mid x); \tag{7.46}$$

this in turn is equivalent to the independence condition $Y_x \perp\!\!\!\perp X$, named "weak ignorability" in Rosenbaum and Rubin (1983).[26]

This definition is pragmatic in that it highlights the reasons economists should be concerned with exogeneity by explicating the policy-analytic benefits of discovering that a variable is exogenous. However, this definition fails to guide an investigator toward

---

[26] We focus the discussion in this section on the causal component of exogeneity, which the econometric literature has unfortunately renamed "superexogeneity" (see Section 5.4.3). Epidemiologists refer to (7.46) as "no-confounding" (see (6.10)). We also postpone discussion of "strong ignorability," defined as the joint independence $\{Y_x, Y_{x'}\} \perp\!\!\!\perp X$, to Chapter 9 (Definition 9.2.3).

verifying, from substantive knowledge of the domain, whether this independence condition holds in any given system, especially when many equations are involved (see Section 11.3.2). To facilitate such judgments, economists (e.g., Koopmans 1950; Orcutt 1952) have adopted the error-based criterion of Definition 5.4.6.

Exogeneity (Error-Based Criterion)

A variable X is exogenous in M relative to Y if X is independent of all error terms that have an influence on Y that is not mediated by X.[27]

This definition is more transparent to human judgment because the reference to error terms tends to focus attention on specific factors, potentially affecting Y, with which scientists are familiar. Still, to judge whether such factors are statistically independent is a difficult mental task unless the independencies considered are dictated by topological considerations that assure their stability. Indeed, the most popular conception of exogeneity is encapsulated in the notion of "common cause"; this may be stated formally as follows.

Exogeneity (Graphical Criterion)

A variable X is exogenous relative to Y if X and Y have no common ancestor in $G(M)$ or, equivalently, if all back-door paths between X and Y are blocked (by colliding arrows).[28]

It is easy to show that the graphical condition implies the error-based condition, which in turn implies the counterfactual (or pragmatic) condition of (7.46). The converse implications do not hold. For example, Figure 6.4 illustrates a case where the graphical criterion fails and both the error-based and counterfactual criteria classify X as exogenous. We argued in Section 6.4 that this type of exogeneity (there called "no confounding") is unstable or incidental, and we have raised the question of whether such cases were meant to be embraced by the definition. If we exclude unstable cases from consideration, then our three-level hierarchy collapses and all three definitions coincide.

Instrumental Variables: Three Definitions

A three-level hierarchy similarly characterizes the notion of instrumental variables (Bowden and Turkington 1984; Pearl 1995c; Angrist et al. 1996), illustrated in Figure 5.9. The traditional definition qualifies a variable Z as *instrumental* (relative to the pair (X, Y)) if (i) Z is independent of all variables (including error terms) that have an influence on Y that is not mediated by X and (ii) Z is not independent of X.

---

[27] Independence relative to *all* errors is sometimes required in the literature (e.g., Dhrymes 1970, p. 169), but this is obviously too strong.

[28] As in Chapter 6 (note 19), the expression "common ancestors" should exclude nodes that have no other connection to Y except through X and should include latent nodes for every pair of dependent errors. Generalization to conditional exogeneity relative to observed covariates is straightforward in all three definitions.

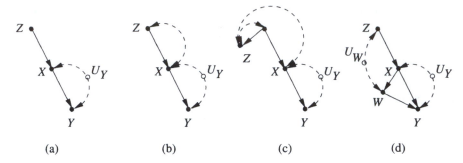

Figure 7.8 Z is a proper instrumental variable in the models of (a), (b), and (c), since it satisfies $Z \perp\!\!\!\perp U_Y$. It is not an instrument in (d) because it is correlated with W, which influences Y.

The counterfactual definition[29] replaces condition (i) with (i'): Z is independent of Y_x. The graphical definition replaces condition (i) with (i''): every unblocked path connecting Z and Y must contain an arrow pointing into X (alternatively, $(Z \perp\!\!\!\perp Y)_{G_{\overline{X}}}$). Figure 7.8 illustrates this definition through examples.

When a set S of covariates is measured, these definitions generalize as follows.

Definition 7.4.1 (Instrument)

A variable Z is an instrument relative to the total effect of X on Y if there exists a set of measurements $S = s$, unaffected by X, such that either of the following criteria holds.

1. **Counterfactual** *criterion*:
   (i) $Z \perp\!\!\!\perp Y_x \,|\, S = s$;
   (ii) $Z \not\!\perp\!\!\!\perp X \,|\, S = s$.

2. **Graphical** *criterion*:
   (i) $(Z \perp\!\!\!\perp Y \,|\, S)_{G_{\overline{X}}}$;
   (ii) $(Z \not\!\perp\!\!\!\perp X \,|\, S)_{G}$.

In concluding this section, I should reemphasize that it is because graphical definitions are insensitive to the values of the model's parameters that graphical vocabulary guides and expresses so well our intuition about causal effects, exogeneity, instruments, confounding, and even (I speculate) more technical notions such as randomness and statistical independence.

---

[29] There is, in fact, no agreed-upon generalization of instrumental variables to nonlinear systems. The definition here, taken from Galles and Pearl (1998), follows by translating the error-based definition into counterfactual vocabulary. Angrist et al. (1996), who expressly rejected all reference to graphs or error terms, assumed two unnecessary restrictions: that Z be ignorable (i.e., randomized; this is violated in Figures 7.8(b) and (c)) and that Z affect X (violated in Figure 7.8(c)). Similar assumptions were made by Heckman and Vytlacil (1999), who used both counterfactuals and structural equation models. Shunning graphs has its hazzards.

7.5 STRUCTURAL VERSUS PROBABILISTIC CAUSALITY

Probabilistic causality is a branch of philosophy that attempts to explicate causal relationships in terms of probabilistic relationships. This attempt was motivated by several ideas and expectations. First and foremost, probabilistic causality promises a solution to the centuries-old puzzle of causal discovery – that is, how humans discover genuine causal relationships from bare empirical observations, free of any causal preconceptions. Given the Humean dictum that all knowledge originates with human experience and the (less compelling but then fashionable) assumption that human experience is encoded in the form of a probability function, it is natural to expect that causal knowledge be reducible to a set of relationships in some probability distribution that is defined over the variables of interest. Second, in contrast to deterministic accounts of causation, probabilistic causality offers substantial cognitive economy. Physical states and physical laws need not be specified in minute detail because they can instead be summarized in the form of probabilistic relationships among macro states so as to match the granularity of natural discourse. Third, probabilistic causality is equipped to deal with the modern (i.e., quantum-theoretical) conception of uncertainty, according to which determinism is merely an epistemic fiction and nondeterminism is the fundamental feature of physical reality.

The formal program of probabilistic causality owes its inception to Reichenbach (1956) and Good (1961), and it has subsequently been pursued by Suppes (1970), Skyrms (1980), Spohn (1980), Otte (1981), Salmon (1984), Cartwright (1989), and Eells (1991). The current state of this program is rather disappointing, considering its original aspirations, but not surprising considering our discussion of Section 1.5. Salmon has abandoned the effort altogether, concluding that "causal relations are not appropriately analyzable in terms of statistical relevance relations" (1984, p. 185); instead, he has proposed an analysis in which "causal processes" are the basic building blocks. More recent accounts by Cartwright and Eells have resolved some of the difficulties encountered by Salmon, but at the price of either complicating the theory beyond recognition or compromising its original goals. The following is a brief account of the major achievements, difficulties, and compromises of probabilistic causality as portrayed in Cartwright (1989) and Eells (1991).

7.5.1 The Reliance on Temporal Ordering

Standard probabilistic accounts of causality assume that, in addition to a probability function P, we are also given the temporal order of the variables in the analysis. This is understandable, considering that causality is an asymmetric relation, whereas statistical relevance is symmetric. Lacking temporal information, it would be impossible to decide which of two dependent variables is the cause and which the effect, since every joint distribution $P(x, y)$ induced by a model in which X is a cause of Y can also be induced by a model in which Y is the cause of X. Thus, any method of inferring that X is a cause of Y must also infer, by symmetry, that Y is a cause of X. In Chapter 2 we demonstrated that, indeed, at least three variables are needed for determining the directionality of arrows in a DAG and, more serious yet, no arrow can be oriented from probability information alone – that is, without the added (causal) assumption of stability or minimality. By imposing the constraint that an effect never precede its cause, the symmetry is broken and causal inference can commence.

The reliance on temporal information has its price, as it excludes a priori the analysis of cases in which the temporal order is not well-defined, either because processes overlap in time or because they (appear to) occur instantaneously. For example, one must give up the prospect of determining (by uncontrolled methods) whether sustained physical exercise contributes to low cholesterol levels or if, conversely, low cholesterol levels enhance the urge to engage in physical exercise. Likewise, the philosophical theory of probabilistic causality would not attempt to distinguish between the claims "tall flag poles cause long shadows" and "long shadows cause tall flag poles" – where, for all practical purposes, the putative cause and effect occur simultaneously.

We have seen in Chapter 2 that some determination of causal directionality can be made from atemporal statistical information, if fortified with the assumption of minimality or stability. These assumptions, however, implicitly reflect generic properties of physical processes – invariance and autonomy (see Section 2.9.1) – that constitute the basis of the structural approach to causality.

7.5.2 The Perils of Circularity

Despite the reliance on temporal precedence, the criteria that philosophers have devised for identifying causal relations suffer from glaring circularity: In order to determine whether an event C is a cause of event E, one must know in advance how other factors are causally related to C and E. Such circularity emerges from the need to define the "background context" under which a causal relation is evaluated, since the intuitive idea that causes should increase the probability of their effects must be qualified by the condition that other things are assumed equal. For example, "studying arithmetic" increases the probability of passing a science test, but only if we keep student age constant; otherwise, studying arithmetic may actually lower the probability of passing the test because it is indicative of young age. Thus, it seems natural to offer the following.

Definition 7.5.1

An event C is causally relevant to E if there is at least one condition F in some background context K such that $P(E \mid C, F) > P(E \mid \neg C, F)$.[30]

But what kind of conditions should we include in the background context? On the one hand, insisting on a complete description of the physical environment would reduce probabilistic causality to deterministic physics (barring quantum-level considerations). On the other hand, ignoring background factors altogether – or describing them too coarsely – would introduce spurious correlations and other confounding effects. A natural compromise

---

[30] The reader can interpret K to be a set of variables and F a particular truth-value assignment to those variables.

is to require that the background context itself be "causally relevant" to the variables in question, but this very move is the source of circularity in the definition of probabilistic causality.

The problem of choosing an appropriate set of background factors is similar to the problem of finding an appropriate adjustment for confounding, as discussed in several previous chapters in connection with Simpson's paradox (e.g., Sections 3.3, 5.1.3, and 6.1). We have seen (e.g., in Section 6.1) that the criterion for choosing an appropriate set of covariates for adjustment cannot be based on probabilistic relationships alone but must rely on causal information. In particular, we must make sure that factors listed as background satisfy the back-door condition, because the inequality in Definition 7.5.1 can always be satisfied by conditioning on some imagined factor F (as in Figure 6.2(c)) that generates spurious associations between C and E. Here we see the emergence of circularity: In order to determine the causal role of C relative to E (e.g., the effect of the drug on recovery), we must first determine the causal role of every factor F (e.g., gender) relative to C and E.

One may try to escape this circularity by conditioning on *all* factors preceding C, but, unfortunately, other factors that cannot be identified through temporal ordering alone must also be weighed. Consider the betting example used in Section 7.2.2. I must bet heads or tails on the outcome of a fair coin toss; I win if I guess correctly and lose if I don't. Naturally, once the coin is tossed (and while the outcome is still unknown), the bet is deemed causally relevant to winning, even though the probability of winning is the same whether I bet heads or tails. In order to reveal the causal relevance of the bet (C), we must include the outcome of the coin (F) in the background context even though F does not meet the common-cause criterion – it does not affect my bet (C), nor is it causally relevant to winning (E) (unless we first declare the bet *is* relevant to winning). Worse yet, we cannot justify including F in the background context by virtue of its occurring earlier than C because whether the coin is tossed before or after my bet is totally irrelevant to the problem at hand. We conclude that temporal precedence alone is insufficient for identifying the appropriate background context, even if we resort to what Eells (1991) called "interacting causes" – namely, (simplified) factors F that (i) are not affected causally by C and (ii) jointly with C (or $\neg C$) increase the probability of E.

Because of the circularity inherent in all definitions of causal relevance, probabilistic causality cannot be regarded as a program for extracting causal relations from temporal–probabilistic information; rather, it should be viewed as a program for validating whether a proposed set of causal relationships is consistent with the available temporal-probabilistic information. More formally, suppose someone gives us a probability distribution P and a temporal order O on a (complete) set of variables V. Furthermore, any pair of variable sets (say, X and Y) in V is annotated by a symbol R or I, where R stands for "causally relevant" and I for "causally irrelevant." Probabilistic causality deals with testing whether the proposed R and I labels are consistent with the pair $\langle P, O \rangle$ and with the doctrine that causes should both precede and increase the probability of their effects.

Currently, the most advanced consistency test is the one based on Eells's (1991) criterion of relevance, which may be translated as follows.

Consistency Test

For each pair of variables labeled $R(X, Y)$, test whether

(i) X precedes Y in O, and

(ii) there exist x, x', y such that $P(y \mid x, z) > P(y \mid x', z)$ for some z in Z, where Z is a set of variables in the background context K such that $I(X, Z)$ and $R(Z, Y)$.

This now raises additional questions.

(a) Is there a consistent label for every pair $\langle P, O \rangle$?

(b) When is the label unique?

(c) Is there a procedure for finding a consistent label when it exists?

Although some insights into these questions are provided by graphical methods (Pearl 1996), the point is that, owing to circularity, the mission of probabilistic causality has been altered: from discovery to consistency testing.

It should also be remarked that the basic program of defining causality in terms of conditionalization, even if it turns out to be successful, is at odds with the natural conception of causation as an oracle for interventions. This program first confounds the causal relation $P(E \mid do(C))$ with epistemic conditionalization $P(E \mid C)$ and then removes spurious correlations through steps of remedial conditionalization, yielding $P(E \mid C, F)$. The structural account, in contrast, defines causation directly in terms of Nature's invariants (i.e., submodel M_x in Definition 7.1.2); see the discussion following Theorem 3.2.2.

7.5.3   Challenging the Closed-World Assumption, with Children

By far the most critical and least defensible paradigm underlying probabilistic causality rests on the assumption that one is in the possession of a probability function on all variables relevant to a given domain. This assumption absolves the analyst from worrying about unmeasured spurious causes that might (physically) affect several variables in the analysis and still remain obscure to the analyst. It is well known that the presence of such "confounders" may reverse or negate any causal conclusion that might be drawn from probabilities. For example, observers might conclude that "bad air" is the cause of malaria if they are not aware of the role of mosquitoes, or that falling barometers are the cause of rain, or that speeding to work is the cause of being late to work, and so on. Because they are unmeasured (or even unsuspected), the confounding factors in such examples cannot be neutralized by conditioning or by "holding them fixed." Thus, taking seriously Hume's program of extracting causal information from raw data entails coping with the problem that the validity of any such information is predicated on the untestable assumption that all relevant factors have been accounted for.

This raises the question of how people ever acquire causal information from the environment and, more specifically, how children extract causal information from experience. The proponents of probabilistic causality who attempt to explain this phenomenon through statistical theories of learning cannot ignore the fact that the child never operates in a closed, isolated environment. Unnoticed external conditions govern the operation of every learning environment, and these conditions often have the potential to confound cause and effect in unexpected and clandestine ways.

Fortunately, that children do not grow in closed, sterile environments does have its advantages. Aside from passive observations, a child possesses two valuable sources of causal information that are not available to the ordinary statistician: manipulative experimentation and linguistic advice. Manipulation subjugates the putative causal event to the sole influence of a known mechanism, thus overruling the influence of uncontrolled factors that might also produce the putative effect. "The beauty of independent manipulation is, of course, that other factors can be kept constant without their being identified" (Cheng 1992). The independence is accomplished by subjecting the object of interest to the whims of one's volition in order to ensure that the manipulation is not influenced by any environmental factor likely to produce the putative effect. Thus, for example, a child can infer that shaking a toy can produce a rattling sound because it is the child's hand, governed solely by the child's volition, that brings about the shaking of the toy and the subsequent rattling sound. The whimsical nature of free manipulation replaces the statistical notion of randomized experimentation and serves to filter sounds produced by the child's actions from those produced by uncontrolled environmental factors.

But manipulative experimentation cannot explain all of the causal knowledge that humans acquire and possess, simply because most variables in our environment are not subject to direct manipulation. The second valuable source of causal knowledge is linguistic advice: explicit causal sentences about the workings of things which we obtain from parents, friends, teachers, and books and which encode the manipulative experience of past generations. As obvious and uninteresting as this source of causal information may appear, it probably accounts for the bulk of our causal knowledge, and understanding how this transference of knowledge works is far from trivial. In order to comprehend and absorb causal sentences such as "The glass broke because you pushed it," the child must already possess a causal schema within which such inputs make sense. To further infer that pushing the glass will make someone angry at you and not at your brother, even though he broke the last glass, requires a truly sophisticated inferential machinery. In most children, this machinery is probably innate (Gopnik et al. 2004).

Note, however, that linguistic input is by and large qualitative; we rarely hear parents explaining to children that placing the glass at the edge of the table increases the probability of breakage by a factor of 2.85. The probabilistic approach to causality embeds such qualitative input in an artificial numerical frame, whereas the structural approach to causality (Section 7.1) builds directly on the qualitative knowledge that we obtain and transmit linguistically.

7.5.4   Singular versus General Causes

In Section 7.2.3 we saw that the distinction between general causes (e.g., "Drinking hemlock causes death") and singular causes (e.g., "Socrates' drinking hemlock caused his death") plays an important role in understanding the nature of explanations. We have also remarked that the notion of singular causation (also known as "token" or "single-event" causation) has not reached an adequate state of conceptualization or formalization in the probabilistic account of causation. In this section we elaborate the nature of these difficulties and conclude that they stem from basic deficiencies in the probabilistic account.

In Chapter 1 (Figure 1.6) we demonstrated that the evaluation of singular causal claims requires knowledge in the form of counterfactual or functional relationships and that such knowledge cannot be extracted from bare statistical data even when obtained under controlled experimentation. This limitation was attributed in Section 7.2.2 to the temporal persistence (or invariance) of information that is needed to sustain counterfactual statements – persistence that is washed out (by averaging) in statistical statements even when enriched with temporal and causally relevant information. The manifestations of this basic limitation have taken an interesting slant in the literature of probabilistic causation and have led to intensive debates regarding the relationships between singular and generic statements (see, e.g., Good 1961; Cartwright 1989; Eells 1991; Hausman 1998).

According to one of the basic tenets of probabilistic causality, a cause should raise the probability of the effect. It is often the case, however, that we judge an event x to be the cause of y when the conditional probability $P(y \mid x)$ is *lower* than $P(y \mid x')$. For example, a vaccine (x) usually decreases the probability of the disease (y), and yet we often say (and can medically verify) that the vaccine itself caused the disease in a given person u. Such reversals would not be problematic to students of structural models, who can interpret the singular statement as saying that "had person u not taken the vaccine (x') then u would still be healthy (y')." The probability of this counterfactual statement $P(Y_{x'} = y' \mid x, y)$ can be high while the conditional probability $P(y \mid x)$ as well as $P(y \mid do(x))$, is low, with all probabilities evaluated from the same structural model (Section 9.2 provides precise relationships between the three quantities). However, this reversal is traumatic to students of probabilistic causation, who mistrust counterfactuals for various reasons – partly because counterfactuals carry an aura of determinism (Kvart 1986, pp. 256–63) and partly because counterfactuals are perceived as resting on a shaky formal foundation "for which we have only the beginnings of a semantics (via the device of measures over possible worlds)" (Cartwright 1983, p. 34).

In order to reconcile the notion of probability increase with that of singular causation, probabilists claim that, if we look hard enough at any given scenario in which x is judged to be a cause of y, then we will always be able to find a subpopulation $Z = z$ in which x raises the probability of y – namely,

$$P(y \mid x, z) > P(y \mid x', z). \tag{7.47}$$

In the vaccine example, we might identify the desired subpopulation as consisting of individuals who are adversely susceptible to the vaccine; by definition, the vaccine would no doubt raise the probability of the disease in that subpopulation. Oddly, only a few philosophers have noticed that factors such as being "adversely susceptible" are defined counterfactually and that, in permitting conditionalization on such factors, one opens a clandestine back door for sneaking determinism and counterfactual information back into the analysis.

Perhaps a less obvious appearance of counterfactuals surfaces in Hesslow's example of the birth-control pill (Hesslow 1976), discussed in Section 4.5.1. Suppose we find that Mrs. Jones is not pregnant and ask whether taking a birth-control pill was the cause of her suffering from thrombosis. The population of nonpregnant women turns out to be too coarse for answering this question unequivocally. If Mrs. Jones belongs to the class of women who would have become pregnant *but for* the pill, then the pill might actually have lowered the probability of thrombosis in her case by preventing her pregnancy. If, on the other hand, she belongs to the class of women who would *not* have become pregnant regardless of the pill, then her taking the pill has surely increased the chance of thrombosis. This example is illuminating because the two classes of test populations do not have established names in the English language (unlike "susceptibility" in the vaccine example) and must be defined explicitly in counterfactual vocabulary. Whether a woman belongs to the former or latter class depends on many social and circumstantial contingencies, which are usually unknown and are not likely to define an invariant attribute of a given person. Still, we recognize the need to consider the two classes separately in evaluating whether the pill was the cause of Mrs. Jones's thrombosis.

Thus we see that there is no escape from counterfactuals when we deal with token-level causation. Probabilists' insistence on counterfactual-free syntax in defining token causal claims has led to subpopulations delineated by none other than counterfactual expressions: "adversely susceptible" in the vaccine example and "would not have become pregnant" in the case of Mrs. Jones.[31]

Probabilists can argue, of course, that there is no need to refine the subclasses $Z = z$ down to deterministic extremes, since one can stop the refinement as soon as one finds a subclass that increases the probability of y, as required in (7.47). This argument borders on the tautological, unless it is accompanied by formal procedures for identifying the test subpopulation $Z = z$ and for computing the quantities in (7.47) from some reasonable model of human knowledge, however hypothetical. Unfortunately, the probabilistic causality literature is silent on questions of procedures and representation.[32]

In particular, probabilists face a tough dilemma in explaining how people search for that rescuing subpopulation z so swiftly and consistently and how the majority of people end up with the same answer when asked whether it was x that caused y. For example (due to Debra Rosen, quoted in Suppes 1970), a tree limb (x) that fortuitously deflects a golf ball is immediately and consistently perceived as "the cause" for the ball finally ending up in the hole, though such collisions generally lower one's chances of reaching the hole (y). Clearly, if there is a subpopulation z that satisfies (7.47) in such examples (and I doubt it ever enters anyone's mind), it must have at least two features.

(1) It must contain events that occur both before and after x. For example, both the angle at which the ball hit the limb and the texture of the grass on which the ball bounced after hitting the limb should be part of z.

---

[31] Cartwright (1989, chap. 3) recognized the insufficiency of observable partitions (e.g., pregnancy) for sustaining the thesis of increased probability, but she did not emphasize the inevitable counterfactual nature of the finer partitions that sustain that thesis. Not incidentally, Cartwright was a strong advocate of excluding counterfactuals from causal analysis (Cartwright 1983, pp. 34–5).

[32] Even Eells (1991, chap. 6) and Shafer (1996), who endeavored to uncover discriminating patterns of increasing probabilities in the actual trajectory of the world leading to y, did not specify what information is needed either to select the appropriate trajectory or to compute the probabilities associated with a given trajectory.

(2) It must depend on x and y. For, surely, a different conditioning set z' would be necessary in (7.47) if we were to test whether the limb caused an alternative consequence y' – say, that the ball stopped two yards short of the hole.

And this brings us to a major methodological inconsistency in the probabilistic approach to causation: If ignorance of x and y leads to the wrong z and if awareness of x and y leads to the correct selection of z, then there must be some process by which people incorporate the occurrence of x and y into their awareness. What could that process be? According to the norms of probabilistic epistemology, evidence is incorporated into one's corpus of knowledge by means of conditionalization. How, then, can we justify excluding from z the very evidence that led to its selection – namely, the occurrence of x and y?

Inspection of (7.47) shows that the exclusion of x and y from z is compelled on syntactic grounds, since it would render $P(y \mid x', z)$ undefined and make $P(y \mid x, z) = 1$. Indeed, in the syntax of probability calculus we cannot ask what the probability of event y *would* be, given that y has in fact occurred – the answer is (trivially) 1. The best we can do is detach ourselves momentarily from the actual world, pretend that we are ignorant of the occurrence of y, and ask for the probability of y under such a state of ignorance. This corresponds precisely to the three steps (abduction, action, and prediction) that govern the evaluation of $P(Y_{x'} = y' \mid x, y)$ (see Theorem 7.1.7), which attains a high value (in our example) and correctly qualifies the tree limb (x) as the cause of making the hole (y). As we see, the desired quantity *can* be expressed and evaluated by ordinary conditionalization on x and y, without explicitly invoking any subpopulation z.[33]

Ironically, by denying counterfactual conditionals, probabilists deprived themselves of using standard conditionals – the very conditionals they were trying to preserve – and were forced to accommodate simple evidential information in roundabout ways. This syntactic barrier that probabilists erected around causation has created an artificial tension between singular and generic causes, but the tension disappears in the structural account. In Section 10.1.1 we show that, by accommodating both standard and counterfactual conditionals (i.e., Y_x), singular and generic causes no longer stand in need of separate analyses. The two types of causes differ merely in the level of scenario-specific information that is brought to bear on a problem, that is, in the specificity of the evidence e that enters the quantity $P(Y_x = y \mid e)$.

7.5.5 Summary

Cartwright (1983, p. 34) listed several reasons for pursuing the probabilistic versus the counterfactual approach to causation:

> [the counterfactual approach] requires us to evaluate the probability of counterfactuals for which we have only the beginnings of a semantics (via the device of measures over possible worlds) and no methodology, much less an account of why the methodology is suited

---

[33] The desired subpopulation z is equal to the set of all u that are mapped into $X(u) = x$, $Y(u) = y$, and $Y_x(u) = y'$.

to the semantics. How do we test claims about probabilities of counterfactuals? We have no answer, much less an answer that fits with our nascent semantics. It would be preferable to have a measure of effectiveness that requires only probabilities over events that can be tested in the actual world in the standard ways.

Examining the progress of the probabilistic approach in the past three decades, it seems clear that Cartwright's aspirations have materialized not in the framework she advocated but rather in the competing framework of counterfactuals, as embodied in structural models. Full characterization of "effectiveness" ("causal effects" in our vocabulary) in terms of "events that can be tested" emerged from Simon's (1953) and Strotz and Wold's (1960) conception of modifiable structural models and culminated in the back-door criterion (Theorem 3.3.2) and to the more general Theorems 3.6.1 and 4.4.1, of which the probabilistic criteria (as in (3.13)) are but crude special cases. The interpretation of singular causation in terms of the counterfactual probability $P(Y_{x'} \neq y \mid x, y)$ has enlisted the support of meaningful formal semantics (Section 7.1) and effective evaluation methodology (Theorem 7.1.7 and Sections 7.1.3–7.2.1), while the probabilistic criterion of (7.47) lingers in vague and procedureless debates. The original dream of rendering causal claims testable was given up in the probabilistic framework as soon as unmeasured entities (e.g., state of the world, background context, causal relevance, susceptibility) were allowed to infiltrate the analysis, and methodologies for answering questions of testability have moved over to the structural–counterfactual framework (see Chapter 9 and Section 11.9).

The ideal of remaining compatible with the teachings of nondeterministic physics seems to be the only viable aspect remaining in the program of probabilistic causation, and this section questions whether maintaining this ideal justifies the sacrifices. It further suggests that the basic agenda of the probabilistic causality program is due for a serious reassessment. If the program is an exercise in epistemology, then the word "probabilistic" is oxymoronic – human perception of causality has remained quasi-deterministic, and these fallible humans are still the main consumers of causal talk. If the program is an exercise in modern physics, then the word "causality" is nonessential – quantum-level causality follows its own rules and intuitions, and another name (perhaps "qua-sality") might be more befitting. However, regarding artificial intelligence and cognitive science, I would venture to predict that robots programmed to emulate the quasi-deterministic macroscopic approximations of Laplace and Einstein would far outperform those built on the correct but counterintuitive theories of Born, Heisenberg, and Bohr.

Acknowledgments

Sections of this chapter are based on the doctoral research of Alex Balke and David Galles. This research has benefitted significantly from the input of Joseph Halpern.

CHAPTER EIGHT

Imperfect Experiments: Bounding Effects and Counterfactuals

*Would that I could discover truth
as easily as I can uncover falsehood.*
                    Cicero (44 B.C.)

Preface

In this chapter we describe how graphical and counterfactual models (Sections 3.2 and 7.1) can combine to elicit causal information from imperfect experiments: experiments that deviate from the ideal protocol of randomized control. A common deviation occurs, for example, when subjects in a randomized clinical trial do not fully comply with their assigned treatment, thus compromising the identification of causal effects. When conditions for identification are not met, the best one can do is derive *bounds* for the quantities of interest – namely, a range of possible values that represents our ignorance about the data-generating process and that cannot be improved with increasing sample size. This chapter demonstrates (i) that such bounds can be derived by simple algebraic methods, (ii) that, despite the imperfection of the experiments, the derived bounds can yield significant and sometimes accurate information on the impact of a policy on the entire population as well as on a particular individual in the study, and (iii) that prior knowledge can be harnessed effectively to obtain Bayesian estimates of those impacts.

8.1 INTRODUCTION

8.1.1 Imperfect and Indirect Experiments

Standard experimental studies in the biological, medical, and behavioral sciences invariably invoke the instrument of randomized control; that is, subjects are assigned at random to various groups (or treatments or programs), and the mean differences between participants in different groups are regarded as measures of the efficacies of the associated programs. Deviations from this ideal setup may take place either by failure to meet any of the experimental requirements or by deliberate attempts to relax these requirements. *Indirect experiments* are studies in which randomized control is either unfeasible or undesirable. In such experiments, subjects are still assigned at random to various groups, but members of each group are simply encouraged (rather than forced) to participate in the program associated with the group; it is up to the individuals to select among the programs.

Recently, use of strict randomization in social and medical experimentation has been questioned for three major reasons.

1. Perfect control is hard to achieve or ascertain. Studies in which treatment is assumed to be randomized may be marred by uncontrolled *imperfect compliance*. For example, subjects experiencing adverse reactions to an experimental drug may decide to reduce the assigned dosage. Alternatively, if the experiment is testing a drug for a terminal disease, a subject suspecting that he or she is in the control group may obtain the drug from other sources. Such imperfect compliance renders the experiment indirect and introduces bias into the conclusions that researchers draw from the data. This bias cannot be corrected unless detailed models of compliance are constructed (Efron and Feldman 1991).

2. Denying subjects assigned to certain control groups the benefits of the best available treatment has moral and legal ramifications. For example, in AIDS research it is difficult to justify placebo programs because those patients assigned to the placebo group would be denied access to potentially life-saving treatment (Palca 1989).

3. Randomization, by its very presence, may influence participation as well as behavior (Heckman 1992). For example, eligible candidates may be wary of applying to a school once they discover that it deliberately randomizes its admission criteria. Likewise, as Kramer and Shapiro (1984) noted, subjects in drug trials may be less likely to participate in randomized trials than in nonexperimental studies, even when the treatments are equally nonthreatening.

Altogether, researchers are beginning to acknowledge that mandated randomization may undermine the reliability of experimental evidence and that experimentation with human subjects often involves – and sometimes *should* involve – an element of self-selection.

This chapter concerns the drawing of inferences from studies in which subjects have final choice of program; the randomization is confined to an indirect *instrument* (or *assignment*) that merely encourages or discourages participation in the various programs. For example, in evaluating the efficacy of a given training program, notices of eligibility may be sent to a randomly selected group of students or, alternatively, eligible candidates may be selected at random to receive scholarships for participating in the program. Similarly, in drug trials, subjects may be given randomly chosen advice on recommended dosage level, yet the final choice of dosage will be determined by the subjects to fit their individual needs.

Imperfect compliance poses a problem because simply comparing the fractions in the treatment and control groups may provide a misleading estimate for how effective the treatment would be if applied uniformly to the population. For example, if those subjects who declined to take the drug are precisely those who would have responded adversely, the experiment might conclude that the drug is more effective than it actually is. In Chapter 3 (see Section 3.5, Figure 3.7(b)), we showed that treatment effectiveness in such studies is actually *nonidentifiable*. That is, in the absence of additional modeling assumptions, treatment effectiveness cannot be estimated from the data without bias, even when the number of subjects in the experiment approaches infinity and even when a record is available of the action and response of each subject.

The question we attempt to answer in this chapter is whether indirect randomization can provide information that allows approximate assessment of the intrinsic merit of a program, as would be measured, for example, if the program were to be extended and mandated uniformly to the population. The analysis presented shows that, given a minimal set of assumptions, such inferences are indeed possible – albeit in the form of *bounds*, rather than precise point estimates, for the causal effect of the program or treatment. These bounds can be used by the analyst to guarantee that the causal effect of a given program must be higher than one measurable quantity and lower than another.

Our most crucial assumption is that, for any given person, the encouraging instrument influences the treatment chosen by that person but has no effect on how that person would respond to the treatment chosen (see the definition of instrumental variables in Section 7.4.5). The second assumption, one which is always made in experimental studies, is that subjects respond to treatment independently of one other. Other than these two assumptions, our model places no constraints on how tendencies to respond to treatments may interact with choices among treatments.

8.1.2   Noncompliance and Intent to Treat

In a popular compromising approach to the problem of imperfect compliance, researchers perform an "intent to treat" analysis in which the control and treatment groups are compared without regard to whether the treatment was actually received.[1] The result of such an analysis is a measure of how well the treatment *assignment* affects the disease, as opposed to the desired measure of how well the treatment *itself* affects the disease. Estimates based on intent-to-treat analyses are valid as long as the experimental conditions perfectly mimic the conditions prevailing in the eventual usage of the treatment. In particular, the experiment should mimic subjects' incentives for receiving each treatment. In situations where field incentives are more compelling than experimental incentives, as is usually the case when drugs receive the approval of a government agency, treatment effectiveness may vary significantly from assignment effectiveness. For example, imagine a study in which (a) the drug has an adverse side-effect on a large segment of the population and (b) only those members of the segment who drop from the treatment "arm" (sub-population) recover. The intent-to-treat analysis will attribute these cases of recovery to the drug because they are part of the intent-to-treat arm, although in reality these cases recovered by *avoiding* the treatment.

Another approach to the problem is to use a correction factor based on an instrumental variables formula (Angrist et al. 1996), according to which the intent-to-treat measure should be divided by the fraction of subjects who comply with the treatment assigned to them. Angrist et al. (1996) showed that, under certain conditions, the corrected formula is valid for the subpopulation of "responsive" subjects – that is, subjects who would have changed treatment status if given a different assignment. Unfortunately, this subpopulation cannot be identified and, more seriously, it cannot serve as a basis for policies

---

[1] This approach is currently used by some FDA agencies to approve new drugs.

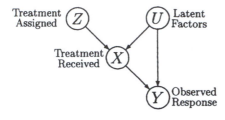

Figure 8.1   Graphical representation of causal dependencies in a randomized clinical trial with partial compliance. Z serves as Instrumental Variable.

involving the entire population because it is instrument-dependent: individuals who are responsive in the study may not remain responsive in the field, where the incentives for obtaining treatment differ from those used in the study. We therefore focus our analysis on the stable aspect of the treatment – the aspect that would remain invariant to changes in compliance behavior.

8.2   BOUNDING CAUSAL EFFECTS WITH INSTRUMENTAL VARIABLES

8.2.1   Problem Formulation: Constrained Optimization

The basic experimental setting associated with indirect experimentation is shown in Figure 8.1, which is isomorphic to Figures 3.7(b) and 5.9. To focus the discussion, we will consider a prototypical clinical trial with partial compliance, although in general the analysis applies to any study in which an instrumental variable (Definition 7.4.1) encourages subjects to choose one program over another.

We assume that Z, X, Y are observed binary variables, where Z represents the (randomized) treatment assignment, X is the treatment actually received, and Y is the observed response. The U term represents all factors, both observed and unobserved, that influence the way a subject responds to treatments; hence, an arrow is drawn from U to Y. The arrow from U to X denotes that the U factors may also influence the subject's choice of treatment X; this dependence may represent a complex decision process standing between the assignment (Z) and the actual treatment (X).

To facilitate the notation, we let z, x, y represent (respectively) the values taken by the variables Z, X, Y, with the following interpretation:

$z \in \{z_0, z_1\}$, z_1 asserts that treatment has been assigned (z_0, its negation);

$x \in \{x_0, x_1\}$, x_1 asserts that treatment has been administered (x_0, its negation); and

$y \in \{y_0, y_1\}$, y_1 asserts a positive observed response (y_0, its negation).

The domain of U remains unspecified and may, in general, combine the spaces of several random variables, both discrete and continuous.

The graphical model reflects two assumptions.

1. The assigned treatment Z does not influence Y directly but rather through the actual treatment X. In practice, any direct effect Z might have on Y would be adjusted for through the use of a placebo.

2. The variables Z and U are marginally independent; this is ensured through the randomization of Z, which rules out a common cause for both Z and U.

These assumptions impose on the joint distribution the decomposition

$$P(y, x, z, u) = P(y \mid x, u)\, P(x \mid z, u)\, P(z)\, P(u), \tag{8.1}$$

which, of course, cannot be observed directly because U is unobserved. However, the marginal distribution $P(y, x, z)$ and, in particular, the conditional distributions

$$P(y, x \mid z) = \sum_u P(y \mid x, u)\, P(x \mid z, u)\, P(u), \quad z \in \{z_0, z_1\} \tag{8.2}$$

are observed,[2] and the challenge is to assess from these distributions the average *change* in Y due to treatment.

Treatment effects are governed by the distribution $P(y \mid do(x))$, which – using the truncated factorization formula of (3.10) – is given by

$$P(y \mid do(x)) = \sum_u P(y \mid x, u)\, P(u); \tag{8.3}$$

here, the factors $P(y \mid x, u)$ and $P(u)$ are the same as those in (8.2). Therefore, if we are interested in the average change in Y due to treatment, then we should compute the *average causal effect*, $\text{ACE}(X \rightarrow Y)$ (Holland 1988), which is given by

$$\begin{aligned}
\text{ACE}(X \rightarrow Y) &= P(y_1 \mid do(x_1)) - P(y_1 \mid do(x_0)) \\
&= \sum_u [P(y_1 \mid x_1, u) - P(y_1 \mid x_0, u)]P(u). \tag{8.4}
\end{aligned}$$

Our task is then to estimate or bound the expression in (8.4) given the observed probabilities $P(y, x \mid z_0)$ and $P(y, x \mid z_1)$, as expressed in (8.2). This task amounts to a constrained optimization exercise of finding the highest and lowest values of (8.4) subject to the equality constraint in (8.2), where the maximization ranges over all possible functions

$$P(u), \ P(y_1 \mid x_0, u), \ P(y_1 \mid x_1, u), \ P(x_1 \mid z_0, u), \ \text{and} \ P(x_1 \mid z_1, u)$$

that satisfy those constraints.

8.2.2 Canonical Partitions: The Evolution of Finite-Response Variables

The bounding exercise described in Section 8.2.1 can be solved using conventional techniques of mathematical optimization. However, the continuous nature of the functions involved – as well as the unspecified domain of U – makes this representation inconvenient for computation. Instead, we can use the observation that U can always be replaced by a finite-state variable such that the resulting model is equivalent with respect to all observations and manipulations of Z, X, and Y (Pearl 1994a).

---

[2] In practice, of course, only a finite sample of $P(y, x \mid z)$ will be observed. But our task is one of identification, not estimation, so we make the large-sample assumption and consider $P(y, x \mid z)$ as given.

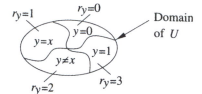

Domain of U

Figure 8.2   The canonical partition of U into four equivalence classes, each inducing a distinct functional mapping from X to Y for any given function $y = f(x, u)$.

Consider the structural equation that connects two binary variables, Y and X, in a causal model:

$$y = f(x, u).$$

For any given u, the relationship between X and Y must be one of four functions:

$$f_0 : y = 0, \quad f_1 : y = x,$$
$$f_2 : y \neq x, \quad f_3 : y = 1. \tag{8.5}$$

As u varies along its domain, regardless of how complex the variation, the only effect it can have on the model is to switch the relationship between X and Y among these four functions. This partitions the domain of U into four *equivalence classes*, as shown in Figure 8.2, where each class contains those points u that correspond to the same function. We can thus replace U by a four-state variable, $R(u)$, such that each state represents one of the four functions. The probability $P(u)$ would automatically translate into a probability function $P(r)$, $r = 0, 1, 2, 3$, that is given by the total weight assigned to the equivalence class corresponding to r. A state-minimal variable like R is called a "response" variable by Balke and Pearl (1994a,b) and a "mapping" variable by Heckerman and Shachter (1995), yet "canonical partition" would be more descriptive.[3]

Because Z, X, and Y are all binary variables, the state space of U divides into 16 equivalence classes: each class dictates two functional mappings, one from Z to X and the other from X to Y. To describe these equivalence classes, it is convenient to regard each of them as a point in the joint space of two four-valued variables R_x and R_y. The variable R_x determines the compliance behavior of a subject through the mapping.

---

[3] In an experimental framework, this partition goes back to Greenland and Robins (1986) and was dubbed "Principal Stratification" by Frangakis and Rubin (2002). In this framework (see Section 7.4.4), u stands for an experimental unit and $R(u)$ corresponds to the potential response of unit u to treatment x. The assumption that each unit (e.g., an individual subject) possesses an intrinsic, seemingly "fatalistic" response function has met with some objections (Dawid 2000), owing to the inherent unobservability of the many factors that might govern an individual response to treatment. The equivalence-class formulation of $R(u)$ mitigates those objections (Pearl 2000) by showing that $R(u)$ evolves naturally and mathematically from any complex system of stochastic latent variables, provided only that we acknowledge their existence through the equation $y = f(x, u)$. Those who invoke quantum-mechanical objections to the latter step as well (e.g., Salmon 1998) should regard the functional relationship $y = f(x, u)$ as an abstract mathematical construct, representing the extreme points (vertices) of the set of conditional probabilities $P(y \mid x, u)$ satisfying the constraints of (8.1) and (8.2).

$$x = f_X(z, r_x) = \begin{cases} x_0 & \text{if } r_x = 0; \\ x_0 & \text{if } r_x = 1 \text{ and } z = z_0, \\ x_1 & \text{if } r_x = 1 \text{ and } z = z_1, \\ x_1 & \text{if } r_x = 2 \text{ and } z = z_0, \\ x_0 & \text{if } r_x = 2 \text{ and } z = z_1; \\ x_1 & \text{if } r_x = 3. \end{cases} \tag{8.6}$$

Imbens and Rubin (1997) call a subject with compliance behavior $r_x = 0, 1, 2, 3$ (respectively) a *never-taker*, a *compiler*, a *defier*, and an *always-taker*. Similarly, the variable R_y determines the response behavior of a subject through the mapping:

$$y = f_Y(x, r_y) = \begin{cases} y_0 & \text{if } r_y = 0; \\ y_0 & \text{if } r_y = 1 \text{ and } x = x_0, \\ y_1 & \text{if } r_y = 1 \text{ and } x = x_1, \\ y_1 & \text{if } r_y = 2 \text{ and } x = x_0, \\ y_0 & \text{if } r_y = 2 \text{ and } x = x_1; \\ y_1 & \text{if } r_y = 3. \end{cases} \tag{8.7}$$

Heckerman and Shachter (1995) call the response behavior $r_y = 0, 1, 2, 3$ (respectively) *never-recover*, *helped*, *hurt*, and *always-recover*.

The correspondence between the states of variable R_y and the counterfactual variables, Y_{x_0} and Y_{x_1}, defined in Section 7.1 (Definition 7.1.4) is as follows:

$$Y_{x_1} = \begin{cases} y_1 & \text{if } r_y = 1 \text{ or } r_y = 3, \\ y_0 & \text{otherwise}; \end{cases}$$

$$Y_{x_0} = \begin{cases} y_1 & \text{if } r_y = 2 \text{ or } r_y = 3, \\ y_0 & \text{otherwise}. \end{cases}$$

In general, response and compliance may not be independent, hence the double arrow $R_x \blacktriangleleft - - \blacktriangleright R_y$ in Figure 8.3. The joint distribution over $R_x \times R_y$ requires 15 independent parameters, and these parameters are sufficient for specifying the model of Figure 8.3, $P(y, x, z, r_x, r_y) = P(y \mid x, r_y) P(x \mid r_x, z) P(z) P(r_x, r_y)$, because Y and X stand in fixed functional relations to their parents in the graph. The causal effect of the treatment can now be obtained directly from (8.7), giving

$$P(y_1 \mid do(x_1)) = P(r_y = 1) + P(r_y = 3), \tag{8.8}$$

$$P(y_1 \mid do(x_0)) = P(r_y = 2) + P(r_y = 3), \tag{8.9}$$

and

$$\text{ACE}(X \to Y) = P(r_y = 1) - P(r_y = 2). \tag{8.10}$$

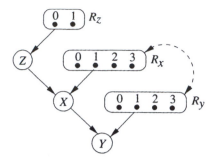

Figure 8.3 A structure equivalent to that of Figure 8.1 but employing finite-state response variables R_z, R_x, and R_y.

8.2.3   Linear Programming Formulation

By explicating the relationship between the parameters of $P(y, x \mid z)$ and those of $P(r_x, r_y)$, we obtain a set of linear constraints needed for minimizing or maximizing $\text{ACE}(X \rightarrow Y)$ given $P(y, x \mid z)$.

The conditional distribution $P(y, x \mid z)$ over the observed variables is fully specified by eight parameters, which will be written as follows:

$$p_{00.0} = P(y_0, x_0 \mid z_0), \quad p_{00.1} = P(y_0, x_0 \mid z_1),$$

$$p_{01.0} = P(y_0, x_1 \mid z_0), \quad p_{01.1} = P(y_0, x_1 \mid z_1),$$

$$p_{10.0} = P(y_1, x_0 \mid z_0), \quad p_{10.1} = P(y_1, x_0 \mid z_1),$$

$$p_{11.0} = P(y_1, x_1 \mid z_0), \quad p_{11.1} = P(y_1, x_1 \mid z_1).$$

The probabilistic constraints

$$\sum_{n=00}^{11} p_{n.0} = 1 \quad \text{and} \quad \sum_{n=00}^{11} p_{n.1} = 1$$

further imply that $\vec{p} = (p_{00.0}, \ldots, p_{11.1})$ can be specified by a point in 6-dimensional space. This space will be referred to as P.

The joint probability $P(r_x, r_y)$ has 16 parameters:

$$q_{jk} \triangleq P(r_x = j, r_y = k), \tag{8.11}$$

where $j, k \in \{0, 1, 2, 3\}$. The probabilistic constraint

$$\sum_{j=0}^{3} \sum_{k=0}^{3} q_{jk} = 1$$

implies that $\vec{q}$ specifies a point in 15-dimensional space. This space will be referred to as Q.

Equation (8.10) can now be rewritten as a linear combination of the Q parameters:

$$\text{ACE}(X \rightarrow Y) = q_{01} + q_{11} + q_{21} + q_{31} - q_{02} - q_{12} - q_{22} - q_{32}. \tag{8.12}$$

Applying (8.6) and (8.7), we can write the linear transformation from a point $\vec{q}$ in Q to a point $\vec{p}$ in P:

$$p_{00.0} = q_{00} + q_{01} + q_{10} + q_{11}, \quad p_{00.1} = q_{00} + q_{01} + q_{20} + q_{21},$$
$$p_{01.0} = q_{20} + q_{22} + q_{30} + q_{32}, \quad p_{01.1} = q_{10} + q_{12} + q_{30} + q_{32},$$
$$p_{10.0} = q_{02} + q_{03} + q_{12} + q_{13}, \quad p_{10.1} = q_{02} + q_{03} + q_{22} + q_{23},$$
$$p_{11.0} = q_{21} + q_{23} + q_{31} + q_{33}, \quad p_{11.1} = q_{11} + q_{13} + q_{31} + q_{33},$$

which can also be written in matrix form as $\vec{p} = \mathbf{R}\vec{q}$.

Given a point $\vec{p}$ in P-space, the strict lower bound on $\mathrm{ACE}(X \to Y)$ can be determined by solving the following linear programming problem.

Minimize $q_{01} + q_{11} + q_{21} + q_{31} - q_{02} - q_{12} - q_{22} - q_{32}$

subject to:

$$\sum_{j=0}^{3}\sum_{k=0}^{3} q_{jk} = 1,$$

$$\mathbf{R}\vec{q} = \vec{p},$$

$$q_{jk} \geq 0 \quad \text{for} \quad j, k \in \{0, 1, 2, 3\}. \tag{8.13}$$

For problems of this size, procedures are available for deriving symbolic expressions for the solution of this optimization exercise (Balke 1995), leading to the following lower bound on the treatment effect:

$$\mathrm{ACE}(X \to Y) \geq \max \left\{ \begin{array}{c} p_{11.1} + p_{00.0} - 1 \\ p_{11.0} + p_{00.1} - 1 \\ p_{11.0} - p_{11.1} - p_{10.1} - p_{01.0} - p_{10.0} \\ p_{11.1} - p_{11.0} - p_{10.0} - p_{01.1} - p_{10.1} \\ -p_{01.1} - p_{10.1} \\ -p_{01.0} - p_{10.0} \\ p_{00.1} - p_{01.1} - p_{10.1} - p_{01.0} - p_{00.0} \\ p_{00.0} - p_{01.0} - p_{10.0} - p_{01.1} - p_{00.1} \end{array} \right\}. \tag{8.14a}$$

Similarly, the upper bound is given by

$$\mathrm{ACE}(X \to Y) \leq \min \left\{ \begin{array}{c} 1 - p_{01.1} - p_{10.0} \\ 1 - p_{01.0} - p_{10.1} \\ -p_{01.0} + p_{01.1} + p_{00.1} + p_{11.0} + p_{00.0} \\ -p_{01.1} + p_{11.1} + p_{00.1} + p_{01.0} + p_{00.0} \\ p_{11.1} + p_{00.1} \\ p_{11.0} - p_{00.0} \\ -p_{10.1} + p_{11.1} + p_{00.1} + p_{11.0} + p_{10.0} \\ -p_{10.0} + p_{11.0} + p_{00.0} + p_{11.1} + p_{10.1} \end{array} \right\}. \tag{8.14b}$$

We may also derive bounds for (8.8) and (8.9) individually (under the same linear constraints), giving:

$$P(y_1 \mid do(x_0)) \geq \max \left\{ \begin{array}{c} p_{10.0} + p_{11.0} - p_{00.1} - p_{11.1} \\ p_{10.1} \\ p_{10.0} \\ p_{01.0} + p_{10.0} - p_{00.1} - p_{01.1} \end{array} \right\},$$

$$P(y_1 \mid do(x_0)) \leq \min \left\{ \begin{array}{c} p_{01.0} + p_{10.0} + p_{10.1} + p_{11.1} \\ 1 - p_{00.1} \\ 1 - p_{00.0} \\ p_{10.0} + p_{11.0} + p_{01.1} + p_{10.1} \end{array} \right\},$$

(8.15)

$$P(y_1 \mid do(x_1)) \geq \max \left\{ \begin{array}{c} p_{11.0} \\ p_{11.1} \\ -p_{00.0} - p_{01.0} + p_{00.1} + p_{11.1} \\ -p_{01.0} - p_{10.0} + p_{10.1} + p_{11.1} \end{array} \right\},$$

$$P(y_1 \mid do(x_1)) \leq \min \left\{ \begin{array}{c} 1 - p_{01.0} \\ 1 - p_{01.0} \\ p_{00.0} + p_{11.0} + p_{10.1} + p_{11.1} \\ p_{10.0} + p_{11.0} + p_{00.1} + p_{11.1} \end{array} \right\}.$$

(8.16)

These expressions give the tightest possible assumption-free[4] bounds on the quantities sought.

8.2.4 The Natural Bounds

The expression for $ACE(X \to Y)$ (equation (8.4)) can be bounded by two simple formulas, each made up of the first two terms in (8.14a) and (8.14b) (Robins 1989; Manski 1990; Pearl 1994a):

$$ACE(X \to Y) \geq P(y_1 \mid z_1) - P(y_1 \mid z_0) - P(y_1, x_0 \mid z_1) - P(y_0, x_1 \mid z_0),$$

$$ACE(X \to Y) \leq P(y_1 \mid z_1) - P(y_1 \mid z_0) + P(y_0, x_0 \mid z_1) + P(y_1, x_1 \mid z_0).$$

(8.17)

Because of their simplicity and wide range of applicability, the bounds given by (8.17) were named the *natural* bounds (Balke and Pearl 1997). The natural bounds guarantee that the causal effect of the actual treatment cannot be smaller than that of the encouragement $(P(y_1 \mid z_1) - P(y_1 \mid z_0))$ by more than the sum of two measurable quantities, $P(y_1, x_0 \mid z_1) + P(y_0, x_1 \mid z_0)$; they also guarantee that the causal effect of the treatment cannot exceed that of the encouragement by more than the sum of two other measurable

---

[4] "Assumption-transparent" might be a better term; we make no assumptions about factors that determine subjects' compliance, but we rely on the assumptions of (i) randomized assignment and (ii) no side effects, as displayed in the graph (e.g., Figure 8.1).

quantities, $P(y_0, x_0 | Z_1) + P(y_1, x_1 | z_0)$. The width of the natural bounds, not surprisingly, is given by the rate of noncompliance: $P(x_1 | z_0) + P(x_0 | z_1)$.

The width of the sharp bounds in (8.14ab) can be substantially narrower, though. In Balke (1995) and Pearl (1995b), it is shown that – even under conditions of 50% noncompliance – these bounds may collapse to a point and thus permit consistent estimation of ACE($X \rightarrow Y$). This occurs whenever (a) the percentage of subjects complying with assignment z_0 is the same as those complying with z_1 and (b) Y and Z are perfectly correlated in at least one treatment arm x (see Table 8.1 in Section 8.5).

Although more complicated than the natural bounds of (8.17), the sharp bounds of (8.14ab) are nevertheless easy to assess once we have the frequency data in the eight cells of $P(y, x | z)$. It can also be shown (Balke 1995) that the natural bounds are optimal when we can safely assume that no subject is *contrarian* – in other words, that no subject would consistently choose a treatment arm contrary to the one assigned.

Note that, if the response Y is continuous, then one can associate y_1 and y_0 with the binary events $Y > t$ and $Y \leq t$ (respectively) and let t vary continuously over the range of Y. Equations (8.15) and (8.16) would then provide bounds on the entire distribution of the treatment effect $P(Y < t | do(x))$.

8.2.5   Effect of Treatment on the Treated (ETT)

Much of the literature assumes that ACE($X \rightarrow Y$) is the parameter of interest, because ACE($X \rightarrow Y$) predicts the impact of applying the treatment uniformly (or randomly) over the population. However, if a policy maker is not interested in introducing new treatment policies but rather in deciding whether to maintain or terminate an existing program under its current incentive system, then the parameter of interest should measure the impact of the treatment *on the treated*, namely, the mean response of the treated subjects compared to the mean response of these same subjects had they not been treated (Heckman 1992). The appropriate formula for this parameter is

$$\begin{aligned}
\text{ETT}(X \rightarrow Y) &= P(Y_{x_1} = y_1 | x_1) - P(Y_{x_0} = y_1 | x_1) \\
&= \sum_u [P(y_1 | x_1, u) - P(y_1 | x_0, u)] P(u | x_1),
\end{aligned} \tag{8.18}$$

which is similar to (8.4) except for replacing the expectation over u with the conditional expectation given $X = x_1$.

The analysis of ETT($X \rightarrow Y$) reveals that, under conditions of *no intrusion* (i.e., $P(x_1 | z_0) = 0$, as in most clinical trials), ETT($X \rightarrow Y$) can be identified precisely (Bloom 1984; Heckman and Robb 1986; Angrist and Imbens 1991). The natural bounds governing ETT($X \rightarrow Y$) in the general case can be obtained by similar means (Pearl 1995b), which yield

$$\begin{aligned}
\text{ETT}(X \rightarrow Y) &\geq \frac{P(y_1 | z_1) - P(y_1 | z_0)}{P(x_1)/P(z_1)} - \frac{P(y_0, x_1 | z_0)}{P(x_1)}, \\
\text{ETT}(X \rightarrow Y) &\leq \frac{P(y_1 | z_1) - P(y_1 | z_0)}{P(x_1)/P(z_1)} + \frac{P(y_1, x_1 | z_0)}{P(x_1)}.
\end{aligned} \tag{8.19}$$

The sharp bounds are presented in Balke (1995, p. 113). Clearly, in situations where treatment may be obtained only by those encouraged (by assignment), we have $P(x_1 \mid z_0) = 0$ and

$$\text{ETT}(X \rightarrow Y) = \frac{P(y_1 \mid z_1) - P(y_1 \mid z_0)}{P(x_1 \mid z_1)}. \tag{8.20}$$

Unlike $\text{ACE}(X \rightarrow Y)$, $\text{ETT}(X \rightarrow Y)$ is not an intrinsic property of the treatment, since it varies with the encouraging instrument. Hence, its significance lies in studies where it is desired to evaluate the efficacy of an existing program on its current participants.

8.2.6 Example: The Effect of Cholestyramine

To demonstrate by example how the bounds for $\text{ACE}(X \rightarrow Y)$ can be used to provide meaningful information about causal effects, consider the Lipid Research Clinics Coronary Primary Prevention Trial data (Program 1984). A portion (covering 337 subjects) of this data was analyzed in Efron and Feldman (1991) and is the focus of this example. Subjects were randomized into two treatment groups of roughly equal size; in one group, all subjects were prescribed cholestyramine (z_1), while subjects in the other group were prescribed a placebo (z_0). Over several years of treatment, each subject's cholesterol level was measured many times, and the average of these measurements was used as the posttreatment cholesterol level (continuous variable C_F). The compliance of each subject was determined by tracking the quantity of prescribed dosage consumed (a continuous quantity).

In order to apply the bounds of (8.17) to data from this study, the continuous data is first transformed, using thresholds, to binary variables representing treatment assignment (Z), received treatment (X), and treatment response (Y). The threshold for dosage consumption was selected as roughly the midpoint between minimum and maximum consumption; the threshold for cholesterol level reduction was set at 28 units. After this "thresholding" procedure, the data samples give rise to the following eight probabilities:[5]

$$P(y_0, x_0 \mid z_0) = 0.919, \quad P(y_0, x_0 \mid z_1) = 0.315,$$

$$P(y_0, x_1 \mid z_0) = 0.000, \quad P(y_0, x_1 \mid z_1) = 0.139,$$

$$P(y_1, x_0 \mid z_0) = 0.081, \quad P(y_1, x_0 \mid z_1) = 0.073,$$

$$P(y_1, x_1 \mid z_0) = 0.000, \quad P(y_1, x_1 \mid z_1) = 0.473.$$

These data represent a compliance rate of

$$P(x_1 \mid z_1) = 0.139 + 0.473 = 0.61,$$

---

[5] We make the large-sample assumption and take the sample frequencies as representing $P(y, x \mid z)$. To account for sample variability, all bounds should be supplemented with confidence intervals and significance levels, as in traditional analyses of controlled experiments. Section 8.5.1 assesses sample variability using Gibbs sampling.

a mean difference (using $P(z_1) = 0.50$) of

$$P(y_1 \mid x_1) - p(y_1 \mid x_0) = \frac{0.473}{0.473 + 0.139} - \frac{0.073 + 0.081}{1 + 0.315 + 0.073} = 0.662,$$

and an encouragement effect (intent to treat) of

$$P(y_1 \mid z_1) - P(y_1 \mid z_0) = 0.073 + 0.473 - 0.081 = 0.465.$$

According to (8.17), ACE$(X \rightarrow Y)$ can be bounded by

$$\text{ETT}(X \rightarrow Y) \geq 0.465 - 0.073 - 0.000 = 0.392,$$

$$\text{ETT}(X \rightarrow Y) \leq 0.465 + 0.315 + 0.000 = 0.780.$$

These are remarkably informative bounds: although 38.8% of the subjects deviated from their treatment protocol, the experimenter can categorically state that, when applied uniformly to the population, the treatment is guaranteed to increase by at least 39.2% the probability of reducing the level of cholesterol by 28 points or more.

The impact of treatment "on the treated" is equally revealing. Using equation (8.20), ETT$(X \rightarrow Y)$ can be evaluated precisely (since $P(x_1 \mid z_0) = 0$):

$$\text{ETT}(X \rightarrow Y) = \frac{0.465}{0.610} = 0.762.$$

In words, those subjects who stayed in the program are much better off than they would have been if not treated: the treatment can be credited with reducing cholesterol levels by at least 28 units in 76.2% of these subjects.

8.3   COUNTERFACTUALS AND LEGAL RESPONSIBILITY

Evaluation of counterfactual probabilities could be enlightening in some legal cases in which a plaintiff claims that a defendant's actions were responsible for the plaintiff's misfortune. Improper rulings can easily be issued without an adequate treatment of counterfactuals (Robins and Greenland 1989). Consider the following hypothetical and fictitious case study, specially crafted in Balke and Pearl (1994a) to accentuate the disparity between causal effects and causal attribution.

The marketer of PeptAid (antacid medication) randomly mailed out product samples to 10% of the households in the city of Stress, California. In a follow-up study, researchers determined for each individual whether they received the PeptAid sample, whether they consumed PeptAid, and whether they developed peptic ulcers in the following month.

The causal structure for this scenario is identical to the partial compliance model given by Figure 8.1, where z_1 asserts that PeptAid was received from the marketer, x_1 asserts that PeptAid was consumed, and y_1 asserts that peptic ulceration occurred. The data showed the following distribution:

$$P(y_0, x_0 \mid z_0) = 0.32, \quad P(y_0, x_0 \mid z_1) = 0.02,$$

$$P(y_0, x_1 \mid z_0) = 0.32, \quad P(y_0, x_1 \mid z_1) = 0.17,$$

$$P(y_1, x_0 \mid z_0) = 0.04, \quad P(y_1, x_0 \mid z_1) = 0.67,$$

$$P(y_1, x_1 \mid z_0) = 0.32, \quad P(y_1, x_1 \mid z_1) = 0.14.$$

These data indicate a high correlation between those who consumed PeptAid and those who developed peptic ulcers:

$$P(y_1 \mid x_1) = 0.50, \quad P(y_1 \mid x_0) = 0.26.$$

In addition, the intent-to-treat analysis showed that those individuals who received the PeptAid samples had a 45% greater chance of developing peptic ulcers:

$$P(y_1 \mid z_1) = 0.81, \quad P(y_1 \mid z_0) = 0.36.$$

The plaintiff (Mr. Smith), having heard of the study, litigated against both the marketing firm and the PeptAid producer. The plaintiff's attorney argued against the producer, claiming that the consumption of PeptAid triggered his client's ulcer and resulting medical expenses. Likewise, the plaintiff's attorney argued against the marketer, claiming that his client would not have developed an ulcer if the marketer had not distributed the product samples.

The defense attorney, representing both the manufacturer and marketer of PeptAid, rebutted this argument, stating that the high correlation between PeptAid consumption and ulcers was attributable to a common factor, namely, pre-ulcer discomfort. Individuals with gastrointestinal discomfort would be much more likely both to use PeptAid and to develop stomach ulcers. To bolster his clients' claims, the defense attorney introduced expert analysis of the data showing that, on average, consumption of PeptAid actually decreases an individual's chances of developing ulcers by at least 15%.

Indeed, the application of (8.14a,b) results in the following bounds on the average causal effect of PeptAid consumption on peptic ulceration:

$$-0.23 \leq \mathrm{ETT}(X \rightarrow Y) \leq -0.15;$$

this proves that PeptAid is beneficial to the population as a whole.

The plaintiff's attorney, though, stressed the distinction between the average treatment effects for the entire population and for the subpopulation consisting of those individuals who, like his client, received the PeptAid sample, consumed it, and then developed ulcers. Analysis of the population data indicated that, had PeptAid not been distributed, Mr. Smith would have had at most a 7% chance of developing ulcers – regardless of any confounding factors such as pre-ulcer pain. Likewise, if Mr. Smith had not consumed PeptAid, he would have had at most a 7% chance of developing ulcers.

The damaging statistics against the marketer are obtained by evaluating the bounds on the counterfactual probability that the plaintiff would have developed a peptic ulcer if he had not received the PeptAid sample, given that he in fact received the sample PeptAid, consumed the PeptAid, and developed peptic ulcers. This probability may be written in terms of the parameters q_{13}, q_{31}, and q_{33} as

$$P(Y_{z_0} = y_1 \mid y_1, x_1, z_1) = \frac{P(r_z = 1)(q_{13} + q_{31} + q_{33})}{P(y_1, x_1, z_1)},$$

since only the combinations $\{r_x = 1, r_y = 3\}$, $\{r_x = 3, r_y = 1\}$, and $\{r_x = 3, r_y = 3\}$ satisfy the joint event $\{X = x_1, Y = y_1, Y_{z_0} = y_1\}$ (see (8.6), (8.7), and (8.11)). Therefore,

$$P(Y_{z_0} = y_1 \mid y_1, x_1, z_1) = \frac{q_{13} + q_{31} + q_{33}}{P(y_1, x_1 \mid z_1)}.$$

This expression is linear in the q parameters and may be bounded using linear programming to give

$$P(Y_{z_0} = y_1 \mid z_1, x_1, y_1) \geq \frac{1}{p_{11.1}} \max \left\{ \begin{array}{c} 0 \\ p_{11.1} - p_{00.0} \\ p_{11.0} - p_{00.1} - p_{10.1} \\ p_{10.0} - p_{01.1} - p_{10.1} \end{array} \right\},$$

$$P(Y_{z_0} = y_1 \mid z_1, x_1, y_1) \leq \frac{1}{p_{11.1}} \min \left\{ \begin{array}{c} p_{11.1} \\ p_{10.0} + p_{11.0} \\ 1 - p_{00.0} - p_{10.1} \end{array} \right\}.$$

Similarly, the damaging evidence against PeptAid's producer is obtained by evaluating the bounds on the counterfactual probability

$$P(Y_{x_0} = y_1 \mid y_1, x_1, z_1) = \frac{q_{13} + q_{33}}{p_{11.1}}.$$

If we minimize and maximize the numerator (subject to (8.13)), we obtain

$$P(Y_{x_0} = y_1 \mid y_1, x_1, z_1) \geq \frac{1}{p_{11.1}} \max \left\{ \begin{array}{c} 0 \\ p_{11.1} - p_{00.0} - p_{11.0} \\ p_{10.0} - p_{01.1} - p_{10.1} \end{array} \right\},$$

$$P(Y_{x_0} = y_1 \mid y_1, x_1, z_1) \leq \frac{1}{p_{11.1}} \min \left\{ \begin{array}{c} p_{11.1} \\ p_{10.0} + p_{11.0} \\ 1 - p_{00.0} - p_{10.1} \end{array} \right\}.$$

Substituting the observed distribution $P(y, x \mid z)$ into these formulas, the following bounds were obtained:

$$0.93 \leq P(Y_{z_0} = y_0 \mid z_1, x_1, y_1) \leq 1.00,$$

$$0.93 \leq P(Y_{x_0} = y_0 \mid z_1, x_1, y_1) \leq 1.00.$$

Thus, at least 93% of the people in the plaintiff's category would not have developed ulcers had they not been encouraged to take PeptAid (z_0) or, similarly, had they not taken PeptAid (x_0). This lends very strong support for the plaintiff's claim that he was adversely affected by the marketer and producer's actions and product.

In Chapter 9 we will continue the analysis of causal attribution in specific events, and we will establish conditions under which the probability of correct attribution can be identified from both experimental and nonexperimental data.

8.4   A TEST FOR INSTRUMENTS

As defined in Section 8.2, our model of imperfect experiment rests on two assumptions: Z is randomized, and Z has no side effect on Y. These two assumptions imply that Z is independent of U, a condition that economists call "exogeneity" and which qualifies Z as an instrumental variable (see Sections 5.4.3 and 7.4.5) relative to the relation between X and Y. For a long time, experimental verification of whether a variable Z is exogenous or instrumental has been thought to be impossible (Imbens and Angrist 1994), since the definition involves unobservable factors (or disturbances, as they are usually called) such as those represented by U.[6] The notion of exogeneity, like that of causation itself, has been viewed as a product of subjective modeling judgment, exempt from the scrutiny of nonexperimental data.

The bounds presented in (8.14a,b) tell a different story. Despite its elusive nature, exogeneity can be given an empirical test. The test is not guaranteed to detect all violations of exogeneity, but it can (in certain circumstances) screen out very bad would-be instruments.

By insisting that each upper bound in (8.14b) be higher than the corresponding lower bound in (8.14a), we obtain the following testable constraints on the observed distribution:

$$P(y_0, x_0 \mid z_0) + P(y_1, x_0 \mid z_1) \leq 1,$$
$$P(y_0, x_1 \mid z_0) + P(y_1, x_1 \mid z_1) \leq 1,$$
$$P(y_1, x_0 \mid z_0) + P(y_0, x_0 \mid z_1) \leq 1,$$
$$P(y_1, x_1 \mid z_0) + P(y_0, x_1 \mid z_1) \leq 1.$$

$$(8.21)$$

If any of these inequalities is violated, the investigator can deduce that at least one of the assumptions underlying our model is violated as well. If the assignment is carefully randomized, then any violation of these inequalities must be attributed to some direct influence that the assignment process has on subjects' responses (e.g., a traumatic experience). Alternatively, if direct effects of Z on Y can be eliminated – say, through an effective use of a placebo – then any observed violation of the inequalities can safely be attributed to spurious correlation between Z and U: namely, to assignment bias and hence loss of exogeneity.

The Instrumental Inequality

The inequalities in (8.21), when generalized to multivalued variables, assume the form

$$\max_{x} \sum_{y} \left[\max_{z} P(y, x \mid z)\right] \leq 1,$$

$$(8.22)$$

---

[6] The tests developed by economists (Wu 1973) merely compare estimates based on two or more instruments and, in case of discrepency, do not tell us objectively which estimate is incorrect.

which is called the *instrumental inequality*. A proof is given in Pearl (1995b,c). Extending the instrumental inequality to the case where Z or Y is continuous presents no special difficulty. If $f(y \mid x, z)$ is the conditional density function of Y given X and Z, then the inequality becomes

$$\int_y \max_z \left[f(y \mid x, z) \, P(x \mid z) \right] dy \leq 1 \quad \forall x. \tag{8.23}$$

However, the transition to a continuous X signals a drastic change in behavior, and led Pearl (1995c) to conjecture that the structure of Figure 8.1 induces no constraint whatsoever on the observed density. The conjecture was proven by Bonet (2001).

From (8.21) we see that the instrumental inequality is violated when the controlling instrument Z manages to produce significant changes in the response variable Y while the treatment X remains constant. Although such changes could in principle be explained by strong correlations among U, X, and Y (since X does not screen off Z from Y), the instrumental inequality sets a limit on the magnitude of the changes.

The similarity of the instrumental inequality to Bell's inequality in quantum physics (Suppes 1988; Cushing and McMullin 1989) is not accidental; both inequalities delineate a class of observed correlations that cannot be explained by hypothesizing latent common causes. The instrumental inequality can, in a sense, be viewed as a generalization of Bell's inequality for cases where direct causal connection is permitted to operate between the correlated observables, X and Y.

The instrumental inequality can be tightened appreciably if we are willing to make additional assumptions about subjects' behavior – for example, that no individual can be discouraged by the encouragement instrument or (mathematically) that, for all u, we have

$$P(x_1 \mid z_1, u) \geq P(x_1 \mid z_0, u).$$

Such an assumption amounts to having no contrarians in the population, that is, no subjects who will consistently choose treatment contrary to their assignment. Under this assumption, the inequalities in (8.21) can be tightened (Balke and Pearl 1997) to yield

$$P(y, x_1 \mid z_1) \geq P(y, x_1 \mid z_0),$$
$$P(y, x_0 \mid z_0) \geq P(y, x_0 \mid z_1) \tag{8.24}$$

for all $y \in \{y_0, y_1\}$. Violation of these inequalities now means either selection bias or direct effect of Z on Y or the presence of defiant subjects.

8.5   A BAYESIAN APPROACH TO NONCOMPLIANCE

8.5.1   Bayesian Methods and Gibbs Sampling

This section describes a general method of estimating causal effects and counterfactual probabilities from finite samples, while incorporating prior knowledge about the population. The method, developed by Chickering and Pearl (1997),[7] is applicable within the Bayesian

---

[7] A similar method, though lacking the graphical perspective, is presented in Imbens and Rubin (1997).

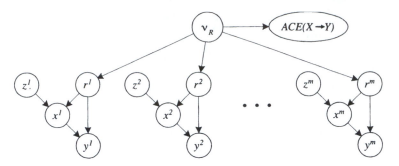

Figure 8.4   Model used to represent the independencies in $P(\{\chi\}) \cup \{v_R\} \cup \{\mathrm{ACE}(X \to Y)\})$.

framework, according to which (i) any unknown statistical parameter can be assigned prior probability and (ii) the estimation of that parameter amounts to computing its posterior distribution, conditioned on the sampled data. In our case the parameter in question is the probability $P(r_x, r_y)$ (or $P(r)$ for short), from which we can deduce $\mathrm{ACE}(X \to Y)$.

If we think of $P(r)$ not as a probability but rather as the fraction v_r of individuals in the population who possess response characteristics given by $R = r$, then the idea of assigning probability to such a quantity would fit the standard philosophy of Bayesian analysis; v_r is a potentially measurable (albeit unknown) physical quantity and can therefore admit a prior probability, one that encodes our uncertainty in that quantity.

Assume there are m subjects in the experiment. We use z^i, x^i, y^i to denote the observed value of Z, X, Y, respectively, for subject i. Similarly, we use r^i to denote the (unobserved) compliance (r_x) and response (r_y) combination for subject i. We use χ^i to denote the triple $\{z^i, x^i\ y^i\}$.

Given the observed data χ from the experiment and a prior distribution over the unknown fractions v_r, our problem is to derive the posterior distribution for $\mathrm{ACE}(X \to Y)$. The posterior distributions of both v_R and $\mathrm{ACE}(X \to Y)$ can be derived using the graphical model shown in Figure 8.4, which explicitly represents the independencies that hold in the joint (Bayesian) distribution defined over the variables $\{\chi, v_R, \mathrm{ACE}(X \to Y)\}$. The model can be understood as m realizations of the response-variable model (Figure 8.3), one for each triple in χ, connected together using the node representing the unknown fractions $v_R = (v_{r_1}, v_{r_2}, \dots, v_{r_{16}})$. The model explicitly represents the assumption that, given the fractions v_R, the probability of a subject belonging to any of the 16 compliance-response subpopulations does not depend on the compliance and response behavior of the other subjects in the experiment. From (8.10), $\mathrm{ACE}(X \to Y)$ is a deterministic function of v_R, and consequently $\mathrm{ACE}(X \to Y)$ is independent of all other variables in the domain once these fractions are known.

In principle, then, estimating $\mathrm{ACE}(X \to Y)$ reduces to the standard inference task of computing the posterior probability for a variable in a fully specified Bayesian network. (The graphical techniques for this inferential computation are briefly summarized in Section 1.2.4.) In many cases, the independencies embodied in the graph can be exploited to render the inference task efficient. Unfortunately, because the r^i are never observed, deriving the posterior distribution for $\mathrm{ACE}(X \to Y)$ is not tractable in our model, even with the given independencies. To obtain an estimate of the posterior distribution of $\mathrm{ACE}(X \to Y)$, an approximation technique known as Gibbs sampling can be

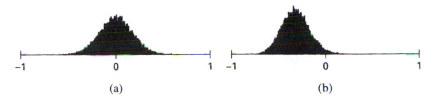

Figure 8.5   (a) The prior distribution of ACE($X \rightarrow Y$) induced by flat priors over the parameters V_{CR}. (b) The distribution of ACE($X \rightarrow Y$) induced by skewed priors over the parameters.

used (Robert and Casella 1999). A graphical version of this technique, called "stochastic simulation," is described in Pearl (1988b, p. 210); the details (as applied to the graph of Figure 8.4) are discussed in Chickering and Pearl (1997). Here we present typical results, in the form of histograms, that demonstrate the general applicability of this technique to problems of causal inference.

8.5.2   The Effects of Sample Size and Prior Distribution

The method takes as input (1) the observed data χ, expressed as the number of cases observed for each of the 8 possible realizations of $\{z, x, y\}$, and (2) a Dirichlet prior over the unknown fractions v_R, expressed in terms of 16 parameters. The system outputs the posterior distribution of ACE($X \rightarrow Y$), expressed in a histogram.

To show the effect of the prior distribution on the output, we present all the results using two different priors. The first is a flat (uniform) distribution over the 16-vector v_R that is commonly used to express ignorance about the domain. The second prior is skewed to represent a strong dependency between the compliance and response characteristics of the subjects. Figure 8.5 shows the distribution of ACE($X \rightarrow Y$) induced by these two prior distributions (in the absence of any data). We see that the skewed prior of Figure 8.5(b) assigns almost all the weight to negative values of ACE($X \rightarrow Y$).

To illustrate how increasing sample size washes away the effect of the prior distribution, we apply the method to simulated data drawn from a distribution $P(x, y \mid z)$ for which ACE is known to be identified. Such a distribution is shown Table 8.1. For this distribution, the resulting upper and lower bounds of (8.14a,b) collapse to a single point: ACE($X \rightarrow Y$) = 0.55.

Figure 8.6 shows the output of the Gibbs sampler when applied to data sets of various sizes drawn from the distribution shown in Table 8.1, using both the flat and the skewed prior. As expected, as the number of cases increases, the posterior distributions become increasingly concentrated near the value 0.55. In general, because the skewed prior for ACE($X \rightarrow Y$) is concentrated further from 0.55 than the uniform prior, more cases are needed before the posterior distribution converges to the value 0.55.

8.5.3   Causal Effects from Clinical Data with Imperfect Compliance

In this section we analyze two clinical data sets obtained under conditions of imperfect compliance. Consider first the Lipid Research Clinics Coronary Primary Prevention data described in Section 8.2.6. The resulting data set (after thresholding) is shown in Table 8.2. Using the large-sample assumption, (8.14a,b) gives the bounds $0.39 \leq$ ACE($X \rightarrow Y$) ≤ 0.78.

Table 8.1. *Distribution Resulting*
in an Identifiable $\text{ACE}(X \rightarrow Y)$

z	x	y	$P(x, y, z)$
0	0	0	0.275
0	0	1	0.0
0	1	0	0.225
0	1	1	0.0
1	0	0	0.225
1	0	1	0.0
1	1	0	0.0
1	1	1	0.275

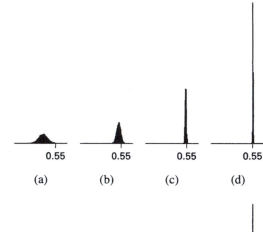

(a) (b) (c) (d)

Figure 8.6 Output histograms for identified treatment effect using two priors,
(a), (b), (c), and (d) show the posteri-
ors for $\text{ACE}(X \rightarrow Y)$ using the flat prior
and data sets that consisted of 10, 100,
1,000 and 10,000 subjects, respectively;
(e), (f), (g), and (h) show the posteriors
for $\text{ACE}(X \rightarrow Y)$ using the skewed prior
with the same respective data sets. (Hor-
izontal lines span the interval $(-1, +1)$.)

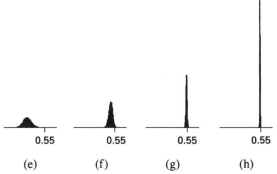

(e) (f) (g) (h)

Figure 8.7 shows posterior densities for $\text{ACE}(X \rightarrow Y)$, based on these data. Rather
remarkably, even with only 337 cases in the data set, both posterior distributions are
highly concentrated within the large-sample bounds of 0.39 and 0.78.

As a second example, we consider an experiment described by Sommer et al. (1986)
that was designed to determine the impact of vitamin A supplementation on childhood
mortality. In the study, 450 villages in northern Sumatra were randomly assigned to participate in a vitamin A supplementation scheme or serve as a control group for one year.
Children in the treatment group received two large doses of vitamin A (x_1), while those

Table 8.2. *Observed Data for the Lipid Study and the Vitamin A Study*

z	x	y	Lipid Study Observations	Vitamin A Study Observations
0	0	0	158	74
0	0	1	14	11,514
0	1	0	0	0
0	1	1	0	0
1	0	0	52	34
1	0	1	12	2,385
1	1	0	23	12
1	1	1	78	9,663

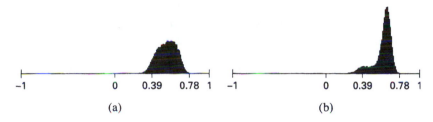

Figure 8.7   Output histograms for the Lipid data: (a) using flat priors; (b) using skewed priors.

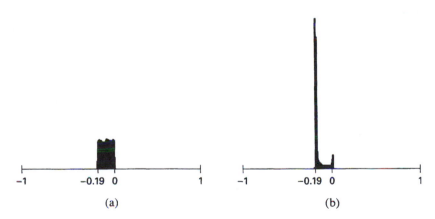

Figure 8.8   Output histograms for the vitamin A data: (a) using flat priors; (b) using skewed priors.

in the control group received no treatment (x_0). After the year had expired, the number of deaths y_0 were counted for both groups. The results of this study are also shown in Table 8.2.

Under the large-sample assumption, the inequalities of (8.14a,b) yield the bounds $-0.19 \leq \text{ACE}(X \to Y) \leq 0.01$. Figure 8.8 shows posterior densities for $\text{ACE}(X \to Y)$, given the data, for two priors. It is interesting to note that, for this study, the choice of the prior distribution has a significant effect on the posterior. This suggests that if the clinician is not very confident in the prior, then a sensitivity analysis should be performed.

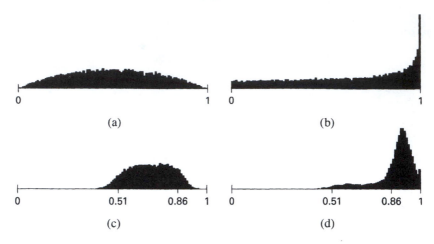

Figure 8.9   Prior ((a) and (b)) and posterior ((c) and (d)) distributions for a subpopulation $f(v_R)$ specified by the counterfactual query: "Would Joe have improved had he taken the drug, given that he did not improve without it?" Part (a) corresponds to the flat prior, (b) to the skewed prior.

In such cases, the asymptotic bounds are more informative than the Bayesian estimates, and the major role of the Gibbs sampler would be to give an indication of the sharpness of the boundaries around those bounds.

8.5.4   Bayesian Estimate of Single-Event Causation

In addition to assessing causal effects, the Bayesian method just described is also capable (with only minor modification) of answering a variety of counterfactual queries concerning individuals with specific characteristics. Queries of this type were analyzed and bounded in Section 8.3 under the large-sample assumption. In this section, we demonstrate a Bayesian analysis of the following query. What is the probability that Joe would have had an improved cholesterol reading had he taken cholestyramine, given that: (1) Joe was in the control group of the Lipid study; (2) Joe took the placebo as prescribed; and (3) Joe's cholesterol level did not improve.

   This query can be answered by running the Gibbs sampler on a model identical to that shown in Figure 8.4, except that the function $\text{ACE}(X \rightarrow Y)$ (equation (8.10)) is replaced by another function of v_R, one that represents our query. If Joe was in the control group and took the placebo, that means he is either a complier or a never-taker. Furthermore, because Joe's cholesterol level did not improve, Joe's response behavior is either never-recover or helped. Consequently, he must be a member of one of the four compliance-response populations:

$$\{(r_x = 0, r_y = 0), (r_x = 0, r_y = 1), (r_x = 1, r_y = 0), (r_x = 1, r_y = 1)\}.$$

Joe would have improved had he taken cholestyramine if and only if his response behavior is helped ($r_y = 1$). It follows that the query of interest is captured by the function

$$f(v_R) = \frac{v_{01} + v_{11}}{v_{01} + v_{02} + v_{11} + v_{12}}.$$

   Figures 8.9(a) and (b) show the prior distribution of $f(v_R)$ that follows from the flat prior and the skewed prior, respectively. Figures 8.9(c) and (d) show the posterior distribution $P(f(v_R \mid \chi))$ obtained from the Lipid data when using the flat prior and the skewed prior, respectively. For reference, the bounds computed under the large-sample assumption are $0.51 \le f(v_R \mid \chi) \le 0.86$.

Thus, despite 39% noncompliance in the treatment group and despite having just 337 subjects, the study strongly supports the conclusion that – given his specific history – Joe would have been better off taking the drug. Moreover, the conclusion holds for both priors.

8.6   CONCLUSION

This chapter has developed causal-analytic techniques for managing one of the major problems in clinical experiments: the assessment of treatment efficacy in the face of imperfect compliance. Estimates based solely on intent-to-treat analysis – as well as those based on instrumental variable formulas – can be misleading, and may even lie entirely outside the theoretical bounds. The general formulas established in this chapter provide instrument-independent guarantees for policy analysis and, in addition, should enable analysts to determine the extent to which efforts to enforce compliance may increase overall treatment effectiveness.

The importance of indirect experimentation and instrumental variables is not confined to studies involving human subjects. Experimental conditions equivalent to those of imperfect compliance occur whenever the variable whose causal effect we seek to assess cannot be manipulated directly yet could be partially influenced by indirect means. Typical applications involve the diagnosis of ongoing processes for which the source of malfunctioning behavior must be identified using indirect means because direct manipulation of suspected sources is either physically impossible or prohibitively expensive.

Methodologically, the message of this chapter has been to demonstrate that, even in cases where causal quantities are not identifiable, reasonable assumptions about the structure of causal relationships in the domain can be harnessed to yield useful quantitative information about the strengths of those relationships. Once such assumptions are articulated in graphical form and re-encoded in terms of canonical partitions, they can be submitted to algebraic methods that yield informative bounds on the quantities of interest. The canonical partition further allows us to supplement structural assumptions with prior beliefs about the population under study and invite Gibbs sampling techniques to facilitate Bayesian estimation of the target quantities.

Acknowledgments

This chapter is based on collaborative works with Alex Balke and David Chickering, and has benefited from discussions with Phil Dawid, James Heckman, Guido Imbens, Steve Jacobsen, Steffen Lauritzen, Charles Manski, and James Robins.

CHAPTER NINE

Probability of Causation: Interpretation and Identification

Come and let us cast lots to find out
who is to blame for this ordeal.
                    Jonah 1:7

Preface

Assessing the likelihood that one event *was the cause* of another guides much of what we understand about (and how we act in) the world. For example, according to common judicial standard, judgment in favor of the plaintiff should be made if and only if it is "more probable than not" that the defendant's action was the *cause* of the plaintiff's damage (or death). But causation has two faces, *necessary* and *sufficient;* which of the two have lawmakers meant us to consider? And how are we to evaluate their probabilities?

This chapter provides formal semantics for the probability that event *x* was a *necessary* or *sufficient* cause (or both) of another event *y*. We then explicate conditions under which the probability of necessary (or sufficient) causation can be learned from statistical data, and we show how data from both experimental and nonexperimental studies can be combined to yield information that neither kind of study alone can provide.

9.1 INTRODUCTION

The standard counterfactual definition of causation (i.e., that *E* would not have occurred were it not for *C*) captures the notion of "necessary cause." Competing notions such as "sufficient cause" and "necessary and sufficient cause" are of interest in a number of applications, and these too can be given concise mathematical definitions in structural model semantics (Section 7.1). Although the distinction between necessary and sufficient causes goes back to J. S. Mill (1843), it has received semiformal explications only in the 1960s – via conditional probabilities (Good 1961) and logical implications (Mackie 1965; Rothman 1976). These explications suffer from basic semantical difficulties,[1] and they do not yield procedures for computing probabilities of causes as those provided by the structural account (Sections 7.1.3 and 8.3).

---

[1] The limitations of the probabilistic account are discussed in Section 7.5; those of the logical account will be discussed in Section 10.1.4.

In this chapter we explore the counterfactual interpretation of necessary and sufficient causes, illustrate the application of structural model semantics to the problem of identifying probabilities of causes, and present, by way of examples, new ways of estimating probabilities of causes from statistical data. Additionally, we argue that necessity and sufficiency are two distinct facets of causation and that both facets should take part in the construction of causal explanations.

Our results have applications in epidemiology, legal reasoning, artificial intelligence (AI), and psychology. Epidemiologists have long been concerned with estimating the probability that a certain case of disease is "attributable" to a particular exposure, which is normally interpreted counterfactually as "the probability that disease would not have occurred in the absence of exposure, given that disease and exposure did in fact occur." This counterfactual notion, which Robins and Greenland (1989) called the "probability of causation," measures how *necessary* the cause is for the production of the effect.[2] It is used frequently in lawsuits, where legal responsibility is at the center of contention (see, e.g., Section 8.3). We shall denote this notion by the symbol PN, an acronym for probability of necessity.

A parallel notion of causation, capturing how *sufficient* a cause is for the production of the effect, finds applications in policy analysis, AI, and psychology. A policy maker may well be interested in the dangers that a certain exposure may present to the healthy population (Khoury et al. 1989). Counterfactually, this notion can be expressed as the "probability that a healthy unexposed individual would have contracted the disease had he or she been exposed," and it will be denoted by PS (probability of sufficiency). A natural extension would be to inquire for the probability of necessary and sufficient causation (PNS) – that is, how likely a given individual is to be affected both ways.

As the examples illustrate, PS assesses the presence of an active causal process capable of producing the effect, while PN emphasizes the absence of alternative processes – not involving the cause in question – that are still capable of explaining the effect. In legal settings, where the occurrence of the cause (x) and the effect (y) are fairly well established, PN is the measure that draws most attention, and the plaintiff must prove that y would not have occurred *but for* x (Robertson 1997). Still, lack of sufficiency may weaken arguments based on PN (Good 1993; Michie 1999).

It is known that PN is in general nonidentifiable, that is, it cannot be estimated from frequency data involving exposures and disease cases (Greenland and Robins 1988; Robins and Greenland 1989). The identification is hindered by two factors.

1. *Confounding* – Exposed and unexposed subjects may differ in several relevant factors or, more generally, the cause and the effect may both be influenced by a third factor. In this case we say that the cause is not *exogenous* relative to the effect (see Section 7.4.5).

---

[2] Greenland and Robins (1988) further distinguish between two ways of measuring probabilities of causation: the first (called "excess fraction") concerns only *whether* the effect (e.g., disease) occurs by a particular time; the second (called "etiological fraction") requires consideration of *when* the effect occurs. We will confine our discussion here to events occurring within a specified time period, or to "all or none" outcomes (such as birth defects) for which the probability of occurrence but not the time to occurrence is important.

2. *Sensitivity to the generative process* – Even in the absence of confounding, probabilities of certain counterfactual relationships cannot be identified from frequency information unless we specify the functional relationships that connect causes and effects. Functional specification is needed whenever the facts at hand (e.g., disease) might be affected by the counterfactual antecedent (e.g., exposure) (see the examples in Sections 1.4, 7.5, and 8.3).

Although PN is not identifiable in the general case, several formulas have nevertheless been proposed to estimate attributions of various kinds in terms of frequencies obtained in epidemiological studies (Breslow and Day 1980; Hennekens and Buring 1987; Cole 1997). Naturally, any such formula must be predicated upon certain implicit assumptions about the data-generating process. Section 9.2 explicates some of those assumptions and explores conditions under which they can be relaxed.[3] It offers new formulas for PN and PS in cases where causes are confounded (with outcomes) but their effects can nevertheless be estimated (e.g., from clinical trials or from auxiliary measurements). Section 9.3 exemplifies the use of these formulas in legal and epidemiological settings, while Section 9.4 provides a general condition for the identifiability of PN and PS when functional relationships are only partially known.

The distinction between necessary and sufficient causes has important implications in AI, especially in systems that generate verbal explanations automatically (see Section 7.2.3). As can be seen from the epidemiological examples, necessary causation is a concept tailored to a specific event under consideration (singular causation), whereas sufficient causation is based on the general tendency of certain event *types* to produce other event types. Adequate explanations should respect both aspects. If we base explanations solely on generic tendencies (i.e., sufficient causation) then we lose important specific information. For instance, aiming a gun at and shooting a person from 1,000 meters away will not qualify as an explanation for that person's death, owing to the very low tendency of shots fired from such long distances to hit their marks. This stands contrary to common sense, for when the shot does hit its mark on that singular day, regardless of the reason, the shooter is an obvious culprit for the consequence. If, on the other hand, we base explanations solely on singular-event considerations (i.e., necessary causation), then various background factors that are normally present in the world would awkwardly qualify as explanations. For example, the presence of oxygen in the room would qualify as an explanation for the fire that broke out, simply because the fire would not have occurred were it not for the oxygen. That we judge the match struck, not the oxygen, to be the actual cause of the fire indicates that we go beyond the singular event at hand (where each factor alone is both necessary and sufficient) and consider situations of the same general type – where oxygen alone is obviously insufficient to start a fire. Clearly, some balance must be struck between the necessary and the sufficient components of causal explanation, and the present chapter illuminates this balance by formally explicating the basic relationships between these two components.

---

[3] A set of sufficient conditions for the identification of etiological fractions are given in Robins and Greenland (1989). These conditions, however, are too restrictive for the identification of PN, which is oblivious to the temporal aspects associated with etiological fractions.

9.2 NECESSARY AND SUFFICIENT CAUSES: CONDITIONS OF IDENTIFICATION

9.2.1 Definitions, Notation, and Basic Relationships

Using the counterfactual notation and the structural model semantics introduced in Section 7.1, we give the following definitions for the three aspects of causation discussed in the introduction.

Definition 9.2.1 (Probability of Necessity, PN)

Let X and Y be two binary variables in a causal model M. Let x and y stand (respectively) for the propositions X = true and Y = true, and let x' and y' denote their complements. The probability of necessity is defined as the expression

$$\text{PN} \triangleq P(Y_{x'} = \textit{false} \mid X = \textit{true}, Y = \textit{true})$$

$$\triangleq P(y'_{x'} \mid x, y). \tag{9.1}$$

In other words, PN stands for the probability of $y'_{x'}$ (that event y would not have occurred in the absence of event x), given that x and y did in fact occur.

Observe the slight change in notation relative to that used in Section 7.1. Lowercase letters (e.g., x and y) denoted values of variables in Section 7.1 but now stand for propositions (or events). Note also the abbreviations y_x for Y_x = true and y'_x for Y_x = false.[4] Readers accustomed to writing "$A > B$" for the counterfactual "B if it were A" can translate (9.1) to read PN $\triangleq P(x' > y' \mid x, y)$.[5]

Definition 9.2.2 (Probability of Sufficiency, PS)

$$\text{PS} \triangleq P(y_x \mid y', x'). \tag{9.2}$$

PS measures the capacity of x to *produce* y, and, since "production" implies a transition from the absence to the presence of x and y, we condition the probability $P(y_x)$ on situations where x and y are both absent. Thus, mirroring the necessity of x (as measured by PN), PS gives the probability that setting x would produce y in a situation where x and y are in fact absent.

Definition 9.2.3 (Probability of Necessity and Sufficiency, PNS)

$$\text{PNS} \triangleq P(y_x, y'_{x'}). \tag{9.3}$$

---

[4] These were proposed by Peyman Meshkat (in class homework) and substantially simplify the derivations.

[5] Definition 9.2.1 generalizes naturally to cases where X and Y are multivalued, say $x \in \{x_1, x_2, \ldots, x_k\}$ and $y \in \{y_1, y_2, \ldots, y_l\}$. We say that event $C = \bigvee_{i \in I} (X = x_i)$ is "counterfactually necessary" for $E = \bigvee_{j \in J} (Y = y_j)$, written $\overline{C} > \overline{E}$, if Y_x falls outside E whenever $X = x$ is outside C. Accordingly, the probability that C was a necessary cause of E is defined as PN $\triangleq P(\overline{C} > \overline{E} \mid C, E)$. For simplicity, however, we will pursue the analysis in the binary case.

PNS stands for the probability that y would respond to x both ways, and therefore measures both the sufficiency and necessity of x to produce y.

Associated with these three basic notions are other counterfactual quantities that have attracted either practical or conceptual interest. We will mention two such quantities but will not dwell on their analyses, since these follow naturally from our treatment of PN, PS, and PNS.

Definition 9.2.4 (Probability of Disablement, PD)

$$\text{PD} \triangleq P(y'_{x'} \mid y). \tag{9.4}$$

PD measures the probability that y would have been prevented if it were not for x; it is therefore of interest to policy makers who wish to assess the social effectiveness of various prevention programs (Fleiss 1981, pp. 75–6).

Definition 9.2.5 (Probability of Enablement, PE)

$$\text{PE} \triangleq P(y_x \mid y').$$

PE is similar to PS, save for the fact that we do not condition on x'. It is applicable, for example, when we wish to assess the danger of an exposure on the entire population of healthy individuals, including those who were already exposed.

Although none of these quantities is sufficient for determining the others, they are not entirely independent, as shown in the following lemma.

Lemma 9.2.6

The probabilities of causation (PNS, PN, *and* PS) *satisfy the following relationship:*

$$\text{PNS} = P(x, y)\text{PN} + P(x', y')\text{PS}. \tag{9.5}$$

Proof

The consistency condition of (7.20), $X = x \Longrightarrow Y_x = Y$, translates in our notation into

$$x \Longrightarrow (y_x = y), \quad x' \Longrightarrow (y_{x'} = y).$$

Hence we can write

$$y_x \wedge y'_{x'} = (y_x \wedge y'_{x'}) \wedge (x \vee x')$$
$$= (y \wedge x \wedge y'_{x'}) \vee (y_x \wedge y' \wedge x').$$

Taking probabilities on both sides and using the disjointness of x and x', we obtain

$$P(y_x, y'_{x'}) = P(y'_{x'}, x, y) + P(y_x, x', y')$$
$$= P(y'_{x'} \mid x, y)P(x, y) + P(y_x \mid x', y')P(x', y'),$$

which proves Lemma 9.2.6. $\qquad\qquad\Box$

To put into focus the aspects of causation captured by PN and PS, it is helpful to characterize those changes in the causal model that would leave each of the two measures invariant. The next two lemmas show that PN is insensitive to the introduction of potential inhibitors of y, while PS is insensitive to the introduction of alternative causes of y.

Lemma 9.2.7

Let PN(x, y) stand for the probability that x is a necessary cause of y. Let $z = y \wedge q$ be a consequence of y that is potentially inhibited by q'. If $q \perp\!\!\!\perp \{X, Y, Y_{x'}\}$, then

$$\text{PN}(x, z) \triangleq P(z'_{x'} \mid x, z) = P(y'_{x'} \mid x, y) \triangleq \text{PN}(x, y).$$

Cascading the process $Y_x(u)$ with the link $z = y \wedge q$ amounts to inhibiting the output of the process with probability $P(q')$. Lemma 9.2.7 asserts that, if q is randomized, we can add such a link without affecting PN. The reason is clear; conditioning on x and z implies that, in the scenario considered, the added link was not inhibited by q'.

Proof of Lemma 9.2.7

We have

$$\text{PN}(x, z) = P(z'_{x'} \mid x, z) = \frac{P(z'_{x'}, x, z)}{P(x, z)}$$

$$= \frac{P(z'_{x'}, x, z \mid q)P(q) + P(z'_{x'}, x, z \mid q')P(q')}{P(z, x, q) + P(z, x, q')}. \tag{9.6}$$

Using $z = y \wedge q$, it follows that

$$q \implies (z = y), \quad q \implies (z'_{x'} = y'_{x'}), \quad \text{and} \quad q' \implies z';$$

therefore,

$$\text{PN}(x, z) = \frac{P(y'_{x'}, x, y \mid q)P(q) + 0}{P(y, x, q) + 0}$$

$$= \frac{P(y'_{x'}, x, y)}{P(y, x)} = P(y'_{x'} \mid x, y) = \text{PN}(x, y). \qquad \square$$

Lemma 9.2.8

Let PS(x, y) stand for the probability that x is a sufficient cause of y, and let $z = y \vee r$ be a consequence of y that may also be triggered by r. If $r \perp\!\!\!\perp \{X, Y, Y_x\}$, then

$$\text{PS}(x, z) \triangleq P(z_x \mid x', z') = P(y_x \mid x', y') \triangleq \text{PS}(x, y).$$

Lemma 9.2.8 asserts that we can add alternative (independent) causes (r) without affecting PS. The reason again is clear; conditioning on the event x' and y' implies that the added causes (r) were not active. The proof of Lemma 9.2.8 is similar to that of Lemma 9.2.7.

Since all the causal measures defined so far invoke conditionalization on y, and since y is presumed to be affected by x, we know that none of these quantities is identifiable from knowledge of the causal diagram $G(M)$ and the data $P(v)$ alone, even under conditions of no-confounding. Moreover, none of these quantities determines the others in the general case. However, simple interrelationships and useful bounds can be derived for these quantities under the assumption of no-confounding, an assumption that we call *exogeneity*.

9.2.2   Bounds and Basic Relationships under Exogeneity

Definition 9.2.9 (Exogeneity)
A variable X is said to be exogenous *relative to Y in model M if and only if*

$$\{Y_x, Y_{x'}\} \perp\!\!\!\perp X. \tag{9.7}$$

In other words, the way Y would potentially respond to conditions x or x' is independent of the actual value of X.

Equation (9.7) is a strong version of those used in Chapter 5 (equation (5.30)) and in Chapter 6 (equation (6.10)) in that it involves the joint variable $\{Y_x, Y_{x'}\}$. This definition was named "strong ignorability" in Rosenbaum and Rubin (1983), and it coincides with the classical error-based criterion for exogeneity (Christ 1966, p. 156; see Section 7.4.5) and with the back-door criterion of Definition 3.3.1. The weaker definition of (5.30) is sufficient for all the results in this chapter except equations (9.11), (9.12), and (9.19), for which strong exogeneity (9.7) is needed.

The importance of exogeneity lies in permitting the identification of $\{P(y_x), P(y_{x'})\}$, the *causal effect* of X on Y, since (using $x \Rightarrow (y_x = y)$)

$$P(y_x) = P(y_x \mid x) = P(y \mid x), \tag{9.8}$$

with similar reduction for $P(y_{x'})$.

Theorem 9.2.10
Under condition of exogeneity, PNS *is bounded as follows:*

$$\max[0, P(y \mid x) - P(y \mid x')] \le \text{PNS} \le \min[P(y \mid x), P(y' \mid x')]. \tag{9.9}$$

Both bounds are sharp in the sense that, for every joint distribution P(x, y), there exists a model y = f(x, u), with u independent of x, that realizes any value of PNS *permitted by the bounds.*

Proof
For any two events A and B, we have the sharp bounds

$$\max[0, P(A) + P(B) - 1] \le P(A, B) \le \min[P(A), P(B)]. \tag{9.10}$$

Equation (9.9) follows from (9.3) and (9.10) using $A = y_x$, $B = y'_{x'}$, $P(y_x) = P(y \mid x)$, and $P(y'_{x'}) = P(y' \mid x')$.   □

Clearly, if exogeneity cannot be ascertained, then PNS is bound by inequalities similar to those of (9.9), with $P(y_x)$ and $P(y'_{x'})$ replacing $P(y \mid x)$ and $P(y' \mid x')$, respectively.

Theorem 9.2.11

Under condition of exogeneity, the probabilities PN, PS, *and* PNS *are related to each other as follows:*

$$\text{PN} = \frac{\text{PNS}}{P(y \mid x)}, \tag{9.11}$$

$$\text{PS} = \frac{\text{PNS}}{P(y' \mid x')}. \tag{9.12}$$

Thus, the bounds for PNS *in (9.9) provide corresponding bounds for* PN *and* PS.

The resulting bounds for PN,

$$\frac{\max\,[0, P(y \mid x) - P(y \mid x')]}{P(y \mid x)} \le \text{PN} \le \frac{\min\,[P(y \mid x), P(y' \mid x')]}{P(y \mid x)}, \tag{9.13}$$

place limits on our ability to identify PN in experimental studies, where exogeneity holds.

Corollary 9.2.12

If x and y occur in an experimental study and if $P(y_x)$ and $P(y_{x'})$ are the causal effects measured in that study, then for any point p in the range

$$\frac{\max\,[0, P(y_x) - P(y_{x'})]}{P(y_x)} \le p \le \frac{\min\,[P(y_x), P(y'_{x'})]}{P(y_x)} \tag{9.14}$$

there exists a causal model M that agrees with $P(y_x)$ and $P(y_{x'})$ and for which PN $= p$.

Other bounds can be established for nonexperimental events if we have data from both experimental and observational studies (as in Section 9.3.4). The nonzero widths of these bounds imply that probabilities of causation cannot be defined uniquely in stochastic (non-Laplacian) models where, for each u, $Y_x(u)$ is specified in probability $P(Y_x(u) = y)$ instead of a single number.[6]

Proof of Theorem 9.2.11

Using $x \implies (y_x = y)$, we can write $x \wedge y_x = x \wedge y$ and so obtain

---

[6] Robins and Greenland (1989), who used a stochastic model of $Y_x(u)$, defined the probability of causation as

$$\text{PN}(u) = [P(y \mid x, u) - P(y \mid x', u)]/P(y \mid x, u)$$

instead of the counterfactual definition in (9.1).

$$\text{PN} = P(y'_{x'} \mid x, y) = \frac{P(y'_{x'}, x, y)}{P(x, y)} \tag{9.15}$$

$$= \frac{P(y'_{x'}, x, y_x)}{P(x, y)} \tag{9.16}$$

$$= \frac{P(y'_{x'}, y_x)P(x)}{P(x, y)} \tag{9.17}$$

$$= \frac{\text{PNS}}{P(y \mid x)}, \tag{9.18}$$

which establishes (9.11). Equation (9.12) follows by identical steps. □

For completeness, we write the relationship between PNS and the probabilities of enablement and disablement:

$$\text{PD} = \frac{P(x)\text{PNS}}{P(y)}, \qquad \text{PE} = \frac{P(x')\text{PNS}}{P(y')}. \tag{9.19}$$

9.2.3   Identifiability under Monotonicity and Exogeneity

Before attacking the general problem of identifying the counterfactual quantities in (9.1)–(9.3), it is instructive to treat a special condition, called *monotonicity,* which is often assumed in practice and which renders these quantities identifiable. The resulting probabilistic expressions will be recognized as familiar measures of causation that often appear in the literature.

Definition 9.2.13 (Monotonicity)
A variable Y is said to be monotonic *relative to variable X in a causal model M if and only if the function $Y_x(u)$ is monotonic in x for all u. Equivalently, Y is monotonic relative to X if and only if*

$$y'_x \wedge y_{x'} = false. \tag{9.20}$$

Monotonicity expresses the assumption that a change from $X = false$ to $X = true$ cannot, under any circumstance, make Y change from true to false.[7] In epidemiology, this assumption is often expressed as "no prevention," that is, no individual in the population can be helped by exposure to the risk factor.

Theorem 9.2.14 (Identifiability under Exogeneity and Monotonicity)
If X is exogenous and Y is monotonic relative to X, then the probabilities PN, PS, *and* PNS *are all identifiable and are given by (9.11)–(9.12), with*

$$\text{PNS} = P(y \mid x) - P(y \mid x'). \tag{9.21}$$

---

[7] Our analysis remains invariant to complementing x or y (or both); hence, the general condition of monotonicity should read: Either $y'_x \wedge y_{x'} = false$ or $y'_{x'} \wedge y_x = false$. For simplicity, however, we will adhere to the definition in (9.20). Note: monotonicity implies that (5.30) entails (9.7).

The r.h.s. of (9.21) is called "risk difference" in epidemiology, and is also misnomered "attributable risk" (Hennekens and Buring 1987, p. 87).

From (9.11) we see that the probability of necessity is identifiable and given by the *excess risk ratio*

$$\text{PN} = \frac{P(y \mid x) - P(y \mid x')}{P(y \mid x)},$$
(9.22)

often misnomered as the "attributable fraction" (Schlesselman 1982), "attributable-rate percent" (Hennekens and Buring 1987, p. 88), or "attributable proportion" (Cole 1997). Taken literally, the ratio presented in (9.22) has nothing to do with attribution, since it is made up of statistical terms and not of causal or counterfactual relationships. However, the assumptions of exogeneity and monotonicity together enable us to translate the notion of attribution embedded in the definition of PN (equation (9.1)) into a ratio of purely statistical associations. This suggests that exogeneity and monotonicity were tacitly assumed by the many authors who proposed or derived (9.22) as a measure for the "fraction of exposed cases that are attributable to the exposure."

Robins and Greenland (1989) analyzed the identification of PN under the assumption of stochastic monotonicity (i.e., $P(Y_x(u) = y) > P(Y_{x'}(u) = y)$) and showed that this assumption is too weak to permit such identification; in fact, it yields the same bounds as in (9.13). This indicates that stochastic monotonicity imposes no constraints whatsoever on the functional mechanisms that mediate between X and Y.

The expression for PS (equation (9.12)) is likewise quite revealing,

$$\text{PS} = \frac{P(y \mid x) - P(y \mid x')}{1 - P(y \mid x')},$$
(9.23)

since it coincides with what epidemiologists call the "relative difference" (Shep 1958), which is used to measure the *susceptibility* of a population to an exposure x. Susceptibility is defined as the proportion of persons who possess "an underlying factor sufficient to make a person contract a disease following exposure" (Khoury et al. 1989). PS offers a formal counterfactual interpretation of susceptibility, which sharpens this definition and renders susceptibility amenable to systematic analysis.

Khoury et al. (1989) recognized that susceptibility in general is not identifiable and derived (9.23) by making three assumptions: no-confounding, monotonicity,[8] and independence (i.e., assuming that susceptibility to exposure is independent of susceptibility to background not involving exposure). This last assumption is often criticized as untenable, and Theorem 9.2.14 assures us that independence is in fact unnecessary; (9.23) attains its validity through exogeneity and monotonicity alone.

Equation (9.23) also coincides with what Cheng (1997) calls "causal power," namely, the effect of x on y after suppressing "all other causes of y." The counterfactual definition of PS, $P(y_x \mid x', y')$, suggests another interpretation of this quantity. It measures the

---

[8] Monotonicity is not mentioned in Khoury et al. (1989), but it must have been assumed implicitly to make their derivations valid.

probability that setting x would produce y in a situation where x and y are in fact absent. Conditioning on y' amounts to selecting (or hypothesizing) only those worlds in which "all other causes of y" are indeed suppressed.

It is important to note, however, that the simple relationships among the three notions of causation (equations (9.11)–(9.12)) hold only under the assumption of exogeneity; the weaker relationship of (9.5) prevails in the general, nonexogenous case. Additionally, all these notions of causation are defined in terms of the global relationships $Y_x(u)$ and $Y_{x'}(u)$, which are too crude to fully characterize the many nuances of causation; the detailed structure of the causal model leading from X to Y is often needed to explicate more refined notions, such as "actual cause" (see Chapter 10).

Proof of Theorem 9.2.14

Writing $y_{x'} \lor y'_{x'} = \text{true}$, we have

$$y_x = y_x \land (y_{x'} \lor y'_{x'}) = (y_x \land y_{x'}) \lor (y_x \land y'_{x'}) \tag{9.24}$$

and

$$y_{x'} = y_{x'} \land (y_x \lor y'_x) = (y_{x'} \land y_x) \lor (y_{x'} \land y'_x) = y_{x'} \land y_x, \tag{9.25}$$

since monotonicity entails $y_{x'} \land y'_x = \text{false}$. Substituting (9.25) into (9.24) yields

$$y_x = y_{x'} \lor (y_x \land y'_{x'}). \tag{9.26}$$

Taking the probability of (9.26) and using the disjointness of $y_{x'}$ and $y'_{x'}$, we obtain

$$P(y_x) = P(y_{x'}) + P(y_x, y'_{x'})$$

or

$$P(y_x, y'_{x'}) = P(y_x) - P(y_{x'}). \tag{9.27}$$

Equation (9.27), together with the assumption of exogeneity (equation (9.8)), establishes equation (9.21). $\square$

9.2.4 Identifiability under Monotonicity and Nonexogeneity

The relations established in Theorems 9.2.10–9.2.14 were based on the assumption of exogeneity. In this section, we relax this assumption and consider cases where the effect of X on Y is confounded, that is, when $P(y_x) \neq P(y \mid x)$. In such cases $P(y_x)$ may still be estimated by auxiliary means (e.g., through adjustment of certain covariates or through experimental studies), and the question is whether this added information can render the probability of causation identifiable. The answer is affirmative.

Theorem 9.2.15

If Y is monotonic relative to X, then PNS, PN, *and* PS *are identifiable whenever the causal effects $P(y_x)$ and $P(y_{x'})$ are identifiable:*

$$\text{PNS} = P(y_x, y'_{x'}) = P(y_x) - P(y_{x'}), \tag{9.28}$$

$$\text{PN} = P(y'_{x'} \mid x, y) = \frac{P(y) - P(y_{x'})}{P(x, y)}, \tag{9.29}$$

$$\text{PS} = P(y_x \mid x', y') = \frac{P(y_x) - P(y)}{P(x', y')}. \tag{9.30}$$

In order to appreciate the difference between equations (9.29) and (9.22), we can expand $P(y)$ and write

$$\text{PN} = \frac{P(y \mid x)P(x) + P(y \mid x')P(x') - P(y_{x'})}{P(y \mid x)P(x)}$$

$$= \frac{P(y \mid x) - P(y \mid x')}{P(y \mid x)} + \frac{P(y \mid x') - P(y_{x'})}{P(x, y)}. \tag{9.31}$$

The first term on the r.h.s. of (9.31) is the familiar excess risk ratio (as in (9.22)) and represents the value of PN under exogeneity. The second term represents the *correction* needed to account for confounding, that is, $P(y_{x'}) \neq P(y \mid x')$.

Equations (9.28)–(9.30) thus provide more refined measures of causation, which can be used in situations where the causal effect $P(y_x)$ can be identified through auxiliary means (see Example 4, Section 9.3.4). It can also be shown that expressions in (9.28)–(9.30) provide lower bounds for PNS, PN, and PS in the general, nonmonotonic case (Tian and Pearl 2000, Section 11.9.2).

Remarkably, since PS and PN must be nonnegative, (9.29)–(9.30) provide a simple necessary test for the assumption of monotonicity:

$$P(y_x) \geq P(y) \geq P(y_{x'}), \tag{9.32}$$

which tightens the standard inequalities (from $x' \wedge y \implies y_{x'}$ and $x \wedge y' \implies y_{x'}$)

$$P(y_{x'}) \geq P(x', y), \qquad P(y'_x) \geq P(x, y'). \tag{9.33}$$

J. Tian has shown that these inequalities are in fact sharp: every combination of experimental and nonexperimental data that satisfies these inequalities can be generated from some causal model in which Y is monotonic in X. That the commonly made assumption of "no prevention" is not entirely exempt from empirical scrutiny should come as a relief to many epidemiologists. Alternatively, if the no-prevention assumption is theoretically unassailable, then (9.32) can be used for testing the compatibility of the experimental and nonexperimental data, that is, whether subjects used in clinical trials are representative of the target population as characterized by the joint distribution $P(x, y)$.

Proof of Theorem 9.2.15
Equation (9.28) was established in (9.27). To prove (9.30), we write

$$P(y_x \mid x', y') = \frac{P(y_x, x', y')}{P(x', y')} = \frac{P(y_x, x', y'_{x'})}{P(x', y')}, \tag{9.34}$$

because $x' \wedge y' = x' \wedge y'_{x'}$ (by consistency). To calculate the numerator of (9.34), we conjoin (9.26) with x' to obtain

$$x' \wedge y_x = (x' \wedge y_{x'}) \vee (y_x \wedge y'_{x'} \wedge x').$$

We then take the probability on both sides, which gives (since $y_{x'}$ and $y'_{x'}$ are disjoint)

$$
\begin{aligned}
P(y_x, y'_{x'}, x') &= P(x', y_x) - P(x', y_{x'}) \\
&= P(x', y_x) - P(x', y) \\
&= P(y_x) - P(x, y_x) - P(x', y) \\
&= P(y_x) - P(x, y) - P(x', y) \\
&= P(y_x) - P(y).
\end{aligned}
$$

Substituting into (9.34), we finally obtain

$$P(y_x \mid x', y') = \frac{P(y_x) - P(y)}{P(x', y')},$$

which establishes (9.30). Equation (9.29) follows via identical steps. $\square$

One common class of models that permits the identification of $P(y_x)$ under conditions of nonexogeneity was exemplified in Chapter 3. It was shown in Section 3.2 (equation (3.13)) that, for every two variables X and Y in a positive Markovian model M, the causal effect $P(y_x)$ is identifiable and is given by

$$P(y_x) = \sum_{pa_X} P(y \mid pa_X, x) P(pa_X), \tag{9.35}$$

where pa_X are (realizations of) the *parents* of X in the causal graph associated with M. Thus, we can combine (9.35) with Theorem 9.2.15 to obtain a concrete condition for the identification of the probability of causation.

Corollary 9.2.16

For any positive Markovian model M, if the function $Y_x(u)$ is monotonic then the probabilities of causation PNS, PS, and PN are identifiable and are given by (9.28)–(9.30), with $P(y_x)$ as given in (9.35).

A broader identification condition can be obtained through the use of the back-door and front-door criteria (Section 3.3), which are applicable to semi-Markovian models. These were further generalized in Galles and Pearl (1995) (see Section 4.3.1) and Tian and Pearl (2002a) (Theorem 3.6.1) and lead to the following corollary.

Corollary 9.2.17

*Let **GP** be the class of semi-Markovian models that satisfy the graphical criterion of Theorem 3.6.1. If $Y_x(u)$ is monotonic, then the probabilities of causation PNS, PS, and*

PN *are identifiable in* **GP** *and are given by* (9.28)–(9.30), *with* $P(y_x)$ *determined by the topology of* **G(M)** *through the algorithm of Tian and Pearl (2002a).*

9.3   EXAMPLES AND APPLICATIONS

9.3.1   Example 1: Betting against a Fair Coin

We must bet heads or tails on the outcome of a fair coin toss; we win a dollar if we guess correctly and lose if we don't. Suppose we bet heads and win a dollar, without glancing at the actual outcome of the coin. Was our bet a necessary cause (or a sufficient cause, or both) for winning?

This example is isomorphic to the clinical trial discussed in Section 1.4.4 (Figure 1.6). Let x stand for "we bet on heads," y for "we win a dollar," and u for "the coin turned up heads." The functional relationship between y, x, and u is

$$y = (x \wedge u) \vee (x' \wedge u'), \tag{9.36}$$

which is not monotonic but, since the model is fully specified, permits us to compute the probabilities of causation from their definitions, (9.1)–(9.3). To exemplify,

$$\text{PN} = P(y'_{x'} \mid x, y) = P(y'_{x'} \mid u) = 1,$$

because $x \wedge y \implies u$ and $Y_{x'}(u) = \text{false}$. In words, knowing the current bet (x) and current win (y) permits us to infer that the coin outcome must have been a head (u), from which we can further deduce that betting tails (x') instead of heads would have resulted in a loss. Similarly,

$$\text{PS} = P(y_x \mid x', y') = P(y_x \mid u) = 1$$

(because $x' \wedge y' \implies u$) and

$$\begin{aligned} \text{PNS} &= P(y_x, y'_{x'}) \\ &= P(y_x, y'_{x'} \mid u)P(u) + P(y_x, y'_{x'} \mid u')P(u') \\ &= 1(0.5) + 0(0.5) = 0.5. \end{aligned}$$

We see that betting heads has 50% chance of being a necessary and sufficient cause of winning. Still, once we win, we can be 100% sure that our bet was necessary for our win, and once we lose (say, on betting tails) we can be 100% sure that betting heads would have been sufficient for producing a win. The empirical content of such counterfactuals is discussed in Section 7.2.2.

It is easy to verify that these counterfactual quantities cannot be computed from the joint probability of X and Y without knowledge of the functional relationship in (9.36), which tells us the (deterministic) policy by which a win or a loss is decided (Section 1.4.4). This can be seen, for instance, from the conditional probabilities and causal effects associated with this example,

$$P(y \mid x) = P(y \mid x') = P(y_x) = P(y_{x'}) = P(y) = \tfrac{1}{2},$$

because identical probabilities would be generated by a random payoff policy in which y is functionally independent of x – say, by a bookie who watches the coin and ignores our bet. In such a random policy, the probabilities of causation PN, PS, and PNS are all zero. Thus, according to our definition of identifiability (Definition 3.2.3), if two models agree on P and do not agree on a quantity Q, then Q is not identifiable. Indeed, the bounds delineated in Theorem 9.2.10 (equation (9.9)) read $0 \le \text{PNS} \le \tfrac{1}{2}$, meaning that the three probabilities of causation cannot be determined from statistical data on X and Y alone, not even in a controlled experiment; knowledge of the functional mechanism is required, as in (9.36).

It is interesting to note that whether the coin is tossed before or after the bet has no bearing on the probabilities of causation as just defined. This stands in contrast with some theories of probabilistic causality (e.g., Good 1961), which attempt to avoid deterministic mechanisms by conditioning all probabilities on "the state of the world just before" the occurrence of the cause in question (x). When applied to our betting story, the intention is to condition all probabilities on the state of the coin (u), but this is not fulfilled if the coin is tossed after the bet is placed. Attempts to enrich the conditioning set with events occurring after the cause in question have led back to deterministic relationships involving counterfactual variables (see Cartwright 1989, Eells 1991, and the discussion in Section 7.5.4).

One may argue, of course, that if the coin is tossed *after* the bet then it is not at all clear what our winnings would be had we bet differently; merely uttering our bet could conceivably affect the trajectory of the coin (Dawid 2000). This objection can be diffused by placing x and u in two remote locations and tossing the coin a split second after the bet is placed but before any light ray could arrive from the betting room to the coin-tossing room. In such a hypothetical situation, the counterfactual statement "our winning would be different had we bet differently" is rather compelling, even though the conditioning event (u) occurs after the cause in question (x). We conclude that temporal descriptions such as "the state of the world just before x" cannot be used to properly identify the appropriate set of conditioning events (u) in a problem; a deterministic model of the mechanisms involved is needed for formulating the notion of "probability of causation."

9.3.2 Example 2: The Firing Squad

Consider again the firing squad of Section 7.1.2 (see Figure 9.1); A and B are riflemen, C is the squad's captain (who is waiting for the court order, U), and T is a condemned prisoner. Let u be the proposition that the court has ordered an execution, x the proposition stating that A pulled the trigger, and y that T is dead. We assume again that $P(u) = \tfrac{1}{2}$, that A and B are perfectly accurate marksmen who are alert and law-abiding, and that T is not likely to die from fright or other extraneous causes. We wish to compute the probability that x was a necessary (or sufficient, or both) cause for y (i.e., we wish to calculate PN, PS, and PNS).

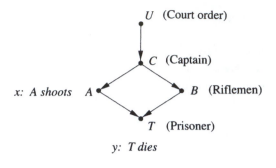

Figure 9.1   Causal relationships in the two-man firing-squad example.

Definitions 9.2.1–9.2.3 permit us to compute these probabilities directly from the given causal model, since all functions and all probabilities are specified, with the truth value of each variable tracing that of U. Accordingly, we can write[9]

$$P(y_x) = P(Y_x(u) = \text{true})P(u) + P(Y_x(u') = \text{true})P(u')$$

$$= \tfrac{1}{2}(1 + 1) = 1. \tag{9.37}$$

Similarly, we have

$$P(y_{x'}) = P(Y_{x'}(u) = \text{true})P(u) + P(Y_{x'}(u') = \text{true})P(u')$$

$$= \tfrac{1}{2}(1 + 0) = \tfrac{1}{2}. \tag{9.38}$$

In order to compute PNS, we must evaluate the probability of the joint event $y'_{x'} \wedge y_x$. Given that these two events are jointly true only when $U = \text{true}$, we have

$$\text{PNS} = P(y_x, y'_{x'})$$

$$= P(y_x, y'_{x'} \mid u)P(u) + P(y_x, y'_{x'} \mid u')P(u')$$

$$= \tfrac{1}{2}(0 + 1) = \tfrac{1}{2}. \tag{9.39}$$

The calculation of PS and PN is likewise simplified by the fact that each of the conditioning events, $x \wedge y$ for PN and $x' \wedge y'$ for PS, is true in only one state of U. We thus have

$$\text{PN} = P(y'_{x'} \mid x, y) = P(y'_{x'} \mid u) = 0.$$

reflecting that, once the court orders an execution (u), T will die (y) from the shot of rifleman B, even if A refrains from shooting (x'). Indeed, upon learning of T's death, we can categorically state that rifleman A's shot was *not* a necessary cause of the death.
Similarly,

$$\text{PS} = P(y_x \mid x', y') = P(y_x \mid u') = 1,$$

---

[9] Recall that $P(Y_x(u') = \text{true})$ involves the submodel M_x, in which X is set to "true" independently of U. Thus, although under condition u' the captain has not given a signal, the potential outcome $Y_x(u')$ calls for hypothesizing that rifleman A pulls the trigger (x) unlawfully.

Table 9.1

	Exposure	
	High (x)	Low (x')
Deaths (y)	30	16
Survivals (y')	69,130	59,010

matching our intuition that a shot fired by an expert marksman would be sufficient for causing the death of T, regardless of the court decision.

Note that Theorems 9.2.10 and 9.2.11 are not applicable to this example because x is not exogenous; events x and y have a common cause (the captain's signal), which renders $P(y \mid x') = 0 \neq P(y_{x'}) = \frac{1}{2}$. However, the monotonicity of Y (in x) permits us to compute PNS, PS, and PN from the joint distribution $P(x, y)$ and the causal effects (using (9.28)–(9.30)), instead of consulting the functional model. Indeed, writing

$$P(x, y) = P(x', y') = \tfrac{1}{2} \tag{9.40}$$

and

$$P(x, y') = P(x', y) = 0, \tag{9.41}$$

we obtain

$$PN = \frac{P(y) - P(y_{x'})}{P(x, y)} = \frac{\frac{1}{2} - \frac{1}{2}}{\frac{1}{2}} = 0 \tag{9.42}$$

and

$$PS = \frac{P(y_x) - P(y)}{P(x', y')} = \frac{1 - \frac{1}{2}}{\frac{1}{2}} = 1, \tag{9.43}$$

as expected.

9.3.3 Example 3: The Effect of Radiation on Leukemia

Consider the following data (Table 9.1, adapted[10] from Finkelstein and Levin 1990) comparing leukemia deaths in children in southern Utah with high and low exposure to radiation from the fallout of nuclear tests in Nevada. Given these data, we wish to estimate the probabilities that high exposure to radiation was a necessary (or sufficient, or both) cause of death due to leukemia.

---

[10] The data in Finkelstein and Levin (1990) are given in "person-year" units. For the purpose of illustration we have converted the data to absolute numbers (of deaths and nondeaths) assuming a ten-year observation period.

Assuming monotonicity – that exposure to nuclear radiation had no remedial effect on any individual in the study – the process can be modeled by a simple disjunctive mechanism represented by the equation

$$y = f(x, u, q) = (x \wedge q) \vee u, \tag{9.44}$$

where u represents "all other causes" of y and where q represents all "enabling" mechanisms that must be present for x to trigger y. Assuming that q and u are both unobserved, the question we ask is under what conditions we can identify the probabilities of causation (PNS, PN, and PS) from the joint distribution of X and Y.

Since (9.44) is monotonic in x, Theorem 9.2.14 states that all three quantities would be identifiable provided X is exogenous; that is, x should be independent of q and u. Under this assumption, (9.21)–(9.23) further permit us to compute the probabilities of causation from frequency data. Taking fractions to represent probabilities, the data in Table 9.1 imply the following numerical results:

$$\text{PNS} = P(y \mid x) - P(y \mid x') = \frac{30}{30 + 69{,}130} - \frac{16}{16 + 59{,}010} = 0.0001625, \tag{9.45}$$

$$\text{PN} = \frac{\text{PNS}}{P(y \mid x)} = \frac{\text{PNS}}{30/(30 + 69{,}130)} = 0.37535, \tag{9.46}$$

$$\text{PS} = \frac{\text{PNS}}{1 - P(y \mid x')} = \frac{\text{PNS}}{1 - 16/(16 + 59{,}010)} = 0.0001625. \tag{9.47}$$

Statistically, these figures mean that:

1. There is a 1.625 in ten thousand chance that a randomly chosen child would both die of leukemia if exposed and survive if not exposed;

2. There is a 37.544% chance that an exposed child who died from leukemia would have survived had he or she not been exposed;

3. There is a 1.625 in ten thousand chance that any unexposed surviving child would have died of leukemia had he or she been exposed.

Glymour (1998) analyzed this example with the aim of identifying the probability $P(q)$ (Cheng's "causal power"), which coincides with PS (see Lemma 9.2.8). Glymour concluded that $P(q)$ is identifiable and is given by (9.23), provided that x, u, and q are mutually independent. Our analysis shows that Glymour's result can be generalized in several ways. First, since Y is monotonic in X, the validity of (9.23) is assured even when q and u are dependent, because exogeneity merely requires independence between x and $\{u, q\}$ jointly. This is important in epidemiological settings, because an individual's susceptibility to nuclear radiation is likely to be associated with susceptibility to other potential causes of leukemia (e.g., natural kinds of radiation).

Second, Theorem 9.2.11 assures us that the relationships among PN, PS, and PNS (equations (9.11)–(9.12)), which Glymour derives for independent q and u, should remain valid even when u and q are dependent.

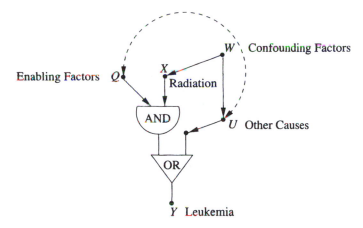

Figure 9.2 Causal relationships in the radiation–leukemia example, where W represents confounding factors.

Finally, Theorem 9.2.15 assures us that PN and PS are identifiable even when x is not independent of $\{u, q\}$, provided only that the mechanism of (9.44) is embedded in a larger causal structure that permits the identification of $P(y_x)$ and $P(y_{x'})$. For example, assume that exposure to nuclear radiation (x) is suspected of being associated with terrain and altitude, which are also factors in determining exposure to cosmic radiation. A model reflecting such consideration is depicted in Figure 9.2, where W represents factors affecting both X and U. A natural way to correct for possible confounding bias in the causal effect of X on Y would be to adjust for W, that is, to calculate $P(y_x)$ and $P(y_{x'})$ using the standard adjustment formula (equation (3.19))

$$P(y_x) = \sum_w P(y \mid x, w)P(w), \qquad P(y_{x'}) = \sum_w P(y \mid x', w)P(w) \qquad (9.48)$$

(instead of $P(y \mid x)$ and $P(y \mid x')$), where the summation runs over levels of W. This adjustment formula, which follows from (9.35), is correct regardless of the mechanisms mediating X and Y, provided only that W represents *all* common factors affecting X and Y (see Section 3.3.1).

Theorem 9.2.15 instructs us to evaluate PN and PS by substituting (9.48) into (9.29) and (9.30), respectively, and it assures us that the resulting expressions constitute consistent estimates of PN and PS. This consistency is guaranteed jointly by the assumption of monotonicity and by the (assumed) topology of the causal graph.

Note that monotonicity as defined in (9.20) is a global property of all pathways between x and y. The causal model may include several nonmonotonic mechanisms along these pathways without affecting the validity of (9.20). However, arguments for the validity of monotonicity must be based on substantive information, since it is not testable in general. For example, Robins and Greenland (1989) argued that exposure to nuclear radiation may conceivably be of benefit to some individuals because such radiation is routinely used clinically in treating cancer patients. The inequalities in (9.32) constitute a statistical test of monotonicity (albeit a weak one) that is based on both experimental and observational studies.

Table 9.2

	Experimental		Nonexperimental	
	x	x'	x	x'
Deaths (y)	16	14	2	28
Survivals (y')	984	986	998	972

9.3.4 Example 4: Legal Responsibility from Experimental and Nonexperimental Data

A lawsuit is filed against the manufacturer of drug x, charging that the drug is likely to have caused the death of Mr. A, who took the drug to relieve symptom S associated with disease D.

The manufacturer claims that experimental data on patients with symptom S show conclusively that drug x may cause only a minor increase in death rates. However, the plaintiff argues that the experimental study is of little relevance to this case because it represents the effect of the drug on *all* patients, not on patients like Mr. A who actually died while using drug x. Moreover, argues the plaintiff, Mr. A is unique in that he used the drug on his own volition, unlike subjects in the experimental study who took the drug to comply with experimental protocols. To support this argument, the plaintiff furnishes nonexperimental data indicating that most patients who chose drug x would have been alive were it not for the drug. The manufacturer counterargues by stating that: (1) counterfactual speculations regarding whether patients would or would not have died are purely metaphysical and should be avoided (Dawid 2000); and (2) nonexperimental data should be dismissed a priori on the grounds that such data may be highly confounded by extraneous factors. The court must now decide, based on both the experimental and nonexperimental studies, what the probability is that drug x was in fact the cause of Mr. A's death.

The (hypothetical) data associated with the two studies are shown in Table 9.2. The experimental data provide the estimates

$$P(y_x) = 16/1000 = 0.016, \tag{9.49}$$

$$P(y_{x'}) = 14/1000 = 0.014; \tag{9.50}$$

the nonexperimental data provide the estimates

$$P(y) = 30/2000 = 0.015, \tag{9.51}$$

$$P(y, x) = 2/2000 = 0.001. \tag{9.52}$$

Substituting these estimates in (9.29), which provides a lower bound on PN (see (11.42)), we obtain

$$PN \geq \frac{P(y) - P(y_{x'})}{P(y, x)} = \frac{0.015 - 0.014}{0.001} = 1.00. \tag{9.53}$$

Thus, the plaintiff was correct; barring sampling errors, the data provide us with 100% assurance that drug x was in fact responsible for the death of Mr. A. Note that a straightforward use of the experimental excess risk ratio would yield a much lower (and incorrect) result:

$$\frac{P(y_x) - P(y_{x'})}{P(y_x)} = \frac{0.016 - 0.014}{0.016} = 0.125. \tag{9.54}$$

Evidently, what the experimental study does not reveal is that, given a choice, terminal patients avoid drug x. Indeed, if there were any terminal patients who would choose x (given the choice), then the control group (x') would have included some such patients (due to randomization) and so the proportion of deaths among the control group $P(y_{x'})$ would have been higher than $P(x', y)$, the population proportion of terminal patients avoiding x. However, the equality $P(y_{x'}) = P(y, x')$ tells us that no such patients were included in the control group; hence (by randomization) no such patients exist in the population at large, and therefore none of the patients who freely chose drug x was a terminal case; all were susceptible to x.

The numbers in Table 9.2 were obviously contrived to represent an extreme case and so facilitate a qualitative explanation of the validity of (9.29). Nevertheless, it is instructive to note that a combination of experimental and nonexperimental studies may unravel what experimental studies alone will not reveal and, in addition, that such combination may provide a necessary test for the adequacy of the experimental procedures. For example, if the frequencies in Table 9.2 were slightly different, they could easily yield a value greater than unity for PN in (9.53) or some other violation of the fundamental inequalities of (9.33). Such violation would indicate an incompatibility of the experimental and nonexperimental groups due, perhaps, to inadequate sampling.

This last point may warrant a word of explanation, lest the reader wonder why two data sets – taken from two separate groups under different experimental conditions – should constrain one another. The explanation is that certain quantities in the two subpopulations are expected to remain invariant to all these differences, provided that the two subpopulations were sampled properly from the population at large. These invariant quantities are simply the causal effects probabilities, $P(y_{x'})$ and $P(y_x)$. Although these counterfactual probabilities were not measured in the observational group, they must (by definition) nevertheless be the same as those measured in the experimental group. The invariance of these quantities is the basic axiom of controlled experimentation, without which *no* inference would be possible from experimental studies to general behavior of the population. The invariance of these quantities implies the inequalities of (9.33) and, if monotonicity holds, (9.32) ensues.

9.3.5   Summary of Results

We now summarize the results from Sections 9.2 and 9.3 that should be of value to practicing epidemiologists and policy makers. These results are shown in Table 9.3, which lists the best estimand of PN (for a nonexperimental event) under various assumptions and various types of data – the stronger the assumptions, the more informative the estimates.

We see that the excess risk ratio (ERR), which epidemiologists commonly equate with the probability of causation, is a valid measure of PN only when two assumptions

Table 9.3. *PN as a Function of Assumptions and Available Data*

Assumptions			Data Available		
Exogeneity	Monotonicity	Additional	Experimental	Observational	Combined
+	+		ERR	ERR	ERR
+	−		bounds	bounds	bounds
−	+	covariate control	—	corrected ERR	corrected ERR
−	+		—	—	corrected ERR
−	−		—	—	bounds

Note: ERR stands for the excess risk ratio, $1 - P(y \mid x')/P(y' \mid x')$; corrected ERR is given in (9.31).

can be ascertained: exogeneity (i.e., no confounding) and monotonicity (i.e., no prevention). When monotonicity does not hold, ERR provides merely a lower bound for PN, as shown in (9.13). (The upper bound is usually unity.) The nonentries (—) in the right-hand side of Table 9.3 represent vacuous bounds (i.e., $0 \leq PN \leq 1$). In the presence of confounding, ERR must be corrected by the additive term $[P(y \mid x') - P(y_{x'})]/P(x, y)$, as stated in (9.31). In other words, when confounding bias (of the causal effect) is positive, PN is higher than ERR by the amount of this additive term. Clearly, owing to the division by $P(x, y)$, the PN bias can be many times higher than the causal effect bias $P(y \mid x') - P(y_{x'})$. However, confounding results only from association between exposure and other factors that affect the outcome; one need not be concerned with associations between such factors and susceptibility to exposure (see Figure 9.2).

The last row in Table 9.3, corresponding to no assumptions whatsoever, leads to vacuous bounds for PN, unless we have combined data. This does not mean, however, that justifiable assumptions *other* than monotonicity and exogeneity could not be helpful in rendering PN identifiable. The use of such assumptions is explored in the next section.

9.4 IDENTIFICATION IN NONMONOTONIC MODELS

In this section we discuss the identification of probabilities of causation without making the assumption of monotonicity. We will assume that we are given a causal model M in which all functional relationships are known, but since the background variables U are not observed, their distribution is not known and the model specification is not complete.

Our first step would be to study under what conditions the function $P(u)$ can be identified, thus rendering the entire model identifiable. If M is Markovian, then the problem can be analyzed by considering each parents–child family separately. Consider any arbitrary equation in M,

$$y = f(pa_Y, u_Y)$$
$$= f(x_1, x_2, \ldots, x_k, u_1, \ldots, u_m), \tag{9.55}$$

where $U_Y = \{U_1, \ldots, U_m\}$ is the set of background (possibly dependent) variables that appear in the equation for Y. In general, the domain of U_Y can be arbitrary, discrete, or continuous, since these variables represent unobserved factors that were omitted from the model. However, since the observed variables are binary, there is only a finite number $(2^{(2^k)})$ of functions from PA_Y to Y and, for any point $U_Y = u$, only one of those functions is realized. This defines a canonical partition of the domain of U_Y into a set S of equivalence classes, where each equivalence class $s \in S$ induces the same function $f^{(s)}$ from PA_Y to Y (see Section 8.2.2). Thus, as u varies over its domain, a set S of such functions is realized, and we can regard S as a new background variable whose values correspond to the set $\{f^{(s)} : s \in S\}$ of functions from PA_Y to Y that are realizable in U_Y. The number of such functions will usually be smaller than $2^{(2^k)}$.[11]

For example, consider the model described in Figure 9.2. As the background variables (Q, U) vary over their respective domains, the relation between X and Y spans three distinct functions:

$$f^{(1)} : Y = \text{true}, \qquad f^{(2)} : Y = \text{false}, \quad \text{and} \quad f^{(3)} : Y = X.$$

The fourth possible function, $Y \neq X$, is never realized because $f_Y(\cdot)$ is monotonic. The cells (q, u) and (q', u) induce the same function between X and Y; hence they belong to the same equivalence class.

If we are given the distribution $P(u_Y)$, then we can compute the distribution $P(s)$, and this will determine the conditional probabilities $P(y \mid pa_Y)$ by summing $P(s)$ over all those functions $f^{(s)}$ that map pa_Y into the value *true*,

$$P(y \mid pa_Y) = \sum_{s \,:\, f^{(s)}(pa_Y)=\text{true}} P(s). \tag{9.56}$$

To ensure model identifiability, it is sufficient that we can invert the process and determine $P(s)$ from $P(y \mid pa_Y)$. If we let the set of conditional probabilities $P(y \mid pa_Y)$ be represented by a vector $\vec{p}$ (of dimensionality 2^k) and $P(s)$ by a vector $\vec{q}$, then (9.56) defines a linear relation between $\vec{p}$ and $\vec{q}$ that can be represented as a matrix multiplication (as in (8.13)),

$$\vec{p} = \mathbf{R}\vec{q}, \tag{9.57}$$

where $\mathbf{R}$ is a $2^k \times |S|$ matrix whose entries are either 0 or 1. Thus, a sufficient condition for identification is simply that $\mathbf{R}$, together with the normalizing equation $\sum_j \vec{q}_j = 1$, be invertible.

In general, $\mathbf{R}$ will *not* be invertible because the dimensionality of $\vec{q}$ can be much larger than that of $\vec{p}$. However, in many cases, such as the "noisy OR" mechanism

$$Y = U_0 \bigvee_{i=1,\ldots,k} (X_i \wedge U_i), \tag{9.58}$$

---

[11] Balke and Pearl (1994a,b) called these S variables "response variables," as in Section 8.2.2; Heckerman and Shachter (1995) called them "mapping variables."

symmetry permits $\vec{q}$ to be identified from $P(y \mid pa_Y)$ even when the exogenous variables $U_0, U_1, \ldots, U_k$ are not independent. This can be seen by noting that every point u for which $U_0 =$ false defines a unique function $f^{(s)}$ because, if T is the set of indices i for which U_i is true, the relationship between PA_Y and Y becomes

$$Y = U_0 \bigvee_{i \in T} X_i \tag{9.59}$$

and, for $U_0 =$ false, this equation defines a distinct function for each T. The number of induced functions is $2^k + 1$, which (subtracting 1 for normalization) is exactly the number of distinct realizations of PA_Y. Moreover, it is easy to show that the matrix connecting $\vec{p}$ and $\vec{q}$ is invertible. We thus conclude that the probability of every counterfactual sentence can be identified in any Markovian model composed of noisy OR mechanisms, regardless of whether the background variables in each family are mutually independent. The same holds, of course, for noisy AND mechanisms or any combination thereof (including negating mechanisms), provided that each family consists of one type of mechanism.

To generalize this result to mechanisms other than noisy OR and noisy AND, we note that – although $f_Y(\cdot)$ in this example was monotonic (in each X_i) – it was the *redundancy* of $f_Y(\cdot)$ and not its monotonicity that ensured identifiability. The following is an example of a monotonic function for which the **R** matrix is not invertible:

$$Y = (X_1 \wedge U_1) \vee (X_2 \wedge U_1) \vee (X_1 \wedge X_2 \wedge U_3).$$

This function represents a noisy OR gate for $U_3 =$ false; it becomes a noisy AND gate for $U_3 =$ true and $U_1 = U_2 =$ false. The number of equivalence classes induced is six, which would require five independent equations to determine their probabilities; the data $P(y \mid pa_Y)$ provide only four such equations.

In contrast, the mechanism governed by the following function, although nonmonotonic, is invertible:

$$Y = \mathrm{XOR}(X_1, \mathrm{XOR}(U_2, \ldots, \mathrm{XOR}(U_{k-1}, \mathrm{XOR}(X_k, U_k)))),$$

where $\mathrm{XOR}(\cdot)$ stands for exclusive OR. This equation induces only two functions from PA_Y to Y:

$$Y = \begin{cases} \mathrm{XOR}\,(X_1, \ldots, X_k) & \text{if } \mathrm{XOR}\,(U_1, \ldots, U_k) = \text{false}, \\ \neg \mathrm{XOR}\,(X_1, \ldots, X_k) & \text{if } \mathrm{XOR}\,(U_1, \ldots, U_k) = \text{true}. \end{cases}$$

A single conditional probability, say $P(y \mid x_1, \ldots, x_k)$, would therefore suffice for computing the one parameter needed for identification: $P[\mathrm{XOR}(U_1, \ldots, U_k) = \text{true}]$.

We summarize these considerations with a theorem.

Definition 9.4.1 (Local Invertibility)
A model M is said to be locally invertible *if, for every variable $V_i \in V$, the set of $2^k + 1$ equations*

$$P(y \mid pa_i) = \sum_{s \,:\, f^{(s)}(pa_i) = true} q_i(s), \tag{9.60}$$

$$\sum_{s} q_i(s) = 1 \tag{9.61}$$

has a unique solution for $q_i(s)$, where each $f_i^{(s)}(pa_i)$ corresponds to the function $f_i(pa_i, u_i)$ induced by u_i in equivalence class s.

Theorem 9.4.2

Given a Markovian model $M = \langle U, V, \{f_i\} \rangle$ in which the functions $\{f_i\}$ are known and the exogenous variables U are unobserved, if M is locally invertible then the probability of every counterfactual sentence is identifiable from the joint probability $P(v)$.

Proof

If (9.60) has a unique solution for $q_i(s)$, then we can replace U with S and obtain an equivalent model as follows:

$$M' = \langle S, V, \{f_i'\} \rangle, \qquad \text{where } f_i' = f_i^{(s)}(pa_i).$$

The model M', together with $q_i(s)$, completely specifies a probabilistic causal model $\langle M', P(s) \rangle$ (owing to the Markov property), from which probabilities of counterfactuals are derivable by definition.   □

Theorem 9.4.2 provides a sufficient condition for identifying probabilities of causation, but of course it does not exhaust the spectrum of assumptions that are helpful in achieving identification. In many cases we might be justified in hypothesizing additional structure on the model – for example, that the U variables entering each family are themselves independent. In such cases, additional constraints are imposed on the probabilities $P(s)$, and (9.60) may be solved even when the cardinality of S far exceeds the number of conditional probabilities $P(y \mid pa_Y)$.

9.5 CONCLUSIONS

This chapter has explicated and analyzed the interplay between the necessary and sufficient components of causation. Using counterfactual interpretations that rest on structural model semantics, we demonstrated how simple techniques of computing probabilities of counterfactuals can be used in computing probabilities of causes, deciding questions of identification, uncovering conditions under which probabilities of causes can be estimated from statistical data, and devising tests for assumptions that are routinely made (often unwittingly) by analysts and investigators.

On the practical side, we have offered several useful tools (partly summarized in Table 9.3) for epidemiologists and health scientists. This chapter formulates and calls attention to subtle assumptions that must be ascertained before statistical measures such as excess risk ratio can be used to represent causal quantities such as attributable risk or probability of causes (Theorem 9.2.14). It shows how data from both experimental and nonexperimental studies can be combined to yield information that neither study alone can reveal (Theorem 9.2.15 and Section 9.3.4). Finally, it provides tests for the commonly made assumption of "no prevention" and for the often asked question of whether a clinical study is representative of its target population (equation (9.32)).

On the conceptual side, we have seen that both the probability of necessity (PN) and probability of sufficiency (PS) play a role in our understanding of causation and that each component has its logic and computational rules. Although the counterfactual concept of necessary cause (i.e., that an outcome would not have occurred "but for" the action) is predominant in legal settings (Robertson 1997) and in ordinary discourse, the sufficiency component of causation has a definite influence on causal thoughts.

The importance of the sufficiency component can be uncovered in examples where the necessary component is either dormant or ensured. Why do we consider striking a match to be a more adequate explanation (of a fire) than the presence of oxygen? Recasting the question in the language of PN and PS, we note that, since both explanations are necessary for the fire, each will command a PN of unity. (In fact, the PN is actually higher for the oxygen if we allow for alternative ways of igniting a spark.) Thus, it must be the sufficiency component that endows the match with greater explanatory power than the oxygen. If the probabilities associated with striking a match and the presence of oxygen are denoted p_m and p_o, respectively, then the PS measures associated with these explanations evaluate to PS(match) $= p_o$ and PS(oxygen) $= p_m$, clearly favoring the match when $p_o \gg P_m$. Thus, a robot instructed to explain why a fire broke out has no choice but to consider both PN and PS in its deliberations.

Should PS enter legal considerations in criminal and tort law? I believe that it should – as does Good (1993) – because attention to sufficiency implies attention to the consequences of one's action. The person who lighted the match ought to have anticipated the presence of oxygen, whereas the person who supplied – or could (but did not) remove – the oxygen is not generally expected to have anticipated match-striking ceremonies.

However, what weight should the law assign to the necessary versus the sufficient component of causation? This question obviously lies beyond the scope of our investigation, and it is not at all clear who would be qualified to tackle the issue or whether our legal system would be prepared to implement the recommendation. I am hopeful, however, that whoever undertakes to consider such questions will find the analysis in this chapter to be of some use. The next chapter combines aspects of necessity and sufficiency in explicating a more refined notion: "actual cause."

Acknowledgments

I am indebted to Sander Greenland for many suggestions and discussions concerning the treatment of attribution in the epidemiological literature and the potential applications of our results in practical epidemiological studies. Donald Michie and Jack Good are responsible for shifting my attention from PN to PS and PNS. Clark Glymour and Patricia Cheng helped to unravel some of the mysteries of causal power theory, and Michelle Pearl provided useful pointers to the epidemiological literature. Blai Bonet corrected omissions from earlier versions of Lemmas 9.2.7 and 9.2.8, and Jin Tian tied it all up in tight bounds.

CHAPTER TEN

The Actual Cause

And now remains
That we find out the cause of this effect,
Or rather say, the cause of this defect,
For this effect defective comes by cause.
Shakespeare (Hamlet II.ii. 100–4)

Preface

This chapter offers a formal explication of the notion of "actual cause," an event recognized as responsible for the production of a given outcome in a specific scenario, as in: "Socrates drinking hemlock was the actual cause of Socrates death." Human intuition is extremely keen in detecting and ascertaining this type of causation and hence is considered the key to constructing explanations (Section 7.2.3) and the ultimate criterion (known as "cause in fact") for determining legal responsibility.

Yet despite its ubiquity in natural thoughts, actual causation is not an easy concept to formulate. A typical example (introduced by Wright 1988) considers two fires advancing toward a house. If fire A burned the house before fire B, we (and many juries nationwide) would surely consider fire A "the actual cause" of the damage, though either fire alone is sufficient (and neither one was necessary) for burning the house. Clearly, actual causation requires information beyond that of necessity and sufficiency; the actual process mediating between the cause and the effect must enter into consideration. But what precisely is a "process" in the language of structural models? What aspects of causal processes define actual causation? How do we piece together evidence about the uncertain aspects of a scenario and so compute probabilities of actual causation?

In this chapter we propose a plausible account of actual causation that can be formulated in structural model semantics. The account is based on the notion of *sustenance,* to be defined in Section 10.2, which combines aspects of necessity and sufficiency to measure the capacity of the cause to maintain the effect despite certain *structural* changes in the model. We show by examples how this account avoids problems associated with the counterfactual dependence account of Lewis (1986) and how it can be used both in generating explanations of specific scenarios and in computing the probabilities that such explanations are in fact correct.

10.1 INTRODUCTION: THE INSUFFICIENCY OF NECESSARY CAUSATION

10.1.1 Singular Causes Revisited

Statements of the type "a car accident was the cause of Joe's death," made relative to a specific scenario, are classified as "singular," "single-event," or "token-level" causal statements. Statements of the type "car accidents cause deaths," when made relative to a type of events or a class of individuals, are classified as "generic" or "type-level" causal claims (see Section 7.5.4). We will call the cause in a single-event statement an *actual cause* and the one in a type-level statement a *general cause*.

The relationship between type and token causal claims has been controversial in the philosophical literature (Woodward 1990; Hitchcock 1995), and priority questions such as "which comes first?" and "can one level be reduced to the other?" (Cartwright 1989; Eells 1991; Hausman 1998) have diverted attention from the more fundamental question: "What tangible claims do type and token statements make about our world, and how is causal knowledge organized so as to substantiate such claims?" The debate has led to theories that view type and token claims as two distinct species of causal relations (as in Good 1961, 1962), each requiring its own philosophical account (see, e.g., Sober 1985; Eells 1991, chap. 6) – "not an altogether happy predicament" (Hitchcock 1997). In contrast, the structural account treats type and token claims as instances of the same species, differing only in the details of the scenario-specific information that is brought to bear on the question. As such, the structural account offers a formal basis for studying the anatomy of the two levels of claims, what information is needed to support each level, and why philosophers have found their relationships so hard to disentangle.

The basic building blocks of the structural account are the functions $\{f_i\}$, which represent lawlike mechanisms and supply information for both type-level and token-level claims. These functions are type-level in the sense of representing generic, counterfactual relationships among variables that are applicable to every hypothetical scenario, not just ones that were realized. At the same time, any specific instantiation of those relationships represents a token-level claim. The ingredients that distinguish one scenario from another are represented in the background variables U. When all such factors are known, $U = u$, we have a "world" on our hands (Definition 7.1.8) – an ideal, full description of a specific scenario in which all relevant details are spelled out and nothing is left to chance or guessing. Causal claims made at the world level would be extreme cases of token causal claims. In general, however, we do not possess the detailed knowledge necessary for specifying a single world $U = u$, and we use a probability $P(u)$ to summarize our ignorance of those details. This takes us to the level of probabilistic causal models $\langle M, P(u) \rangle$ (Definition 7.1.6). Causal claims made on the basis of such models, with no reference to the actual scenario, would be classified as type-level claims. Causal effects assertions, such as $P(Y_x = y) = p$, are examples of such claims, for they express the general tendency of x to bring about y, as judged over all potential scenarios.[1] In most cases, however, we possess partial information about the scenario at hand – for example, that Joe died, that he was in a car accident, and perhaps that he drove a sports car and suffered a head injury. The totality of such episode-specific information is called "evidence" (e) and can be used to update $P(u)$ into $P(u \mid e)$. Causal claims derived from the model $\langle M, P(u \mid e) \rangle$ represent token claims of varying shades, depending on the specificity of e.

---

[1] Occasionally, causal effect assertions can even be made on the basis of an incomplete probabilistic model, where only $G(M)$ and $P(v)$ are given – this is the issue of identification (Chapter 3). But no token-level statement can be made on such basis alone without some knowledge of $\{f_i\}$ or $P(u)$ (assuming, of course, that x and y are known to have occurred).

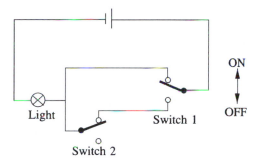

Figure 10.1   Switch 1 (and not switch 2) is perceived to be causing the light, though neither is necessary.

Thus, the distinction between type and token claims is a matter of degree in the structural account. The more episode-specific evidence we gather, the closer we come to the ideals of token claims and actual causes. The notions of PS and PN (the focus of Chapter 9) represent intermediate points along this spectrum. Probable sufficiency (PS) is close to a type-level claim because the actual scenario is not taken into account and is, in fact, excluded from consideration. Probable necessity (PN) makes some reference to the actual scenario, albeit a rudimentary one (i.e., that x and y are true). In this section we will attempt to come closer to the notion of actual cause by taking additional information into consideration.

10.1.2   Preemption and the Role of Structural Information

In Section 9.2, we alluded to the fact that both PN and PS are global (i.e., input–output) features of a causal model, depending only on the function $Y_x(u)$ and not on the structure of the process mediating between the cause (x) and the effect (y). That such structure plays a role in causal explanation is seen in the following example.

Consider an electrical circuit consisting of a light bulb and two switches, as shown in Figure 10.1. From the user's viewpoint, the light responds symmetrically to the two switches; either switch is sufficient to turn the light on. Internally, however, when switch 1 is on it not only activates the light but also disconnects switch 2 from the circuit, rendering it inoperative. Consequently, with both switches on, we would not hesitate to proclaim switch 1 as the "actual cause" of the current flowing in the light bulb, knowing as we do that switch 2 can have no effect whatsoever on the electric pathway in this particular state of affairs. There is nothing in PN and PS that could possibly account for this asymmetry; each is based on the response function $Y_x(u)$ and is therefore oblivious to the internal workings of the circuit.

This example is representative of a class of counterexamples, involving *preemption*, that were brought up against Lewis's counterfactual account of causation. It illustrates how an event (e.g., switch 1 being on) can be considered a cause although the effect persists in its absence. Lewis's (1986) answer to such counterexamples was to modify the counterfactual criterion and let x be a cause of y as long as there exists a *counterfactual dependence chain* of intermediate variables between x to y; that is, the output of every link in the chain is counterfactually dependent on its input. Such a chain does not exist for switch 2 because, given the current state of affairs (i.e., both switches being on), no

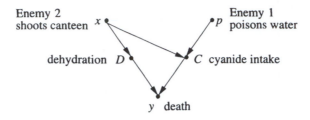

Figure 10.2   Causal relationships in the desert traveler example.

part of the circuit would be affected (electrically) by turning switch 2 on or off. This can be shown more clearly in the following example.

> **Example 10.1.1** (The Desert Traveler–after P. Suppes) A desert traveler T has two enemies. Enemy 1 poisons T's canteen, and enemy 2, unaware of enemy 1's action, shoots and empties the canteen. A week later, T is found dead and the two enemies confess to action and intention. A jury must decide whose action was the *actual cause* of T's death.

Let x and p be (respectively) the propositions "enemy 2 shot" and "enemy 1 poisoned the water," and let y denote "T is dead." In addition to these events we will also use the intermediate variable C (connoting cyanide) and D (connoting dehydration), as shown in Figure 10.2. The functions $f_i(pa_i, u)$ are not shown explicitly in Figure 10.2, but they are presumed to determine the value of each child variable from those of its parent variables in the graph, in accordance with the usual understanding of the story:[2]

$$c = px',$$

$$d = x, \tag{10.1}$$

$$y = c \vee d.$$

When we substitute c and d into the expression for y, we obtain a simple disjunction

$$y = x \vee px' \equiv x \vee p, \tag{10.2}$$

which is deceiving in its symmetry.

    Here we see in vivid symbols the role played by structural information. Although it is true that $x \vee x'p$ is logically equivalent to $x \vee p$, the two are not structurally equivalent; $x \vee p$ is completely symmetric relative to exchanging x and p, whereas $x \vee x'p$ tells us that, when x is true, p has no effect whatsoever – not only on y, but also on any of the intermediate conditions that could potentially affect y. It is this asymmetry that makes us proclaim x and not p to be the cause of death.

    According to Lewis, the difference between x and p lies in the nature of the chains that connect each of them to y. From x, there exists a causal chain $x \rightarrow d \rightarrow y$ such that every element is counterfactually dependent on its antecedent. Such a chain does not exist from p to y because, when x is true, the chain $p \rightarrow c \rightarrow y$ is *preempted* (at c);

---

[2] For simplicity, we drop the "$\wedge$" symbol in the rest of this chapter.

that is, c is "stuck" at false regardless of p. Put another way, although x does not satisfy the counterfactual test for causing y, one of its consequences (d) does; given that x and p are true, y would be false were it not for d.

Lewis's chain criterion retains the connection between causation and counterfactuals, but it is rather ad hoc; after all, why should the existence of a counterfactual dependence chain be taken as a defining test for a concept as crucial as "actual cause," by which we decide the guilt or innocence of defendants in a court of law? The basic counterfactual criterion does embody a pragmatic rationale; we would not wish to punish a person for a damage that could not have been avoided, and we would like to encourage people to watch for circumstances where their actions could make a substantial difference. However, once the counterfactual dependence between the action and the consequence is destroyed by the presence of another cause, what good is it to insist on intermediate counterfactual dependencies along a chain that connects them?

10.1.3   Overdetermination and Quasi-Dependence

Another problem with Lewis's chain is its failure to capture cases of simultaneous disjunctive causes. For example, consider the firing squad in Figure 9.1, and assume that riflemen A and B shot together and killed the prisoner. Our intuition regards each of the riflemen as a *contributory* actual cause of the death, though neither rifleman passes the counterfactual test and neither supports a counterfactual dependence chain in the presence of the other.

This example is representative of a condition called *overdetermination*, which presents a tough challenge to the counterfactual account. Lewis answered this challenge by offering yet another repair of the counterfactual criterion. He proposed that chains of counterfactual dependence should be regarded as intrinsic to the process (e.g., the flight of the bullet from A to D) and that the disappearance of dependence due to peculiar surroundings (e.g., the flight of the bullet from B to D) should not be considered an intrinsic loss of dependence; we should still count such a process as *quasi-dependent* "if only the surroundings were different" (Lewis 1986, p. 206).

Hall (2004) observed that the notion of quasi-dependence raises difficult questions: "First, what exactly is a process? Second, what does it mean to say that one process is 'just like' another process in its intrinsic character? Third, how exactly do we 'measure the variety of the surroundings'?" We will propose an answer to these questions using an object called a *causal beam* (Section 10.3.1), which can be regarded as a structural–semantic explication of the notion of a "process." We will return to chains and beams and to questions of preemption and overdetermination in Section 10.2, after a short excursion into Mackie's approach, which also deals with the problem of actual causation – though from a different perspective.

10.1.4   Mackie's INUS Condition

The problems we encountered in the previous section are typical of many attempts by philosophers to give a satisfactory logical explication to the notion of single-event causation (here, "actual causation"). These attempts seem to have started with Mill's observation that no cause is truly sufficient or necessary for its effect (Mill 1843, p. 398). The numerous accounts subsequently proposed – based on more elaborate combinations of sufficiency and necessity conditions – all suffer from insurmountable difficulties (Sosa and Tooley 1993, pp. 1–8). Mackie's treatment (1965) appears to be the earliest attempt to offer a semiformal explication of "actual causation" within this logical framework; his solution, known as the INUS condition, became extremely popular.

The INUS condition states that an event C is perceived to be the cause of event E if C is "an *insufficient* but *necessary* part of a condition which is itself *unnecessary* but *sufficient* for the result" (Mackie 1965).[3] Although attempts to give INUS precise formulation (including some by Mackie 1980) have not resulted in a coherent proposal (Sosa and Tooley 1993, pp. 1–8), the basic idea behind INUS is appealing: If we can think of $\{S_1, S_2, S_3, \ldots\}$ as a collection of every minimally sufficient set of conditions (for E), then event C is an INUS condition for E if it is a conjunct of some S_i. Furthermore, C is considered a *cause* of E if C is an INUS condition for E and if, under the circumstances, C was sufficient for one of those S_i. Thus, for example, if E can be written in disjunctive normal form as

$$E = AB \lor CD,$$

then C is an INUS condition by virtue of being a member of a disjunct, CD, which is minimal and sufficient for E. Thus C would be considered a cause of E if D were present on the occasion in question.[4]

This basic intuition is shared by researchers from many disciplines. Legal scholars, for example, have advocated a relation called NESS (Wright 1988), standing for "necessary element of sufficient set," which is a rephrasing of Mackie's INUS condition in a simpler mnemonic. In epidemiology, Rothman (1976) proposed a similar criterion – dubbed "sufficient components" – for recognizing when an exposure is said to cause a disease: "We say that the exposure E causes disease if a sufficient cause that contains E is the first sufficient cause to be completed" (Rothman and Greenland 1998, p. 53). Hoover (1990, p. 218) related the INUS condition to causality in econometrics: "Any variable that causes another in Simon's sense may be regarded as an INUS condition for that other variable."

However, all these proposals suffer from a basic flaw: the language of logical necessity and sufficiency is inadequate for explicating these intuitions (Kim 1971). Similar conclusions are implicit in the analysis of Cartwright (1989, pp. 25–34), who starts out enchanted with INUS's intuition and ends up having to correct INUS's mistakes.

The basic flaw of the logical account stems from the lack of a syntactic distinction between formulas that represent stable mechanisms (or "dispositional relations," to use Mackie's terminology) and those that represent circumstantial conditions. The simplest manifestation of this limitation can be seen in contraposition: "A implies B"

---

[3] The two negations and the two "buts" in this acronym make INUS one of the least helpful mnemonics in the philosophical literature. Simplified, it should read: "a necessary element in a sufficient set of conditions, NESS" (Wright 1988).

[4] Mackie (1965) also required that every disjunct of E that does not contain C as a conjunct be absent, but this would render Mackie's definition identical to the counterfactual test of Lewis. I use a broader definition here to allow for simultaneous causes and overdetermination; see Mackie (1980, pp. 43–7).

is logically equivalent to "not B implies not A," which renders not-B an INUS-cause of not-A. This is counterintuive; from "disease causes a symptom" we cannot infer that eliminating a symptom will cause the disappearance of the disease. The failure of contraposition further entails problems with transduction (inference through common causes): if a disease D causes two symptoms, A and B, then curing symptom A would entail (in the logical account of INUS) the disappearance of symptom B.

Another set of problems stems from syntax sensitivity. Suppose we apply Mackie's INUS condition to the firing squad story of Figure 9.1. If we write the conditions for the prisoner's death as:

$$D = A \lor B,$$

then A satisfies the INUS criterion, and we can plausibly conclude that A was a cause of D. However, substituting $A = C$, which is explicit in our model, we obtain

$$D = C \lor B,$$

and suddenly A no longer appears as a conjunct in the expression for D. Shall we conclude that A was not a cause of D? We can, of course, avoid this disappearance by forbidding substitutions and insisting that A remain in the disjunction together with B and C. But then a worse problems ensues: in circumstances where the captain gives a signal (C) and both riflemen fail to shoot, the prisoner will still be deemed dead. In short, the structural information conveying the flow of influences in the story cannot be encoded in standard logical syntax – the intuitions of Mackie, Rothman, and Wright must be reformulated.

Finally, let us consider the desert traveler example, where the traveler's death was expressed in (10.2) as

$$y = x \lor x'p.$$

This expression is not in minimal disjunctive normal form because it can be rewritten as

$$y = x \lor p,$$

from which one would obtain the counterintuitive result that x and p are equal partners in causing y. If, on the other hand, we permit nonminimal expressions like $y = x \lor x'p$, then we might as well permit the equivalent expression $y = xp' \lor p$, from which we would absurdly conclude that not poisoning the water (p') would be a cause for our traveler's misfortune, provided someone shoots the canteen (x).

We return now to structural analysis, in which such syntactic problems do not arise. Dispositional information is conveyed through structural or counterfactual expressions (e.g., $v_i = f_i(pa_i, u)$) in which u is generic, whereas circumstantial information is conveyed through prepositional expressions (e.g., $X(u) = x$)) that refer to one specific world $U = u$. Structural models do not permit arbitrary transformations and substitutions, even when truth values are preserved. For example, substituting the expression for c in $y = d \lor c$ would not be permitted if c (cyanide intake) is understood to be governed by a separate mechanism, independent of that which governs y.

Using structural analysis, we will now propose a formal setting that captures the intuitions of Mackie and Lewis. Our analysis will be based on an aspect of causation called *sustenance,* which combines elements of sufficiency and necessity and also takes structural information into account.

10.2 PRODUCTION, DEPENDENCE, AND SUSTENANCE

The probabilistic concept of causal sufficiency, PS (Definition 9.2.2), suggests a way of rescuing the counterfactual account of causation. Consider again the symmetric over-determination in the firing-squad example. The shot of each rifleman features a PS value of unity (see (9.43)), because each shot would cause the prisoner's death in a state u' in which the prisoner is alive. This high PS value supports our intuition that each shot is an actual cause of death, despite a low PN value (PN = 0). Thus, it seems plausible to argue that our intuition gives some consideration to sufficiency, and that we could formulate an adequate criterion for actual causation using the right mixture of PN and PS components.

Similar expectations are expressed in Hall (2004). In analyzing problems faced by the counterfactual approach, Hall made the observation that there are two concepts of causation, only one of which is captured by the counterfactual account, and that failure to capture the second concept may well explain its clashes with intuition. Hall calls the first concept "dependence" and the second "production." In the firing-squad example, intuition considers each shot to be an equal "producer" of death. In contrast, the counterfactual account tests for "dependence" only, and it fails because the state of the prisoner does not "depend" on either shot alone.

The notions of dependence and production closely parallel those of necessity and sufficiency, respectively. Thus, our formulation of PS could well provide the formal basis for Hall's notion of production and serve as a step toward the formalization of actual causation. However, for this program to succeed, a basic hurdle must first be overcome: productive causation is oblivious to scenario-specific information (Pearl 1999), as can be seen from the following considerations.

The ***dependence*** aspect of causation appeals to the necessity of a cause x in maintaining the effect y in the face of certain contingencies, which otherwise will negate y (Definition 9.2.1):

$$X(u) = x, \quad Y(u) = y, \quad Y_{x'}(u) = y'. \tag{10.3}$$

The ***production*** aspect, on the other hand, appeals to the capacity of a cause (x) to bring about the effect (y) in a situation (u') where both are absent (Definition 9.2.2):

$$X(u') = x', \quad Y(u') = y', \quad Y_x(u') = y. \tag{10.4}$$

Comparing these two definitions, we note a peculiar feature of production: To test production, we must step outside our world momentarily, imagine a new world u' with x and y absent, apply x, and see if y sets in. Therefore, the sentence "x produced y" can be true only in worlds u' where x and y are false, and thus it appears (a) that nothing could possibly explain (by consideration of production) any events that did materialize in the actual world and (b) that evidence gathered about the actual world u could not be brought to bear on the hypothetical world u' in which production is defined.

To overcome this hurdle, we resort to an aspect of causation called sustenance, which enriches the notion of dependence with features of production while remaining in a world

u in which both x and y are true. Sustenance differs from dependence in the type of contingencies against which x is expected to protect y. Whereas the contingencies considered in (10.3) are "circumstantial" – that is, evolving from a set of circumstances $U = u$ that are specific to the scenario at hand – we now insist that x will maintain y against contingencies that evolve from *structural* modification of the model itself (Pearl 1998b).

Definition 10.2.1 (Sustenance)
Let W be a set of variables in V, and let w, w' be specific realizations of these variables. We say that x causally sustains y in u relative to contingencies in W if and only if

   (i) $X(u) = x$;

   (ii) $Y(u) = y$;   (10.5)

   (iii) $Y_{xw}(u) = y$ *for all w; and*

   (iv) $Y_{x'w'}(u) = y' \neq y$ *for some $x' \neq x$ and some w'.*

The sustenance feature of (10.5) is expressed in condition (iii), $Y_{xw}(u) = y$, which requires that x *alone* be sufficient for maintaining y. It reads: If we set X to its actual value (x) in u then, even if W is set to any value (w) that is different from the actual, Y will still retain its actual value (y) in u. Condition (iv), $Y_{x'w'}(u) = y'$, attributes to $X = x$ the "responsibility" for sustaining $Y = y$ under such adverse conditions; if we set X to some other value (x'), then Y will relinquish its current value (y) under at least one setting $W = w'$. Put together, (iii) and (iv) imply that there exists a setting $W = w'$ in which x is both necessary and sufficient for y.

Is sustenance a reasonable requirement to impose on an "actual cause"? Consider again the two bullets that caused the prisoner's death in Figure 9.1. We consider A to be an actual cause of D in this scenario, because A would have sustained D "alone," even in the absence of B. But how do we express, formally, the absence of B in our scenario $U = u$, given that B did in fact occur? If we wish to (hypothetically) suppress B within the context created by u, then we must use a structural contingency and imagine that B is made false by some external intervention (or "miracle") that violates the rule $B = C$; for example, that rifleman B was prevented from shooting by some mechanical failure. We know perfectly well that such failure did not occur, yet we are committed to contemplating such failures by the very act of representing our story in the form of a multistage causal model, as in Figure 9.1.

Recalling that every causal model stands not for just one but for a whole set of models, one for each possible state of the $do(\cdot)$ operator, contemplating interventional contingencies is an intrinsic feature of every such model. In other words, the autonomy of the mechanisms in the model means that each mechanism advertises its possible breakdown, and these breakdowns signal contingencies against which causal explanations should operate. It is reasonable, therefore, that we build such contingencies into the definition of actual causation, which is a form of explanation.

The choice of W in Definition 10.2.1 should be made with caution. We obviously cannot permit W to include all variables that mediate between X and Y, for this would preclude any x from ever sustaining y. More seriously, by not restricting W we run the risk of removing genuine preemptions and turning noncauses into causes. For example, by choosing $W = \{X\}$ and $w' = 0$ in the desert traveler story (Figure 10.2), we turn enemy 1 into the actual cause of death, contrary to intuition and contrary to the actual scenario (which excludes cyanide intake). The notion of "causal beam" (Pearl 1998b) is devised to make the choice of W minimally disruptive to the actual scenario.[5]

10.3 CAUSAL BEAMS AND SUSTENANCE-BASED CAUSATION

10.3.1 Causal Beams: Definitions and Implications

We start by considering a causal model M, as defined in Section 7.1, and selecting a subset S of *sustaining* parent variables for each family and each u. Recall that the arguments of the functions $\{f_i\}$ in a causal model were assumed to be minimal in some sense, since we have pruned from each f_i all redundant arguments and retained only those called pa_i that render $f_i(pa_i, u)$ nontrivial (Definition 7.1.1). However, in that definition we were concerned with nontriviality relative to all possible u; further pruning is feasible when we are situated at a particular state $U = u$.

To illustrate, consider the function $f_i = ax_1 + bux_2$. Here $PA_i = \{X_1, X_2\}$, because there is always some value of u that would make f_i sensitive to changes in either x_1 or x_2. However, given that we are in a state for which $u = 0$, we can safely consider X_2 to be a trivial argument, replace f_i with $f_i^0 = ax_1$, and consider X_1 as the only *essential* argument of f_i'. We shall call f_i^0 the *projection* of f_i on $u = 0$; more generally, we will consider the projection of the entire model M by replacing every function in $\{f_i\}$ with its projection relative to a specific u and a specific value of its nonessential part. This leads to a new model, which we call *causal beam*.

Definition 10.3.1 (Causal Beam)
For model $M = \langle U, V, \{f_i\}\rangle$ and state $U = u$, a causal beam *is a new model $M_u = \langle u, V, \{f_i^u\}\rangle$ in which the set of functions f_i^u is constructed from $\{f_i\}$ as follows.*

1. *For each variable $V_i \in V$, partition PA_i into two subsets, $PA_i = S \cup \bar{S}$, where S (connoting "sustaining") is any subset of PA_i satisfying*[6]

   $$f_i(S(u), \bar{s}, u) = f_i(S(u), \bar{s}', u) \quad \text{for all } \bar{s}'. \tag{10.6}$$

   In words, S is any set of PA_i sufficient to entail the actual value of $V_i(u)$, regardless of how we set the other members of PA_i.

2. *For each variable $V_i \in V$, find a subset W of $\bar{S}$ for which there exists some realization $W = w$ that renders the function $f_i(s, \bar{S}_w(u), u)$ nontrivial in s; that is,*

   $$f_i(s', \bar{S}_w(u), u) \neq V_i(u) \text{ for some } s'.$$

---

[5] Halpern and Pearl (1999) permit the choice of any set W such that its complement, $Z = V - W$, is sustained by x; that is, $Z_{xw}(u) = Z(u)$ for all w.

[6] Pearl (1998b) required that S be minimal, but this restriction is unnecessary for our purposes (though all our examples will invoke minimally sufficient sets). As usual, we use lowercase letters (e.g., $s, \bar{s}$) to denote specific realizations of the corresponding variables (e.g., $S, \bar{S}$) and use $S_x(u)$ to denote the realization of S under $U = u$ and $do(X = x)$. Of course, each parent set PA_i would have a distinct partition $PA_i = S_i \cup \bar{S}_i$, but we drop the i index for clarity.

Here, $\overline{S}$ should not intersect the sustaining set of any other variable $V_j, j \neq i$. (Likewise, setting $W = w$ should not contradict any such setting elsewhere.)

3. *Replace $f_i(s, \overline{s}, u)$ by its projection $f_i^u(s)$, which is given by*

$$f_i^u(s) = f_i(s, \overline{S}_w(u), u). \tag{10.7}$$

Thus the new parent set of V_i becomes $PA_i^u = S$, and every f^u function is responsive to its new parent set S.

Definition 10.3.2 (Natural Beam)

A causal beam M_u is said to be natural if condition 2 of Definition 10.3.1 is satisfied with $W = \emptyset$ for all $V_i \in V$.

In words, a natural beam is formed by "freezing" all variables outside the sustaining set at their actual values, $\overline{S}(u)$, thus yielding the projection $f_i^u(s) = f_i(s, \overline{S}(u), u)$.

Definition 10.3.3 (Actual Cause)

We say that event $X = x$ was an actual cause of $Y = y$ in a state u (abbreviated "x caused y") if and only if there exists a natural beam M_u such that

$$Y_x = y \quad in \; M_u \tag{10.8}$$

and

$$Y_{x'} \neq y \quad in \; M_u \quad for \; some \; x' \neq x. \tag{10.9}$$

Note that (10.8) is equivalent to

$$Y_x(u) = y, \tag{10.10}$$

which is implied by $X(u) = x$ and $Y(u) = y$. But (10.9) ensures that, after "freezing the trivial surroundings" represented by $\overline{S}$, $Y = y$ would not be sustained by some value x' of X.

Definition 10.3.4 (Contributory Cause)

We say that x is a contributory cause of y in a state u if and only if there exists a causal beam, but no natural beam, that satisfies (10.8) and (10.9).

In summary, the causal beam can be interpreted as a theory that provides a sufficient and nontrivial explanation for each actual event $V_i(u) = v_i$ under a hypothetical freezing of some variables ($\overline{S}$) by the $do(\cdot)$ operator. Using this new theory, we subject the event $X = x$ to a counterfactual test and check whether Y would change if X were not x. If a change occurs in Y when freezing takes place at the actual values of $\overline{S}$ (i.e., $W = \emptyset$), we say that "x was an actual cause of y." If changes occur only under a freeze state that is removed from the actual state (i.e., $W \neq \emptyset$), we say that "x was a contributory cause of y."

> **Remark:** Although W was chosen to make V_i responsive to S, this does not guarantee that $S(u)$ is necessary and sufficient for $V_i(u)$ because local responsiveness

does not preclude the existence of another state $s'' \neq S(u)$ for which $f_i^u(s'') = V_i(u)$. Thus, (10.8) does not guarantee that x is both necessary and sufficient for y. That is the reason for the final counterfactual test in (10.9). It would be too restrictive to require that w render f^u nontrivial for every s of S; such a W may not exist. If (10.8)–(10.9) are satisfied, then $W = w$ represents some hypothetical modification of the world model under which x is both sufficient and necessary for y.

Remarks on Multivariate Events: Although Definitions 10.3.3 and 10.3.4 apply to univariate as well as multivariate causes and effects, some refinements are in order when X and Y consist of sets of variables.[7] If the effect considered, E, is any Boolean function of a set $Y = \{Y_1, \ldots, Y_k\}$ of variables, then (10.8) should apply to every member Y_i of Y, and (10.9) should be modified to read $Y_{x'} \implies \neg E$ instead of $Y_{x'} \neq y$. Additionally, if X consists of several variables, then it is reasonable to demand that X be minimal – in other words, to demand that no subset of those variables passes the test of (10.8)–(10.9). This requirement strips X from irrelevant, overspecified details. For example, if drinking poison qualifies as the actual cause of Joe's death then, awkwardly, drinking poison and sneezing would also pass the test of (10.8)–(10.9) and qualify as the cause of Joe's death. Minimality removes "sneezing" from the causal event $X = x$.

Incorporating Probabilities and Evidence

Suppose that the state u is uncertain and that the uncertainty is characterized by the probability $P(u)$. If e is the evidence available in the case, then the probability that x caused y can be obtained by summing up the weight of evidence $P(u \mid e)$ over all states u in which the assertion "x caused y" is true.

Definition 10.3.5 (Probability of Actual Causation)
Let U_{xy} be the set of states in which the assertion "x is an actual cause of y" is true (Definition 10.3.2), and let U_e be the set of states compatible with the evidence e. The probability that x caused y in light of evidence e, denoted $P(caused(x, y \mid e))$, is given by the expression

$$P(caused\,(x, y \mid e)) = \frac{P(U_{xy} \cap U_e)}{P(U_e)}. \tag{10.11}$$

10.3.2   Examples: From Disjunction to General Formulas

Overdetermination and Contributory Causes

Contributory causation is typified by cases where two actions concur to bring about an event yet either action, operating alone, would still have brought about the event. In such cases the model consists of just one mechanism, which connects the effect E to the two

---

[7] These were formulated by Joseph Halpern in the context of the definition presented in Halpern and Pearl (1999).

actions through a simple disjunction: $E = A_1 \vee A_2$. There exists no natural beam to qualify either A_1 or A_2 as an actual cause of E. If we fix either A_1 or A_2 at its current value (namely, true), then E will become a trivial function of the other action. However, if we deviate from the current state of affairs and set A_2 to false (i.e., forming a beam with $W = \{A_2\}$ and setting W to false), then E would then become responsive to A_1 and so pass the counterfactual test of (10.9).

This example illustrates the sense in which the beam criterion encapsulates Lewis's notion of quasi-dependence. Event E can be considered quasi-dependent on A_1 if we agree to test such dependence in a hypothetical submodel created by the $do(A_2 = \text{false})$ operator. In Section 10.2 we argued that such a hypothetical test – though it conflicts with the current scenario u – is implicitly written into the charter of every causal model. A causal beam may thus be considered a formal explication of Lewis's notion of a quasi-dependent process, and the combined sets W represent the "peculiar surroundings" of the process that (when properly modified) renders $X = x$ necessary for $Y = y$.

Disjunctive Normal Form

Consider a single mechanism characterized by the Boolean function

$$y = f(x, z, r, h, t, u) = xz \vee rh \vee t,$$

where (for simplicity) the variables X, Z, R, H, T are assumed to be causally independent of each other (i.e., none is a descendant of another in the causal graph $G(M)$). We next illustrate conditions under which x would qualify as a contributory or an actual cause for y.

First, consider a state $U = u$ where all variables are true:

$$X(u) = Z(u) = R(u) = H(u) = T(u) = Y(u) = \text{true}.$$

In this state, every disjunct represents a minimal set of sustaining variables. In particular, taking $S = \{X, Z,\}$ we find that the projection $f^u(x, z) = f(x, z, R(u), H(u), T(u))$ becomes trivially true. Thus, there is no natural beam M_u, and x could not be the actual cause of y. Feasible causal beams can be obtained by using $w = \{r', t'\}$ or $w = \{h', t'\}$, where primes denote complementation. Each of these two choices yields the projection $f^u(x, z) = xz$. Clearly, M_u meets the conditions of (10.8) and (10.9), thus certifying x as a contributory cause of y.

Using the same argument, it is easy to see that, at a state u' for which

$$X(u') = Z(u') = \text{true} \quad \text{and} \quad R(u') = T(u') = \text{false},$$

a natural beam exists; that is, a nontrivial projection $f^{u'}(x, z) = xz$ is realized by setting the redundant $(\bar{S})$ variables R, H, and T to their actual values in u'. Hence, x qualifies as an actual cause of y.

This example illustrates how Mackie's intuition for the INUS condition can be explicated in the structural framework. It also illustrates the precise roles played by structural (or "dispositional") knowledge (e.g., $f_i(pa_i, u)$) and circumstantial knowledge ($X(u) = \text{true}$), which were not clearly distinguished by the strictly logical account.

The next example illustrates how the INUS condition generalizes to arbitrary Boolean functions, especially those having several minimal disjunctive normal forms.

Single Mechanism in General Boolean Form

Consider the function

$$y = f(x, z, h, u) = xz' \lor x'z \lor xh', \tag{10.12}$$

which has the equivalent form

$$y = f(x, z, h, u) = xz' \lor x'z \lor zh'. \tag{10.13}$$

Assume, as before, that (a) we consider a state u in which X, Z, and H are true and (b) we inquire as to whether the event $x : X =$ true caused the event $y : Y =$ false. In this state, the only sustaining set is $S = \{X, Z, R\}$, because no choice of two variables (valued at this u) would entail $Y =$ false regardless of the third. Since $\bar{S}$ is empty, the choice of beam is unique: $M_u = M$, for which $y = f^u(x, z, h) = xz' \lor x'z \lor xh'$. This M_u passes the counterfactual test of (10.9), because $f^u(x', z, h) =$ true; we therefore conclude that x was an actual cause of y. Similarly, we can see that the event $H =$ true was an actual cause of $Y =$ false. This follows directly from the counterfactual test

$$Y_h(u) = \text{false} \quad \text{and} \quad Y_{h'}(u) = \text{true}.$$

Because Definitions 10.3.3 and 10.3.4 rest on semantical considerations, identical conclusions would be obtained from any logically equivalent form of f (not necessarily in minimal disjunctive form) – as long as f represents a single mechanism. In simple, single-mechanism models, the beam criterion can therefore be considered the semantical basis behind the INUS intuition. The structure-sensitive aspects of the beam criterion will surface in the next two examples, where models of several layers are considered.

10.3.3   Beams, Preemption, and the Probability of Single-Event Causation

In this section we apply the beam criterion to a probabilistic version of the desert traveler example. This will illustrate (i) how structural information is utilized in problems involving preemption and (ii) how we can compute the probability that one event "was the actual cause of another," given a set of observations.

Consider a modification of the desert traveler example in which we do not know whether the traveler managed to drink any of the poisoned water before the canteen was emptied. To model this uncertainty, we add a bivalued variable U that indicates whether poison was drunk ($u = 0$) or not ($u = 1$). Since U affects both D and C, we obtain the structure shown in Figure 10.3. To complete the specification of the model, we need to assign functions $f_i(pa_i, u)$ to the families in the diagram and a probability distribution $P(u)$. To formally complete the model, we introduce the dummy background variables U_X and U_P, which represent the factors behind the enemies' actions.

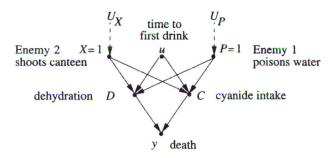

Figure 10.3 Causal relationships for the probabilistic desert traveler.

The usual understanding of the story yields the following functional relationships:

$$c = p(u' \lor x'),$$
$$d = x(u \lor p'),$$
$$y = c \lor d,$$

together with the evidential information

$$X(u_X) = 1, \qquad P(u_P) = 1.$$

(We assume that T will not survive with an empty canteen (x) even after drinking un-poisoned water before the shot ($p'u'$).)

In order to construct the causal beam M_u, we examine each of the three functions and form their respective projections on u. For example, for $u = 1$ we obtain the functions shown in (10.1), for which the (minimal) sustaining parent sets are: X (for C), X (for D), and D (for Y). The projected functions become

$$c = x',$$
$$d = x, \tag{10.14}$$
$$y = d,$$

and the beam model $M_{u=1}$ is natural; its structure is depicted in Figure 10.4. To test whether x (or p) was the cause of y, we apply (10.8)–(10.9) and obtain

$$Y_x = 1 \text{ and } Y_{x'} = 0 \text{ in } M_{u=1},$$
$$Y_p = 1 \text{ and } Y_{p'} = 1 \text{ in } M_{u=1}. \tag{10.15}$$

Thus, enemy 2 shooting at the container (x) is classified as the actual cause of T's death (y), whereas enemy 1 poisoning the water (p) was not the actual cause of y.

Next, consider the state $u = 0$, which denotes the event that our traveler reached for a drink before enemy 2 shot at the canteen. The graph corresponding to $M_{u=0}$ is shown in Figure 10.5 and gives

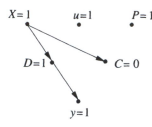

Figure 10.4   Natural causal beam representing the state $u = 1$.

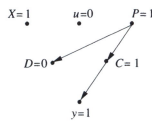

Figure 10.5   Natural causal beam representing the state $u = 0$.

$$Y_x = 1 \text{ and } Y_{x'} = 1 \text{ in } M_{u=0}, \tag{10.16}$$

$$Y_p = 1 \text{ and } Y_{p'} = 0 \text{ in } M_{u=0}.$$

Thus, in this state of affairs we classify enemy 1's action to be the actual cause of T's death, while enemy 2's action is not considered the cause of death.

If we do not know which state prevailed, $u = 1$ or $u = 0$, then we must settle for the *probability* that x caused y. Likewise, if we observe some evidence e reflecting on the probability $P(u)$, such evidence would yield (see (10.11))

$$P(\text{caused}(x, y \mid e)) = P(u = 1 \mid e)$$

and

$$P(\text{caused}(p, y \mid e)) = P(u = 0 \mid e).$$

For example, a forensic report confirming "no cyanide in the body" would rule out state $u = 0$ in favor of $u = 1$, and the probability of x being the cause of y becomes 100%. More elaborate probabilistic models are analyzed in Pearl (1999).

10.3.4   Path-Switching Causation

Example 10.3.6 Let x be the state of a two-position switch. In position 1 ($x = 1$), the switch turns on a lamp ($z = 1$) and turns off a flashlight ($w = 0$). In position 0 ($x = 0$), the switch turns on the flashlight ($w = 1$) and turns off the lamp ($z = 0$). Let $Y = 1$ be the proposition that the room is lighted.

The causal beams M_u and $M_{u'}$ associated with the states in which the switch is in position 1 and 2 (respectively) are shown in the graphs of Figure 10.6. Once again, M_u entails

$X(u) = 1$

$z = 1$ $w = 0$

$y = 1$

$X(u') = 0$

$z = 0$ $w = 1$

$y = 1$

Figure 10.6 Natural beams that represent path switching in Example 10.3.6.

$Y_x = 1$ and $Y_{x'} = 0$. Likewise $M_{u'}$ entails $Y_x = 1$ and $Y_{x'} = 0$. Thus "switch in position 1" and "switch in position 2" are *both* considered actual causes for "room is lighted," although neither is a necessary cause.

This example further highlights the subtlety of the notion of "actual cause"; changing X from 1 to 0 merely changes the course of the causal pathway while keeping its source and destination the same. Should the current switch position ($X = 1$) be considered the actual cause of (or an "explanation of") the light in the room? Although $X = 1$ enables the passage of electric current through the lamp and is in fact the only mechanism currently sustaining light, one may argue that it does not deserve the title "cause" in ordinary conversation. It would be odd to say, for instance, that $X = 1$ was the cause of spoiling an attempted burglary. However, recalling that causal explanations earn their value in the abnormal circumstances created by structural contingencies, the possibility of a malfunctioning flashlight should enter our mind whenever we designate it as a separate mechanism in the model. Keeping this contingency in mind, it should not be too odd to name the switch position as a cause of spoiling the burglary.

10.3.5   Temporal Preemption

Consider the example mentioned in the preface of this chapter, in which two fires are advancing toward a house. If fire A burned the house before fire B, then we would consider fire A "the actual cause" of the damage, even though fire B would have done the same were it not for A. If we simply write the structural model as

$$H = A \lor B,$$

where H stands for "house burns down," then the beam method would classify each fire as an equally contributory cause, which is counterintuitive – fire B is not regarded as having made any contribution to H.

This example is similar to yet differs from the desert traveler; here, the way in which one cause preempts the other is more subtle in that the second cause becomes ineffective only because the effect has already happened. Hall (2004) regards this sort of preemption as equivalent to ordinary preemption, and he models it by a causal diagram in which H, once activated, inhibits its own parents. Such inhibitory feedback loops lead to irreversible behavior, contrary to the unique-solution assumption of Definition 7.1.1.

A more direct way of expressing the fact that a house, once burned, will remain burned even when the causes of fire disappear is to resort to dynamic causal models (as in Figure 3.3), in which variables are time-indexed. Indeed, it is impossible to capture temporal relationships such as "arriving first" by using the static causal models defined in Section 7.1; instead, dynamic models must be invoked.

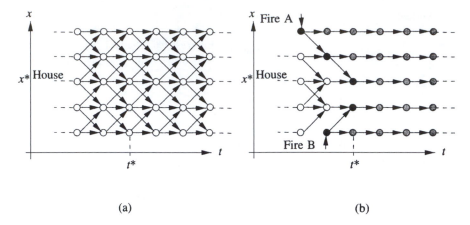

(a)   (b)

Figure 10.7   (a) Causal diagram associated with the dynamic model of (10.17). (b) Causal beam associated with starting fire A and fire B at different times, showing no connection between fire B and the state of the house at $x = x^*$.

Let the state of the fire $V(x, t)$ at location x and time t take on three values: g (for green), f (for on fire), and b (for burned). The dynamic structural equations characterizing the propagation of fire can then be written (in simplified form) as:

$$V(x, t) = \begin{cases} f & \text{if } V(x, t - 1) = g \text{ and } V(x - 1, t - 1) = f, \\ f & \text{if } V(x, t - 1) = g \text{ and } V(x + 1, t - 1) = f, \\ b & \text{if } V(x, t - 1) = b \text{ and } V(x, t - 1) = f, \\ g & \text{otherwise.} \end{cases} \qquad (10.17)$$

The causal diagram associated with this model is illustrated in Figure 10.7(a), designating three parents for each variable $V(x, t)$: the previous state $V(x + 1, t - 1)$ of its northern neighbor, the previous state $V(x - 1, t - 1)$ of its southern neighbor, and the previous state $V(x, t - 1)$ at location x. The scenario emanating from starting fire A and fire B one time unit apart (corresponding to actions $do(V(x^* + 2, t^* - 2) = f)$ and $do(V(x^* - 2, t^* - 1) = f)$) is shown in Figure 10.7(b). Black and grey bullets represent, respectively, space–time regions in states f (on fire) and b (burned). This beam is both natural and unique, as can be seen from (10.17). The arrows in Figure 10.7(b) represent a natural beam constructed from the (unique) minimally sufficient sets S at each family. The state of the parent set S that this beam assigns to each variable constitutes an event that is both necessary and sufficient for the actual state of that variable (assuming variables in $\bar{S}$ are frozen at their actual values).

Applying the test of (10.9) to this beam, we find that a counterfactual dependence exists between the event $V(x^* - 2, t^* - 2) = f$ (representing the start of fire A) and the sequence $V(x^*, t)$, $t > t^*$ (representing the state of the house through time). No such dependence exists for fire B. On that basis, we classify fire A as the actual cause of the house fire. Remarkably, the common intuition of attributing causation to an event that hastens the occurrence of the effect is seen to be a corollary of the beam test in the spatiotemporal representation of the story. However, this intuition cannot serve as the defining principle for actual causation, as suggested by Paul (1998). In our story, for example, each fire alone did not hasten (or delay, or change any property of) the following event: E = the owner of the house did not enjoy breakfast the next day. Yet we still consider fire A, not B, to be the actual cause of E, as predicted by the beam criterion.

The conceptual basis of this criterion can be illuminated by examining the construction of the minimal beam shown in Figure 10.7(b). The pivotal step in this construction lies in the space–time region (x^*, t^*), which represents the house at the arrival of fire. The variable representing the state of the house at that time, $V(x^*, t^*)$, has a two-parent sustaining set, $S = \{V(x^* + 1, t^* - 1) \text{ and } V(x^*, t^* - 1)\}$, with values f and g, respectively. Using (10.17), we see that the south parent $V(x^* - 1, t^* - 1)$ is redundant, because the value of $V(x^*, t^*)$ is determined (at f) by the current values of the other two parents. Hence, this parent can be excluded from the beam, rendering $V(x^*, t^*)$ dependent on fire A. Moreover, since the value of the south parent is g, that parent cannot be part of any minimally sustaining set, thus ensuring that $V(x^*, t^*)$ is independent of fire B. (We could, of course, add this parent to S, but $V(x^*, t^*)$ would remain independent of fire B.) The next variable to examine is $V(x^*, t^* + 1)$, with parents $V(x^* - 1, t^*)$, $V(x^*, t^*)$, and $V(x^* - 1, t^*)$ valued at $b, f,$ and f, respectively. From (10.17), the value f of the middle parent is sufficient to ensure the value b for the child variable; hence this parent qualifies as a singleton sustaining set, $S = \{V(x^*, t^*)\}$, which permits us to exclude the other two parents from the beam and so render the child dependent on fire A (through S) but not on fire B. The north and south parents are not, in themselves, sufficient for sustaining the current value (b) of the child node (fires at neighboring regions can cause the house to catch fire but not to become immediately "burned"); hence we must keep the middle parent in S and, in so doing, we render all variables $V(x^*, t), t > t^*$, independent of fire B.

We see that sustenance considerations lead to the intuitive results through two crucial steps: (1) permitting the exclusion (from the beam) of the south parent of every variable $V(x^*, t), t > t^*$, thus maintaining the dependence of $V(x^*, t)$ on fire A; and (2) requiring the inclusion (in any beam) of the middle parent of every variable $V(x^*, t), t > t^*$, thus preventing the dependence of $V(x^*, t)$ on fire B. Step (1) corresponds to selecting the intrinsic process from cause to effect and then suppressing the influence of its nonintrinsic surrounding. Step (2) prevents the growth of causal processes beyond their intrinsic boundaries.

10.4   CONCLUSIONS

We have seen that the property of sustenance (Definition 10.2.1), as embodied in the beam test (Definition 10.3.3), is the key to explicating the notion of actual causation (or "cause in fact," in legal terminology); this property should replace the "but for" test in cases involving multistage scenarios with several potential causes. Sustenance captures the capacity of the putative cause to maintain the value of the effect in the face of structural contingencies and includes the counterfactual test of necessity as a special case, with structural contingencies suppressed (i.e., $W = \emptyset$). We have argued that (a) it is the structural rather than circumstantial contingencies that convey the true meaning of causal claims and (b) these structural contingencies should therefore serve as the basis for causal explanation. We further demonstrated how explanations based on such contingencies resolve difficulties that have plagued the counterfactual account of single-event causation – primarily difficulties associated with preemption, overdetermination, temporal preemption, and switching causation.

Sustenance, however, does not totally replace production, the second component of sufficiency – that is, the capacity of the putative cause to produce the effect in situations where the effect is absent. In the match–oxygen example (see Section 9.5), for instance, oxygen and a lit match each satisfy the sustenance test of Definition 10.3.3 (with $W = \emptyset$ and $\bar{S} = \emptyset$); hence, each factor would qualify as an actual cause of the observed fire. What makes oxygen an awkward explanation in this case is not its ineptness at sustaining fire against contingencies (the contingency set W is empty) but rather its inability to produce fire in the most common circumstance that we encounter, $U = u'$, in which a match is not struck (and a fire does not break out).

This argument still does not tell us why we should consider such hypothetical circumstances ($U = u'$) in the match–oxygen story and not, say, in any of the examples considered in this chapter, where sustenance ruled triumphantly. With all due respect to the regularity and commonality of worlds $U = u'$ in which a match is not struck, those are nevertheless contrary-to-fact worlds, since a fire did break out. Why, then, should one travel to such a would-be world when issuing an explanation for events (fire) in the actual world?

The answer, I believe, lies in the pragmatics of the explanation sought. The tacit target of explanation in the match–oxygen story is the question: "How could the fire have been prevented?" In view of this target, we have no choice but abandon the actual world (in which fire broke out) and travel to one ($U = u'$) in which agents are still capable of preventing this fire.[8]

A different pragmatics motivates the causal explanation in the switch–light story of Example 10.3.6. Here one might be more concerned with keeping the room lit, and the target question is: "How can we ensure that the room remains lit in the face of unforeseen contingencies?" Given this target, we might as well remain in the comfort of our factual world, $U = u$, and apply the criterion of sustenance rather than production.

It appears that pragmatic issues surrounding our quest for explanation are the key to deciding which facet of causation should be used, and that the mathematical formulation of this pragmatics is a key step toward the automatic generation of adequate explanations. Unfortunately, I must now leave this task for future investigation.

Acknowledgments

My interest in the topic of actual causation was kindled by Don Michie, who spent many e-mail messages trying to convince me that (1) the problem is not trivial and (2) Good's

---

[8] Herbert Simon has related to me that a common criterion in accident liability cases, often applied to railroad crossing accidents, is the "last clear chance" doctrine: the person liable for a collision is the one who had the last clear chance of avoiding it.

Figure 10.8   An example showing the need for beam refinement.

(1961, 1962) measures of causal tendency can be extended to handle individual events. He succeeded with regard to (1), and this chapter is based on a seminar given at UCLA (in the spring of 1998) in which "actual causation" was the main topic. I thank the seminar participants, Ray Golish, Andrew Lister, Eitan Mendelowitz, Peyman Meshkat, Igor Roizen, and Jin Tian for knocking down two earlier attempts at beams and sustenance and for stimulating discussions leading to the current proposal. Discussions with Clark Glymour, Igal Kvart, Jim Woodward, Ned Hall, Herbert Simon, Gary Schwartz, and Richard Baldwin sharpened my understanding of the philosophical and legal issues involved. Subsequent collaboration with Joseph Halpern helped to polish these ideas further and led to the more general and declarative definition of actual cause reported in Halpern and Pearl (2000).

Postscript for the Second Edition

Halpern and Pearl (2001a,b) discovered a need to refine the causal beam definition of Section 10.3.3. They retained the idea of defining actual causation by counterfactual dependency in a world perturbed by contingencies, but permitted a wider set of contingencies.

To see that the causal beam definition requires refinement, consider the following example.

> **Example 10.4.1** A vote takes place, involving two people. The measure Y is passed if at least one of them votes in favor. In fact, both of them vote in favor, and the measure passes.

This version of the story is identical to the disjunctive scenario discussed in Section 10.1.3, where we wish to proclaim each favorable vote, $V_1 = 1$ and $V_2 = 1$, a contributing cause of $Y = 1$.

However, suppose there is a voting machine that tabulates the votes. Let M represent the total number of votes recorded by the machine. Clearly $M = V_1 + V_2$ and $Y = 1$ iff $M \geq 1$. Figure 10.8 represents this more refined version of the story.

In this scenario, the beam criterion no longer qualifies $V_1 = 1$ as a contributing cause of $P = 1$, because V_2 cannot be labeled "inactive" relative to M, hence we are not at liberty to set the contingency $V_2 = 0$ and test the counterfactual dependency of Y on V_1 as we did in the simple disjunctive case.

A refinement that properly handles such counterexamples was proposed in Halpern and Pearl (2001a,b) but, unfortunately, Hopkins and Pearl (2002) showed that the constraints on the contingencies were too liberal. This led to a further refinement (Halpern and Pearl 2005a,b) and to the definition given below:

Definition 10.4.2 (Actual Causation) (Halpern and Pearl 2005)
$X = x$ is an actual cause of $Y = y$ in a world $U = u$ if the following three conditions hold:

  AC1. $X(u) = x$, $Y(u) = y$

  AC2. *There is a partition of V into two subsets, Z and W, with $X \subseteq Z$ and a setting x' and w of the variables in X and W, respectively, such that if $Z(u) = z^*$, then both of the following conditions hold:*

   (a) $Y_{x',w} \neq y$.

   (b) $Y_{x,w,z^*} = y$ *for all subsets W' of W and all subsets Z' of Z, with the setting w of W' and z^* of Z' equal to the setting of those variables in $W = w$ and $Z = z^*$, respectively.*

  AC3. *W is minimal; no subset of X satisfies conditions AC1 and AC2.*

The assignment $W = w$ acts as a contingency against which $X = x$ is given the counterfactual test, as expressed in AC2(a).

  AC2 (b) limits the choice of contingencies. Roughly speaking, it says that if the variables in X are reset to their original values, then $Y = y$ must hold, even under the contingency $W = w$ and even if some variables in Z are given their original values (i.e., the values in z^*).

  In the case of the voting machine, if we identify $W = w$ with $V_2 = 0$, and $Z = z^*$ with $V_1 = 1$, we see that $V_i = 1$ qualifies as a cause under AC2; we no longer require that M remains invariant to the contingency $V_2 = 0$; the invariance of $Y = 1$ suffices.

  This definition, though it correctly solves most problems posed in the literature (Hiddleston 2005; Hall 2007; Hitchcock 2007, 2008), still suffers from one deficiency; it must rule out certain contingencies as unreasonable. Halpern (2008) has offered a solution to this problem by appealing to the notion of "normality" in default logic (Spohn 1988; Kraus et al. 1990; Pearl 1990b); only those contingencies should be considered which are at the same level of "normality" as their counterparts in the actual world.

CHAPTER ELEVEN

Reflections, Elaborations, and Discussions with Readers

As X-rays are to the surgeon,
graphs are for causation.
The author

In this chapter, I reflect back on the material covered in Chapters 1 to 10, discuss issues that require further elaboration, introduce new results obtained in the past eight years, and answer questions of general interest posed to me by readers of the first edition. These range from clarification of specific passages in the text, to conceptual and philosophical issues concerning the controversial status of causation, how it is taught in classrooms and how it is treated in textbooks and research articles.

The discussions follow roughly the order in which these issues are presented in the book, with section numbers indicating the corresponding chapters.

11.1 CAUSAL, STATISTICAL, AND GRAPHICAL VOCABULARY

11.1.1 Is the Causal–Statistical Dichotomy Necessary?
***Question to Author* (from many readers)**

Chapter 1 (Section 1.5) insists on a sharp distinction between statistical and causal concepts; the former are definable in terms of a joint distribution function (of observed variables), the latter are not. Considering that many concepts which the book classifies as "causal" (e.g., "randomization," "confounding," and "instrumental variables") are commonly discussed in the statistical literature, is this distinction crisp? Is it necessary? Is it useful?

Author Answer

The distinction is crisp,[1] necessary, and useful, and, as I tell audiences in all my lectures: "If you get nothing out of this lecture except the importance of keeping statistical and causal concepts apart, I would consider it a success." Here, I would dare go even further:

---

[1] The basic distinction has been given a variety of other nomenclatures, e.g., descriptive vs. etiological, associational vs. causal, empirical vs. theoretical, observational vs. experimental, and many others. I am not satisfied with any of these surrogates, partly because they were not as crisply defined, partly because their boundaries got blurred through the years, and partly because the concatenation "nonstatistical" triggers openness to new perspectives.

"If I am remembered for no other contribution except for insisting on the causal–statistical distinction, I would consider my scientific work worthwhile."

The distinction is embarrassingly crisp and simple, because it is based on the fundamental distinction between statics and kinematics. Standard statistical analysis, typified by regression, estimation, and hypothesis-testing techniques, aims to assess parameters of a static distribution from samples drawn of that distribution. With the help of such parameters, one can infer associations among variables, estimate the likelihood of past and future events, as well as update the likelihood of events in light of new evidence or new measurements. These tasks are managed well by standard statistical analysis so long as experimental conditions remain the same. Causal analysis goes one step further; its aim is to infer not only the likelihood of events under static conditions, but also the dynamics of events under *changing conditions,* for example, changes induced by treatments or external interventions, or by new policies or new experimental designs.

This distinction implies that causal and statistical concepts do not mix. There is nothing in the joint distribution of symptoms and diseases to tell us that curing the former would or would not cure the latter. More generally, there is nothing in a distribution function to tell us how that distribution would differ if external conditions were to change – say, from observational to experimental setup – because the laws of probability theory do not dictate how one property of a distribution ought to change when another property is modified. This information must be provided by extra assumptions that identify what in the distribution remains invariant when the specified modification takes place. The sum total of these extra assumptions is what we call "causal knowledge."

These considerations imply that the slogan "correlation does not imply causation" can be translated into a useful principle: behind every causal conclusion there must lie some causal assumption that is not discernible from the distribution function.

Take the concept of randomization – why is it not statistical? Assume we are given a bivariate density function $f(x,y)$, and we are told that one of the variables is randomized; can we tell which one it is by just examining $f(x, y)$? Of course not; therefore, following our definition, randomization is a causal, not a statistical concept. Indeed, *every* randomized experiment is based on external *intervention*; that is, subjects are "forced" to take one treatment or another in accordance with the experimental protocol, regardless of their natural inclination. The presence of intervention immediately qualifies the experimental setup, as well as all relationships inferred from that setup, as causal.

Note, however, that the purpose of the causal–statistical demarcation line (as stated in Section 1.4, p. 40) is not to exclude causal concepts from the province of statistical analysis but, rather, to encourage investigators to treat causal concepts distinctly, with the proper set of mathematical and inferential tools. Indeed, statisticians were the first to conceive of randomized experiments, and have used them successfully since the time of Fisher (1926). However, both the assumptions and conclusions in those studies were kept implicit, in the mind of ingenious investigators; they did not make their way into the mathematics. For example, one would be extremely hard pressed to find a statistics textbook, even at the graduate level, containing a mathematical proof that randomization indeed produces unbiased estimates of the quantities we wish estimated – i.e., efficacy of treatments or policies.

As a related example, very few statistics teachers today can write down a formula stating that "randomized experiments prove drug x_1 to be twice as effective as drug x_2."

Of course, they can write: $P(y \mid x_1)/P(y \mid x_2) = 2$ (y being the desirable outcome), but then they must keep in mind that this ratio applies to a specific randomized condition, and should not be confused with likelihood ratios prevailing in observational studies. Scientific progress requires that such distinctions be expressed mathematically.[2]

The most important contribution of causal analysis in the past two decades has been the emergence of mathematical languages in which not merely the data, but the experimental design itself can be given mathematical description. Such description is essential, in fact, if one wishes the results of one experiment to serve as premises in another, or to predict outcomes of one design from data obtained under another, or merely to decide if we are in possession of sufficient knowledge to render such cross-design predictions possible.

Is the Distinction Necessary?

Science thrives on distinctions, especially those that do not readily mix. The distinction between rational and irrational numbers, for example, is extremely important in number theory, for it spares us futile efforts to define the latter through some arithmetic operations on the former. The same can be said about the distinctions between prime, composite, algebraic, and transcendental numbers. Logicians, among them George Boole (1815–1864) and Augustus De Morgan (1806–1871), wasted half a century trying to prove syllogisms of first-order logic (e.g., all men are mortal) using the machinery of propositional logic; the distinction between the two was made crisp only at the end of the nineteenth century.

A similar situation occurred in the history of causality. Philosophers have struggled for half a century trying to reduce causality to probabilities (Section 7.5) and have gotten nowhere, except for traps such as "evidential decision theory" (Section 4.1). Epidemiologists have struggled for half a century to define "confounding" in the language of associations (Chapter 6, pp. 183, 194). Some are still struggling (see Section 11.6.4). This effort could have been avoided by appealing to first principles: If confounding were a statistical concept, we would have been able to identify confounders from features of nonexperimental data, adjust for those confounders, and obtain unbiased estimates of causal effects. This would have violated our golden rule: behind any causal conclusion there must be some causal assumption, untested in observational studies. That epidemiologists did not recognize in advance the futility of such attempts is a puzzle that can have only two explanations: they either did not take seriously the causal–statistical divide, or were afraid to classify "confounding" – a simple, intuitive concept – as "nonstatistical."

Divorcing simple concepts from the province of statistics - the most powerful formal language known to empirical scientists – can be traumatic indeed. Social scientists have been laboring for half a century to evaluate public policies using statistical analysis, anchored in regression techniques, and only recently have confessed, with great disappointment, what should have been recognized as obvious in the 1960's: "Regression analyses typically do nothing more than produce from a data set a collection of conditional means and conditional variances" (Berk 2004, p. 237). Economists have gone through a

---

[2] The potential-outcome approach of Neyman (1923) and Rubin (1974) does offer a notational distinction, by writing $P(Y_{x_1} = y)/P(Y_{x_2} = y) = 2$ for the former, and $P(y \mid x_1)/P(y \mid x_2) = 2$ for the latter. However, the opaqueness of this notation and the incomplete state of its semantics (see Sections 3.6.3 and 11.3.2) have prevented it from penetrating classrooms, textbooks, and laboratories.

similar trauma with the concept of exogeneity (Section 5.4.3). Even those who recognized that a strand of exogeneity (i.e., superexogeneity) is of a causal variety came back to define it in terms of distributions (Maddala 1992; Hendry 1995) – crossing the demarcation line was irresistible. And we understand why; defining concepts in term of prior and conditional distributions – the ultimate oracles of empirical knowledge – was considered a mark of scientific prudence. We know better now.

Is the Distinction Useful?

I am fairly confident that today, enlightened by failed experiments in philosophy, epidemiology, and economics, no reputable discipline would waste half a century chasing after a distribution-based definition of another causal concept, however tempted by prudence or intuition. Today, the usefulness of the demarcation line lies primarily in helping investigators trace the assumptions that are needed to support various types of scientific claims. Since every claim invoking causal concepts must rely on some judgmental premises that invoke causal vocabulary, and since causal vocabulary can only be formulated in causally distinct notation, the demarcation line provides notational tools for identifying the judgmental assumptions to which every causal claim is vulnerable.

Statistical assumptions, even untested, are testable in principle, given a sufficiently large sample and sufficiently fine measurements. Causal assumptions, in contrast, cannot be verified even in principle, unless one resorts to experimental control. This difference stands out in Bayesian analysis. Though the priors that Bayesians commonly assign to statistical parameters are untested quantities, the sensitivity to these priors tends to diminish with increasing sample size. In contrast, sensitivity to prior causal assumptions – say, that treatment does not change gender – remains high regardless of sample size.

This makes it doubly important that the notation we use for expressing causal assumptions be meaningful and unambiguous so that scientists can clearly judge the plausibility or inevitability of the assumptions articulated.

How Does One Recognize Causal Expressions in the Statistical Literature?

Those versed in the potential-outcome notation (Neyman 1923; Rubin 1974; Holland 1988) can recognize such expressions through the subscripts that are attached to counterfactual events and variables, e.g., $Y_x(u)$ or Z_{xy}. (Some authors use parenthetical expressions, e.g., $Y(x, u)$ or $Z(x, y)$.) (See Section 3.6.3 for semantics.)

Alternatively, this book also uses expressions of the form $P(Y = y \mid do(X = x))$ or $P(Y_x = y)$ to denote the probability (or frequency) that event $(Y = y)$ would occur if treatment condition $X = x$ were enforced uniformly over the population. (Clearly, $P(Y = y \mid do(X = x))$ is equivalent to $P(Y_x = y)$.) Still a third formal notation is provided by graphical models, where the arrows represent either causal influences, as in Definition 1.3.1, or functional (i.e., counterfactual) relationships, as in Figure 1.6(c).

These notational devices are extremely useful for detecting and tracing the causal premises with which every causal inference study must commence. Any causal premise that is cast in standard probability expressions, void of graphs, counterfactual subscripts, or $do(*)$ operators, can safely be discarded as inadequate. Consequently, any article describing an empirical investigation that does not commence with expressions involving graphs, counterfactual subscripts, or $do(*)$ can safely be proclaimed as inadequately written.

$x \longrightarrow r \longrightarrow s \longrightarrow t \longleftarrow u \longleftarrow v \longrightarrow y$

Figure 11.1   A graph containing a collider at t.

While this harsh verdict may condemn valuable articles in the empirical literature to the province of inadequacy, it can save investigators endless hours of confusion and argumentation in deciding whether causal claims from one study are relevant to another. More importantly, the verdict should encourage investigators to visibly explicate causal premises, so that they can be communicated unambiguously to other investigators and invite professional scrutiny, deliberation, and refinement.

11.1.2   *d*-Separation without Tears (Chapter 1, pp. 16–18)

At the request of many who have had difficulties switching from algebraic to graphical thinking, I am including a gentle introduction to *d*-separation, supplementing the formal definition given in Chapter 1, pp. 16–18.

Introduction

d-separation is a criterion for deciding, from a given causal graph, whether a set X of variables is independent of another set Y, given a third set Z. The idea is to associate "dependence" with "connectedness" (i.e., the existence of a connecting path) and "independence" with "unconnectedness" or "separation." The only twist on this simple idea is to define what we mean by "connecting path," given that we are dealing with a system of directed arrows in which some vertices (those residing in Z) correspond to measured variables, whose values are known precisely. To account for the orientations of the arrows we use the terms "*d*-separated" and "*d*-connected" (*d* connotes "directional"). We start by considering separation between two singleton variables, x and y; the extension to sets of variables is straightforward (i.e., two sets are separated if and only if each element in one set is separated from every element in the other).

Unconditional Separation

Rule 1: x and y are *d*-connected if there is an unblocked path between them.

By a "path" we mean any consecutive sequence of edges, disregarding their directionalities. By "unblocked path" we mean a path that can be traced without traversing a pair of arrows that collide "head-to-head." In other words, arrows that meet head-to-head do not constitute a connection for the purpose of passing information; such a meeting will be called a "collider."

Example 11.1.1 The graph in Figure 11.1 contains one collider, at t. The path $x - r - s - t$ is unblocked, hence x and t are *d*-connected. So also is the path $t - u - v - y$, hence t and y are *d*-connected, as well as the pairs u and y, t and v, t and u, x and s, etc. However, x and y are not *d*-connected; there is no way of tracing a path from x to y without traversing the collider at t. Therefore, we conclude that x and y are *d*-separated, as well as x and v, s and u, r and u, etc. (In linear models, the ramification is that the covariance terms corresponding to these pairs of variables will be zero, for every choice of model parameters.)

Figure 11.2   The set $Z = \{r, v\}$ d-separates x from t and t from y.

$x \longrightarrow (r) \longrightarrow s \longrightarrow t \longleftarrow u \longleftarrow v \longrightarrow y$

Figure 11.3 s and y are d-connected given p, a descendant of the collider t.

Blocking by Conditioning

Motivation: When we measure a set Z of variables, and take their values as given, the conditional distribution of the remaining variables changes character; some dependent variables become independent, and some independent variables become dependent. To represent this dynamic in the graph, we need the notion of "conditional d-connectedness" or, more concretely, "d-connectedness, conditioned on a set Z of measurements."

Rule 2: x and y are d-connected, conditioned on a set Z of nodes, if there is a collider-free path between x and y that traverses no member of Z. If no such path exists, we say that x and y are d-separated by Z. We also say then that every path between x and y is "blocked" by Z.

Example 11.1.2 Let Z be the set $\{r, v\}$ (marked by circles in Figure 11.2). Rule 2 tells us that x and y are d-separated by Z, and so also are x and s, u and y, s and u, etc. The path $x - r - s$ is blocked by Z, and so also are the paths $u - v - y$ and $s - t - u$. The only pairs of unmeasured nodes that remain d-connected in this example, conditioned on Z, are s and t and u and t. Note that, although t is not in Z, the path $s - t - u$ is nevertheless blocked by Z, since t is a collider, and is blocked by Rule 1.

Conditioning on Colliders

Motivation: When we measure a common effect of two independent causes, the causes become dependent, because finding the truth of one makes the other less likely (or "explained away," p. 17), and refuting one implies the truth of the other. This phenomenon (known as Berkson paradox, or "explaining away") requires a slightly special treatment when we condition on colliders (representing common effects) or on any descendant of a collider (representing evidence for a common effect).

Rule 3: If a collider is a member of the conditioning set Z, or has a descendant in Z, then it no longer blocks any path that traces this collider.

Example 11.1.3 Let Z be the set $\{r, p\}$ (again, marked with circles in Figure 11.3). Rule 3 tells us that s and y are d-connected by Z, because the collider at t has a descendant (p) in Z, which unblocks the path $s - t - u - v - y$. However, x and u are still d-separated by Z, because although the linkage at t is unblocked, the one at r is blocked by Rule 2 (since r is in Z).

This completes the definition of d-separation, and readers are invited to try it on some more intricate graphs, such as those shown in Chapter 1, Figure 1.3.

Typical application: Consider Example 11.1.3. Suppose we form the regression of y on p, r, and x,

$$y = c_1 p + c_2 r + c_3 x + \epsilon,$$

and wish to predict which coefficient in this regression is zero. From the discussion above we can conclude immediately that c_3 is zero, because y and x are d-separated given p and r, hence y is independent of x given p and r, or, x cannot offer any information about y once we know p and r. (Formally, the partial correlation between y and x, conditioned on p and r, must vanish.) c_1 and c_2, on the other hand, will in general not be zero, as can be seen from the graph: $Z = \{r, x\}$ does not d-separate y from p, and $Z = \{p, x\}$ does not d-separate y from r.

Remark on correlated errors: Correlated exogenous variables (or error terms) need no special treatment. These are represented by bi-directed arcs (double-arrowed), and their arrowheads are treated as any other arrowheads for the purpose of path tracing. For example, if we add to the graph in Figure 11.3 a bi-directed arc between x and t, then y and x will no longer be d-separated (by $Z = \{r, p\}$), because the path $x - t - u - v - y$ is d-connected – the collider at t is unblocked by virtue of having a descendant, p, in Z.

11.2   REVERSING STATISTICAL TIME (CHAPTER 2, pp. 58–59)

Question to Author:

Keith Markus requested a general method of achieving time reversal by changing coordinate systems or, in the specific example of equation (2.3), a general method of solving for the parameters a, b, c, and d to make the statistical time run opposite to the physical time (p. 59).

Author's Reply:

Consider any two time-dependent variables $X(t)$ and $Y(t)$. These may represent the position of two particles in one dimension, temperature and pressure, sales and advertising budget, and so on.

Assume that temporal variation of $X(t)$ and $Y(t)$ is governed by the equations:

$$\begin{aligned} X(t) &= \alpha X(t-1) + \beta Y(t-1) + \epsilon(t) \\ Y(t) &= \gamma X(t-1) + \delta Y(t-1) + \eta(t), \end{aligned} \tag{11.1}$$

with $\epsilon(t)$ and $\eta(t)$ being mutually and serially uncorrelated noise terms.

In this coordinate system, we find that the two components of the current state, $X(t)$ and $Y(t)$, are uncorrelated conditioned on the components of the previous state, $X(t-1)$ and $Y(t-1)$. Simultaneously, the components of the current state, $X(t)$ and $Y(t)$, are correlated conditioned on the components of the future state, $X(t+1)$ and $Y(t+1)$. Thus, according to Definition 2.8.1 (p. 58), the statistical time coincides with the physical time.

Now let us rotate the coordinates using the transformation

$$\begin{aligned} X'(t) &= aX(t) + bY(t) \\ Y'(t) &= cX(t) + dY(t). \end{aligned} \tag{11.2}$$

338 Reflections, Elaborations, and Discussions with Readers

The governing physical equations remain the same as equation (11.1), but, written in the new coordinate system, they read

$$X'(t) = \alpha'X'(t-1) + \beta'Y(t-1) + \epsilon'(t)$$
$$Y'(t) = \gamma'X'_i(t-1) + \delta'Y'(t-1) + \eta'(t). \tag{11.3}$$

The primed coefficients can be obtained from the original (unprimed) coefficients by matrix multiplication. Likewise, we have:

$$\epsilon'(t) = a\epsilon(t) + b\eta(t)$$
$$\eta'(t) = c\epsilon(t) + d\eta(t).$$

Since $\epsilon(t)$ and $\eta(t)$ are uncorrelated, $\epsilon'(t)$ and $\eta'(t)$ will be correlated, and we no longer have the condition that the components of the current state, $X'(t)$ and $Y'(t)$, are uncorrelated conditioned on the components of the previous state, $X'(t-1)$ and $Y'(t-1)$. Thus, the statistical time (if there is one) no longer runs along the physical time.

Now we need to show that we can choose the parameters a, b, c, and d in such a way as to have the statistical time run opposite to the physical time, namely, to make the components of the current state, $X'(t)$ and $Y'(t)$, uncorrelated conditioned on the components of the future state, $X'(t+1)$ and $Y'(t+1)$.

By inverting equation (11.3) we can express $X'(t-1)$ and $Y'(t-1)$ in terms of linear combinations of $X'(t)$, $Y'(t)$, $\epsilon'(t)$, and $\eta'(t)$. Clearly, since $e(t)$ and $h(t)$ are uncorrelated, we can choose a, b, c, d in such a way that the noise term appearing in the $X'(t-1)$ equation is uncorrelated with the one appearing in the $Y'(t-1)$ equation. (This is better demonstrated in matrix calculus.)

Thus, the general principle for selecting the alternative coordinate system is to diagonalize the noise correlation matrix in the reverse direction.

I hope that readers will undertake the challenge of testing the Temporal Bias Conjecture (p. 59):

> "In most natural phenomenon, the physical time coincides with at least one statistical time."

Alex Balke (personal communication) tried to test it with economic time series, but the results were not too conclusive, for lack of adequate data. I still believe the conjecture to be true, and I hope readers will have better luck.

11.3 ESTIMATING CAUSAL EFFECTS

11.3.1 The Intuition behind the Back-Door Criterion (Chapter 3, p. 79)
Question to Author:

In the definition of the back-door condition (p. 79, Definition 3.3.1), the exclusion of X's descendants (Condition (i)) seems to be introduced as an after fact, just because we get into trouble if we don't. Why can't we get it from first principles; first define sufficiency of Z in terms of the goal of removing bias, and then show that, to achieve this goal, we neither want nor need descendants of X in Z.

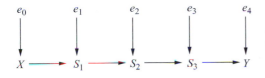

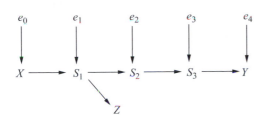

Figure 11.4 Showing the noise factors on the path from X to Y.

Figure 11.5 Conditioning on Z creates dependence between X and e_1, which biases the estimated effect of X on Y.

Author's Answer:

The exclusion of descendants from the back-door criterion is indeed based on first principles, in terms of the goal of removing bias. The principles are as follows: We wish to measure a certain quantity (causal effect) and, instead, we measure a dependency $P(y \mid x)$ that results from all the paths in the diagram; some are spurious (the back-door paths), and some are genuinely causal (the directed paths from X to Y). Thus, to remove bias, we need to modify the measured dependency and make it equal to the desired quantity. To do this systematically, we condition on a set Z of variables while ensuring that:

1.   We block all spurious paths from X to Y,

2.   We leave all directed paths unperturbed,

3.   We create no new spurious paths.

Principles 1 and 2 are accomplished by blocking all back-door paths and only those paths, as articulated in condition (ii). Principle 3 requires that we do not condition on descendants of X, even those that do not block directed paths, because such descendants may create new spurious paths between X and Y. To see why, consider the graph

$$X \to S_1 \to S_2 \to S_3 \to Y.$$

The intermediate variables, $S_1, S_2,\ldots,$ (as well as Y) are affected by noise factors $e_0, e_1, e_2,\ldots$ which are not shown explicitly in the diagram. However, under magnification, the chain unfolds into the graph in Figure 11.4.

Now imagine that we condition on a descendant Z of S_1 as shown in Figure 11.5. Since S_1 is a collider, this creates dependency between X and e_1 which is equivalent to a back-door path

$$X \leftrightarrow e_1 \to S_1 \to S_2 \to S_3 \to Y.$$

By principle 3, such paths should not be created, for it introduces spurious dependence between X and Y.

Note that a descendant Z of X that is not also a descendant of some S_i escapes this exclusion; it can safely be conditioned on without introducing bias (though it may decrease the efficiency of the associated estimator of the causal effect of X on Y). Section

11.3.3 provides an alternative proof of the back-door criterion where the need to exclude descendants of X is even more transparent.

It is also important to note that the danger of creating new bias by adjusting for wrong variables can threaten randomized trials as well. In such trials, investigators may wish to adjust for covariates despite the fact that, asymptotically, randomization neutralizes both measured and unmeasured confounders. Adjustment may be sought either to improve precision (Cox 1958, pp. 48–55), or to match imbalanced samples, or to obtain covariate-specific causal effects. Randomized trials are immune to adjustment-induced bias when adjustment is restricted to pre-treatment covariates, but adjustment for post-treatment variables may induce bias by the mechanism shown in Figure 11.5 or, more severely, when correlation exists between the adjusted variable Z and some factor that affects outcome (e.g., e_4 in Figure 11.5).

As an example, suppose treatment has a side effect (e.g., headache) in patients who are predisposed to disease Y. If we wish to adjust for disposition and adjust instead for its proxy, headache, a bias would emerge through the spurious path: treatment $\rightarrow$ headache $\leftarrow$ predisposition $\rightarrow$ disease. However, if we are careful never to adjust for any consequence of treatment (not only those that are on the causal pathway to disease), no bias will emerge in randomized trials.

Further Questions from This Reader:

This explanation for excluding descendants of X is reasonable, but it has two shortcomings:

1. It does not address cases such as

    $$X \leftarrow C \rightarrow Y \rightarrow F,$$

    which occur frequently in epidemiology, and where tradition permits the adjustment for $Z = \{C, F\}$.

2. The explanation seems to redefine confounding and sufficiency to represent something different from what they have meant to epidemiologists in the past few decades. Can we find something in graph theory that is closer to their traditional meaning?

Author's Answer

1. Epidemiological tradition permits the adjustment for $Z = (C, F)$ for the task of testing whether X has a causal effect on Y, but not for estimating the magnitude of that effect. In the former case, while conditioning on F creates a spurious path between C and the noise factor affecting Y, that path is blocked upon conditioning on C. Thus, conditioning on $Z = \{C, F\}$ leaves X and Y independent. If we happen to measure such dependence in any stratum of Z, it must be that the model is wrong, i.e., either there is a direct causal effect of X on Y, or some other paths exist that are not shown in the graph.

    Thus, if we wish to test the (null) hypothesis that there is no causal effect of X on Y, adjusting for $Z = \{C, F\}$ is perfectly legitimate, and the graph shows it (i.e., C and F are nondescendant of X). However, adjusting for Z is not legitimate for assessing the causal effect of X on Y when such effect is suspected,

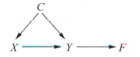

Figure 11.6   Graph applicable for accessing the effect of X on Y.

because the graph applicable for this task is given in Figure 11.6; F becomes a descendant of X, and is excluded by the back-door criterion.

2. If the explanation of confounding and sufficiency sounds at variance with traditional epidemiology, it is only because traditional epidemiologists did not have proper means of expressing the operations of blocking or creating dependencies. They might have had a healthy intuition about dependencies, but graphs translate this intuition into a formal system of closing and opening paths.

We should also note that before 1985, causal analysis in epidemiology was in a state of confusion, because the healthy intuitions of leading epidemiologists had to be expressed in the language of associations – an impossible task. Even the idea that confounding stands for "bias," namely, a "difference between two dependencies, one that we wish to measure, the other that we do measure," was resisted by many (see Chapter 6), because they could not express the former mathematically.[3]

Therefore, instead of finding "something in graph language that is closer to traditional meaning," we can do better: explicate what that "traditional meaning" ought to have been.

In other words, traditional meaning was informal and occasionally misguided, while graphical criteria are formal and mathematically proven.

Chapter 6 (pp. 183, 194) records a long history of epidemiological intuitions, some by prominent epidemiologists, that have gone astray when confronted with questions of confounding and adjustment (see Greenland and Robins 1986; Wickramaratne and Holford 1987; Weinberg 1993). Although most leading epidemiologists today are keenly attuned to modern developments in causal analysis, (e.g., Glymour and Greenland 2008), epidemiological folklore is still permeated with traditional intuitions that are highly suspect. (See Section 6.5.2.)

In summary, graphical criteria, as well as principles 1–3 above, give us a sensible, friendly, and unambiguous interpretation of the "traditional meaning of epidemiological concepts."

11.3.2   Demystifying "Strong Ignorability"

Researchers working within the confines of the potential-outcome language express the condition of "zero bias" or "no-confounding" using an independence relationship called

---

[3] Recall that Greenland and Robins (1986) were a lone beacon of truth for many years, and even they had to resort to the "black-box" language of "exchangeability" to define "bias," which discouraged intuitive interpretations of confounding (see Section 6.5.3). Indeed, it took epidemiologists another six years (Weinberg 1993) to discover that adjusting for factors affected by the exposure (as in Figure 11.5) would introduce bias.

"strong ignorability" (Rosenbaum and Rubin 1983). Formally, if X is a binary treatment (or action), strong ignorability is written as:

$$\{Y(0), Y(1)\} \perp\!\!\!\perp X \mid Z, \tag{11.4}$$

where $Y(0)$ and $Y(1)$ are the (unobservable) potential outcomes under actions $do(X = 0)$ and $do(X = 1)$, respectively (see equation (3.51) for definition), and Z is a set of measured covariates. When "strong ignorability" holds, Z is *admissible,* or *deconfounding*, that is, treatment effects can be estimated without bias using the adjustment estimand, as shown in the derivation of equation (3.54).

Strong ignorability, as the derivation shows, is a convenient syntactic tool for manipulating counterfactual formulas, as well as a convenient way of formally assuming admissibility (of Z) without having to justify it. However, as we have noted several times in this book, hardly anyone knows how to apply it in practice, because the counterfactual variables $Y(0)$ and $Y(1)$ are unobservable, and scientific knowledge is not stored in a form that allows reliable judgment about conditional independence of counterfactuals. It is not surprising, therefore, that "strong ignorability" is used almost exclusively as a surrogate for the assumption "Z is admissible," that is,

$$P(y \mid do(x)) = \sum_z P(y \mid z, x)P(z), \tag{11.5}$$

and rarely, if ever, as a criterion to protect us from bad choices of Z.[4]

Readers enlightened by graphical models would recognize immediately that equation (11.4) must mirror the back-door criterion (p. 79, Definition 3.3.1), since the latter too entails admissibility. This recognition allows us not merely to pose equation (11.4) as a claim, or an assumption, but also to reason about the cause–effect relationships that render it valid.

The question arises, however, whether the variables $Y(0)$ and $Y(1)$ could be represented in the causal graph in a way that would allow us to test equation (11.4) by graphical means, using d-separation. In other words, we seek a set W of nodes such that Z would d-separate X from W if and only if Z satisfies equation (11.4).

The answer follows directly from the rules of translation between graphs and potential outcome (Section 3.6.3). According to this translation, $\{Y(0), Y(1)\}$ represents the sum total of all exogenous variables, latent as well as observed, which can influence Y through paths that avoid X. The reason is as follows: according to the structural definition of $\{Y(0), Y(1)\}$ (equation (3.51)), $Y(0)$ (similarly $Y(1)$) represents the value of Y under a condition where all arrows entering X are severed, and X is held constant at $X = 0$. Statistical variations of $Y(0)$ would therefore be governed by all exogenous ancestors of Y in the mutilated graphs with the arrows entering X removed.

In Figure 11.4, for example, $\{Y(0), Y(1)\}$ will be represented by the exogenous variables $\{e_1, e_2, e_3, e_4\}$. In Figure 3.4, as another example, $\{Y(0), Y(1)\}$ will be represented by the noise factors (not shown in the graph) that affect variables X_4, X_1, X_2, X_5, and X_6. However,

---

[4] In fact, in the rare cases where "strong ignorability" is used to guide the choice of covariates, the guidelines issued are wrong or inaccurate, perpetuating myths such as: "there is no reason to avoid adjustment for a variable describing subjects before treatment," "a confounder is any variable associated with both treatment and disease," and "strong ignorability requires measurement of all covariates related to both treatment and outcome" (citations withheld to spare embarrassment).

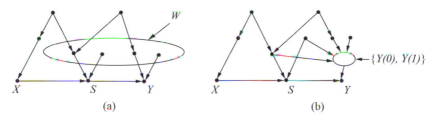

Figure 11.7   Graphical interpretation of counterfactuals $\{Y(0), Y(l)\}$ in the "strong ignorability" condition.

since variables X_4 and X_5 summarize (for Y) the variations of their ancestors, a sufficient set for representing $\{Y(0), Y(l)\}$ would be X_4, X_1 and the noise factors affecting Y and X_6.

In summary, the potential outcomes $\{Y(0), Y(l)\}$ are represented by the observed and unobserved parents[5] of all nodes on paths from X to Y. Schematically, we can represent these parents as in Figure 11.7(a). It is easy to see that, with this interpretation of $\{Y(0), Y(l)\}$, a set of covariates Z d-separates W from X if and only if Z satisfies the back-door criterion.

It should be noted that the set of observable variables designated W in Figure 11.7(a) are merely surrogates of the unobservable counterfactuals $\{Y(0), Y(l)\}$ for the purpose of confirming conditional independencies (e.g., equation (11.4)) in the causal graph (via d-separation.) A more accurate allocation of $\{Y(0), Y(l)\}$ is given in Figure 11.7(b), where they are shown as (dummy) parents of Y that are functions of, though not identical to, the actual (observable) parents of Y and S.

Readers versed in structural equation modeling would recognize the graphical representations $\{Y(0), Y(1)\}$ as a refinement of the classical economentric notion of "disturbance," or "error term" (in the equation for Y), and "strong ignorability" as the requirement that, for X to be "exogenous," it must be independent of this "disturbance" (see Section 5.4.3). This notion fell into ill repute in the 1970s (Richard 1980) together with the causal interpretation of econometric equations, and I have predicted its re-acceptance (p. 170) in view of the clarity that graphical models shine on the structural equation formalism. Figure 11.7 should further help this acceptance.

Having translated "strong ignorability" into a simple separation condition in a model that encodes substantive process knowledge should demystify the nebulous concept of "strong ignorability" and invite investigators who speak "ignorability" to benefit from its graphical interpretation.

This interpretation permits researchers to understand what conditions covariates must fulfill before they eliminate bias, what to watch for and what to think about when covariates are selected, and what experiments we can do to test, at least partially, if we have the knowledge needed for covariate selection. Section 11.3.4 exemplifies such considerations.

One application where the symbiosis between the graphical and counterfactual frameworks has been useful is in estimating the effect of treatments on the treated: ETT $= P(Y_{x'} = y \mid x)$ (see Sections 8.2.5 and 11.9.1). This counterfactual quantity (e.g., the probability that a treated person would recover if not treated, or the rate of disease among the exposed, had the exposure been avoided) is not easily analyzed in the do-calculus notation. The counterfactual notation, however, allows us to derive a

---

[5] The reason for explicitly including latent parents is explained in Section 11.3.1.

useful conclusion: Whenever a set of covariates Z exists that satisfies the back-door criterion, ETT can be estimated from observational studies. This follows directly from

$$(Y \perp\!\!\!\perp X \mid Z)_{G_{\underline{X}}} \implies Y_{x'} \perp\!\!\!\perp X \mid Z,$$

which allows us to write

$$
\begin{aligned}
\text{ETT} &= P(Y_{x'} = y \mid x) \\
&= \sum_z P(Y_{x'} = y \mid x, z) P(z \mid x) \\
&= \sum_z P(Y_{x'} = y \mid x', z) P(z \mid x) \\
&= \sum_z P(y \mid x', z) P(z \mid x).
\end{aligned}
$$

The graphical demystification of "strong ignorability" also helps explain why the probability of causation $P(Y_{x'} = y' \mid x, y)$ and, in fact, any counterfactual expression conditioned on y, would not permit such a derivation and is, in general, non-identifiable (see Chapter 9).

11.3.3 Alternative Proof of the Back-Door Criterion

The original proof of the back-door criterion (Theorem 3.3.2) used an auxiliary intervention node F (Figure 3.2) and was rather indirect. An alternative proof is presented below, where the need for restricting Z to nondescendants of X is transparent.

Proof of the Back-Door Criterion

Consider a Markovian model G in which T stands for the set of parents of X. From equation (3.13), we know that the causal effect of X on Y is given by

$$P(y \mid \hat{x}) = \sum_{t \in T} P(y \mid x, t) P(t). \tag{11.6}$$

Now assume some members of T are unobserved. We seek another set Z of observed variables, to replace T so that

$$P(y \mid \hat{x}) = \sum_{z \in Z} P(y \mid x, z) P(z). \tag{11.7}$$

It is easily verified that (11.7) follow from (11.6) if Z satisfies:

(i) $(Y \perp\!\!\!\perp T \mid X, Z)$

(ii) $(X \perp\!\!\!\perp Z \mid T)$.

Indeed, conditioning on Z, (i) permits us to rewrite (11.6) as

$$P(y \mid \hat{x}) = \sum_t P(t) \sum_z P(y \mid z, x) P(z \mid t, x),$$

and (ii) further yields $P(z \mid t, x) = P(z \mid t)$, from which (11.7) follows.

It is now a purely graphical exercise to prove that the back-door criterion implies (i) and (ii). Indeed, (ii) follows directly from the fact that Z consists of nondescendants of X, while the blockage of all back-door paths by Z implies $(Y \perp\!\!\!\perp T \mid X, Z)_G$, hence (i). This follows from observing that any path from Y to T in G that is unblocked by $\{X, Z\}$ can be extended to a back-door path from Y to X, unblocked by Z.

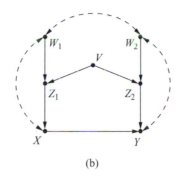

$$(a)\qquad\qquad\qquad\qquad (b)$$

Figure 11.8   (a) $S_1 = \{Z_1, W_2\}$ and $S_2 = \{Z_2, W_1\}$ are each admissible yet not satisfying C_1 or C_2. (b) No subset of $C = \{Z_1, Z_2, W_1, W_2, V\}$ is admissible.

On Recognizing Admissible Sets of Deconfounders

Note that conditions (i) and (ii) allow us to recognize a set Z as *admissible* (i.e., satisfying equation (11.7)) starting from any other admissible T, not necessarily the parents of X. The parenthood status of T was used merely to established (11.6) but played no role in replacing T with Z to establish (11.7). Still, starting with the parent set T has the unique advantage of allowing us to recognize *every* other admissible set Z via (i) and (ii). For any other starting set, T, there exists an admissible Z that does not satisfy (i) and (ii). For an obvious example, choosing X's parents for Z would violate (i) and (ii) because no set can d-separate X from its parents as would be required by (i).

Note also that conditions (i) and (ii) are purely statistical, invoking no knowledge of the graph or any other causal assumption. It is interesting to ask, therefore, whether there are general independence conditions, similar to (i) and (ii), that connect any two admissible sets, S_1 and S_2. A partial answer is given by the Stone–Robins criterion (page 187) for the case where S_1 is a subset of S_2; another is provided by the following observation.

Define two subsets, S_1 and S_2, as *c-equivalent* ("c" connotes "confounding") if the following equality holds:

$$\sum_{s_1} P(y \mid x, s_1)P(s_1) = \sum_{s_2} P(y \mid x, s_2)P(s_2). \tag{11.8}$$

This equality guarantees that, if adjusted for, sets S_1 and S_2 would produce the same bias relative to estimating the causal effect of X on Y.

Claim: A sufficient condition for c-equivalence of S_1 and S_2 is that either one of the following two conditions holds:

$$C_1 : X \perp\!\!\!\perp S_2 \mid S_1 \qquad \text{and} \qquad Y \perp\!\!\!\perp S_1 \mid S_2, X$$
$$C_2 : X \perp\!\!\!\perp S_1 \mid S_2 \qquad \text{and} \qquad Y \perp\!\!\!\perp S_2 \mid S_1, X.$$

C_1 permits us to derive the right-hand side of equation (11.8) from the left-hand side, while C_2 permits us to go the other way around. Therefore, if S_1 is known to be admissible, the admissibility of S_2 can be confirmed by either C_1 or C_2. This broader condition allows us, for example, to certify $S_2 = PA_X$ as admissible from any other admissible set S_1, since condition C_2 would be satisfied by any such choice.

This broader condition still does not characterize *all* c-equivalent pairs, S_1 and S_2. For example, consider the graph in Figure 11.8(a), in which each of $S_1 = \{Z_1, W_2\}$ and

$S_2 = \{Z_2, W_2\}$ is admissible (by virtue of satisfying the back-door criterion), hence S_1 and S_2 are c-equivalent. Yet neither C_1 nor C_2 holds in this case.

A natural attempt would be to impose the condition that S_1 and S_2 each be c-equivalent to $S_1 \cup S_2$ and invoke the criterion of Stone (1993) and Robins (1997) for the required set-subset equivalence. The resulting criterion, while valid, is still not complete; there are cases where S_1 and S_2 are c-equivalent yet not c-equivalent to their union. A theorem by Pearl and Paz (2008) broadens this condition using irreducible sets.

Having given a conditional-independence characterization of c-equivalence does not solve, of course, the problem of identifying admissible sets; the latter is a causal notion and cannot be given statistical characterization.

The graph depicted in Figure 11.8(b) demonstrates the difficulties commonly faced by social and health scientists. Suppose our target is to estimate $P(y\,|\,do(x))$ given measurements on $\{X, Y, Z_1, Z_2, W_1, W_2, V\}$, but having no idea of the underlying graph structure. The conventional wisdom is to start with all available covariates $C = \{Z_1, Z_2, W_1, W_2, V\}$, and test if a proper subset of C would yield an equivalent estimand upon adjustment. Statistical methods for such reduction are described in Greenland et al. (1999b), Geng et al. (2002), and Wang et al. (2008). For example, $\{Z_1, V\}$, $\{Z_2, V\}$, or $\{Z_1, Z_2\}$ can be removed from C by successively applying conditions C_1 and C_2. This reduction method would produce three irreducible subsets, $\{Z_1, W_1, W_2\}$, $\{Z_2, W_1, W_2\}$, and $\{V, W_1, W_2\}$, all c-equivalent to the original covariate set C. However, none of these subsets is admissible for adjustment, because none (including C) satisfies the back-door criterion. While a theorem due to Tian et al. (1998) assures us that any c-equivalent subset of a set C can be reached from C by a step-at-a-time removal method, going through a sequence of c-equivalent subsets, the problem of covariate selection is that, lacking the graph structure, we do not know which (if any) of the many subsets of C is admissible. The next subsection discusses how external knowledge, as well as more refined analysis of the data at hand, can be brought to bear on the problem.

11.3.4 Data vs. Knowledge in Covariate Selection

What then can be done in the absence of a causal graph? One way is to postulate a plausible graph, based on one's understanding of the domain, and check if the data refutes any of the statistical claims implied by that graph. In our case, the graph of Figure 11.8(b) advertises several such claims, cast as conditional independence constraints, each associated with a missing arrow in the graph:

$$V \perp\!\!\!\perp X \mid Z_1, W_1 \qquad V \perp\!\!\!\perp W_2 \qquad Z_2 \perp\!\!\!\perp W_1 \mid W_2$$
$$V \perp\!\!\!\perp Y \mid X, Z_2, W_2 \qquad Z_1 \perp\!\!\!\perp Z_2 \mid V, W_1, W_2 \qquad Z_1 \perp\!\!\!\perp W_2 \mid W_1$$
$$V \perp\!\!\!\perp W_1 \qquad X \perp\!\!\!\perp Z_2 \mid Z, W_1, W_2 \qquad X \perp\!\!\!\perp \{V, Z_2\} \mid Z_1, W_1, W_2.$$

Satisfying these constraints does not establish, of course, the validity of the causal model postulated because, as we have seen in Chapter 2, alternative models may exist which satisfy the same independence constraints yet embody markedly different causal structures, hence, markedly different admissible sets and effect estimands. A trivial example would be a complete graph, with arbitrary orientation of arrows which, with a clever choice of parameters, can emulate any other graph. A less trivial example, one that is not sensitive to choice of parameters, lies in the class of equivalent structures, in

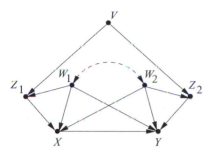

Figure 11.9   A model that is dependence-wise indistinguishable from that of Figure 11.8 (b), in which the irreducible sets $\{Z_1, W_1, W_2\}$, $\{W_1, W_2, V\}$, and $\{W_1, W_2, Z_2\}$ are admissible.

which all conditional independencies emanate from graph separations. The search techniques developed in Chapter 2 provide systematic ways of representing all equivalent models compatible with a given set of conditional independence relations.

For example, the model depicted in Figure 11.9 is indistinguishable from that of Figure 11.8(b), in that it satisfies all the conditional independencies implied by the latter, and no others.[6] However, in contrast to Figure 11.8(b), the sets $\{Z_1, W_1, W_2\}$, $\{V, W_1, W_2\}$, and $\{Z_2, W_1, W_2\}$ are admissible. Adjusting for the latter would remove bias if the correct model is Figure 11.9 and might produce bias if the correct model is Figure 11.8(b).

Is there a way of telling the two models apart? Although the notion of "observational equivalence" precludes discrimination by statistical means, substantive causal knowledge may provide discriminating information. For example, the model of Figure 11.9 can be ruled out if we have good reasons to believe that variable W_2 cannot have any influence on X (e.g., it may occur *later* than X,) or that W_1 could not possibly have direct effect on Y.

The power of graphs lies in offering investigators a transparent language to reason about, to discuss the plausibility of such assumptions and, when consensus is not reached, to isolate differences of opinion and identify what additional observations would be needed to resolve differences. This facility is lacking in the potential-outcome approach where, for most investigators, "strong ignorability" remains a mystical black box.

In addition to serving as carriers of substantive judgments, graphs also offer one the ability to reject large classes of models without testing each member of the class. For example, all models in which V and W_1 are the sole parents of X, thus rendering $\{V, W_1\}$ (as well as C) admissible, could be rejected at once if the condition $X \perp\!\!\!\perp Z_1 \mid V, W_1$ does not hold in the data.

In Chapter 3, for example, we demonstrated how the measurement of an additional variable, mediating between X and Y, was sufficient for identifying the causal effect of X on Y. This facility can also be demonstrated in Figure 11.8(b); measurement of a variable Z judged to be on the pathway between X and Y would render $P(y \mid do(x))$ identifiable and estimable through equation (3.29). This is predicated, of course, on Figure 11.8(b) being the correct data-generating model. If, on the other hand, it is Figure 11.9 that represents the correct model, the causal effect would be given by

$$P(y \mid do(x)) = \sum_{pa_X} P(y \mid pa_X, x) P(pa_X)$$

$$= \sum_{z_1, w_1, w_2} P(y \mid x, z_1, w_1, w_2) \, P(z_1, w_1, w_2),$$

---

[6] Semi-Markovian models may also be distinguished by functional relationships that are not expressible as conditional independencies (Verma and Pearl 1990; Tian and Pearl 2002b; Shpitser and Pearl 2008). We do not consider these useful constraints in this example.

348   **Reflections, Elaborations, and Discussions with Readers**

which might or might not agree with equation (3.29). In the latter case, we would have good reason to reject the model in Figure 11.9 as inconsistent, and seek perhaps additional measurements to confirm or refute Figure 11.8(b).

Auxiliary experiments may offer an even more powerful discriminatory tool than auxiliary observations. Consider variable W_1 in Figure 11.8(b). If we could conduct a controlled experiment with W_1 randomized, instead of X, the data obtained would enable us to estimate the causal effect of X on Y with no bias (see Section 3.4.4). At the very least, we would be able to discern whether W_1 is a parent of X, as in Figure 11.9, or an indirect ancestor of X, as in Figure 11.8(b).

In an attempt to adhere to traditional statistical methodology, some causal analysts have adopted a method called "sensitivity analysis" (e.g., Rosenbaum 2002, pp. 105–170), which gives the impression that causal assumptions are not invoked in the analysis. This, of course, is an illusion. Instead of drawing inferences by assuming the absence of certain causal relationships in the model, the analyst tries such assumptions and evaluates how strong alternative causal relationships must be in order to explain the observed data. The result is then submitted to a judgment of plausibility, the nature of which is no different from the judgments invoked in positing a model like the one in Figure 11.9. In its richer setting, sensitivity analysis amounts to loading a diagram with causal relationships whose strength is limited by plausibility judgments and, given the data, attempting to draw conclusions without violating those plausibility constraints. It is a noble endeavor, which thus far has been limited to problems with a very small number of variables. The advent of diagrams promises to expand the applicability of this method to more realistic problems.

11.3.5 Understanding Propensity Scores

The method of propensity score (Rosenbaum and Rubin 1983), or propensity score matching (PSM), is the most developed and popular strategy for causal analysis in observational studies. It is not emphasized in this book, because it is an estimation method, designed to deal with the variability of finite samples, but does not add much to our understanding of the asymptotic, large-sample limits, which is the main focus of the book. However, due to the prominence of the propensity score method in causal analysis, and recent controversies surrounding its usage, we devote this section to explain where it falls in the grand scheme of graphical models, admissibility, identifiability, bias reduction, and the statistical vs. causal dichotomy.

The method of propensity score is based on a simple, yet ingenious, idea of purely statistical character. Assuming a binary action (or treatment) X, and an arbitrary set S of measured covariates, the propensity score $L(s)$ is the probability that action $X = 1$ will be chosen by a participant with characteristics $S = s$, or

$$L(s) = P(X = 1 \mid S = s). \tag{11.9}$$

What Rosenbaum and Rubin showed is that, viewing $L(s)$ as a function of S, hence, as a random variable, X and S are independent given $L(s)$, that is, $X \perp\!\!\!\perp S \mid L(s)$. In words, all units that map into the same value of $L(s)$ are comparable, or "balanced," in the sense that, within each stratum of L, treated and untreated units have the same distribution of characteristics S.[7]

---

[7] This independence emanates from the special nature of the function $L(s)$ and is not represented in the graph, i.e., if we depict L as a child of S, L would not in general d-separate S from X.

To see the significance of this result, let us assume, for simplicity, that $L(s)$ can be estimated separately and approximated by discrete strata $L = \{l_1, l_2, \ldots, l_k\}$. The conditional independence $X \perp\!\!\!\perp S \mid L(s)$, together with the functional mapping $S \to L$, renders S and L c-equivalent in the sense defined in Section 11.3.3, equation (11.8), namely, for any Y,

$$\sum_s P(y \mid s, x)P(s) = \sum_l P(y \mid l, x)P(l). \qquad (11.10)$$

This follows immediately by writing:[8]

$$\sum_l P(y \mid l, x)P(l) = \sum_s \sum_l P(y \mid l, s, x)P(l)P(s \mid l, x)$$
$$= \sum_s \sum_l P(y \mid s, x)P(l)P(s \mid l)$$
$$= \sum_s P(y \mid s, x)P(s).$$

Thus far we have not mentioned any causal relationship, nor the fact that Y is an outcome variable and that, eventually, our task would be to estimate the causal effect of X on Y. The c-equivalence of S and L merely implies that, if for any reason one wishes to estimate the "adjustment estimand" $\sum_s P(y \mid s, x)P(s)$, with S and Y two arbitrary sets of variables, then, instead of summing over a high-dimensional set S, one might as well sum over a one-dimensional vector $L(s)$. The asymptotic estimate, in the limit of a very large sample, would be the same in either method.

This c-equivalence further implies – and this is where causal inference first comes into the picture – that if one chooses to approximate the causal effect $P(y \mid do(x))$ by the adjustment estimand $E_s P(y \mid s, x)$, then, asymptotically, the same approximation can be achieved using the estimand $E_l P(y \mid l, x)$, where the adjustment is performed over the strata of L. The latter has the advantage that, for finite samples, each of the strata is less likely to be empty and each is likely to contain both treated and untreated units.

The method of propensity score can thus be seen as an efficient estimator of the adjustment estimand, formed by an arbitrary set of covariates S; it makes no statement regarding the appropriateness of S, nor does it promise to correct for any confounding bias, or to refrain from creating new bias where none exists.

In the special case where S is admissible, that is,

$$P(y \mid do(x)) = E_s P(y \mid s, x), \qquad (11.11)$$

L would be admissible as well, and we would then have an unbiased estimand of the causal effect,[9]

$$P(y \mid do(x)) = E_l P(y \mid l, x),$$

accompanied by an efficient method of estimating the right-hand side. Conversely, if S is inadmissible, L would be inadmissible as well, and all we can guarantee is that the bias produced by the former would be faithfully and efficiently reproduced by the latter.

---

[8] This also follows from the fact that condition C_2 is satisfied by the substitution $S_1 = S$ and $S_2 = L(s)$.
[9] Rosenbaum and Rubin (1983) proved the c-equivalence of S and L only for admissible S, which is unfortunate; it gives readers the impression that the propensity score matching somehow contributes to bias reduction.

The Controversy Surrounding Propensity Score

Thus far, our presentation of propensity score leaves no room for misunderstanding, and readers of this book would find it hard to understand how a controversy could emerge from an innocent estimation method which merely offers an efficient way of estimating a statistical quantity that sometimes does, and sometimes does not, coincide with the causal quantity of interest, depending on the choice of S.

But a controversy has developed recently, most likely due to the increased popularity of the method and the strong endorsement it received from prominent statisticians (Rubin 2007), social scientists (Morgan and Winship 2007; Berk and de Leeuw 1999), health scientists (Austin 2007), and economists (Heckman 1992). The popularity of the method has in fact grown to the point where some federal agencies now expect program evaluators to use this approach as a substitute for experimental designs (Peikes et al. 2008). This move reflects a general tendency among investigators to play down the cautionary note concerning the required admissibility of S, and to interpret the mathematical proof of Rosenbaum and Rubin as a guarantee that, in each strata of L, matching treated and untreated subjects somehow eliminates confounding from the data and contributes therefore to overall bias reduction. This tendency was further reinforced by empirical studies (Heckman et al. 1998; Dehejia and Wahba 1999) in which agreement was found between propensity score analysis and randomized trials, and in which the agreement was attributed to the ability of the former to "balance" treatment and control groups on important characteristics. Rubin has encouraged such interpretations by stating: "This application uses propensity score methods to create subgroups of treated units and control units ... as if they had been randomized. The collection of these subgroups then 'approximate' a randomized block experiment with respect to the observed covariates" (Rubin 2007).

Subsequent empirical studies, however, have taken a more critical view of propensity score, noting with disappointment that a substantial bias is sometimes measured when careful comparisons are made to results of clinical studies (Smith and Todd 2005; Luellen et al. 2005; Peikes et al. 2008).

But why would anyone play down the cautionary note of Rosenbaum and Rubin when doing so would violate the golden rule of causal analysis: No causal claim can be established by a purely statistical method, be it propensity scores, regression, stratification, or any other distribution-based design. The answer, I believe, rests with the language that Rosenbaum and Rubin used to formulate the condition of admissibility, i.e., equation (11.11). The condition was articulated in the restricted language of potential-outcome, stating that the set S must render X "strongly ignorable," i.e., $\{Y_1, Y_0\} \perp\!\!\!\perp X \mid S$. As stated several times in this book, the opacity of "ignorability" is the Achilles' heel of the potential-outcome approach – no mortal can apply this condition to judge whether it holds even in simple problems, with all causal relationships correctly specified, let alone in partially specified problems that involve dozens of variables.[10]

---

[10] Advocates of the potential outcome tradition are invited to inspect Figure 11.8(b) (or any model, or story, or toy-example of their choice) and judge whether any subset of C renders X "strongly ignorable." This could easily be determined, of course, by the back-door criterion, but, unfortunately, graphs are still feared and misunderstood by some of the chief advocates of the potential-outcome camp (e.g., Rubin 2004, 2008b, 2009).

The difficulty that most investigators experience in comprehending what "ignorability" means, and what judgment it summons them to exercise, has tempted them to assume that it is automatically satisfied, or at least is likely to be satisfied, if one includes in the analysis as many covariates as possible. The prevailing attitude is that adding more covariates can cause no harm (Rosenbaum 2002, p. 76) and can absolve one from thinking about the causal relationships among those covariates, the treatment, the outcome and, most importantly, the confounders left unmeasured (Rubin 2009).

This attitude stands contrary to what students of graphical models have learned, and what this book has attempted to teach. The admissibility of S can be established only by appealing to the causal knowledge available to the investigator, and that knowledge, as we know from graph theory and the back-door criterion, makes bias reduction a non-monotonic operation, i.e., eliminating bias (or imbalance) due to one confounder may awaken and unleash bias due to dormant, unmeasured confounders. Examples abound (e.g., Figure 6.3) where adding a variable to the analysis not only is not needed, but would introduce irreparable bias (Pearl 2009, Shrier 2009, Sjölander 2009).

Another factor inflaming the controversy has been the general belief that the bias-reducing potential of propensity score methods can be assessed experimentally by running case studies and comparing effect estimates obtained by propensity scores to those obtained by controlled randomized experiments (Shadish and Cook 2009).[11] This belief is unjustified because the bias-reducing potential of propensity scores depends critically on the specific choice of S or, more accurately, on the cause–effect relationships among variables inside and outside S. Measuring significant bias in one problem instance (say, an educational program in Oklahoma) does not preclude finding zero bias in another (say, crime control in Arkansas), even under identical statistical distributions $P(x, s, y)$.

With these considerations in mind, one is justified in asking a social science type question: What is it about propensity scores that has inhibited a more general understanding of their promise and limitations?

Richard Berk, in *Regression Analysis: A Constructive Critique* (Berk 2004), recalls similar phenomena in social science, where immaculate ideas were misinterpreted by the scientific community: "I recall a conversation with Don Campbell in which he openly wished that he had never written Campbell and Stanley (1966). The intent of the justly famous book, *Experimental and Quasi-Experimental Designs for Research,* was to contrast randomized experiments to quasi-experimental approximations and to strongly discourage the latter. Yet the apparent impact of the book was to legitimize a host of quasi-experimental designs for a wide variety of applied social science. After I got to know Dudley Duncan late in his career, he said that he often thought that his influential book on path analysis, *Introduction to Structural Equation Models* was a big mistake. Researchers had come away from the book believing that fundamental policy questions about social inequality could be quickly and easily answered with path analysis." (p. xvii)

---

[11] Such beliefs are encouraged by valiant statements such as: "For dramatic evidence that such an analysis can reach the same conclusion as an exactly parallel randomized experiment, see Shadish and Clark (2006, unpublished)" (Rubin 2007).

I believe that a similar cultural phenomenon has evolved around propensity scores.

It is not that Rosenbaum and Rubin were careless in stating the conditions for success. Formally, they were very clear in warning practitioners that propensity scores work only under "strong ignorability" conditions. However, what they failed to realize is that it is not enough to warn people against dangers they cannot recognize; to protect them from perilous adventures, we must also give them eyeglasses to spot the threats, and a meaningful language to reason about them. By failing to equip readers with tools (e.g., graphs) for recognizing how "strong ignorability" can be violated or achieved, they have encouraged a generation of researchers (including federal agencies) to assume that ignorability either holds in most cases, or can be made to hold by clever designs.

11.3.6 The Intuition behind *do*-Calculus
Question to Author Regarding Theorem 3.4.1:

In the inference rules of *do*-calculus (p. 85), the subgraph $G_{\overline{X}}$, represents the distribution prevailing under the operation $do(X = x)$, since all direct causes of X are removed. What distribution does the submodel $G_{\underline{X}}$ represent, with the direct effects of X removed?

Author's Reply:

The graph $G_{\underline{X}}$ represents the hypothetical act of "holding constant" all children of X. This severs all directed paths from X to Y, while leaving all back-door paths intact. So, if X and Y are *d*-connected in that graph, it must be due to (unblocked) confounding paths between the two. Conversely, if we find a set Z of nodes that *d*-separate X from Y in that graph, we are assured that Z blocks all back-door paths in the original graph. If we further condition on variables Z, we are assured, by the back-door criterion, that we have neutralized all confounders and that whatever dependence we measure after such conditioning must be due to the causal effect of X on Y, free of confoundings.

11.3.7 The Validity of *G*-Estimation

In Section 3.6.4 we introduced the *G*-estimation formula (3.63), together with the counterfactual independency (3.62), $(Y(x) \perp\!\!\!\perp X_k \mid \overline{L}_k, \overline{X}_{k-1} = \overline{x}_{k-1})$, which Robins proved to be a sufficient condition for (3.63). In general, condition (3.62) is both overrestrictive and lacks intuitive basis. A more general and intuitive condition leading to (3.63) is derived in (4.5) (p. 122), which reads as follows:

(3.62*) General Condition for g-Estimation (Sequential Deconfounding)
$P(y \mid g = x)$ is identifiable and is given by (3.63) if every action-avoiding back-door path from X_k to Y is blocked by some subset L_k of nondescendants of X_k. (By "action-avoiding" we mean a path containing no arrows entering an X variable later than X_k.)

This condition bears several improvements over (3.62), as demonstrated in the following three examples.

Example 11.3.1 Figure 11.10 demonstrates cases where the *g*-formula (3.63) is valid with a subset L_k of the past but not with the entire past. Assuming U_1 and U_2 are

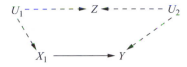

Figure 11.10   Conditioning on the entire past $L_1 = Z$ would invalidate g-estimation.

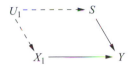

Figure 11.11 g-estimation is rendered valid by including a non-predecessor S.

unobserved, and temporal order: U_1, Z, X_1, U_2, Y, we see that both (3.62) and (3.62*), hence (3.63), are satisfied with $L_1 = 0$, while taking the whole past $L_1 = Z$ would violate both.

Example 11.3.2 Figure 11.11 demonstrates cases where defining L_k as the set of "nondescendants" of X_k (as opposed to temporal predecessors of X_k) broadens (3.62). Assuming temporal order: U_1, X_1, S, Y, both (3.62) and (3.62*) are satisfied with $L_1 = S$, but not with $L_1 = 0$.

Example 11.3.3 (constructed by Ilya Shpitser in response to Eliezer Yudkowsky) Figure 11.12 demonstrates cases where (3.62) is not satisfied even with the new interpretation of L_k, but the graphical condition (3.62*) is. It is easy to see that (3.62*) is satisfied; all back-door action-avoiding paths from X_1 to Y are blocked by $\{X_0, Z\}$. At the same time, it is possible to show (using the Twin Network Method, p. 213) that $Y(x_0, x_1)$ is not independent of X_1, given Z and X_0. (In the twin network model there is a d-connected path from X_1 to $Y(x_0, x_1)$, as follows: $X_1 \leftarrow Z \leftrightarrow Z* \rightarrow Y*$. Therefore, (3.62) is not satisfied for $Y(x_0, x_1)$ and X_1.)

This example is another demonstration of the weakness of the potential-outcome language initially taken by Robins in deriving (3.63). The counterfactual condition (3.62) that legitimizes the use of the g-estimation formula is opaque, evoking no intuitive support. Epidemiologists who apply this formula are doing so under no guidance of substantive medical knowledge. Fortunately, graphical methods are rapidly making their way into epidemiological practice (Greenland et al. 1999a; Robins 2001; Hernán et al. 2002; Greenland and Brumback 2002; Kaufman et al. 2005; Petersen et al. 2006; VanderWeele and Robins 2007) as more and more researchers begin to understand the assumptions behind g-estimation. With the added understanding that structural equation models subsume, unify, and underlie the graphical, counterfactual, potential outcome and sufficient-component (Rothman 1976) approaches to causation,[12] epidemiology stands a good chance of becoming the first discipline to fully liberate itself from past dogmas and "break through our academically enforced reluctance to think directly about causes (Weinberg 2007)."

---

[12] This unification has not been sufficiently emphasized by leading epidemiologists (Greenland and Brumback 2002), economists (Heckman and Vytlacil 2007), and social scientists (Morgan and Winship 2007), not to mention statisticians (Cox and Wermuth 2004; Rubin 2005). With all due respect to multiculturalism, all approaches to causation are variants or abstractions of the structural theory presented in this book (Chapter 7).

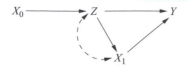

Figure 11.12   A graph for which g-estimation is valid while Robins' condition (3.62) is violated.

11.4   POLICY EVALUATION AND THE *do*-OPERATOR

11.4.1   Identifying Conditional Plans (Section 4.2, p. 113)
Question to Author:

Section 4.2 of the book (p. 113) gives an identification condition and estimation formula for the effect of a conditional action, namely, the effect of an action $do(X = g(z))$ where $Z = z$ is a measurement taken prior to the action. Is this equation generalizable to the case of several actions, i.e., conditional plan?

The difficulty seen is that this formula was derived on the assumption that X does not change the value of Z. However, in a multiaction plan, some actions in X could change observations Z that guide future actions. We do not have notation for distinguishing post-intervention from pre-intervention observations. Absent such notation, it is not clear how conditional plans can be expressed formally and submitted to the *do*-calculus for analysis.

Author's Reply (with Ilya Shpitser):

A notational distinction between post-intervention pre-intervention observations is introduced in Chapter 7 using the language of counterfactuals. The case of conditional plans, however, can be handled without resorting to richer notation. The reason is that the observations that dictate the choice of an action are not changed by that action, while those that have been changed by previous actions are well captured by the $P(y \mid do(x), z)$ notation.

To see that this is the case, however, we will first introduce counterfactual notation, and then show that it can be eliminated from our expression. We will use bold letters to denote sets, and normal letters to denote individual elements. Also, capital letters will denote random variables, and small letters will denote possible values these variables could attain. We will write Y_x to mean 'the value Y attains if we set variables X to values x.' Similarly, Y_{Xg} is taken to mean 'the value Y attains if we set variables X to whatever values they would have attained under the stochastic policy g.' Note that Y_x and Y_{Xg} are both random variables, just as the original variable Y.

Say we have a set of K action variables X that occur in some temporal order. We will indicate the time at which a given variable is acted on by a superscript, so a variable X^i occurs before X^j if $i < j$. For a given X^i, we denote $X^{<i}$ to be the set of action variables preceding X^i.

We are interested in the probability distribution of a set of outcome variables Y, under a policy that sets the values of each $X^i \in X$ to the output of functions g_i (known in advance) which pay attention to some set of prior variables Z_i, as well as the previous interventions on $X^{<i}$. At the same time, the variables Z^i are themselves affected by previous interventions. To define this recursion appropriately, we use an inductive definition. The base case is $X_g^1 = g_1(Z_1)$. The inductive case is $X_g^i = g_i(Z_{X_g^{<i}}^i, X_g^{<i})$. Here the

subscript g represents the policy we use, in other words, $g = \{g_i \mid i = 1, 2, ..., K\}$. We can now write the quantity of interest:

$$P(Y_{X_g} = y) = P(Y = y \mid do(X^1 = X_g^1), do(X^2 = X_g^2), ..., do(X^K = X_g^K)).$$

Let $Z_g = \cup_i Z_{X_g}^{i,<i}$. The key observation here is that if we observe Z_g to take on particular values, X_g collapse to unique values as well because X_g is a function of Z_g. We let $x_z = \{x_z^1, ..., x_z^K\}$ be the values attained by X_g in the situation where Z_g has been observed to equal $z = \{z_1, ..., z_K\}$. We note here that if we know z, we can compute x_z in advance, because the functions g_i are fixed in advance and known to us. However, we don't know what values Z_g might obtain, so we use case analysis to consider all possible value combinations. We then obtain:

$$P(Y_{X_g} = y) = \Sigma z^1, ..., z^K P(Y = y \mid do(X = x_z), Z^1 = z^1, ..., Z^K_{x_z < K} = z^K)$$

$$P(Z^1 = z^1, ..., Z^K_{x_z < K} = z^K \mid do(X = x_z)).$$

Here we note that Z_i cannot depend on subsequent interventions. So we obtain

$$\sum_z P(Y = y \mid do(X = x_z), Z^1_{x_z} = z^1, ..., Z^K_{x_z} = z^K) P(Z^1 = z^1, ..., Z^K = z^K \mid do(X = x_z)).$$

Now we note that the subscripts in the first and second terms are redundant, since the $do(x_z)$ already implies such subscripts for all variables in the expression. Thus we can rewrite the target quantity as

$$\sum_z P(Y = y \mid do(X = x_z), Z^1 = z^1, ..., Z^K = z^K) P(Z^1 = z^1, ..., Z^K = z^K do(X = x_z))$$

or, more succinctly,

$$\sum_z P(y \mid do(x_z), z) P(z \mid do(x_z)).$$

We see that we can compute this expression from $P(y \mid do(x))$, z) and $P(z \mid do(x))$, where Y, X, Z are disjoint sets.

To summarize, though conditional plans are represented naturally by nested counterfactual expressions, their identification can nevertheless be reduced to identification of conditional interventional distributions of the form $P(y \mid do(x), z)$ (possibly with z being empty). Complete conditions for identifying these quantities from a joint distribution in a given graph G are given in Shpitser and Pearl (2006a,b).

11.4.2 The Meaning of Indirect Effects

Question to Author:

I am teaching a course in latent variable modeling (to biostatistics and other public health students) and was yesterday introducing path analysis concepts, including direct and indirect effects.

I showed how to calculate indirect effects by taking the product of direct paths. Then a student asked about how to interpret the indirect effect, and I gave the answer that I always give, that the indirect effect ab (in the simple model of Fig. 11.13) is the effect that a change in X has on Y through its relationship with Z.

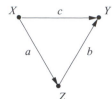

Figure 11.13   Demonstrating an indirect effect of X on Y via Z.

After chewing on this for a second, the student asked the following:

Student: "The interpretation of the b path is: b is the increase we would see in Y given a unit increase in Z while holding X fixed, right?"

Me: "That's right."

Student: "Then what is being held constant when we interpret an indirect effect?"

Me: "Not sure what you mean."

Student: "You said the interpretation of the indirect effect ab is: ab is the increase we would see in Y given a one unit increase in X through its causal effect on Z. But since b (the direct effect from Z to Y) requires X to be held constant, how can it be used in a calculation that is also requiring X to change one unit."

Me: "Hmm. Very good question. I'm not sure I have a good answer for you. In the case where the direct path from X to Y is zero, I think we have no problem, since the relationship between Z and Y then has nothing to do with X. But you are right, here if "c" is nonzero then we must interpret b as the effect of Z on Y when X is held constant. I understand that this sounds like it conflicts with the interpretation of the ab indirect effect, where we are examining what a change in X will cause. How about I get back to you. As I have told you before, the calculations here aren't hard, its trying to truly understand what your model means that's hard."

Author's Reply:

Commend your student on his/her inquisitive mind. The answer can be formulated rather simply (see Section 4.5.5, which was appended to the second edition):

> The indirect effect of X on Y is the increase we would see in Y while holding X constant and increasing Z to whatever value Z would attain under a unit increase of X.

Incidentally, the definition of b (the direct effect of Z on Y) does not "require X to be held constant"; it requires merely that the increase in Z be produced by intervention, and not in response to other variations in the system. See discussion on p. 97 and equation (5.24) (p. 161).

Author's Afterthought:

This question represents one of several areas where standard education in structural equation models (SEM) can stand reform. While some SEM textbooks give a cursory mention of the interpretation of structural parameters as effect coefficients, this interpretation is not taken very seriously by authors, teachers, and students. Writing in 2008, I find that the bulk of SEM education still focuses on techniques of statistical estimation and model fitting, and one can hardly find a serious discussion of what the model means, once it is fitted and estimated (see Section 11.5.3 for SEM survival kit).[13]

The weakness of this educational tradition surfaces when inquisitive students ask questions that deviate slightly from standard LISREL routines, the answers to which hinge on the causal interpretation of structural coefficients and structural equations. For example:

1. Why should we define the total effect the way we do? (i.e., as the sum of products of certain direct effects). Is this an arbitrary definition, or is it compelled by the causal interpretation of the path coefficients?

2. Why should we define the indirect effect as the difference between the total and direct effects?

3. How can we define direct effect in nonlinear systems or in systems involving dichotomous variables?

4. How should we, in a meaningful way, define effects in systems involving feedback loops (i.e., reciprocal causation) so as to avoid the pitfalls of erroneous definitions quoted in SEM textbooks? (see p. 164)

5. Would our assessment of direct and total effects remain the same if we were to take some measurements prior to implementing the action whose effect we attempt to estimate?

Readers will be pleased to note that these questions can be given formal answers, as in Sections 4.5.4, 4.5.5, 11.5.2, 11.5.3, and 11.7.1.

On a personal note, my interest in direct and indirect effects was triggered by a message from Jacques Hagenaars, who wrote (September 15, 2000): "Indirect effects do occupy an important place in substantive theories. Many social science theories 'agree' on the input (background characteristics) and output (behavioral) variables, but differ exactly with regard to the intervening mechanisms. To take a simple example, we know that the influence of Education on Political Preferences is mediated through 'economic status' (higher educated people get the better jobs and earn more money) and through a 'cultural mechanism' (having to do with the contents of the education and the accompanying socialization processes at school). We need to know and separate the nature and consequences of these two different processes, that is, we want to know the signs and the magnitudes of the indirect effects. In the parametric linear version of structural equation models, there exists a 'calculus of path coefficients' in which we can write total effects in terms of direct and several indirect effects. But this is not possible in the general nonparametric cases and not, e.g., in the log-linear parametric version. For systems of logic models there does not exist a comparable 'calculus of path coefficients' as has been remarked long ago. However, given its overriding theoretical importance, the issue of indirect effects cannot be simply neglected."

Stimulated by these comments, and armed with the notation of nested counterfactuals, I set out to formalize the legal definition of hiring discrimination given on page 147, and

---

[13] The word "causal" does not appear in the index of any of the post-2000 SEM textbooks that I have examined.

was led to the results of Sections 4.5.4 and 4.5.5, some already anticipated by Robins and Greenland (1992). Enlightened by these results, I was compelled and delighted to retract an earlier statement made on page 165 of the first edition of *Causality:* "indirect effects lack intrinsic operational meaning" because they cannot be isolated using the $do(x)$ operator. While it is true that indirect effects cannot be isolated using the $do(x)$ operator, they do possess intrinsic operational meaning. Policy-making implications of direct and indirect effects are further exemplified in Pearl (2001) and Petersen et al. (2006).

11.4.3   Can $do(x)$ Represent Practical Experiments?
Question to Author:

L.B.S., from the University of Arizona, questioned whether the $do(x)$ operator can represent realistic actions or experiments: "Even an otherwise perfectly executed randomized experiment may yield perfectly misleading conclusions. A good example is a study involving injected vitamin E as a treatment for incubated children at risk for retrolental fibroplasia. The randomized experiment indicated efficacy for the injections, but it was soon discovered that the actual effective treatment was opening the pressurized, oxygen-saturated incubators several times per day to give the injections, thus lowering the barometric pressure and oxygen levels in the blood of the infants (Leonard, *Major Medical Mistakes*). Any statistical analysis would have been misleading in that case."

S.M., from Georgia Institute of Technology, adds:

"Your example of the misleading causal effect shows the kind of thing that troubles me about the $do(x)$ concept. You $do(x)$ or don't $do(x)$, but it may be something else that covaries with $do(x)$ that is the cause and not the $do(x)$ per se."

Author's Reply:

Mathematics deals with ideal situations, and it is the experimenter's job to make sure that the experimental conditions approximate the mathematical ideal as closely as possible. The $do(x)$ operator stands for doing $X = x$ in an ideal experiment, where X and X alone is manipulated, not any other variable in the model.

In your example of the vitamin E injection, there is another variable being manipulated together with X, namely, the incubator cover, Z, which turns the experiment into a $do(x, z)$ condition instead of $do(x)$. Thus, the experiment was far from ideal, and far even from the standard experimental protocol, which requires the use of a placebo. Had a placebo been used (to approximate the requirement of the $do(x)$ operator), the result would not have been biased.

There is no way a model can predict the effect of an action unless one specifies what variables in the model are affected by the action, and how. The $do(x)$ operator is a mathematical device that helps us specify explicitly and formally what is held constant, and what is free to vary in any given experiment. The do-calculus then helps us predict the logical ramifications of such specifications, assuming they are executed faithfully, assuming we have a valid causal model of the environment, and assuming we have data from other experiments conducted under well-specified conditions.

11.4.4 Is the *do*(x) Operator Universal?

Question to Author (from Bill Shipley)

In most experiments, the external manipulation consists of adding (or subtracting) some amount from X without removing preexisting causes of X. For example, adding 5 kg/h of fertilizer to a field, adding 5 mg/l of insulin to subjects, etc. Here, the preexisting causes of the manipulated variable still exert effects, but a new variable (M) is added.

The problem that I see with the $do(x)$ operator as a general operator of external manipulation is that it requires two things: (1) removing any preexisting causes of x and (2) setting x to some value. This corresponds to some types of external manipulation, but not to all (or even most) external manipulations. I would introduce an $add(x = n)$ operator, meaning "add, external to the preexisting causal process, an amount 'n' to x." Graphically, this consists of augmenting the preexisting causal graph with a new edge, namely, $M - n \rightarrow X$. Algebraically, this would consist of adding a new term -n- as a cause of X (Shipley 2000b).

Author's Answer:

In many cases, your "additive intervention" indeed represents the only way we can intervene on a variable X; in others, it may represent the actual policy we wish evaluated. In fact, the general notion of intervention (p. 113) permits us to replace the equation of X by any other equation that fits the circumstances, not necessarily a constant $X = x$.

What you are proposing corresponds to replacing the old equation of X, $x = f(pa_X)$, with a new equation: $x = f(pa_X) + n$. This replacement can be represented using "instrumental variables," since it is equivalent to writing $x = f(paX) + I$ (where I is an instrument) and varying I from 0 to n.

There are three points to notice:

1. The additive manipulation *can* be represented in the *do*( ) framework – we merely apply the *do*( ) operator to the instrument I, and not to X itself. This is a different kind of manipulation that needs to be distinguished from $do(x)$ because, as is shown below, the effect on Y may be different.

2. In many cases, scientists are not satisfied with estimating the effect of the instrument on Y, but are trying hard to estimate the effect of X itself, which is often more meaningful or more transportable to other situations. (See p. 261 for discussion of the effect of "intention to treat.")

3. Consider the nonrecursive example where LISREL fails $y = bx + e_1 + I$, $x = ay + e_2$, (p. 164). If we interpret "total effects" as the response of Y to a unit change of the instrument I, then LISREL's formula obtains: The effect of I on Y is $b/(1 - ab)$. However, if we adhere to the notion of "per unit change in X," we get back the *do*-formula: The effect of X on Y is b, not $b/(1 - ab)$, even though the manipulation is done through an instrument. In other words, we change I from 0 to 1 and observe the changes in X and in Y; if we divide the change in Y by the change in X, we get b, not $b/(1 - ab)$.

To summarize: Yes, additive manipulation is sometimes what we need to model, and it can be done in the $do(x)$ framework using instrumental variables. We still need to distinguish, though, between the effect of the instrument and the effect of X. The former is not stable (p. 261), the latter is. LISREL's formula corresponds to the effect of an instrument, not to the effect of X.

Bill Shipley Further Asked:

Thanks for the clarification. It seems to me that the simplest, and most straightforward, way of modeling and representing manipulations of a causal system is to simply (1) modify the causal graph of the unmanipulated system to represent the proposed manipulation, (2) translate this new graph into structural equations, and (3) derive predictions (including conditional predictions) from the resulting equations; this is how I have treated the notion in my book. Why worry about $do(x)$ at all? In particular, one can model quite sophisticated manipulations this way. For instance, one might well ask what would happen if one added an amount z to some variable x in the causal graph, in which z is dependent on some other variable in the graph.

Author's Reply:

The method you are proposing, to replace the current equation $x = f(pa_X)$ with $x = g(f(pa_X), I, z)$, requires that we know the functional forms of f and g, as in linear systems or, alternatively, that the parents of X are observed, as in the Process Control example on page 74. These do not hold, however, in the non-parametric, partially observable settings of Chapters 3 and 4, which might render it impossible to predict the effect of the proposed intervention from data gathered prior to the intervention, a problem we called *identification*. Because pre-intervention statistics is not available for variable I, and f is unknown, there are semi-Markovian cases where $P(y \mid do(x))$ is identifiable while $P(y \mid do(x = g(f(pa_X), I, z)))$ is not; each case must be analyzed on its own merits. It is important, therefore, to impose certain standards on this vast space of potential interventions, and focus attention on those that could illuminate others.

Science thrives on standards, because standards serve (at least) two purposes: communication and theoretical focus. Mathematicians, for example, have decided that the derivative operator "dy/dx" is a nice standard for communicating information about change, so that is what we teach in calculus, although other operators might also serve the purpose, for example, xdy/dx or $(dy/dx)/y$, etc. The same applies to causal analysis:

1. **Communication**: If we were to eliminate the term "treatment effect" from epidemiology, and replace it with detailed descriptions of how the effect was measured, we would practically choke all communication among epidemiologists. A standard was therefore established: what we measure in a controlled, randomized experiment will be called "treatment effect"; the rest will be considered variations on the theme. The "*do*-operator" represents this standard faithfully.

   The same goes for SEM. Sewall Wright talked about "effect coefficients" and established them as the standard of "direct effect" in path analysis (before it got molested with regressional jargon), with the help of which more elaborate effects can be constructed. Again, the "*do*-operator" is the basis for defining this standard.

2. **Theoretical focus:** Many of the variants of manipulations can be reduced to "*do*," or to several applications of "*do*." Theoretical results established for "*do*"

are then applicable to those variants. Example: your "add *n*" manipulation is expressible as "*do*" on an instrument. Another example: questions of identification for expressions involving "*do*" are applicable to questions of identification for more sophisticated effects. On page 113, for example, we show that if the expression $P(y \mid do(x), z)$ is identifiable, then so also is the effect of conditional actions $P(y \mid do(x = g(z) \text{ if } Z = z))$. The same goes for many other theoretical results in the book; they were developed for the "*do*"-operator, they borrow from each other, and they are applicable to many variants.

Finally, the "surgical" operation underlying the *do*-operator provides the appropriate formalism for interpreting counterfactual sentences (see p. 204), and counterfactuals are abundant in scientific discourse (see pp. 217–19). I have yet to see any competing candidate with comparable versatility, generality, formal power, and (not the least) conceptual appeal.

11.4.5 Causation without Manipulation!!!
Question to Author

In the analysis of direct effects, Section 4.5 invokes an example of sex discrimination in school admission and, in several of the formulas, gender is placed under the "hat" symbol or, equivalently, as an argument of the *do*-operator. How can gender be placed after the *do*-operator when it is a variable that cannot be manipulated?

Author's Reply

Since Holland coined the phrase "No Causation without Manipulation" (Holland 1986), many good ideas have been stifled or dismissed from causal analysis. To suppress talk about how gender causes the many biological, social, and psychological distinctions between males and females is to suppress 90% of our knowledge about gender differences.

Surely we have causation without manipulation. The moon causes tides, race causes discrimination, and sex causes the secretion of certain hormones and not others. Nature is a society of mechanisms that relentlessly sense the values of some variables and determine the values of others; it does not wait for a human manipulator before activating those mechanisms.

True, manipulation is one way (albeit a crude one) for scientists to test the workings of mechanisms, but it should not in any way inhibit causal thoughts, formal definitions, and mathematical analyses of the mechanisms that propel the phenomena under investigation. It is for that reason, perhaps, that scientists invented counterfactuals; it permits them to state and conceive the realization of antecedent conditions without specifying the physical means by which these conditions are established.

The purpose of the "hat" symbol in Definition 4.5.1 is not to stimulate thoughts about possible ways of changing applicants' gender, but to remind us that any definition concerned with "effects" should focus on causal links and filter out spurious associations from the quantities defined. True, in the case of gender, one can safely replace $P(y \mid do(female))$ with $P(y \mid female)$, because the mechanism determining gender can safely be assumed to be independent of the background factors that influence Y (thus ensuring no confounding). But as a general definition, and even as part of an instructive example, mathematical expressions concerned with direct effects and sex discrimination should maintain the hat symbol. If nothing else, placing "female" under the "hat" symbol should help propagate the long-overdue counter-slogan: "Causation without manipulation? You bet!"

11.4.6 Hunting Causes with Cartwright

In her book *Hunting Causes and Using Them* (Cambridge University Press, 2007), Nancy Cartwright expresses several objections to the $do(x)$ operator and the "surgery" semantics on which it is based (p. 72, p. 201). In so doing, she unveils several areas in need of systematic clarification; I will address them in turn.

Cartwright description of surgery goes as follows:

> Pearl gives a precise and detailed semantics for counterfactuals. But what is the semantics a semantics of? The particular semantics Pearl develops is unsuited to a host of natural language uses of counterfactuals, especially those for planning and evaluation of the kind I have been discussing. That is because of the special way in which he imagines that the counterfactual antecedent will be brought about: by a precision incision that changes exactly the counterfactual antecedent and nothing else (except what follows causally from just that difference). But when we consider implementing a policy, this is not at all the question we need to ask. For policy and evaluation we generally want to know what would happen were the policy really set in place. And whatever we know about how it might be put in place, the one thing we can usually be sure of is that it will not be by a precise incision of the kind Pearl assumes.
>
> Consider for example Pearl's axiom of composition, which he proves to hold in all causal models – given his characterization of a causal model and his semantics for counterfactuals. This axiom states that 'if we force a variable (W) to a value w that it would have had without our intervention, then the intervention will have no effect on other variables in the system' (p. 229). This axiom is reasonable if we envisage interventions that bring about the antecedent of the counterfactual in as minimal a way as possible. But it is clearly violated in a great many realistic cases. Often we have no idea whether the antecedent will in fact obtain or not, and this is true even if we allow that the governing principles are deterministic. We implement a policy to ensure that it will obtain – and the policy may affect a host of changes in other variables in the system, some envisaged and some not. (Cartwright 2007, pp. 246–7)

Cartwright's objections can thus be summarized in three claims; each will be addressed separately.

1. In most studies we need to predict the effect of nonatomic interventions.
2. For policy evaluation "we generally want to know what would happen were the policy really set in place," but, unfortunately, "the policy may affect a host of changes in other variables in the system, some envisaged and some not."
3. Because practical policies are nonatomic, they cannot be evaluated from the atomic semantics of the $do(x)$ calculus even if we *could* envisage the variables that are affected by the policy.

Let us start with claim (2) – the easiest one to disprove. This objection is identical to the one discussed in Section 11.4.3, the answer to which was: "There is no way a model can predict the effect of an action unless one specifies correctly what variables in the model are affected by the action, and how." In other words, under the state of ignorance described in claim (2) of Cartwright, a policy evaluation study must end with a trivial answer: There is not enough information, hence, anything can happen. It is like pressing an unknown button in the dark, or trying to solve two equations with three unknowns. Moreover, the *do*-calculus can be used to test whether the state of ignorance in any given situation should justify such a trivial answer. Thus, it would be a mistake to assume that serious policy evaluation studies are conducted under such a state of ignorance; all policy analyses I have seen commence by assuming knowledge of the variables affected by the policy, and expressing that knowledge formally.

Claim (1) may apply in some cases, but certainly not in most; in many studies our goal is not to predict the effect of the crude, nonatomic intervention that we are about to implement but, rather, to evaluate an ideal, atomic policy that cannot be implemented given the available tools, but that represents nevertheless a theoretical relationship that is pivotal for our understanding of the domain.

An example will help. Smoking cannot be stopped by any legal or educational means available to us today; cigarette advertising can. That does not stop researchers from aiming to estimate "the effect of smoking on cancer," and doing so from experiments in which they vary the instrument – cigarette advertisement – not smoking.

The reason they would be interested in the atomic intervention $P(cancer \mid do(smoking))$ rather than (or in addition to) $P(cancer \mid do(advertising))$ is that the former represents a stable biological characteristic of the population, uncontaminated by social factors that affect susceptibility to advertisement. With the help of this stable characteristic one can assess the effects of a wide variety of practical policies, each employing a different smoking-reduction instrument.

Finally, claim (3) is demonstratively disproved in almost every chapter of this book. What could be more nonatomic than a policy involving a sequence of actions, each chosen by a set of observations Z which, in turn, are affected by previous actions (see Sections 4.4 and 11.4.1)? And yet the effect of implementing such a complex policy can be predicted using the "surgical" semantics of the *do*-calculus in much the same way that properties of complex molecules can be predicted from atomic physics.

I once challenged Nancy Cartwright (Pearl 2003a), and I would like to challenge her again, to cite a single example of a policy that cannot either be specified and analyzed using the $do(x)$ operators, or proven "unpredictable" (e.g., pressing an unknown button in the dark), again using the calculus of $do(x)$ operators.

Ironically, shunning mathematics based on ideal atomic intervention may condemn scientists to ineptness in handling realistic non-atomic interventions.

Science and mathematics are full of auxiliary abstract quantities that are not directly measured or tested, but serve to analyze those that are. Pure chemical elements do not exist in nature, yet they are indispensable to the understanding of alloys and compounds. Negative numbers (let alone imaginary numbers) do not exist in isolation, yet they are essential for the understanding of manipulations on positive numbers.

The broad set of problems tackled (and solved) in this book testifies that, invariably, questions about interventions and experimentation, ideal as well as non-ideal, practical as well as epistemological, can be formulated precisely and managed systematically using the atomic intervention as a primitive notion.

11.4.7 The Illusion of Nonmodularity

In her critique of the *do*-operator, Cartwright invokes yet another argument – the failure of modularity, which allegedly plagues most mechanical and social systems.

In her words:

> "When Pearl talked about this recently at LSE he illustrated this requirement with a Boolean input-output diagram for a circuit. In it, not only could the entire input for each variable be changed independently of that for each other, so too could each Boolean component of that input. But most arrangements we study are not like that. They are rather like a toaster or a carburetor."

At this point, Cartwright provides a four-equation model of a car carburetor and concludes:

> The gas in the chamber is the result of the pumped gas and the gas exiting the emulsion tube. How much each contributes is fixed by other factors: for the pumped gas both the amount of airflow and a parameter a, which is partly determined by the geometry of the chamber; and for the gas exiting the emulsion tube, by a parameter a', which also depends on the geometry of the chamber. The point is this. In Pearl's circuit-board, there is one distinct physical mechanism to underwrite each distinct causal connection. But that is incredibly wasteful of space and materials, which matters for the carburetor. One of the central tricks for an engineer in designing a carburetor is to ensure that one and the same physical design – for example, the design of the chamber – can underwrite or ensure a number of different causal connections that we need all at once.

> Just look back at my diagrammatic equations, where we can see a large number of laws all of which depend on the same physical features – the geometry of the carburetor. So no one of these laws can be changed on its own. To change any one requires a redesign of the carburetor, which will change the others in train. By design the different causal laws are harnessed together and cannot be changed singly. So modularity fails. (Cartwright 2007, pp. 15–16)

Thus, for Cartwright, a set of equations that share parameters is inherently nonmodular; changing one equation means modifying at least one of its parameters, and if this parameter appears in some other equation, it must change as well, in violation of modularity.

Heckman (2005, p. 44) makes similar claims: "Putting a constraint on one equation places a restriction on the entire set of internal variables." "Shutting down one equation might also affect the parameters of the other equations in the system and violate the requirements of parameter stability."

Such fears and warnings are illusory. Surgery, and the whole semantics and calculus built around it, does not assume that in the physical world we have the technology to incisively modify the mechanism behind each structural equation while leaving all others unaltered. Symbolic modularity does not assume physical modularity. Surgery is a symbolic operation which makes no claims about the physical means available to the experimenter, or about invisible connections that might exist between the mechanisms involved.

Symbolically, one can surely change one equation without altering others and proceed to define quantities that rest on such "atomic" changes. Whether the quantities defined in this manner correspond to changes that can be physically realized is a totally different question that can only be addressed once we have a formal description of the interventions available to us. More importantly, shutting down an equation does not necessarily mean meddling with its parameters; it means overruling that equation, namely, leaving the equation intact but lifting the outcome variable from its influence.

A simple example will illustrate this point.

Assume we have two objects under free-fall conditions. The respective accelerations, a_1 and a_2, of the two objects are given by the equations:

$$a_1 = g \tag{11.12}$$
$$a_2 = g, \tag{11.13}$$

where g is the earth's gravitational pull. The two equations share a parameter, g, and appear to be nonmodular in Cartwright's sense; there is indeed no physical way of changing the gravitational force on one object without a corresponding change on the other. However, this does not mean that we cannot intervene on object 1 without touching object 2. Assume we grab object 1 and bring it to a stop. Mathematically, the intervention amounts to replacing equation (11.12) by

$$a_1 = 0 \tag{11.14}$$

while leaving equation (11.13) intact. Setting g to zero in equation (11.12) is a symbolic surgery that does not alter g in the physical world but, rather, sets a_1 to 0 by bringing object 1 under the influence of a new force, f, emanating from our grabbing hand. Thus, equation (11.14) is a result of two forces:

$$a_1 = g + f/m_1, \tag{11.15}$$

where $f = -gm_1$, which is identical to (11.14).

This same operation can be applied to Cartwright carburetor; for example, the gas outflow can be fixed without changing the chamber geometry by installing a flow regulator at the emulsion tube. It definitely applies to economic systems, where human agents are behind most of the equations; the left-hand side of the equations can be fixed by exposing agents to different information, rather than by changing parameters in the physical world. A typical example emerges in job discrimination cases (Section 4.5.3). To test the "effect of gender on hiring" one need not physically change the applicant's gender; it is enough to change the employer's awareness of the applicant's gender. I have yet to see an example of an economic system which is not modular in the sense described here.

This operation of adding a term to the right-hand side of an equation to ensure constancy of the left-hand side is precisely how Haavelmo (1943) envisioned surgery in economic settings. Why his wisdom disappeared from the teachings of his disciples in 2008 is one of the great mysteries of economics (see Hoover (2004)); my theory remains (p. 138) that it all happened due to a careless choice of notation which crumbled under the ruthless invasion of statistical thinking in the early 1970s.

More on the confusion in econometrics and the reluctance of modern-day econometricians to reinstate Haavelmo's wisdom is to be discussed in Section 11.5.4.

11.5   CAUSAL ANALYSIS IN LINEAR STRUCTURAL MODELS

11.5.1   General Criterion for Parameter Identification (Chapter 5, pp. 149–54)
Question to Author:

The parameter identification method described in Section 5.3.1 rests on repetitive applications of two basic criteria: (1) the single-door criterion of Theorem 5.3.1, and (2) the back-door criterion of Theorem 5.3.2. This method may require appreciable bookkeeping in combining results from various segments of the graph. Is there a single graphical criterion of identification that unifies the two theorems and thus avoids much of the bookkeeping involved?

Author's Reply:

A unifying criterion is described in the following lemma (Pearl 2004):

Lemma 11.5.1 (*Graphical identification of direct effects*)
Let c stand for the path coefficient assigned to the arrow $X \rightarrow Y$ in a causal graph G. Parameter c is identified if there exists a pair (W, Z), where W is a single node in G (not excluding W = X), and Z is a (possibly empty) set of nodes in G, such that:

1. *Z consists of nondescendants of Y,*

2. *Z d-separates W from Y in the graph G_c formed by removing $X \rightarrow Y$ from G,*

3. *W and X are d-connected, given Z, in G_c.*

Moreover, the estimand induced by the pair (W, Z) is given by:

$$c = \frac{cov(Y, W \mid Z)}{cov(X, W \mid Z)}.$$

The intuition is that, conditional on Z, W acts as an instrumental variable relative to $X \rightarrow Y$. See also McDonald (2002a).

More general identification methods are reported in Brito and Pearl (2002a,b,c; 2006).

11.5.2   The Causal Interpretation of Structural Coefficients
Question to Author:

In response to assertions made in Sections 5.1 and 5.4 that a correct causal interpretation is conspicuously absent from SEM books and papers, including all 1970–99 texts in economics, two readers wrote that the "unit-change" interpretation is common and well accepted in the SEM literature. L.H. from the University of Alberta wrote:

> Page 245 of L. Hayduk, *Structural Equation Modeling with LISREL: Essentials and Advances*, 1987, [states] that a slope can be interpreted as: the magnitude of the change in *y* that would be predicted to accompany a unit change in *x* with the other variables in the equation left untouched at their original values.

> O.D. Duncan, *Introduction to Structural Equation Models* (1975) pages 1 and 2 are pretty clear on *b* as causal. More precisely, it says that "a change of one unit in *x* …produces a change of *b* units in *y*" (page 2). I suspect that H. M. Blalock's book

'Causal models in the social sciences,' and D. Heise's book 'Causal analysis' probably speak of *b* as causal.

S.M., from Georgia Tech, concurs:
Heise, author of *Causal Analysis* (1975), regarded the *b* of causal equations to be how
much a unit change in a cause produced an effect in an effect variable. This is a well-
accepted idea.

Author's Reply:

The "unit change" idea appears, sporadically and informally, in several SEM publications, yet, similarly to counterfactuals in econometrics (Section 11.5.5) it has not been
operationalized through a precise mathematical definition.

The paragraph cited above (from Hayduk 1987) can serve to illustrate how the
unit change idea is typically introduced in the SEM literature and how it *should be*
introduced using modern understanding of causal modeling. The original paragraph
reads:

The interpretation of structural coefficients as "effect coefficients" originates with
ordinary regression equations like

$$X_0 = a + b_1 X_1 + b_2 X_2 + b_3 X_3 + e$$

for the effects of variables X_1, X_2, and X_3 on variable X_0. We can interpret the estimate
of b_i as the magnitude of the change in X_0 that would be predicted to accompany a
unit change INCREASE in X_1 with X_2 and X_3 left untouched at their original values.
We avoid ending with the phrase "held constant" because this phrase must be abandoned for models containing multiple equations, as we shall later see. Parallel interpretations are appropriate for b_2 and b_3. (Hayduk 1987, p. 245)

This paragraph illustrates how two basic distinctions are often conflated in the SEM
literature. The first is the distinction between structural coefficients and regressional
(or statistical) estimates of those coefficients. We rarely find the former defined independently of the latter – a confusion that is rampant and especially embarrassing in
econometric texts. The second is the distinction between "held constant" and "left
untouched" or "found to remain constant," for which the $do(x)$ operator was devised.
(Related distinctions: "doing" versus "seeing" and "interventional change" versus "natural change.")

To emphasize the centrality of these distinctions I will now propose a concise revision of Hayduk's paragraph:

Proposed Revised Paragraph

The interpretation of structural coefficients as "effect coefficients" bears some resemblance to, but differs fundamentally from, the interpretation of coefficients in regression
equations like

$$X_0 = a + b_1 X_1 + b_2 X_2 + b_3 X_3 + e. \tag{11.16}$$

If (11.16) is a regression equation, then b_1 stands for the change in X_0 that would be
predicted to accompany a unit change in X_1 in those situations where X_2 and X_3 remain

constant at their original values. We formally express this interpretation using conditional expectations:

$$b_1 = E(X_0 \mid x_1 + 1, x_2, x_3) - E(X_0 \mid x_1, x_2, x_3)$$
$$\quad = R_{X_0 X_1 \cdot X_2 X_3}. \tag{11.17}$$

Note that, as a regression equation, (11.16) is claimless; i.e., it cannot be falsified by any experiment and, from (11.17), e is automatically rendered uncorrelated with X_1, X_2, and X_3.

In contrast, if equation (11.16) represents a structural equation, it makes empirical claims about the world (e.g., that other variables in the system do not affect X_0 once we hold X_1, X_2, and X_3 fixed), and the interpretation of b_1 must be modified in two fundamental ways. First, the phrase "a unit change in X_1" must be qualified to mean "a unit interventional change in X_1," thus ruling out changes in X_1 that are produced by other variables in the model (possibly correlated with e). Second, the phrase "where X_2 and X_3 remain constant" must be abandoned and replaced by the phrase "if we hold X_2 and X_3 constant," thus ensuring constancy even when X_2 is affected by X_1.

Formally, these two modifications are expressed as:

$$b_1 = E(X_0 \mid do(x_1 + 1, x_2, x_3)) - E(X_0 \mid do(x_1, x_2, x_3)). \tag{11.18}$$

The phrase "left untouched at their original values" may lead to ambiguities. Leaving variables untouched permits those variables to vary (e.g., in response to the unit increase in X_1 or other influences), in which case the change in X_0 would correspond to the total effect

$$E(X_0 \mid do(x_1 + 1)) - E(X_0 \mid do(x_1)) \tag{11.19}$$

or to the marginal conditional expectation

$$E(X_0 \mid x_1 + 1) - E(X_0 \mid x_1), \tag{11.20}$$

depending on whether the change in X_1 is interventional or observational. None of (11.19) or (11.20) matches the meaning of b_1 in equation (11.16), regardless of whether we treat (11.16) as a structural or a regression equation.

The interpretation expressed in (11.18) holds in *all* models, including those containing multiple equations, recursive and nonrecursive, regardless of whether e is correlated with other variables in the model and regardless of whether X_2 and X_3 are affected by X_1. In contrast, expression (11.17) coincides with (11.18) only under very special circumstances (defined by the single-door criterion of Theorem 5.3.1). It is for this reason that we consider (11.18), not (11.17), to be an "interpretation" of b_1; (11.17) interprets the "regression estimate" of b_1 (which might well be biased), while (11.18) interprets b_1 itself.

11.5.3 Defending the Causal Interpretation of SEM (or, SEM Survival Kit)
Question to Author:

J. Wilson from Surrey, England, asked about ways of defending his Ph.D. thesis before examiners who do not approve of the causal interpretation of structural equation models (SEM). He complained about "the complete lack of emphasis in PhD programmes on

Figure 11.14 The graph underlying equations (11.21)–(11.22).

how to defend causal interpretations and policy implications in a viva when SEM is used …
if only causality had been fully explained at the beginning of the programme, then each
of the 70,000 words used in my thesis would have been carefully measured to defend
first the causal assumptions, then the data, and finally the interpretations … (I wonder
how widespread this problem is?) Back to the present and urgent task of trying to satisfy
the examiners, especially those two very awkward Stat Professors – they seem to be
trying to outdo each other in nastiness."

Author's Reply:

The phenomenon that you complain about is precisely what triggered my writing of
Chapter 5 – the causal interpretation of SEM is still a mystery to most SEMs researchers,
leaders, educators, and practitioners. I have spent hours on SEMNET Discussion List
trying to rectify the current neglect, but it is only students like yourself who can turn
things around and help reinstate the causal interpretation to its central role in SEM
research.

   As to your concrete question – how to defend the causal interpretation of SEM
against nasty examiners who oppose such interpretation – permit me to assist by
sketching a hypothetical scenario in which you defend the causal interpretation of your
thesis in front of a hostile examiner, Dr. EX. (Any resemblance to Dr. EX is purely coin-
cidental.)

A Dialogue with a Hostile Examiner
or
SEM Survival Kit

For simplicity, let us assume that the model in your thesis consists of just two-equations,

$$y = bx + e_1 \tag{11.21}$$
$$z = cy + e_2, \tag{11.22}$$

with e_2 uncorrelated with x. The associated diagram is given in Figure 11.14. Let us further assume that the target of your thesis was to estimate parameter c, that you have estimated c satisfactorily to be $c = 0.78$ using the best SEM methods, and that you have
given a causal interpretation to your finding.

   Now your nasty examiner, Dr. EX, arrives and questions your interpretation.

   Dr. EX: What do you mean by "c has a causal interpretation"?

   You: I mean that a unit change in y will bring about a c units change in $E(Z)$.

   Dr. EX: The words "change" and "bring about" make me uncomfortable; let's be
scientific. Do you mean $E(Z \mid y) = cy + a$??? I can understand this last expression,
because the conditional expectation of Z given y, $E(Z \mid y)$, is well defined mathematically, and I know how to estimate it from data. But "change" and "bring about" is jargon to me.

   You: I actually mean "change," not "an increase in conditional expectation," and by
"change" I mean the following: If we had the physical means of fixing y at some

constant y_1, and of changing that constant from y_1 to y_2, then the observed change in $E(Z)$ would be $c(y_2 - y_1)$.

Dr. EX: Well, well, aren't we getting a bit metaphysical here? I never heard about "fixing" in my statistics classes.

You: Oh, sorry, I did not realize you have statistics background. In that case, let me rephrase my interpretation a bit, to read as follows: If we had the means of conducting a controlled randomized experiment, with y randomized, then if we set the control group to y_1 and the experimental group to y_2, the observed difference in $E(Z)$ would be $E(Z_2) - E(Z_1) = c(y_2 - y_1)$ regardless of what values y_1 and y_2 we choose. (Z_1 and Z_2 are the measurements of z under the control and experimental groups, respectively.)[14]

Dr. EX: That sounds much closer to what I can understand. But I am bothered by a giant leap that you seem to be making. Your data was nonexperimental, and in your entire study you have not conducted a single experiment. Are you telling us that your SEM exercise can take data from an observational study, do some LISREL analysis on it, and come up with a prediction of what the outcome of a controlled randomized experiment will be? You've got to be kidding!! Do you know how much money can be saved nationwide if we could replace experimental studies with SEM magic?

You: This is not magic, Dr. EX, it is plain logic. The input to my LISREL analysis was more than just nonexperimental data. The input consisted of two components: (1) data, (2) causal assumptions; my conclusion logically follows from the two. The second component is absent in standard experimental studies, and that is what makes them so expensive.

Dr. EX: What kind of assumptions? "Causal"? I never heard of such strangers. Can you express them mathematically the way we normally express assumptions – say, in the form of conditions on the joint density, or properties of the covariance matrix?

You: Causal assumptions are of a different kind; they cannot be written in the vocabulary of density functions or covariance matrices. Instead, they are expressed in my causal model.

Dr. EX: Looking at your model, equations (11.21)–(11.22), I do not see any new vocabulary; all I see is equations.

You: These are not ordinary algebraic equations, Dr. EX. These are "structural equations," and if we read them correctly, they convey a set of assumptions with which you are familiar, namely, assumptions about the outcomes of hypothetical randomized experiments conducted on the population – we call them "causal" or "modeling" assumptions, for want of better words, but they can be understood as assumptions about the behavior of the population under various randomized experiments.

Dr. EX: Wait a minute! Now that I begin to understand what your causal assumptions are, I am even more puzzled than before. If you allow yourself to make assumptions about the behavior of the population under randomized experiments, why go through the trouble of conducting a study? Why not make the assumption directly that in a randomized experiment, with y randomized, the observed difference in $E(Z)$ should be $c'(y_2 - y_1)$, with c' just any convenient number, and save yourself agonizing months of data collection and analysis. He who believes your other untested assumptions should also believe your $E(Z_2) - E(Z_1) = c'(y_2 - y_1)$ assumption.

---

[14] Just in case Dr. EX asks: "Is that the only claim?" you should add: Moreover, I claim that the distribution of the random variable $Z_1 - cy_1$ will be the same as that of the variable $Z_2 - cy_2$.

You: Not so, Dr. EX. The modeling assumptions with which my program begins are much milder than the assertion $E(Z_2) - E(Z_1) = 0.78(y_2 - y_1)$ with which my study concludes. First, my modeling assumptions are qualitative, while my conclusion is quantitative, making a commitment to a specific value of $c = 0.78$. Second, many researchers (including you, Dr. EX) would be prepared to accept my assumptions, not my conclusion, because the former conforms to commonsense understanding and general theoretical knowledge of how the world operates. Third, the majority of my assumptions can be tested by experiments that do not involve randomization of y. This means that if randomizing y is expensive, or infeasible, we still can test the assumptions by controlling other, less formidable variables. Finally, though this is not the case in my study, modeling assumptions often have some statistical implications that can be tested in nonexperimental studies, and, if the test turns out to be successful (we call it "fit"), it gives us further confirmation of the validity of those assumptions.

Dr. EX: This is getting interesting. Let me see some of those "causal" or modeling assumptions, so I can judge how mild they are.

You: That's easy, have a look at our model, Figure 11.14, where

z —   student's score on the final exam,
y —   number of hours the student spent on homework,
x —   weight of homework (as announced by the teacher) in the final grade.

When I put this model down on paper, I had in mind two randomized experiments, one where x is randomized (i.e., teachers assigning weight at random), the second where the actual time spent on homework (y) is randomized. The assumptions I made while thinking of those experiments were:

1. Linearity and exclusion for y: $E(Y_2) - E(Y_1) = b(x_2 - x_1)$, with b unknown (Y_2 and Y_1 are the time that would be spent on homework under announced weights x_2 and x_1, respectively.) Also, by excluding z from the equation, I assumed that the score z would not affect y, because z is not known at the time y is decided.

2. Linearity and exclusion for z: $E(Z_2) - E(Z_1) = c(y_2 - y_1)$ for all x, with c unknown. In words, x has no effect on z, except through y.

In addition, I made qualitative assumptions about unmeasured factors that govern x under nonexperimental conditions; I assumed that there are no common causes for x and z.

Do you, Dr. EX, see any objection to any of these assumptions?

Dr. EX: Well, I agree that these assumptions are milder than a blunt, unsupported declaration of your thesis conclusion, $E(Z_2) - E(Z_1) = 0.78(y_2 - y_1)$, and I am somewhat amazed that such mild assumptions can support a daring prediction about the actual effect of homework on score (under experimental setup). But I am still unhappy with your common cause assumption. It seems to me that a teacher who emphasizes the importance of homework would also be an inspiring, effective teacher, so e_2 (which includes factors such as quality of teaching) should be correlated with x, contrary to your assumption.

You: Dr. EX, now you begin to talk like an SEM researcher. Instead of attacking the method and its philosophy, we are beginning to discuss substantive issues – e.g., whether it is reasonable to assume that a teacher's effectiveness is uncorrelated with the weight

Figure 11.15 Statistically equivalent models to Figure 11.14.

that the teacher assigns to homework. I personally have had great teachers that could not care less about homework, and conversely so.

But this is not what my thesis is about. I am not claiming that teachers' effectiveness is uncorrelated with how they weigh homework; I leave that to other researchers to test in future studies (or might it have been tested already?). All I am claiming is: Those researchers who are willing to accept the assumption that teachers' effectiveness is uncorrelated with how they weigh homework will find it interesting to note that this assumption, coupled with the data, logically implies the conclusion that an increase of one homework-hour per day causes an (average) increase of 0.78 grade points in student's score. And this claim can be verified empirically if we are allowed a controlled experiment with randomized amounts of homework (y).

Dr. EX: I am glad you do not insist that your modeling assumptions are true; you merely state their plausibility and explicate their ramifications. I cannot object to that. But I have another question. You said that your model does not have any statistical implications, so it cannot be tested for fitness to data. How do you know that? And doesn't this bother you?

You: I know it by just looking at the graph and examining the missing links. A criterion named d-separation (see Section 11.1.2, "d-separation without tears") permits students of SEM to glance at a graph and determine whether the corresponding model implies any constraint in the form of a vanishing partial correlation between variables. Most statistical implications (though not all) are of this nature. The model in our example does not imply any constraint on the covariance matrix, so it can fit perfectly any data whatsoever. We call this model "saturated," a feature that some SEM researchers, unable to shake off statistical-testing traditions, regard as a fault of the model. It isn't. Having a saturated model at hand simply means that the investigator is not willing to make implausible causal assumptions, and that the mild assumptions he/she is willing to make are too weak to produce statistical implications. Such a conservative attitude should be commended, not condemned. Admittedly, I would be happy if my model were not saturated – say, if e_1 and e_2 were uncorrelated. But this is not the case at hand; common sense tells us that e_1 and e_2 are correlated, and it also shows in the data. I tried assuming $cov(e_1, e_2) = 0$, and I got terrible fit. Am I going to make unwarranted assumptions just to get my model "knighted" as "nonsaturated"? No! I would rather make reasonable assumptions, get useful conclusions, and report my results side by side with my assumptions.

Dr. EX: But suppose there is another saturated model, based on equally plausible assumptions, yet leading to a different value of c. Shouldn't you be concerned with the possibility that some of your initial assumptions are wrong, hence that your conclusion $c = 0.78$ is wrong? There is nothing in the data that can help you prefer one model over the other.

You: I am concerned indeed, and, in fact, I can immediately enumerate the structures of all such competing models; the two models in Figure 11.15 are examples, and many more. (This too can be done using the *d*-separation criterion; see pp. 145–8.) But note that the existence of competing models does not in any way weaken my earlier stated claim: "Researchers who accept the qualitative assumptions of model M are compelled to accept the conclusion $c = 0.78$." This claim remains logically invincible. Moreover, the claim can be further refined by reporting the conclusions of each contending model, together with the assumptions underlying that model. The format of the conclusion will then read:

If you accept assumption set A_1, then $c = c_1$ is implied,
If you accept assumption set A_2, then $c = c_2$ is implied,

and so on.

Dr. EX: I see, but still, in case we wish to go beyond these conditional statements and do something about deciding among the various assumption sets, are there no SEM methods to assist one in this endeavor? We, in statistics, are not used to facing problems with two competing hypotheses that cannot be submitted to some test, however feeble.

You: This is a fundamental difference between statistical data analysis and SEM. Statistical hypotheses, by definition, are testable by statistical methods. SEM models, in contrast, rest on *causal* assumptions, which, also by definition (see p. 39), cannot be given statistical tests. If our two competing models are saturated, we know in advance that there is nothing more we can do but report our conclusions in a conditional format, as listed above. If, however, the competition is among equally plausible yet statistically distinct models, then we are facing the century-old problem of model selection, where various selection criteria such as AIC have been suggested for analysis. However, the problem of model selection is now given a new, causal twist – our mission is not to maximize fitness, or to maximize predictive power, but rather to produce the most reliable estimate of causal parameters such as c. This is a new arena altogether (see Pearl 2004).

Dr. EX: Interesting. Now I understand why my statistician colleagues got so totally confused, mistrustful, even antagonistic, upon encountering SEM methodology (e.g., Freedman 1987; Holland 1988; Wermuth 1992). One last question. You started talking about randomized experiments only after realizing that I am a statistician. How would you explain your SEM strategy to a nonstatistician?

You: I would use plain English and say: "If we have the physical means of fixing y at some constant y_1, and of changing that constant from y_1 to y_2, then the observed change in $E(Z)$ would be $c(y_2 - y_1)$." Most people understand what "fixing" means, because this is on the mind of policy makers. For example, a teacher interested in the effect of homework on performance does not think in terms of randomizing homework. Randomization is merely an indirect means for predicting the effect of fixing.

Actually, if the person I am talking to is really enlightened (and many statisticians are), I might even resort to counterfactual vocabulary and say, for example, that a student who scored z on the exam after spending y hours on homework would have scored $z + c$ had he/she spent $y + 1$ hours on homework. To be honest, this is what I truly had in mind when writing the equation $z = cy + e_2$, where e_2 stood for all other characteristics of the student that were not given variable names in our model and that are not affected by y. I did not even think about $E(Z)$, only about z of a typical student. Counterfactuals are the most precise linguistic tool we have for expressing the meaning of scientific relations. But I refrain from mentioning counterfactuals when I talk to statisticians because, and this is regrettable, statisticians tend to suspect deterministic concepts, or concepts that are not immediately testable, and counterfactuals are such concepts (Dawid 2000; Pearl 2000).

Dr. EX: Thanks for educating me on these aspects of SEM. No further questions.
You: The pleasure is mine.

11.5.4   Where Is Economic Modeling Today? – Courting Causes with Heckman

Section 5.2 of this book decries the decline in the understanding of structural equation modeling in econometric in the past three decades (see also Hoover 2003, "Lost Causes") and attributes this decline to a careless choice of notation which blurred the essential distinction between algebraic and structural equations. In a series of articles (Heckman 2000, 2003, 2005; Heckman and Vytlacil 2007), James Heckman has set out to overturn this perception, reclaim causal modeling as the central focus of economic research, and reestablish economics as an active frontier in causal analysis. This is not an easy task by any measure. To adopt the conceptual and technical advances that have emerged in neighboring disciplines would amount to admitting decades of neglect in econometrics, while to dismiss those advances would necessitate finding them econometric surrogates. Heckman chose the latter route, even though most modern advances in causal modeling are rooted in the ideas of economists such as Haavelmo (1943), Marschak (1950), and Strotz and Wold (1960).

One step in Heckman's program was to reject the *do*-operator and the "surgery" semantics upon which it is based, thus depriving economists of the structural semantics of counterfactuals developed in this book (especially Chapter 7), which unifies traditional econometrics with the potential-outcome approach. Heckman's reasons for rejecting surgery are summarized thus:

> Controlled variation in external (forcing) variables is the key to defining causal effects in nonrecursive models … Pearl defines a causal effect by 'shutting one equation down' or performing 'surgery' in his colorful language. He implicitly assumes that 'surgery,' or shutting down an equation in a system of simultaneous equations, uniquely fixes one outcome or internal variable (the consumption of the other person in my example). In general, it does not. Putting a constraint on one equation places a restriction on the entire set of internal variables. In general, no single equation in a system of simultaneous equations uniquely determines any single outcome variable. Shutting down one equation might also affect the parameters of the other equations in the system and violate the requirements of parameter stability. (Heckman and Vytlacil 2007)

Clearly, Heckman's objections are the same as Cartwright's (Section 11.4.6):

1. Ideal surgery may be technically infeasible,
2. Economic systems are nonmodular.

We have repudiated these objections in four previous subsections (11.4.3–11.4.6) which readers can easily reapply to deconstruct Heckman's arguments. It is important to reemphasize, though, that, as in the case of Cartwright, these objections emanate from conflating the task of definition (of counterfactuals) with those of identification and practical estimation, a frequent confusion among researchers which Heckman (2005) sternly warns readers to avoid.

This conflation is particularly visible in Heckman's concern that "shutting down one equation might also affect the parameters of the other equations in the system." In the physical world, attempting to implement the conditions dictated by a "surgery" may sometimes affect parameters in other equations, and, as we shall see, the same applies to Heckman's proposal of "external variation." However, we are dealing here with symbolic, not physical, manipulations. Our task is to formulate a meaningful mathematical definition of "the causal effect of one variable on another" in a symbolic system called a "model." This permits us to manipulate symbols at will, while ignoring the technical feasibility of these manipulations. Implementational considerations need not enter the discussion of *definition*.

A New Definition of Causal Effects: "External Variation"

Absent surgery semantics, Heckman and Vytlacil (HV) set out to configure a new definition of causal effects, which, hopefully, would be free of the faults they discovered in the surgery procedure, by basing it on "external-variations," instead of shutting down equations. It is only unfortunate that their new definition, the cornerstone of their logic of counterfactuals, is not given an explicit formal exposition: it is relegated to a semiformal footnote (HV, p. 77) that even a curious and hard-working reader would find difficult to decipher. The following is my extrapolation of HV's definition as it applies to multi-equations and nonlinear systems.

Given a system of equations:

$$Y_i = f_i(Y, X, U) \; i = 1, 2, \ldots, n,$$

where X and U are sets of observed and unobserved external variables, respectively, the causal effect of Y_j on Y_k is computed in four steps:

1. Choose any member X_t of X that appears in f_j. If none exists, exit with failure.

2. If X_t appears in any other equation as well, consider excluding it from that equation (e.g., set its coefficient to zero if the equation is linear or replace X_t by a constant).[15]

3. Solve for the reduced form

    $$Y_i = g_i(X, U) \; i = 1, 2, \ldots, n \qquad (11.23)$$

    of the resulting system of equations.

4. The causal effect of Y_j on Y_k is given by the partial derivative:

    $$dY_k/dY_j = dg_k/dX_t : dg_j/dX_t. \qquad (11.24)$$

Example 11.5.2 Consider a system of three equations:

$$Y_1 = aY_2 + cY_3 + eX + U_1$$
$$Y_2 = bY_1 + X + U_2$$
$$Y_3 = dY_1 + U_3.$$

---

[15] It is not clear what conditions (if any) would forbid one from setting $e = 0$, in example 11.5.2, or ignoring X altogether and adding a dummy variable X' to the second equation. HV give the impression that deciding on whether e can be set to 0 requires deep understanding of the problem at hand; if this is their intention, it need not be.

Needed: the causal effect of Y_2 on Y_1.

The system has one external variable, X, which appears in the first two equations. If we can set $e = 0$, x will appear in the equation of Y_2 only, and we can then proceed to Step 3 of the "external variation" procedure. The reduced form of the modified model yields:

$$dY_1/dX = a/(1 - ba - cd) \quad dY_2/dX = (1 - cd)/(1 - ab - cd),$$

and the causal effect of Y_1 on Y_2 calculates to:

$$dY_1/dY_2 = a/(1 - cd).$$

In comparison, the surgery procedure constructs the following modified system of equations:

$$
\begin{aligned}
Y_1 &= aY_2 + cY_3 + eX + U_1 \\
Y_2 &= y_2 \\
Y_3 &= dY_1 + U_3,
\end{aligned}
$$

from which we obtain for the causal effect of Y_2 on Y_1;

$$dY_1/dy_2 = a/(1 - cd),$$

an expression identical to that obtained from the "external variation" procedure.

It is highly probable that the two procedures always yield identical results, which would bestow validity and conceptual clarity on the "external variation" definition.

11.5.5 External Variation versus Surgery

In comparing their definition to the one provided by the surgery procedure, HV write (p. 79): "Shutting down an equation or fiddling with the parameters … is not required to define causality in an interdependent, nonrecursive system or to identify causal parameters. The more basic idea is exclusion of different external variables from different equations which, when manipulated, allow the analyst to construct the desired causal quantities."

I differ with HV on this issue. I believe that "surgery" is the more basic idea, more solidly motivated, and more appropriate for policy evaluation tasks. I further note that basing a definition on exclusion and external variation suffers from the following flaws:

1. In general, "exclusion" involves the removal of a variable from an equation and amounts to "fiddling with the parameters." It is, therefore, a form of "surgery" – a modification of the original system of equations – and would be subject to the same criticism one may raise against "surgery." Although we have refuted such criticism in previous sections, we should nevertheless note that if it ever has a grain of validity, the criticism would apply equally to both methods.

2. The idea of relying exclusively on external variables to reveal internal cause–effect relationships has its roots in the literature on *identification* (e.g., as in the studies of "instrumental variables") when such variables act as "nature's experiments." This restriction, however, is unjustified in the context

of *defining* causal effect, since "causal effects" are meant to quantify effects produced by *new* external manipulations, not necessarily those shown explicitly in the model and not necessarily those operating in the data-gathering phase of the study. Moreover, every causal structural equation model, by its very nature, provides an implicit mechanism for emulating such external manipulations, via surgery.

Indeed, most policy evaluation tasks are concerned with *new* external manipulations which exercise direct control over endogenous variables. Take, for example, a manufacturer deciding whether to double the current price of a given product after years of letting the price track the cost, i.e., *price* = *f*(*cost*). Such a decision amounts to removing the equation *price* = *f*(*cost*) from the model at hand (i.e., the one responsible for the available data) and replacing it with a constant equal to the new price. This removal emulates faithfully the decision under evaluation, and attempts to circumvent it by appealing to "external variables" are artificial and hardly helpful.

As another example, consider the well-studied problem (Heckman 1992) of evaluating the impact of terminating an educational program for which students are admitted based on a set of qualifications. The equation *admission* = *f*(*qualifications*) will no longer hold under program termination, and no external variable can simulate the new condition (i.e., *admission* = 0) save for one that actually neutralizes (or "ignores," or "shuts down") the equation *admission* = *f*(*qualifications*).

It is also interesting to note that the method used in Haavelmo (1943) to define causal effects is mathematically equivalent to surgery, not to external variation. Instead of replacing the equation $Y_j = f_j(Y, X, U)$ with $Y_j = y_j$, as would be required by surgery, Haavelmo writes $Y_j = f_j(Y, X, U) + x_j$, where X_j is chosen so as to make Y_j constant, $Y_j = y_j$. Thus, since X_j liberates Y_j from any residual influence of $f_j(Y, X, U)$, Haavelmo's method is equivalent to that of surgery. Heckman's method of external variation leaves Y_j under the influence f_j.

3. Definitions based on external variation have the obvious flaw that the target equation may not contain any observable external variable. In fact, in many cases the set of observed external variables in the system is empty (e.g., Figure 3.5). Additionally, a definition based on a ratio of two partial derivatives does not generalize easily to nonlinear systems with discrete variables. Thus, those who seriously accept Heckman's definition would be deprived of the many identification techniques now available for instrumentless models (see Chapters 3 and 4) and, more seriously yet, would be unable to even ask whether causal effects are identified in any such model – identification questions are meaningless for undefined quantities.

    Fortunately, liberated by the understanding that definitions can be based on purely symbolic manipulations, we can modify Heckman's proposal and *add* fictitious external variables to any equation we desire. The added variables can then serve to define causal effects in a manner similar to the steps in equations (11.23) and (11.24) (assuming continuous variables). This brings us closer to surgery, with the one basic difference of leaving Y_j under the influence of $f_j(Y, X, U)$.

Having argued that definitions based on "external variation" are conceptually ill-motivated, we now explore whether they can handle noncausal systems of equations.

Equation Ambiguity in Noncausal Systems

Several economists (Leroy 2002; Neuberg 2003; Heckman and Vytlacil 2007) have criticized the *do*-operator for its reliance on *causal*, or directional, structural equations, where we have a one-to-one correspondence between variables and equations. HV voice this criticism thus: "In general, no single equation in a system of simultaneous equations uniquely determines any single outcome variable" (Heckman and Vytlacil 2007, p. 79).

One may guess that Heckman and Vytlacil refer here to systems containing nondirectional equations, namely, equations in which the equality sign does not stand for the non-symmetrical relation "is determined by" or "is caused by" but for symmetrical algebraic equality. In econometrics, such noncausal equations usually convey equilibrium or resource constraints; they impose equality between the two sides of the equation but do not endow the variable on the left-hand side with the special status of an "outcome" variable.

The presence of nondirectional equations creates ambiguity in the surgical definition of the counterfactual Y_x, which calls for replacing the equation *determining* X with the constant equation $X = x$. If X appears in several equations, and if the position of X in the equation is arbitrary, then each one of those equations would be equally qualified for replacement by $X = x$, and the value of Y_x (i.e., the solution for Y after replacement) would be ambiguous.

Note that symmetrical equalities differ structurally from reciprocal causation in directional nonrecursive systems (i.e., systems with feedback, as in Figure 7.4), since, in the latter, each variable is an "outcome" of precisely one equation. Symmetrical constraints can nevertheless be modeled as the solution of a dynamic feedback system in which equilibrium is reached almost instantaneously (Lauritzen and Richardson 2002; Pearl 2003a).

Heckman and Vytlacil create the impression that equation ambiguity is a flaw of the surgery definition and does not plague the exclusion-based definition. This is not the case. In a system of nondirectional equations, we have no way of knowing which external variable to exclude from which equation to get the right causal effect.

For example: Consider a nonrecursive system of two equations that is discussed in HV, p. 75:

$$Y_1 = a_1 + c_{12}Y_2 + b_{11}X_1 + b_{12}X_2 + U_1 \tag{11.25}$$

$$Y_2 = a_2 + c_{21}Y_1 + b_{21}X_1 + b_{22}X_2 + U_2. \tag{11.26}$$

Suppose we move Y_1 to the l.h.s. of (11.26) and get:

$$Y_1 = [a_2 - Y_2 + b_{21}X_1 + b_{22}X_2 + U_2]/c_{21}. \tag{11.27}$$

To define the causal effect of Y_1 on Y_2, we now have a choice of excluding X_2 from (11.25) or from (11.27). The former yields c_{12}, while the latter yields $1/c_{21}$. We see that the ambiguity we have in choosing an equation for surgery translates into ambiguity in choosing an equation and an external variable for exclusion.

Methods of breaking this ambiguity were proposed by Simon (1953) and are discussed on pages 226–8.

Summary – Economic Modeling Reinvigorated

The idea of constructing causal quantities by exclusion and manipulation of external variables, while soundly motivated in the context of identification problems, has no logical basis when it comes to model-based definitions. Definitions based on surgery, on the other hand, enjoy generality, semantic clarity, and computational simplicity.

So, where does this leave econometric modeling? Is the failure of the "external variable" approach central or tangential to economic analysis and policy evaluation?

In almost every one of his recent articles James Heckman stresses the importance of counterfactuals as a necessary component of economic analysis and the hallmark of econometric achievement in the past century. For example, the first paragraph of the HV article reads: "they [policy comparisons] require that the economist construct counterfactuals. Counterfactuals are required to forecast the effects of policies that have been tried in one environment but are proposed to be applied in new environments and to forecast the effects of new policies." Likewise, in his *Sociological Methodology* article (2005), Heckman states: "Economists since the time of Haavelmo (1943, 1944) have recognized the need for precise models to construct counterfactuals... The econometric framework is explicit about how counterfactuals are generated and how interventions are assigned..."

And yet, despite the proclaimed centrality of counterfactuals in econometric analysis, a curious reader will be hard pressed to identify even one econometric article or textbook in the past 40 years in which counterfactuals or causal effects are formally defined. Needed is a procedure for computing the counterfactual $Y(x, u)$ in a well-posed, fully specified economic model, with X and Y two arbitrary variables in the model. By rejecting Haavelmo's definition of $Y(x, u)$, based on surgery, Heckman commits econometrics to another decade of division and ambiguity, with two antagonistic camps working in almost total isolation.

Economists working within the potential-outcome framework of the Neyman-Rubin model take counterfactuals as primitive, unobservable variables, totally detached from the knowledge encoded in structural equation models (e.g., Angrist 2004; Imbens 2004). Even those applying propensity score techniques, whose validity rests entirely on the causal assumption of "ignorability," or unconfoundedness, rarely know how to confirm or invalidate that assumption using structural knowledge (see Section 11.3.5). Economists working within the structural equation framework (e.g., Kennedy 2003; Mittelhammer et al. 2000; Intriligator et al. 1996) are busy estimating parameters while treating counterfactuals as metaphysical ghosts that should not concern ordinary mortals. They trust leaders such as Heckman to define precisely what the policy implications are of the structural parameters they labor to estimate, and to relate them to what their colleagues in the potential-outcome camp are doing.[16]

The surgery semantics (pp. 98–102) and the causal theory entailed by it (Chapters 7–10) offer a simple and precise unification of these two antagonistic and narrowly focused schools of econometric research – a theorem in one approach entails a theorem in the other, and vice versa. Economists will do well resurrecting the basic

---

[16] Notably, the bibliographical list in the comprehensive review article by economist Hoover (2008) is almost disjoint from those of economists Angrist (2004) and Imbens (2004) – the cleavage is culturally deep.

ideas of Haavelmo (1943), Marschak (1950), and Strotz and Wold (1960) and re-
invigorating them with the logic of graphs and counterfactuals presented in this book.

For completeness, I reiterate here explicitly (using parenthetical notation) the two
fundamental connections between counterfactuals and structural equations.

1. The structural definition of counterfactuals is:

$$Y_M(x, u) = Y_{M_x}(u).$$

   Read: For any model M and background information u, the counterfactual conditional "Y if X had been x" is given by the solution for Y in submodel M_x (i.e.,
   the mutilated version of M with the equation determining X replaced by $X = x$).

2. The empirical claim of the structural equation $y = f(x, e(u))$ is:

$$Y(x, z, u) = f(x, e(u)),$$

   for any set Z not intersecting X or Y.

   Read: Had X and Z been x and z, respectively, Y would be $f(x, e(u))$, independently of z, and independently of other equations in the model.

11.6   DECISIONS AND CONFOUNDING (CHAPTER 6)

11.6.1   Simpson's Paradox and Decision Trees
Nimrod Megiddo (IBM Almaden) Wrote:

"I do not agree that 'causality' is the key to resolving the paradox (but this is also a
matter of definition) and that tools for looking at it did not exist twenty years ago.
Coming from game theory, I think the issue is not difficult for people who like to
draw decision trees with 'decision' nodes distinguished from 'chance' nodes.

I drew two such trees [Figure 11.16(a) and (b)] which I think clarify the correct decision in different circumstances."

Author's Reply:

The fact that you have constructed two different decision trees for the same input tables
implies that the key to the construction was not in the data, but in some information you
obtained from the story behind the data. What is that information?

The literature of decision tree analysis has indeed been in existence for at least fifty
years, but, to the best of my knowledge, it has not dealt seriously with the problem posed
above: "What information do we use to guide us into setting up the correct decision tree?"

We agree that giving a robot the frequency tables *alone* would not be sufficient for
the task. But what else would Mr. Robot (or a statistician) need? Changing the story
from F = "*female*" to F = "*blood pressure*" seems to be enough for people, because
people understand informally the distinct roles that gender and blood pressure play in the
scheme of things. Can we characterize these roles formally, so that our robot would be
able to construct the correct decision tree?

My proposal: give the robot (or a statistician or a decision-tree expert) a pair (T, G),
where T is the set of frequency tables and G is a causal graph, and, lo and behold, the

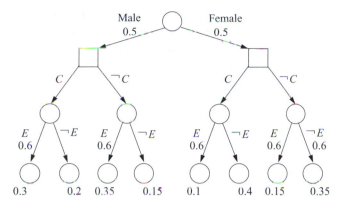

Figure 11.16(a)   Decision tree corresponding to Figure 11.17(a). Given Male, $-C$ is better than C $(0.35 > 0.3)$. Given Female, $-C$ is also better $(0.15 > 0.1)$. Unconditionally, with *any* probability p for Male and $1 - p$ for Female, again, $-C$ is better than C $(0.35p + 0.15(1 - p) > 0.3p = 0.1(1 - p))$.

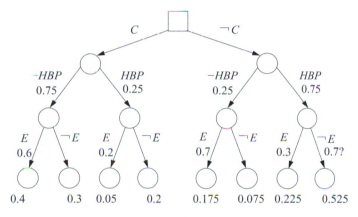

Figure 11.16(b)   Decision tree corresponding to Figure 11.17(b). The tree can be compressed as shown in Figure 11.16(c).

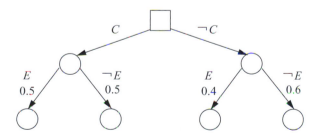

Figure 11.16(c)   Compressed decision tree. Obviously, C is better than $-C$ (since $0.5 > 0.4$).

robot would be able to set up the correct decision tree automatically. This is what I mean in saying that the resolution of the paradox lies in causal considerations. Moreover, one can go further and argue: "If the information in (T, G) is sufficient, why not skip the construction of a decision tree altogether, and get the right answer directly from (T, G)?" This is the gist of Chapters 3–4 in this book. Can the rich literature on decision analysis benefit from the more economical encoding of decision problems in the syntax of (T, G)? The introduction of influence diagrams (Howard and Matheson 1981; Pearl 2005) was a step in this direction, and, as Section 4.1.2 indicates, the second step might not be too far off. While an influence diagram is parameterized for a specific choice of decision and chance variables, a causal diagram is not specific to any such choice; it contains the information for parameterizing all influence diagrams (and decision trees) for any choice of decision variables and any choice of chance variables that one may observe at decision time.

More recently, Dawid (2002) has developed a hybrid representation, combining influence diagrams and Bayesian networks, by attaching to some variables in the latter a decision node (as in Figure 3.2) representing a potential intervention, capable of switching from an observational (idle) to interventional (active) mode (as in equation (3.8)).

11.6.2 Is Chronological Information Sufficient for Decision Trees?

Megiddo Wrote Back:

"The term 'causality' introduces into the problem issues that do not have to be there, such as determinism, free will, cause and effect, etc. What does matter is a specification that, in the outcome fixing process, fixing the value of variable X occurs before fixing the value of a variable Y, and Y depends on X. You like to call this situation a causality relation. Of course in a mathematical theory you can choose any name you like, but then people are initially tempted to develop some intuition, which may be wrong due to the external meaning of the name you choose. The interpretation of this intuition outside the mathematical model often has real-life implications that may be wrong, for example, that X really causes Y. The decision tree is a simple way to demonstrate the additional chronological information, and simple directed graphs can of course encode that information more concisely. When you have to define precisely what these graphs mean, you refer to a fuller description like the trees. So, in summary, my only objection is to the use of the word 'causality' and I never had doubts that chronological order information was crucial to a correct decision making based on past data."

Author's Reply

1. Before the 21st century, there was some danger in attributing to certain mathematical relations labels such as "causal," which were loaded with intuition, mystery, and controversy. This is no longer the case – the mystery is gone, and the real-life implication of this interpretation is not wrong; X *really* causes Y.

   Moreover, if in addition to getting the mathematics right, one is also interested in explicating those valuable intuitions, so that we can interpret them more precisely and even teach them to robots, then there is no escape but to label those relationships with whatever names they currently enjoy in our language, namely, "causal."

2. There is more that enters a decision tree than chronological and dependence information. For example, the chronological and dependence information that is conveyed by Figure 11.17(c) is identical to that of Figure 11.17(a) (assuming F occurs before C), yet (c) calls for a different decision tree (and yields a different conclusion), because the dependence between F and Y is "causal" in (a) and associational in (c). Thus, causal considerations must supplement chronological

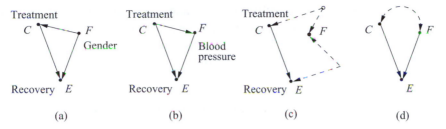

Figure 11.17 Graphs demonstrating the insufficiency of chronological information.

and dependence information if we are to construct correct decision trees and to load their branches with correct probabilities.

As a thought experiment, imagine that we wish to write a program that automatically constructs decision trees from stories like those in Figure 11.17(a)-(b)-(c). The program is given the empirical frequency tables and is allowed to ask us questions about chronological and dependence relationships among C, E, and F, but is not allowed to use any causal vocabulary. Would the program be able to distinguish between (a) and (c)? The answer is: No. If one ignores causal considerations and attends, as you suggest, to chronological and dependence information alone, then the decision tree constructed for Figure 11.17(c) would be identical to the one drawn for $F = gender$ (Figure 11.16(a)), and the wrong conclusion would ensue, namely, that the drug (C) is harmful for both $F = true$ and $F = false$ patients. This is wrong, because the correct answer is that the drug is beneficial to the population as a whole (as in the blood pressure example), hence it must be beneficial for either $F = true$ or $F = false$ patients (or both). The decision tree hides this information and yields the wrong results.

The error stems from attributing the wrong probabilities to the branches of the decision tree. For example, the leftmost branch in the tree is assigned a probability $P(E \mid C, \neg F) = 0.6$, which is wrong; the correct probability should be $P(E \mid do(C), \neg F)$, which, in the case of Figure 11.17(c), cannot be determined from the graph and the tables. We do know, however, that $P(E \mid do(C)) = 0.5$ and $P(E \mid do(\neg C)) = 0.4$, so either equation (6.4) or equation (6.5) must be violated.

The conventional wisdom in decision analysis is to assign the tree branches conditional probabilities of the type $P(E \mid action, Z)$, where Z is the information available at decision time. Practitioners should be warned against following this convention blindly. The correct assignment should be, of course, $P(E \mid do(action), Z)$ which is estimable either from experimental studies or when the causal graph renders this quantity identifiable.

3. I do not agree that in order to define precisely what causal graphs mean, we must "refer to a fuller description like the trees." In Section 7.1 we find a formal definition of causal graphs as a collection of functions, and this definition invokes no decision trees (at least not explicitly). Thus, a causal graph has meaning of its own, independent of the many decision trees that the graph may help us construct. By analogy, it would be awkward (though not mathematically wrong) to

say that the meaning of a differential equation (say, for particle motion) lies in the set of trajectories that obey that equation; the meaning of each term in a differential equation can be defined precisely, from first principles, without having a clue about the solution to the equation.

11.6.3 Lindley on Causality, Decision Trees, and Bayesianism

Question to Author (from Dennis Lindley):

If your assumption, that controlling X at x is equivalent to removing the function for X and putting $X = x$ elsewhere, is applicable, then it makes sense. What I do not understand at the moment is the relevance of this to decision trees. At a decision node, one conditions on the quantities known at the time of the decision. At a random node, one includes all relevant uncertain quantities under known conditions. Nothing more than the joint distributions (and utility considerations) are needed. For example, in the medical case, the confounding factor may either be known or not at the time the decision about treatment is made, and this determines the structure of the tree. Where causation may enter is when the data are used to assess the probabilities needed in the tree, and it is here that Novick and I used exchangeability. The Bayesian paradigm makes a sharp distinction between probability as belief and probability as frequency, calling the latter, chance. If I understand causation, it would be reasonable that our concept could conveniently be replaced by yours in this context.

Author's Reply:

Many decision analysts take the position that causality is not needed because: "Nothing more than the joint distributions (and utility considerations) are needed" (see discussion in section 11.6.2). They are right and wrong. While it is true that probability distributions are all that is needed, the distributions needed, $P(y \mid do(x), see(z))$, are of a peculiar kind; they are not derivable from the joint distribution $P(y, x, z)$ which governs our data, unless we supplement the data with causal knowledge, such as that provided by a causal graph.

Your next sentence says it all:

> Where causation may enter is when the data are used to assess the probabilities needed in the tree,...

I take the frequency/belief distinction to be tangential to discussions of causality. Let us assume that the tables in Simpson's story were not frequency tables, but summaries of one's subjective beliefs about the occurrence of various joint events, (C, E, F), $(C, E, \neg F)$... etc. My assertion remains that this summary of beliefs is not sufficient for constructing our decision tree. We need also to assess our belief in the hypothetical event "E would occur if a decision $do(C)$ is taken," and, as we have seen, temporal information alone is insufficient for deriving this assessment from the tabulated belief summaries, hence, we cannot construct the decision tree from this belief summary; we need an extra ingredient, which I call "causal" information and you choose to call "exchangeability" – I would not quarrel about nomenclature except to note that, if we trust human beings to provide us with reliable information, we should also respect the vocabulary in which they cast that information, and, much to the disappointment of prudent academics, that vocabulary is unmistakably causal.

Update Question to Author:

Your point about probability and decision trees is well taken and I am in agreement with what you say here; a point that I had not appreciated before. Thank you. Let me rephrase the argument to see whether we are in agreement. In handling a decision tree it is easy to see what probabilities are needed to solve the problem. It is not so easy to see how these might be assessed numerically. However, I do not follow how causality completely resolves the issue. Nor, of course, does exchangeability.

Author's Reply:

I am very glad that we have narrowed the problem down to simple and concrete issues: (1) how to assess the probabilities needed for a decision tree, (2) where those probabilities come from, (3) how those probabilities can be encoded economically, and perhaps even (4) whether those probabilities must comply with certain rules of internal coherence, especially when we construct several decision trees, differing in the choice of decision and chance nodes.

The reason I welcome this narrowing of the problem is that it would greatly facilitate my argument for explicit distinction between causal and probabilistic relationships. In general, I have found Bayesian statisticians to be the hardest breed of statisticians to convince of the necessity of this distinction. Why? Because whereas classical statisticians are constantly on watch against assumptions that cannot be substantiated by hard data, Bayesians are more permissive in this regard, and rightly so. However, by licensing human judgment as a legitimate source of information, Bayesians have become less meticulous in keeping track of the character and origin of that information. Your earlier statement describes the habitual, uncritical use of conditiong among Bayesian researchers, which occasionally lures the unwary into blind alleys (e.g., Rubin (2009)):

> At a decision node, one conditions on the quantities known at the time of the decision. At a random node, one includes all relevant uncertain quantities under known conditions. Nothing more than the joint distributions (and utility considerations) are needed.

As Newcomb's paradox teaches us (see Section 4.1), it is not exactly true that "at a decision node, one conditions on the quantities known at the time of the decision" – at least some of the "conditionings" need to be replaced with "doing." If this were not the case, then all decision trees would turn into a joke; "patients should avoid going to the doctor to reduce the probability that one is seriously ill (Skyrms 1980, p. 130); workers should never hurry to work, to reduce the probability of having overslept; students should not prepare for exams, lest this would prove them behind in their studies; and so on. In short, all remedial actions should be banished lest they increase the probability that a remedy is indeed needed." (Chapter 4, p. 108).

But even after escaping this "conditioning" trap, the Bayesian philosopher does not see any difference between assessing probabilities for branches emanating from decision nodes and those emanating from chance nodes. For a Bayesian, both assessments are probability assessments. That the former involves a mental simulation of hypothetical experiment while the latter involves the mental envisioning of passive observations is of minor significance, because Bayesians are preoccupied with the distinction between probability as belief and probability as frequency.

This preoccupation renders them less sensitive to the fact that beliefs come from a variety of sources and that it is important to distinguish beliefs about outcomes of experiments from beliefs about outcomes of passive observations (Pearl 2001a, "Bayesianism and Causality"). First, the latter are robust, the former are not; that is, if our subjective beliefs about passive observations are erroneous, the impact of the error will get washed out in due time, as the number of observations increases. This is not the case for belief about outcomes of experiments; erroneous priors will produce a permanent bias regardless of sample size. Second, beliefs about outcomes of experiments cannot be articulated in the grammar of probability calculus; therefore, in order to formally express such beliefs and coherently combine them with data, probability calculus must be enriched with causal notation (i.e., graphs, $do(x)$, or counterfactuals).

So far, I have found two effective ways to win the hearts of Bayesians, one involving the notion of "economy" (see my discussion in Section 11.6.2), the other the notion of "coherence."

Given a set of n variables of interest, there is a huge number of decision trees that can conceivably be constructed from these variables, each corresponding to a different choice of temporal ordering and a different choice of decision nodes and chance nodes from those variables. The question naturally arises, how can a decision maker ensure that probability assessments for all these decision trees be reproducible? Surely we cannot assume that human explicitly store all these potential decision trees in their heads. For reproducibility, we must assume that all these assessments must be derived from some economical representation of knowledge about decisions and chance events. Causal relationships can thus be viewed as an economical representation from which decision trees are constructed. Indeed, as I wrote to Megiddo (Section 11.6.2), if we were in need of instructing a robot to construct such decision trees on demand, in accordance with the available knowledge and belief, our best approach would be to feed the robot a pair of inputs (G, P), where G is a causal graph and P is our joint distribution over the variables of interest (subjective distribution, if we were Bayesian). With the help of this pair of objects, the robot should be able to construct consistently all the decision trees required, for any partition of the variables into decision and chance nodes, and replicate the parameters on the branches. This is one way a Bayesian could appreciate causality without offending the traditional stance that "it is nothing more than the joint distributions…"

The second approach involves "coherence." Coherence is something of which Bayesians are very proud, because DeFinetti, Savage, and others have labored hard to construct qualitative axioms that prevent probability judgments from being totally whimsical, and that compel beliefs to conform to the calculus of probability.

We can ask the Bayesian philosopher to tell us whether judgments about joint probabilities, say $P(x, y)$, should in some way cohere with judgments about decision-based probabilities, say $P(y \mid do(x))$, which quantify the branch emanating from a decision node with two alternatives $do(X = x)$ and $do(X = x')$. We can then ask the Bayesian whether these probabilities should bear any connection to the usual conditional probabilities, $P(y \mid x)$, namely, the probability assessed for outcome $Y = y$ that emanates (in some other decision tree) from a chance event $X = x$.

It should not be too hard to convince our Bayesian that these two assessments could not be totally arbitrary, but must obey some restrictions of coherence. For example, the inequality $P(y \mid do(x)) \geq P(y, x)$ should be obeyed for all events x and y.[17] Moreover, coherence restrictions of this kind are automatically satisfied whenever $P(y \mid do(x))$ is derived from a causal network according to the rules of Chapter 3. These two arguments should be inviting for a Bayesian to start drawing mathematical benefits from causal calculus, while maintaining caution and skepticism, and, as they say in the Talmud:

> "*From benefits comes understanding*"
> (free translation of "*mitoch shelo lishma, ba lishma*" (Talmud, Psahim, 50b)).

Bayesians will eventually embrace causal vocabulary, I have no doubt.

11.6.4 Why Isn't Confounding a Statistical Concept?

In June 2001, I received two anonymous reviews of my paper "Causal Inference in the Health Sciences" (Pearl 2001c). The questions raised by the reviewers astounded me, for they reminded me of the archaic way some statisticians still think about causality and of the immense educational effort that still lies ahead. In the interest of contributing to this effort, I am including my reply in this chapter. Related discussion on the causal–statistical distinction is presented in Section 11.1.

Excerpts from Reviewers' Comments:

Reviewer 1.

"The contrast between statistical and causal concepts is overdrawn. Randomization, instrumental variables, and so forth have clear statistical definitions. ... [the paper urges] 'that any systematic approach to causal analysis must require new mathematical notation.' This is false: there is a long tradition of informal – but systematic and successful – causal inference in the medical sciences."

Reviewer 2.

"The paper makes many sweeping comments which rely on distinguishing 'statistical' and 'causal' concepts ... Also, included in the list of causal (and therefore, according to the paper, non-statistical) concepts is, for example, confounding, which is solidly founded in standard, frequentist statistics. Statisticians are inclined to say things like 'U is a potential confounder for examining the effect of treatment X on outcome Y when both U and X and U and Y are not independent. So why isn't confounding a statistical concept?' ... If the author wants me to believe this, he's going to have to show at least one example of how the usual analyses fail."

---

[17] This inequality follows from (3.52) or (9.33). A complete characterization of coherence constraints is given in Tian, Kang, and Pearl (2006). As an example, for any three variables X, Y, Z, coherence dictates: $P(y \mid do(x, z)) - P(y, x \mid do(z)) - P(y, z \mid do(x)) + P(x, y, z) \geq 0$. If the structure of a causal graph is known, the conditions of Definition 1.3.1 constitute a complete characterization of all coherence requirements.

Author's Response:

Reviewer #1 seems to advocate an informal approach to causation (whatever that means; it reminds me of informal statistics before Bernoulli), and, as such, he/she comes to this forum with set ideas against the basic aims and principles of my paper. Let history decide between us.

The target of this paper are readers of Reviewer #2's persuasion, who attempt to reconcile the claims of this paper (occasionally sweeping, I admit) with traditional statistical wisdom. To this reviewer I have the following comments.

You question the usefulness of my proposed demarcation line between statistical and causal concepts. Let me try to demonstrate this usefulness by considering the example that you bring up: confounding. You write that "confounding is solidly founded in standard, frequentist statistics." and that statisticians are inclined to say things like "U is a potential confounder for examining the effect of treatment X on outcome Y when both U and X and U and Y are not independent. So why isn't confounding a statistical concept?"

Chapter 6 of my book goes to great lengths explaining why this definition fails on both sufficiency and necessity tests, and why all variants of this definition must fail by first principles. I will bring just a couple of examples to demonstrate the point; additional discussion is provided in Section 11.3.3. Consider a variable U that is affected by both X and Y – say, one that turns 1 whenever both X and Y reach high levels. U satisfies your criterion, and yet U is not a confounder for examining the effect of treatment X on outcome – in fact, U can safely be ignored in our analysis. (The same goes for any variable U whose effect on Y is mediated by X, like Z in Figure 2 of my paper.) As a second example, consider a variable U that resides "on the causal pathway" from X to Y. This variable too satisfies your criterion yet it is not a confounder – generations of statisticians have been warned (see Cox 1958) not to adjust for such variables. One might argue that your definition is merely a necessary, but not sufficient condition for confounding. But this too fails. Chapter 6 (pp. 185–6) describes examples where there is no variable that satisfies your definition, and still the effect of treatment X on outcome Y is confounded.

One can also construct an example (Figure 6.5) where U is a confounder (i.e., must be adjusted to remove effect bias), and still U is not associated with either X or Y.

I am not the first to discover discrepancies between confounding and its various statistical "definitions." Miettinen and Cook and Robins and Greenland have been arguing this point in epidemiology since the mid-1980s – to no avail. Investigators continue to equate collapsibility with no-confounding and continue to adjust for the wrong variables (Weinberg 1993). Moreover, the popular conception that any important concept (e.g., randomization, confounding, instrumental variables) *must* have a statistical definition is so deeply entrenched in the health sciences that even today, 15 months past the publication of my book, people with the highest qualifications and purest of intentions continue to ask: "So why isn't confounding a statistical concept?"

I believe that any attempt to correct this tradition would necessarily sound sweeping, and nothing but a sweeping campaign can ever eradicate these misconceptions about confounding, statistics, and causation. Statistical education is firmly in the hands of people of Reviewer 1's persuasion. And people with your quest for understanding are rarely given a public forum to ask, "So why isn't confounding a statistical concept?"

The same argument applies to the concepts of "randomization" and "instrumental variables." (Ironically, Reviewer #1 states authoritatively that these concepts "have clear statistical definitions." I would hand him a bivariate distribution $f(x, y)$ and ask to tell us if X is randomized.) Any definition of these concepts must invoke causal vocabulary, undefined in terms of distributions – there is simply no escape.

And this brings me to reemphasize the usefulness of the statistical–causal demarcation line, as defined in Section 1.5. Those who recognize that concepts such as randomization, confounding, and instrumental variables must rest on causal information would be on guard to isolate and explicate the causal assumptions underlying studies invoking these concepts. In contrast, those who ignore the causal–statistical distinction (e.g., Reviewer #1) will seek no such explication, and will continue to turn a blind eye to "how the usual analysis fails."

11.7   THE CALCULUS OF COUNTERFACTUALS

11.7.1   Counterfactuals in Linear Systems

We know that, in general, counterfactual queries of the form $P(Y_x = y \mid e)$ may or may not be empirically identifiable, even in experimental studies. For example, the probability of causation, $P(Y_x = y \mid x', y')$, is in general not identifiable from either observational or experimental studies (p. 290, Corollary 9.2.12). The question we address in this section is whether the assumption of linearity renders counterfactual assertions more empirically grounded. The answer is positive:

Claim A. Any counterfactual query of the form $E(Y_x \mid e)$ is empirically identifiable in linear causal models, with e an arbitrary evidence.

Claim B. Whenever the causal effect T is identified, $E(Y_x \mid e)$ is identified as well.

Claim C. $E(Y_x \mid e)$ is given by

$$E(Y_x \mid e) = E(Y \mid e) + T[x - E(X \mid e)], \tag{11.28}$$

where T is the total effect coefficient of X on Y, i.e.,

$$T = dE[Y_x]/dx = E(Y \mid do(x + 1)) - E(Y \mid do(x)). \tag{11.29}$$

Claim A is not surprising. It has been established in generality by Balke and Pearl (1994b) where $E(Y_x \mid e)$ is given explicit expression in terms of the covariance matrix and the structural coefficients; the latter are empirically identifiable.

Claim B renders $E(Y_x \mid e)$ observationally identifiable as well, in all models where the causal effect $E(Y_x)$ is identifiable.

Claim C offers an intuitively compelling interpretation of $E(Y_x \mid e)$ that reads as follows: Given evidence e, to calculate $E(Y_x \mid e)$ (i.e., the expectation of Y under the hypothetical assumption that X were x, rather than its current value), first calculate the best estimate of Y conditioned on the evidence e, $E(Y \mid e)$, then add to it whatever change is expected in Y when X undergoes a forced transition from its current best estimate, $E(X \mid e)$, to its hypothetical value $X = x$. That last addition is none other than the effect coefficient T times the expected change in X, i.e., $T[x - E(X \mid e)]$.

Note: Equation (11.28) can also be written in $do(x)$ notation as

$$E(Y_x|e) = E(Y|e) + E(Y|do(x)) - E[Y|do(X = E(X|e))].\qquad(11.30)$$

Proof:

(With help from Ilya Shpitser)

Assume, without loss of generality, that we are dealing with a zero-mean model. Since the model is linear, we can write the relation between X and Y as:

$$Y = TX + I + U,\qquad(11.31)$$

where T is the total effect of X on Y, given in (11.29); I represents terms containing other variables in the model, nondescendants of X; and U represents exogenous variables.

It is always possible to bring the function determining Y into the form (11.31) by recursively substituting the functions for each r.h.s. variable that has X as an ancestor, and grouping all the X terms together to form TX. Clearly, T is the Wright-rule sum of the path costs originating from X and ending in Y (Wright 1921).

From (11.31) we can write:

$$Y_x = Tx + I + U,\qquad(11.32)$$

since I and U are not affected by the hypothetical change from $X = x$, and, moreover,

$$E(Y_x|e) = Tx + E(I + U|e),\qquad(11.33)$$

since x is a constant.

The last term in (11.33) can be evaluated by taking expectations on both sides of (11.31), giving:

$$E(I + U|e) = E(Y|e) - TE(X|e),\qquad(11.34)$$

which, substituted into (11.33), yields

$$E(Y_x|e) = Tx + E(Y|e) - E(X|e)\qquad(11.35)$$

and proves our target formula (11.28). □

Three special cases of e are worth noting:

Example 11.7.1 $e : X = x', Y = y'$ (the linear equivalent of the probability of causation, Chapter 9). From (11.28) we obtain directly

$$E(Y_x \mid Y = y', X = x') = y' + T(x - x').$$

This is intuitively compelling. The hypothetical expectation of Y is simply the observed value, y', plus the anticipated change in Y due to the change $x - x'$ in X.

Example 11.7.2 $e : X = x'$ (the effect of treatment on the treated, Chapter 8.2.5).

$$\begin{aligned} E(Y_x \mid X = x') &= E(Y|x') + T(x - x') \\ &= rx' + T(x - x') \\ &= rx' + E(Y|do(x)) - E(Y|do(x')), \end{aligned}\qquad(11.36)$$

where r is the regression coefficient of Y on X.

Example 11.7.3 $e : Y = y'$ (e.g., the expected income Y of those who currently earn $Y = y'$ if we were to mandate x hours of training each month).

$$E(Y_x \mid Y = y') = y' + T[x - E(X \mid y')]$$
$$= y' + E(Y \mid do(x)) - E[Y \mid do(X = r'y')], \qquad (11.37)$$

where r' is the regression coefficient of X on Y.

Example 11.7.4 Consider the nonrecursive price-demand model of p. 215, equations (7.9)–(7.10):

$$q = b_1 p + d_1 i + u_1$$
$$p = b_2 q + d_2 w + u_2. \qquad (11.38)$$

Our counterfactual problem (p. 216) reads: Given that the current price is $P = p_0$, what would be the expected value of the demand Q if we were to control the price at $P = p_1$?

Making the correspondence $P = X, Q = Y, e = \{P = p_0, i, w\}$, we see that this problem is identical to Example 11.7.2 above (effect of treatment on the treated), subject to conditioning on i and w. Hence, since $T = b_1$, we can immediately write

$$E(Q_{p_1} \mid p_0, i, w) = E(Y \mid p_0, i, w) + b_1(p_1 - p_0)$$
$$= r_p p_0 + r_i i + r_w w + b_1(p_1 - p_0), \qquad (11.39)$$

where r_p, r_i, and r_w are the coefficients of P, i and w, respectively, in the regression of Q on P, i, and w.

Equation (11.39) replaces equation (7.17) on page 217. Note that the parameters of the price equation, $p = b_2 q + d_2 w + u_2$, enter (11.39) only via the regression coefficients. Thus, they need not be calculated explicitly in cases where they are estimated directly by least square.

> **Remark:** Example 11.7.1 is not really surprising; we know that the probability of causation is empirically identifiable under the assumption of monotonicity (p. 293). But examples 11.7.2 and 11.7.3 trigger the following conjecture:

Conjecture:

Any counterfactual query of the form $P(Y_x \mid e)$ is empirically identifiable when Y is monotonic relative to X.

It is good to end on a challenging note.

11.7.2   The Meaning of Counterfactuals
Question to Author:

I have a hard time understanding what counterfactuals are actually useful for. To me, they seem to be answering the wrong question. In your book, you give at least a couple of different reasons for when one would need the answer to a counterfactual question, so let me tackle these separately:

1.   Legal questions of responsibility. From your text, I infer that the American legal system says that a defendant is guilty if he or she caused the plaintiff's

misfortune. But in my mind, the law is clearly flawed. Responsibility should rest with the predicted outcome of the defendant's action, not with what actually happened. A doctor should not be held responsible if he administers, for a serious disease, a drug which cures 99.99999% of the population but kills 0.00001%, even if he was unlucky and his patient died. If the law is based on the counterfactual notion of responsibility then the law is seriously flawed, in my mind.

2. The use of context in decision making. On p. 217, you write "At this point, it is worth emphasizing that the problem of computing counterfactual expectations is not an academic exercise; it represents in fact the typical case in almost every decision-making situation." I agree that context is important in decision making, but do not agree that we need to answer counterfactual questions.

   In decision making, the thing we want to estimate is $P(future \mid do(action), see(context))$. This is of course a regular do-probability, not a counterfactual query. So why do we need to compute counterfactuals?

3. In the latter part of your book, you use counterfactuals to define concepts such as 'the cause of X' or 'necessary and sufficient cause of Y'. Again, I can understand that it is tempting to mathematically define such concepts since they are in use in everyday language, but I do not think that this is generally very helpful. Why do we need to know 'the cause' of a particular event? Yes, we are interested in knowing 'causes' of events in the sense that they allows us to predict the future, but this is again a case of point (2) above.

   To put it in the most simplified form, my argument is the following: Regardless of whether we represent individuals, businesses, organizations, or government, we are constantly faced with decisions of how to act (and these are the only decisions we have!). What we want to know is, what will likely happen if we act in particular ways. So what we want to know is $P(future \mid do(action), see(context))$. We do not want or need the answers to counterfactuals.

Where does my reasoning go wrong?

Author's Reply:

1. Your first question doubts the wisdom of using single-event probabilities, rather than population probabilities, in deciding legal responsibility. Suppose there is a small percentage of patients who are allergic to a given drug, and the manufacturer nevertheless distributes the drug with no warning about possible allergic reactions. Wouldn't we agree that when an allergic patient dies he is entitled to compensation? Normally, drug makers take insurance for those exceptional cases, rather than submit the entire population to expensive tests prior to taking the drug – it pays economically. The physician, of course, is exonerated from guilt, for he/she just followed accepted practice. But the law makes sure that someone pays for the injury if one can prove that, counterfactually, the specific death in question would not have occurred had the patient not taken the drug.

2. Your second question deals with decisions that are conditioned on the results of observations. Or, as you put it: "In decision making, the thing we want to estimate is $P(future \mid do(action), see(context))$."

The problem is that, to make this prediction properly, we need a sequential, time-indexed model, where "future" variables are clearly distinct from the "context" variables. Often, however, we are given only a static model, in which the "context" variable are shown as *consequences* of the "action," and then using the expression $P(y \mid do(x), z)$ would be inappropriate; it represents the probability of $Y = y$ given that we do $X = x$ and *later* observe $Z = z$, while what is needed is the probability of $Y = y$ given that we first observe $Z = z$ and then do $X = x$. Counterfactuals give us a way of expressing the desired probability, by writing

$$P = P(y_x \mid z),$$

which stands for the probability that $Y = y$ would occur had X been x, given that $Z = z$ is currently observed. Note that $P = P(y_x \mid z) = P(y \mid do(x), z)$ if Z is a non-descendant of X in the static model.

An example will be instructive. Suppose an engineer draws a circuit diagram M containing a chain of gates $X \rightarrow Y \rightarrow Z$. At time t_1 we observe $Z(t_1)$, and we want to know the causal effect of $X(t_2)$ on $Y(t_3)$ conditioned on $Z(t_1)$. We can do this exercise through *do*-calculus, with all the necessary time indices, if we replicate the model M and assign each time slot t_i, a model M_i, showing the relationships among $X(t_i)$, $Y(t_i)$, $Z(t_i)$ as well as to previous variables, then compute $P(y(t_3) \mid do(x(t_2)), z(t_1))$. But we can do it much better in the static model, using the counterfactuals $P(Y_x = y \mid z)$. The static model saves us miles and miles of drawing the sequential model equivalent, and counterfactuals enable us to take advantage of this savings. It is an admirable invention. Once can argue, of course, that if counterfactual claims are merely "conversational shorthand" (p. 118) for scientific predictions in the sequential model equivalent, then they are not needed at all. But this would stand contrary to the spirit of scientific methods, where symbolic economy plays a crucial role. In principle, multiplication is not needed in algebra – we can live with addition alone and add a number to itself as many times as needed. But science would not have advanced very far without multiplication. Same with counterfactuals.

3.   In the final analysis, the reason we want to know "causes of events" is indeed to allow us to better predict and better act in the future. But we do not always know in advance under what future circumstances our knowledge will be summoned to help in making decisions. The reason we need to know the cause of a specific accident may be multifaceted: to warn the public against the use of similar vehicles, to improve maintenance procedures of certain equipment, to calculate the cost of taking remedial action, to cause similar accidents in enemy territories, and many more. Each of these applications may demand different information from our model, so "causes of events" may be a useful way of registering our experience so as to amortize it over the wide spectrum of possible applications.

11.7.3   *d*-Separation of Counterfactuals
Question to Author:

I am trying to generalize the twin network method of Figure 7.3 to cases where counterfactuals involve more than two possible worlds. Consider the causal model $X \rightarrow Z \rightarrow Y$,

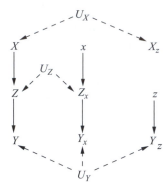

Figure 11.18   Triple network drawn to test equation (11.40).

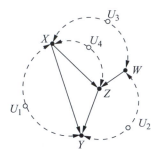

Figure 11.19   Graph in which $Y_{xz} \perp\!\!\!\perp Z_{x*} \mid W$ holds by virtue of the equality $W_{x*} = W$.

and assume we wish to test whether the assertion

$$Y_x \perp\!\!\!\perp X \mid Y_z, Z_x, Y \qquad\qquad (11.40)$$

is true in the model. I would naively construct the following "triple network":

The left part corresponds to a world in which no intervention is imposed; the middle part to a world in which $do(X = x)$ is imposed; and in the right part, $do(Z = z)$ is imposed. In this network (11.40) does not follow from d-separation, since the path from Y_x to X is open by conditioning on Y. Is there anything wrong with this way of generalizing the twin network method?

Author's Reply (with Ilya Shpitser):

Your generalization of the twin network to more than two worlds is correct, and so is your conclusion; Y_x is not independent of X given Y_z, Z_x, Y. In fact, a recent paper (Shpitser and Pearl 2007) formulates this graphical representation of counterfactuals in multiple worlds under the rubric "counterfactual graphs."

Follow-up Question to Author:

Your answer helped me understand the twin network method, as well as the one used in the paper "Direct and Indirect Effects" (Pearl 2001c), but it raises a new question: why does $Y_{xz} \perp\!\!\!\perp Z_{x*} \mid W$ hold in the graph of Figure 11.19, as claimed in (Pearl 2001)? If one draws a "triple network," it is obvious that the path between Y_{xz} and Z_{x*} is not blocked by W; the latter is situated in a totally different world.

Author's Second Reply:

The independence $Y_{xz} \perp\!\!\!\perp Z_{x*} \mid W$ actually holds in the graph shown in Figure 11.19.

This is because Y_{xz} is separated from Z_{x*} by the variable W_{x*}, in the "triple network" that you mentioned. The license to replace W with W_{x*} is obtained from rule 3 of *do*-calculus, which implies $W_{x*} = W$, since X is a nondescendant of X. This points to an important refinement needed in the twin network generalization: causal axioms may entail certain equality constraints among seemingly distinct counterfactual variables, and these hidden equalities need to be considered when we apply *d*-separation to counterfactual networks. A systematic way of encoding and managing these equalities is presented in Shpitser and Pearl (2007).

11.8   INSTRUMENTAL VARIABLES AND NONCOMPLIANCE

11.8.1   Tight Bounds under Noncompliance (Question to Author)

I am referring to the way you improved Manski's bounds on treatment effects when we have imperfect compliance. Which information does your approach exploit that the one by Manski does not? What is the intuition behind it?

Author's Reply:

We used the same information and same assumptions as Manski, and we derived the tight bounds using linear-programming analysis in the 16-dimensional space defined by the canonical partition of U (Balke and Pearl 1994a, 1995a). Manski, evidently, either did not aim at getting tight bounds, or was not aware of the power of partitioning U into its equivalence classes. Recall, this partition was unknown to economists before Frangakis and Rubin (2002) popularized it, under the rubric "principal stratification."

Manski's bounds, as I state on page 269, are tight under certain conditions, e.g., no contrarians. This means that one can get narrower bounds *only* when there are contrarians in the population, as in the examples discussed in Pearl (1995b). It is shown there how data representing the presence of contrarians can provide enough information to make the bounds collapse to a point. That article also gives an intuitive explanation of how this can happen.

It is important to mention at this point that the canonical partition conception, coupled with the linear programming method developed in Balke and Pearl (1994a, 1995a,b), has turned into a powerful analytical tool in a variety of applications. Tian and Pearl (2000) applied it to bound probabilities of causation; Kaufman et al. (2005) and Cai et al. (2008) used it to bound direct effects in the presence of confounded mediation, and, similarly, Imai et al. (2008) used it to bound natural direct and indirect effects. The closed-form expressions derived by this method enable researchers to assess what features of the distribution are critical for narrowing the widths of the bounds.

Rubin (2004), in an independent exploration, attempted to apply canonical partitions to the analysis of direct and indirect effects within the traditional potential-outcome framework but, lacking the graphical and structural perspectives, was led to conclude that such effects are "ill-defined" and "more deceptive than helpful." I believe readers of this book, guided by the structural roots of potential-outcome analysis, will reach more positive conclusions (see Sections 4.5 and 11.4.2).

11.9   MORE ON PROBABILITIES OF CAUSATION

Looking back, eight years past the first publication of *Causality*, I consider the results obtained in Chapter 9 to be a triumph of counterfactual analysis, dispelling all fears and trepidations that some researchers express concerning the empirical content of counterfactuals (Dawid 2000; Pearl 2000). It demonstrates that a quantity PN which at first glance appears to be hypothetical, ill-defined, untestable, and hence unworthy of scientific analysis is nevertheless definable, testable, and, in certain cases, even identifiable. Moreover, the fact that, under certain combination of data, and making no assumptions whatsoever, an important legal claim such as "the plaintiff would be alive had he not taken the drug" can be ascertained *with probability one* is truly astonishing.

11.9.1   Is "Guilty with Probability One" Ever Possible?

I have presented the example of Section 9.3.4 in dozens of talks at conferences and universities, and, invariably, the result PN ≥ 1 of equation (9.53) meets with universal disbelief; how can we determine, from frequency data, that the defendant is guilty with probability one, i.e., that Mr. A would *definitely* be alive today had he not taken the drug. Professor Stephen Fienberg attended two of these talks, and twice shook his head with: "It can't be true." To me, this reaction constitutes a solid proof that counterfactual analysis can yield nontrivial results, and hence that it is real, meaningful, and useful; metaphysical analysis would not evoke such resistance.

What causes people to disbelieve this result are three puzzling aspects of the problem:

1. that a hypothetical, generally untestable quantity can be ascertained with probability one under certain conditions;

2. that frequency tables which, individually, do not reveal a substantial effect of the drug imply a perfect susceptibility to the drug when taken together; and

3. that a property of an untested individual can be assigned a probability one, on the basis of data taken from a sampled population.

The first puzzle is not really surprising for students of science who take seriously the benefits of logic and mathematics. Once we give a quantity formal semantics we essentially define its relation to the data, and it is not inconceivable that data obtained under certain conditions would sufficiently constrain that quantity, to a point where it can be determined exactly.

This benefit of formal counterfactual analysis can in fact be demonstrated in a much simpler example. Consider the effect of treatment on the treated, $P(y_{x'} \mid x)$ (Section 8.2.5). In words, $P(y_{x'} \mid x)$ stands for the chances that a treated patient, $X = x$, would survive for a time period $Y = y$ had he/she not been treated ($X = x'$). This counterfactual quantity too seems to defy empirical measurement, because we can never rerun history and deny treatment for those who received it. Yet, for binary X, we can write

$$P(y_{x'}) = P(y_{x'} \mid x')P(x') + P(y_{x'} \mid x)P(x)$$

and derive

$$P(y_{x'} \mid x) = [P(y_{x'}) - P(y, x')]/P(x).$$

In other words, $P(y_{x'} \mid x)$ is reducible to empirically estimable quantities; $P(y_{x'}) = P(y \mid do(x'))$ is estimable in experimental studies and the other quantities in observational studies. Moreover, if data support the equality $P(y_{x'}) = P(y, x')$, we can safely conclude that a treated patient would have zero chance of survival had the treatment not been taken. Those who mistrust counterfactual analysis a priori, as a calculus dealing with undefined quantities, would never enjoy the discovery that some of those quantities are empirically definable. Logic, when gracious, can rerun history for us.

The second puzzle was given intuitive explanation in the paragraph following equation (9.54).

The third puzzle is the one that gives most people a shock of disbelief. For a statistician, in particular, it is a rare case to be able to say anything certain about a specific individual who was not tested directly. This emanates from two factors. First, statisticians normally deal with finite samples, the variability of which rules out certainty in any claim, not merely about an individual but also about any property of the underlying distribution. This factor, however, should not enter into our discussion, for we have been assuming infinite samples throughout. (Readers should imagine that the numbers in Table 9.2 stand for millions.)

The second factor emanates from the fact that, even when we know a distribution precisely, we cannot assign a definite probabilistic estimate to a property of a specific individual drawn from that distribution. The reason is, so the argument goes, that we never know, let alone measure, all the anatomical and psychological variables that determine an individual's behavior, and, even if we knew, we would not be able to represent them in the crude categories provided by the distribution at hand. Thus, because of this inherent crudeness, the sentence "Mr. A would be dead" can never be assigned a probability of one (or, in fact, any definite probability).

This argument, advanced by Freedman and Stark (1999), is incompatible with the way probability statements are used in ordinary discourse, for it implies that every probability statement about an individual must be a statement about a restricted subpopulation that shares *all* the individual's characteristics. Taken to the extreme, such a restrictive interpretation would insist on characterizing the plaintiff in minute detail, and would reduce PN to zero or one when all relevant details were accounted for. It is inconceivable that this interpretation underlies the intent of judicial standards. By using the wording "more probable than not," lawmakers have instructed us to ignore specific features for which data is not available, and to base our determination on the most specific features for which reliable data is available. In our example, two properties of Mr. A were noted: (1) that he died and (2) that he chose to use the drug; these were properly taken into account in bounding PN. If additional properties of Mr. A become known, and deemed relevant (e.g., that he had red hair, or was left-handed), these too could, in principle, be accounted for by restricting the analysis to data representing the appropriate subpopulations. However, in the absence of such data, and knowing in advance that we will never be able to match *all* the idiosyncratic properties of Mr. A, the lawmakers' specification must be interpreted relative to the properties at hand.

11.9.2   Tightening the Bounds on Probabilities of Causation

Systematic work by Jin Tian (Tian and Pearl 2000) has improved on the results of Chapter 9 in several ways. First, Tian showed that for most of these results, the assumption of strong exogeneity, equation (9.10), can be replaced by weak exogeneity:

$$Y_x \perp\!\!\!\perp X \quad \text{and} \quad Y_{x'} \perp\!\!\!\perp X.$$

Second, the estimands obtained under the assumption of monotonicity (Definition 9.2.13) constitute *lower bounds* when monotonicity is not assumed. Finally, the bounds derived in Chapter 9 are *sharp*, that is, they cannot be improved without strengthening the assumptions.

Of particular interest are the bounds obtained when data from both experimental and nonexperimental studies are available, and no other assumptions are made. These read:

$$\max \begin{Bmatrix} 0 \\ P(y_x) - P(y_{x'}) \\ P(y) - P(y_{x'}) \\ P(y_x) - P(y) \end{Bmatrix} \leq PNS \leq \min \begin{Bmatrix} P(y_x) \\ P(y'_{x'}) \\ P(x, y) + P(x', y') \\ P(y_x) - P(y_{x'}) + P(x, y') + P(x', y) \end{Bmatrix} \tag{11.41}$$

$$\max \begin{Bmatrix} 0 \\ \dfrac{P(y) - P(y_{x'})}{P(x, y)} \end{Bmatrix} \leq PN \leq \min \begin{Bmatrix} 1 \\ \dfrac{P(y'_{x'}) - P(x', y')}{P(x, y)} \end{Bmatrix} \tag{11.42}$$

$$\max \begin{Bmatrix} 0 \\ \dfrac{P(y_x) - P(y)}{P(x', y')} \end{Bmatrix} \leq PS \leq \min \begin{Bmatrix} 1 \\ \dfrac{P(y_x) - P(x, y)}{P(x', y')} \end{Bmatrix} \tag{11.43}$$

It is worth noting that, in drug-related litigation, it is not uncommon to obtain data from both experimental and observational studies. The former is usually available at the manufacturer or the agency that approved the drug for distribution (e.g., FDA), while the latter is easy to obtain by random surveys of the population. In such cases, the standard lower bound used by epidemiologists to establish legal responsibility, the Excess Risk Ratio (equation (9.22)), can be substantially improved by using the lower bound of equation (11.42). Likewise, the upper bound of equation (11.42) can be used to exonerate the drug maker from legal responsibility. Cai and Kuroki (2006) analyzed the statistical properties of PNS, PN, and PS.

Also noteworthy is the fact that in tasks of abduction, i.e., reasoning to the best explanation, PN is the most reasonable measure to use in deciding among competing explanations of the event $Y = y$. In such applications, the bounds given in (11.42) can be computed from a causal Bayesian network model, where $P(y_x)$ and $P(y_{x'})$ are computable through equation (9.48).

Acknowledgments

I thank all readers who sent in questions and contributed in this way to this chapter. These include: David Bessler, Nimrod Megiddo, David Kenny, Keith A. Markus, Jos

Lehmann, Dennis Lindley, Jacques A. Hagenaars, Jonathan Wilson, Stan Mulaik, Bill Shipley, Nozer D. Singpurwalla, Les Hayduk, Erich Battistin, Sampsa Hautaniemi, Melanie Wall, Susan Scott, Patrik Hoyer, Joseph Halpern, Phil Dawid, Sander Greenland, Arvid Sjolander, Eliezer S. Yudkowsky, UCLA students in CS262Z (Seminar in Causality, Spring 2006), and the UCLA Epidemiology Class – EPIDEM 200C.

I similarly thank all reviewers of the first edition of the book and the editors who helped bring these reviews to the attention of their readers. These include: *Choice* (Byerly 2000), *Structural Equation Modeling* (Shipley 2000a), *Chance* (McDonald 2001), *Technometrics* (Zelterman 2001), *Mathematical Reviews* (Lawry 2001), *Politische Vierteljahrsschrlft* (Didelez and Pigeot 2001), *Technological Forecasting & Social Change* (Payson 2001), *British Journal for the Philosophy of Science* (Gillics 2001), *Human Biology* (Chakraborty 2001), *The Philosophical Review* (Hitchcock 2001), *Intelligence* (O'Rourke 2001), *Journal of Marketing Research* (Rigdon 2002), *Tijdschrlft Voor* (Decock 2002), *Psychometrika* (McDonald 2002b), *International Statistical Review* (Lindley 2002), *Journal of Economic Methodology* (Leroy 2002), *Statistics in Medicine* (Didelez 2002), *Journal of Economic Literature* (Swanson 2002), *Journal of Mathematical Psychology* (Butler 2002), *IIE Transactions* (Gursoy 2002), *Royal Economic Society* (Hoover 2003), *Econometric Theory* (Neuberg 2003), *Economica* (Abbring 2003), *Economics and Philosophy* (Woodward 2003), *Sociological Methods and Research* (Morgan 2004), *Review of Social Economy* (Boumans 2004), *Journal of the American Statistical Association* (Hadlock 2005), and *Artificial Intelligence* (Kyburg 2005).

Thanks also go to the contributors to UCLA's *Causality* blog (http://www.mii.ucla.edu/causality/) and to William Hsu, the blog's curator.

A special word of appreciation and admiration goes to Dennis Lindley, who went to the trouble of studying my ideas from first principles, and allowed me to conclude that readers from the statistics-based sciences would benefit from this book. I am fortunate that our paths have crossed and that I was able to witness the intellect, curiosity, and integrity of a true gentleman.

This chapter could not have been written without the support and insight of Jin Tian, Avin Chen, Carlo Brito, Blai Bonet, Mark Hopkins, Ilya Shpitser, Azaria Paz, Manabu Kuroki, Zhihong Cai, Kaoru Mulvihill, and all members of the Cognitive System Laboratory at UCLA who, in the past six years, continued to explore the green pastures of causation while I was summoned to mend a world that took the life of my son Daniel (murdered by extremists in Karachi, Pakistan, 2002). These years have convinced me, beyond a shadow of doubt, that the forces of reason and enlightenment will win over fanaticism and inhumanity.

Finally, I dedicate this edition to my wife, Ruth, for her strength, love, and guidance throughout our ordeal, to my daughters, Tamara and Michelle, and my grandchildren, Leora, Torri, Adam, Ari, and Evan for being with me through this journey and making it meaningful and purposeful.

The Art and Science of Cause and Effect

*A public lecture delivered November 1996 as part of
the UCLA Faculty Research Lectureship Program*

The topic of this lecture is causality – namely, our awareness of what causes what in the world and why it matters.

Though it is basic to human thought, causality is a notion shrouded in mystery, controversy, and caution, because scientists and philosophers have had difficulties defining when one event *truly causes* another.

We all understand that the rooster's crow does not cause the sun to rise, but even this simple fact cannot easily be translated into a mathematical equation.

Today, I would like to share with you a set of ideas which I have found very useful in studying phenomena of this kind. These ideas have led to practical tools that I hope you will find useful on your next encounter with cause and effect.

It is hard to imagine anyone here who is *not* dealing with cause and effect.

Whether you are evaluating the impact of bilingual education programs or running an experiment on how mice distinguish food from danger or speculating about why Julius Caesar crossed the Rubicon or diagnosing a patient or predicting who will win the presidential election, you are dealing with a tangled web of cause–effect considerations.

The story that I am about to tell is aimed at helping researchers deal with the complexities of such considerations, and to clarify their meaning.

This lecture is divided into three parts.

I begin with a brief historical sketch of the difficulties that various disciplines have had with causation.

Next I outline the ideas that reduce or eliminate several of these historical difficulties.

Finally, in honor of my engineering background, I will show how these ideas lead to simple practical tools, which will be demonstrated in the areas of statistics and social science.

In the beginning, as far as we can tell, causality was not problematic.

The urge to ask *why* and the capacity to find causal explanations came very early in human development.

The Bible, for example, tells us that just a few hours after tasting from the tree of knowledge, Adam is already an expert in causal arguments.

When God asks: "Did you eat from that tree?"

This is what Adam replies: "The woman whom you gave to be with me, She handed me the fruit from the tree; and I ate."

Eve is just as skillful: "The serpent deceived me, and I ate."

The thing to notice about this story is that God did not ask for explanation, only for the facts – it was Adam who felt the need to explain. The message is clear: causal explanation is a man-made concept.

Another interesting point about the story: explanations are used exclusively for passing responsibilities.

Indeed, for thousands of years explanations had no other function. Therefore, only Gods, people, and animals could cause things to happen, not objects, events, or physical processes.

Natural events entered into causal explanations much later because, in the ancient world, events were simply *predetermined*.

Storms and earthquakes were *controlled* by the angry gods [slide 2] and could not in themselves assume causal responsibility for the consequences.

Even an erratic and unpredictable event such as the roll of a die [3] was not considered a *chance* event but rather a divine message demanding proper interpretation.

One such message gave the prophet Jonah the scare of his life when he was identified as God's renegade and was thrown overboard [4].

Quoting from the book of Jonah: "And the sailors said: 'Come and let us cast lots to find out who is to blame for this ordeal.' So they cast lots and the lot fell on Jonah."

Obviously, on this luxury Phoenician cruiser, "casting lots" was used not for recreation but for communication – a one-way modem for processing messages of vital importance.

In summary, the agents of causal forces in the ancient world were either deities, who cause things to happen for a purpose, or human beings and animals, who possess free will, for which they are punished and rewarded.

This notion of causation was naive, but clear and unproblematic.

The problems began, as usual, with engineering; when machines had to be constructed to do useful jobs [5].

As engineers grew ambitious, they decided that the earth, too, can be moved [6], but not with a single lever.

Systems consisting of many pulleys and wheels [7], one driving another, were needed for projects of such magnitude.

And, once people started building multistage systems, an interesting thing happened to causality – *physical objects began acquiring causal character*.

When a system like that broke down, it was futile to blame God or the operator – instead, a broken rope or a rusty pulley were more useful

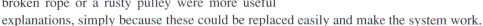

explanations, simply because these could be replaced easily and make the system work.

At that point in history, Gods and humans ceased to be the sole agents of causal forces – lifeless objects and processes became partners in responsibility.

A wheel turned and stopped *because* the wheel preceding it turned and stopped – the human operator became secondary.

Not surprisingly, these new agents of causation took on some of the characteristics of their predecessors – Gods and humans.

Natural objects became not only carriers of credit and blame but also carriers of force, will, and even purpose.

Aristotle regarded explanation in terms of a *purpose* to be the only complete and satisfactory explanation for why a thing is what it is.

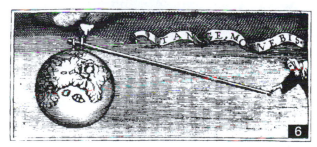

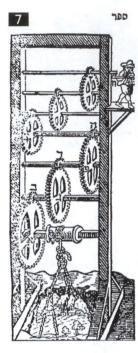

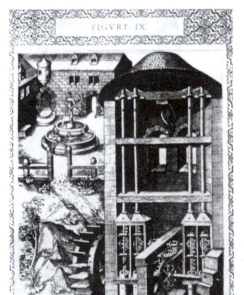

He even called it a *final cause* – namely, the final aim of scientific inquiry.

From that point on, causality served a dual role: *causes* were the targets of credit and blame on one hand and the carriers of physical flow of control on the other.

This duality survived in relative tranquility [8] until about the time of the Renaissance, when it encountered conceptual difficulties.

What happened can be seen on the title page [9] of Recordes's book "The Castle of Knowledge," the first science book in English, published in 1575.

The wheel of fortune is turned, not by the wisdom of God, but by the ignorance of man.

And, as God's role as the final cause was taken over by human knowledge, the whole notion of causal explanation came under attack.

The erosion started with the work of Galileo [10].

Most of us know Galileo as the man who was brought before the inquisition and imprisoned [11] for defending the heliocentric theory of the world.

But while all that was going on, Galileo also managed to quietly engineer the most profound revolution that science has ever known.

This revolution, expounded in his 1638 book "Discorsi" [12], published in Leyden, far from Rome, consists of two maxims:

One, description first, explanation second – that is, the "how" precedes the "why"; and

Two, description is carried out in the language of mathematics; namely, equations.

Ask not, said Galileo, whether an object falls because it is pulled from below or pushed from above.

Ask how well you can predict the time it takes for the object to travel a certain distance, and how that time will vary from object to object and as the angle of the track changes.

Moreover, said Galileo, do not attempt to answer such questions in the qualitative and slippery nuances of human language; say it in the form of mathematical equations [13].

It is hard for us to appreciate today how strange that idea sounded in 1638, barely 50 years after the introduction of algebraic notation by Vieta. To proclaim algebra the *universal* language of science would sound today like proclaiming Esperanto the language of economics.

Why would Nature agree to speak algebra? Of all languages?

But you can't argue with success.

The distance traveled by an object turned out indeed to be proportional to the square of the time.

Even more successful than predicting outcomes of experiments were the computational aspects of algebraic equations.

They enabled engineers, for the first time in history, to ask "how to" questions in addition to "what if" questions.

In addition to asking: "What if we narrow the beam, will it carry the load?", they began to ask more difficult questions: "How to shape the beam so that it *will* carry the load?" [14]

This was made possible by the availability of methods for solving equations.

The algebraic machinery does not discriminate among variables; instead of predicting behavior in terms of parameters, we can turn things around and solve for the parameters in terms of the desired behavior.

Let us concentrate now on Galileo's first maxim – "description first, explanation second" – because that idea was taken very seriously by the scientists and changed the character of science from speculative to empirical.

Physics became flooded with empirical laws that were extremely useful.

Snell's law [15], Hooke's law, Ohm's law, and Joule's law are examples of purely empirical generalizations that were discovered and used long before they were explained by more fundamental principles.

Philosophers, however, were reluctant to give up the idea of causal explanation and continued to search for the origin and justification of those successful Galilean equations.

For example, Descartes ascribed cause to *eternal truth.*

Liebniz made cause a self-evident logical law.

Finally, about one hundred years after Galileo, a Scottish philosopher by the name of David Hume [16] carried Galileo's first maxim to an extreme [17].

Hume argued convincingly that the *why* is not merely second to the *how*, but that the *why* is totally superfluous as it is subsumed by the *how*.

On page 156 of Hume's "Treatise of Human Nature" [18], we find the paragraph that shook up causation so thoroughly that it has not recovered to this day.

I always get a kick reading it: "Thus we remember to have seen that species of object we call *flame*, and to have felt that species of sensation we call *heat*. We likewise call to mind their constant conjunction in all past instances. Without any farther ceremony, we call the one *cause* and the other *effect*, and infer the existence of the one from that of the other."

Thus, causal connections according to Hume are the product of observations. Causation is a learnable habit of the mind, almost as fictional as optical illusions and as transitory as Pavlov's conditioning.

DISCORSI
E
DIMOSTRAZIONI
MATEMATICHE,
intorno à due nuoue scienze
Attenenti alla
MECANICA & i MOVIMENTI LOCALI;
del Signor
GALILEO GALILEI LINCEO,
Filosofo e Matematico primario del Sereniſſimo
Grand Duca di Toſcana.
Con vna Appendice del centro di grauità d'alcuni Solidi.

IN LEIDA,
Appreſſo gli Elſevirii. M. D. C. XXXVIII.

It is hard to believe that Hume was not aware of the difficulties inherent in his proposed recipe.

He knew quite well that the rooster crow *stands* in constant conjunction to the sunrise, yet it does not *cause* the sun to rise.

He knew that the barometer reading *stands* in constant conjunction to the rain but does not *cause* the rain.

Today these difficulties fall under the rubric of *spurious correlations,* namely "correlations that do not imply causation."

Now, taking Hume's dictum that all knowledge comes from experience encoded in the mind as correlation, and our observation that correlation does not imply causation, we are led into our first riddle of causation: How do people *ever* acquire knowledge of *causation?*

We saw in the rooster example that regularity of succession is not sufficient; what *would* be sufficient?

What patterns of experience would justify calling a connection "causal"?

Moreover: What patterns of experience *convince* people that a connection is "causal"?

If the first riddle concerns the *learning* of causal connection, the second concerns its usage: What *difference* would it make if I told you that a certain connection is or is not causal?

Continuing our example, what difference would it make if I told you that the rooster does cause the sun to rise?

This may sound trivial.

The obvious answer is that knowing "what causes what" makes a big difference in how we act.

If the rooster's crow causes the sun to rise, we could make the night shorter by waking up our rooster earlier and making him crow – say, by telling him the latest rooster joke.

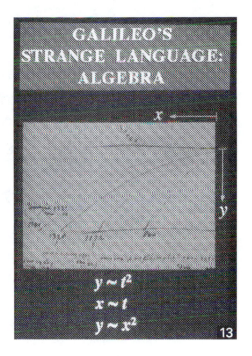

y ∼ t²
x ∼ t
y ∼ x²

But this riddle is *not* as trivial as it seems.

If causal information has an empirical meaning beyond regularity of succession, then that information should show up in the laws of physics.

But it does not!

The philosopher Bertrand Russell made this argument [19] in 1913:

"All philosophers," says Russell, "imagine that causation is one of the fundamental axioms of science, yet oddly enough, in advanced sciences, the word 'cause' never occurs.... The law of causality, I believe, is a relic of bygone age, surviving, like the monarchy, only because it is erroneously supposed to do no harm."

Another philosopher, Patrick Suppes, who argued for the importance of causality, noted that:

"There is scarcely an issue of 'Physical Review' that does not contain at least one article using either 'cause' or 'causality' in its title."

What we conclude from this exchange is that physicists talk, write, and think one way and formulate physics in another.

Such bilingual activity would be forgiven if causality was used merely as a convenient communication device – a shorthand for expressing complex patterns of physical relationships that would otherwise take many equations to write.

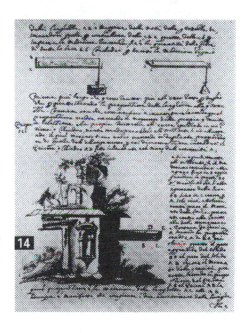

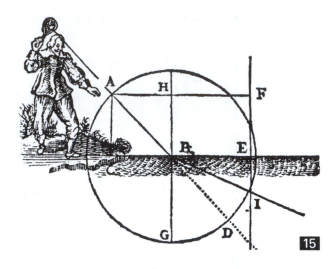

15

After all! Science is full of abbreviations: We use "multiply x by 5" instead of "add x to itself 5 times"; we say "density" instead of "the ratio of weight to volume."

Why pick on causality?

"Because causality is different," Lord Russell would argue. "It could not possibly be an abbreviation, because the laws of physics are all symmetrical, going both ways, while causal relations are unidirectional, going from cause to effect."

Take, for instance, Newton's law:

$$f = ma.$$

The rules of algebra permit us to write this law in a wild variety of syntactic forms, all meaning the same thing – that if we know any two of the three quantities, the third is determined.

Yet, in ordinary discourse we say that force causes acceleration – not that acceleration causes force, and we feel very strongly about this distinction.

Likewise, we say that the ratio f/a helps us *determine* the mass, not that it *causes* the mass.

Such distinctions are not supported by the equations of physics, and this leads us to ask whether the whole causal vocabulary is purely metaphysical, "surviving, like the monarchy ...".

Fortunately, very few physicists paid attention to Russell's enigma. They continued to write equations in the office and talk cause–effect in the cafeteria; with astonishing success they smashed the atom, invented the transistor and the laser.

The same is true for engineering.

But in another arena the tension could not go unnoticed, because in that arena the demand for distinguishing causal from other relationships was very explicit.

This arena is statistics.

The story begins with the discovery of correlation, about one hundred years ago.

16

Francis Galton [20], inventor of fingerprinting and cousin of Charles Darwin, quite understandably set out to prove that talent and virtue run in families.

Galton's investigations drove him to consider various ways of measuring how properties of one class of individuals or objects are related to those of another class.

In 1888, he measured the length of a person's forearm and the size of that person's head and asked to what degree one of these quantities can predict the other [21].

He stumbled upon the following discovery: If you plot one quantity against the other and scale the two axes properly, then the slope of the best-fit line has some nice mathematical properties. The slope is 1 only when one quantity can predict the other precisely; it is zero whenever the prediction is no better than a random guess; and, most remarkably, the slope is the same no matter if you plot X against Y or Y against X.

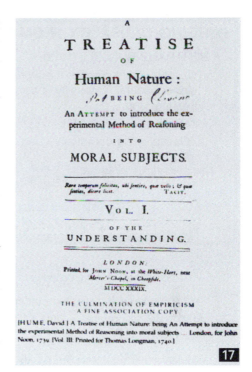

"It is easy to see," said Galton, "that co-relation must be the consequence of the variations of the two organs being partly due to common causes."

Here we have, for the first time, an objective measure of how two variables are "related" to each other, based strictly on the data, clear of human judgment or opinion.

Galton's discovery dazzled one of his disciples, Karl Pearson [22], now considered to be one of the founders of modern statistics.

Pearson was 30 years old at the time, an accomplished physicist and philosopher about to turn lawyer, and this is how he describes, 45 years later [23], his initial reaction to Galton's discovery:

"I felt like a buccaneer of Drake's days. ...

"I interpreted ... Galton to mean that there was a category broader than causation, namely correlation, of which causation was only the limit, and that this new conception of correlation brought psychology, anthropology, medicine, and sociology in large parts into the field of mathematical treatment."

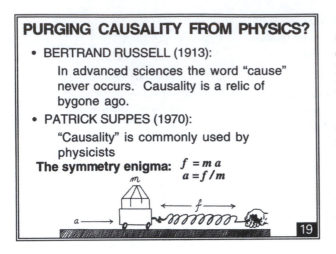

PURGING CAUSALITY FROM PHYSICS?

- BERTRAND RUSSELL (1913):

  In advanced sciences the word "cause" never occurs. Causality is a relic of bygone ago.

- PATRICK SUPPES (1970):

  "Causality" is commonly used by physicists

The symmetry enigma: $f = m\,a$
$a = f/m$

`19`

Now, Pearson has been described as a man "with the kind of drive and determination that took Hannibal over the Alps and Marco Polo to China."

When Pearson felt like a buccaneer, you can be sure he gets his bounty.

The year 1911 saw the publication of the third edition of his book "The Grammar of Science." It contained a new chapter titled "Contingency and Correlation – The Insufficiency of Causation," and this is what Pearson says in that chapter:

"Beyond such discarded fundamentals as 'matter' and 'force' lies still another fetish amidst the inscrutable arcana of modern science, namely, the category of cause and effect."

And what does Pearson substitute for the archaic category of cause and effect? You wouldn't believe your ears: *contingency tables* [24].

"Such a table is termed a contingency table, and the ultimate scientific statement of description of the relation between two things can always be thrown back upon such a contingency table. . . .

"Once the reader realizes the nature of such a table, he will have grasped the essence of the conception of association between cause and effect."

Thus, Pearson categorically denies the need for an independent concept of causal relation beyond correlation.

`20`

He held this view throughout his life and, accordingly, did not mention causation in *any* of his technical papers.

His crusade against animistic concepts such as "will" and "force" was so fierce and his rejection of determinism so absolute that he *exterminated* causation from statistics before it had a chance to take root.

It took another 25 years and another strong-willed person, Sir Ronald Fisher [25], for statisticians to formulate the randomized experiment – the only scientifically proven method of testing causal relations from data, and to this day, the one and only causal concept permitted in mainstream statistics.

And that is roughly where things stand today.

If we count the number of doctoral theses, research papers, or textbook pages written on causation, we get the impression that Pearson still rules statistics.

The "Encyclopedia of Statistical Science" devotes twelve pages to correlation but only two pages to causation – and spends one of those pages demonstrating that "correlation does not imply causation."

Let us hear what modern statisticians say about causality.

Philip Dawid, the current editor of "Biometrika" (the journal founded by Pearson), admits: "Causal inference is one of the most important, most subtle, and most neglected of all the problems of statistics."

Terry Speed, former president of the Biometric Society (whom you might remember as an expert witness at the O. J. Simpson murder trial), declares: "Considerations of causality should be treated as they have always been treated in statistics: preferably not at all but, if necessary, then with very great care."

Sir David Cox and Nanny Wermuth, in a book published just a few months ago, apologize as follows: "We did not in this book use the words *causal* or *causality*.... Our reason for caution is that it is rare that firm conclusions about causality can be drawn from one study."

This position of caution and avoidance has paralyzed many fields that look to statistics for guidance, especially economics and social science.

A leading social scientist stated in 1987: "It would be very healthy if more researchers abandon thinking of and using terms such as cause and effect."

Can this state of affairs be the work of just one person? Even a buccaneer like Pearson?

I doubt it.

But how else can we explain why statistics, the field that has given the world such powerful concepts as the testing of hypothesis and the

23

design of experiment, would give up so early on causation?

One obvious explanation is, of course, that causation is much harder to measure than correlation.

Correlations can be estimated directly in a single uncontrolled study, while causal conclusions require controlled experiments.

But this is too simplistic; statisticians are not easily deterred by difficulties, and children manage to learn cause–effect relations *without* running controlled experiments.

The answer, I believe, lies deeper, and it has to do with the official language of statistics – namely, the language of probability.

This may come as a surprise to some of you, but the word *cause* is not in the vocabulary of probability theory; we cannot express in the language of probabilities the sentence, *mud does not cause rain* – all we can say is that the two are mutually correlated or dependent – meaning that if we find one, we can expect the other.

Naturally, if we lack a language to express a certain concept explicitly, we can't expect to develop scientific activity around that concept.

Scientific development requires that knowledge be transferred reliably from one study to another and, as Galileo showed 350 years ago, such transference requires the precision and computational benefits of a formal language.

I will soon come back to discuss the importance of language and notation, but first I wish to conclude this historical survey with a tale from another field in which causation has had its share of difficulty.

This time it is computer science – the science of symbols – a field that is relatively new yet one that has placed a tremendous emphasis on language and notation and therefore may offer a useful perspective on the problem.

When researchers began to encode causal relationships using computers, the two riddles of causation were awakened with renewed vigor.

24 CONTINGENCY AND CORRELATION 159

B_1 occurs n_{p_1}, B_2 occurs n_{p_2} times, and so on. We thus are able to obtain a general distribution of B's for each class of A that we can form, and were we to go through the whole population, N, of A's in this manner we should obtain a table of the following kind :—

TYPE OF A OBSERVED

		A_1	A_2	A_3	...	...	A_p	...	...	...	Total.
TYPE OF B OBSERVED	B_1	n_{11}	n_{21}	n_{31}	...	...	n_{p1}	...	...	...	n_{a1}
	B_2	n_{12}	n_{22}	n_{32}	...	...	n_{p2}	...	...	...	n_{a2}
	B_3	n_{13}	n_{23}	n_{33}	...	...	n_{p3}	...	...	...	n_{a3}
	...	...	...	...	...	...	...	...	...	...	...
	...	...	...	...	...	...	...	...	...	...	...
	B_q	n_{1q}	n_{2q}	n_{3q}	...	...	n_{pq}	...	...	...	n_{aq}
	...	...	...	...	...	...	...	...	...	...	...
	...	...	...	...	...	...	...	...	...	...	...
	Total	n_{1b}	n_{2b}	n_{3b}	...	...	n_{pb}	...	...	...	N

Put yourself in the shoes of this robot [26] who is trying to make sense of what is going on in a kitchen or a laboratory.

Conceptually, the robot's problems are the same as those faced by an economist seeking to model the national debt or an epidemiologist attempting to understand the spread of a disease.

Our robot, economist, and epidemiologist all need to track down cause–effect relations from the environment, using limited actions and noisy observations.

This puts them right at Hume's first riddle of causation: *how*?

The second riddle of causation also plays a role in the robot's world.

Assume we wish to take a shortcut and teach our robot all we know about cause and effect in this room [27].

How should the robot organize and make use of this information?

Thus, the two philosophical riddles of causation are now translated into concrete and practical questions:

How should a robot acquire causal information through interaction with its environment? How should a robot process causal information received from its creator–programmer?

Again, the second riddle is not as trivial as it might seem. Lord Russell's warning that causal relations and physical equations are incompatible now surfaces as an apparent flaw in logic.

For example, when given the information, "If the grass is wet, then it rained" and "If we break this bottle, the grass will get wet," the computer will conclude "If we break this bottle, then it rained" [28].

The swiftness and specificity with which such programming bugs surface have made Artificial Intelligence programs an ideal laboratory for studying the fine print of causation.

This brings us to the second part of the lecture: how the second riddle of causation can be solved by combining equations with graphs, and how this solution makes the first riddle less formidable.

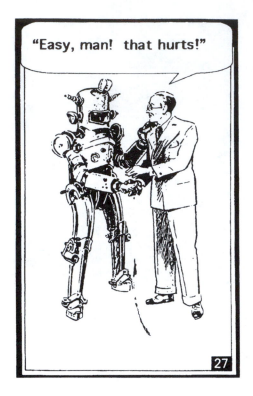

"Easy, man! that hurts!"

27

The overriding ideas in this solution are:

First – treating causation as a summary of behavior under interventions; and

Second – using equations and graphs as a mathematical language within which causal thoughts can be represented and manipulated.

And to put the two together, we need a *third* concept: treating interventions as a surgery over equations.

Let us start with an area that uses causation extensively and never had any trouble with it: engineering.

Here is an engineering drawing [29] of a circuit diagram that shows cause–effect relations among the signals in the circuit. The circuit consists of *and* gates and *or* gates, each performing some logical function between input and output. Let us examine this diagram closely, since its simplicity and familiarity are very deceiving. This diagram is, in fact, one of the greatest marvels of science. It is capable of conveying more information than millions of algebraic equations or probability functions or logical expressions. What makes this diagram so much more powerful is the ability to predict not merely how the circuit behaves under normal conditions but also how the circuit will behave under millions of *abnormal* conditions. For example, given this circuit diagram, we can easily tell what the output will be if some input changes from 0 to 1. This is normal and can easily be expressed by a simple input–output equation. Now comes the abnormal part. We can also tell what the output will be when we set Y to 0 (zero), or tie it to X, or change this *and* gate to an *or* gate, or when we perform any of the millions of combinations of these operations. The designer of this circuit did not anticipate or even consider such weird interventions, yet, miraculously, we can predict their consequences. How? Where does this representational power come from?

It comes from what early economists called *autonomy*. Namely, the gates in this diagram represent independent mechanisms – it is easy to change one without changing the other. The diagram takes advantage of this independence and

CAUSATION AS A PROGRAMMER'S NIGHTMARE

Input: 1. "If the grass is wet, then it rained"

            2. "If we break this bottle, the grass will get wet"

Output: "If we break this bottle, then it rained"

28

describes the normal functioning of the circuit *using precisely those building blocks that will remain unaltered under intervention.*

My colleagues from Boelter Hall are surely wondering why I stand here before you blathering about an engineering triviality as if it were the eighth wonder of the world. I have three reasons for doing this. First, I will try to show that there is a lot of unexploited wisdom in practices that engineers take for granted.

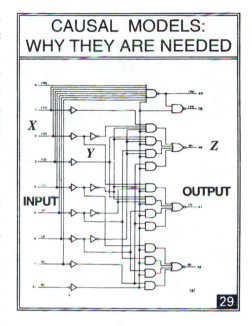

Second, I am trying to remind economists and social scientists of the benefits of this diagrammatic method. They have been using a similar method on and off for over 75 years, called structural equation modeling and path diagrams, but in recent years they have allowed algebraic convenience to suppress the diagrammatic representation, together with its benefits. Finally, these diagrams capture, in my opinion, the very essence of causation – the ability to predict the consequences of abnormal eventualities and new manipulations. In S.Wright's diagram [30], for example, it is possible to predict what coat pattern the guinea-pig litter is likely to have if we change environmental factors, shown here as input (*E*), or even genetic factors, shown as intermediate nodes between parents and offsprings (*H*). Such predictions cannot be made on the basis of algebraic or correlational analysis.

Viewing causality this way explains why scientists pursue causal explanations with such zeal and why attaining a causal model is accompanied by a sense of gaining "deep understanding" and "being in control."

Deep understanding [31] means knowing not merely how things behaved yesterday but also how things will behave under new hypothetical circumstances, control being one such circumstance. Interestingly, when we have such understanding we feel "in control" even if we have no practical way of controlling things. For example, we have no practical way to control celestial motion, and still the theory of gravitation gives us a feeling of understanding and control, because it provides a blueprint for hypothetical control. We can predict the effect on tidal waves of unexpected new events – say, the moon being hit by a meteor or the gravitational constant suddenly diminishing by a

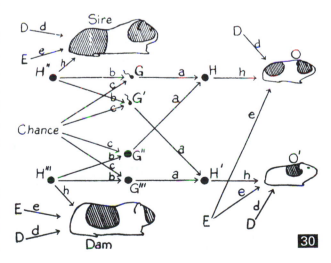

DEEP UNDERSTANDING = HOW THINGS WORK WHEN TAKEN APART

factor of 2 – and, just as important, the gravitational theory gives us the assurance that ordinary manipulation of earthly things will *not* control tidal waves. It is not surprising that causal models are viewed as the litmus test for distinguishing deliberate reasoning from reactive or instinctive response. Birds and monkeys may possibly be trained to perform complex tasks such as fixing a broken wire, but that requires trial-and-error training. Deliberate reasoners, on the other hand, can anticipate the consequences of new manipulations *without ever trying* those manipulations.

Let us magnify [32] a portion of the circuit diagram so that we can understand why the diagram can predict outcomes that equations cannot. Let us also switch from logical gates to linear equations (to make everyone here more comfortable), and assume we are dealing with a system containing just two components: a multiplier and an adder. The *multiplier* takes the input and multiplies it by a factor of 2; the *adder* takes its input and adds a 1 to it. The equations describing these two components are given here on the left.

But are these equations *equivalent* to the diagram on the right? Obviously not! If they were, then let us switch the variables around, and the resulting two equations should be equivalent to the circuit shown below. But these two circuits are different. The top one tells us that if we physically manipulate Y it will affect Z, while the bottom one shows that manipulating Y will affect X and will have no effect on Z. Moreover, performing some additional algebraic operations on our equations, we can obtain two new equations, shown at the bottom, which point to no structure *at all*; they simply represent two constraints on three variables without telling us how they influence each other.

Let us examine more closely the mental process by which we determine the effect of physically manipulating Y – say, setting Y to 0 [33].

Clearly, when we set Y to 0, the relation between X and Y is no longer given by the multiplier – a

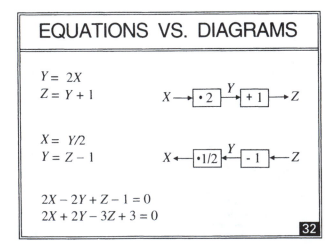

EQUATIONS VS. DIAGRAMS

$Y = 2X$
$Z = Y + 1$

$X \longrightarrow \boxed{\cdot 2} \xrightarrow{Y} \boxed{+1} \longrightarrow Z$

$X = Y/2$
$Y = Z - 1$

$X \longleftarrow \boxed{\cdot 1/2} \xleftarrow{Y} \boxed{-1} \longleftarrow Z$

$2X - 2Y + Z - 1 = 0$
$2X + 2Y - 3Z + 3 = 0$

new mechanism now controls Y, in which X has no say. In the equational representation, this amounts to replacing the equation $Y = 2X$ by a new equation $Y = 0$ and solving a new set of equations, which gives $Z = 1$. If we perform this surgery on the lower pair of equations, representing the lower model, we get of course a different solution. The second equation will need to be replaced, which will yield $X = 0$ and leave Z unconstrained.

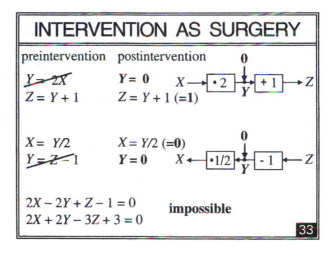

INTERVENTION AS SURGERY

preintervention	postintervention	
$Y = 2X$	$Y = 0$	
$Z = Y + 1$	$Z = Y + 1 \ (=1)$	
$X = Y/2$	$X = Y/2 \ (=0)$	
$Y = Z - 1$	$Y = 0$	
$2X - 2Y + Z - 1 = 0$	**impossible**	
$2X + 2Y - 3Z + 3 = 0$		

33

We now see how this model of intervention leads to a formal definition of causation: "Y is a cause of Z if we can change Z by manipulating Y, namely, if after surgically removing the equation for Y, the solution for Z will depend on the new value we substitute for Y." We also see how vital the diagram is in this process. *The diagram tells us which equation is to be deleted when we manipulate Y.* That information is totally washed out when we transform the equations into algebraically equivalent form, as shown at the bottom of the screen. From this pair of equations alone, it is impossible to predict the result of setting Y to 0, because we do not know what surgery to perform – there is no such thing as "the equation for Y."

In summary, *intervention amounts to a surgery on equations* (guided by a diagram) and *causation means predicting the consequences of such a surgery.*

This is a universal theme that goes beyond physical systems. In fact, the idea of modeling interventions by "wiping out" equations was first proposed in 1960 by an *economist*, Herman Wold, but his teachings have all but disappeared from the economics literature. History books attribute this mysterious disappearance to Wold's personality, but I tend to believe that the reason goes deeper: Early econometricians were very careful mathematicians; they fought hard to keep their algebra clean and formal, and they could not agree to have it contaminated by gimmicks such as diagrams. And as we see on the screen, the surgery operation makes no mathematical sense without the diagram, as it is sensitive to the way we write the equations.

Before expounding on the properties of this new mathematical operation, let me demonstrate how useful it is for clarifying concepts in statistics and economics.

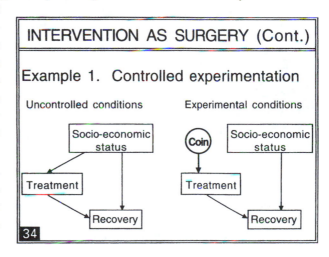

INTERVENTION AS SURGERY (Cont.)

Example 1. Controlled experimentation

Uncontrolled conditions

Experimental conditions

34

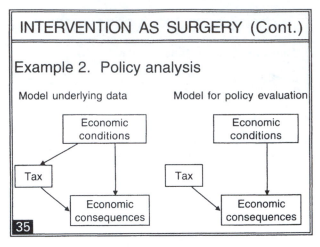

Why do we prefer controlled experiment over uncontrolled studies? Assume we wish to study the effect of some drug treatment on recovery of patients suffering from a given disorder. The mechanism governing the behavior of each patient is similar in structure to the circuit diagram we saw earlier. Recovery is a function of both the treatment and other factors, such as socioeconomic conditions, life style, diet, age, et cetera. Only one such factor is shown here [34].

Under uncontrolled conditions, the choice of treatment is up to the patients and may depend on the patients' socioeconomic backgrounds. This creates a problem, because we can't tell if changes in recovery rates are due to treatment or to those background factors. What we wish to do is compare patients of like backgrounds, and that is precisely what Fisher's *randomized experiment* accomplishes. How? It actually consists of two parts, randomization and *intervention*.

Intervention means that we change the natural behavior of the individual: we separate subjects into two groups, called treatment and control, and we convince the subjects to obey the experimental policy. We assign treatment to some patients who, under normal circumstances, will not seek treatment, and we give a placebo to patients who otherwise would receive treatment. That, in our new vocabulary, means *surgery* – we are severing one functional link and replacing it with another. Fisher's great insight was that connecting the new link to a random coin flip *guarantees* that the link we wish to break is actually broken. The reason is that a random coin is assumed to be unaffected by anything we can measure on a macroscopic level – including, of course, a patient's socioeconomic background.

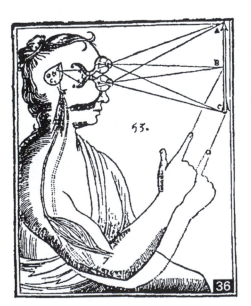

This picture provides a meaningful and formal rationale for the universally accepted procedure of randomized trials. In contrast, our next example uses the surgery idea to point out inadequacies in widely accepted procedures.

The example [35] involves a government official trying to evaluate the economic consequences of some policy – say, taxation. A deliberate decision to raise or lower taxes is a surgery on the model of the economy because it modifies the conditions prevailing when the model was built. Economic models are built on the basis of data taken over some period of time, and during this period of time taxes were lowered and raised in response to some economic conditions or political pressure. However, when we *evaluate* a policy, we wish to compare alternative policies under the *same* economic conditions – namely, we wish to sever this link that, in the past, has tied policies to those conditions. In this setup, it is of course impossible to connect our policy to a coin toss and run a controlled experiment; we do not have the time for that, and we might ruin the economy before the experiment is over. Nevertheless the analysis that we *should conduct* is to infer the behavior of this mutilated model from data governed by a nonmutilated model.

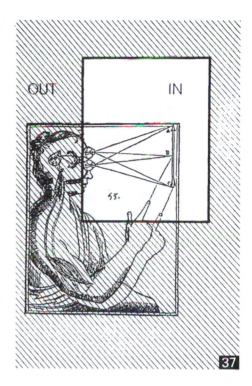

I said *should conduct* because you will not find such analysis in any economics textbook. As I mentioned earlier, the surgery idea of Herman Wold was stamped out of the economics literature in the 1970s, and all discussions on policy analysis that I could find assume that the mutilated model prevails throughout. That taxation is under government control at the time of evaluation is assumed to be sufficient for treating taxation as an exogenous variable throughout, when in fact taxation is an endogenous variable during the model-building phase and turns exogenous only when evaluated. Of course, I am not claiming that reinstating the surgery model would enable the government to balance its budget overnight, but it is certainly something worth trying.

Let us now examine how the surgery interpretation resolves Russell's enigma concerning the clash between the directionality of causal relations and the symmetry of physical equations. The equations of physics are indeed symmetrical, but when we compare the phrases "*A* causes *B*" versus "*B* causes *A*," we are not talking about a single set of equations. Rather, we are comparing two world models, represented by two different sets of equations: one in which the equation for *A* is surgically removed; the other where the equation for *B* is removed. Russell would probably stop us at this point and ask: "How can you talk about *two* world models when in fact there is only one world model, given by all the equations of physics put together?" The answer is: *yes*. If you wish to

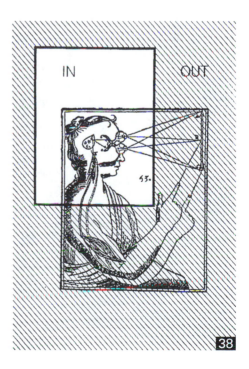

FROM PHYSICS TO CAUSALITY

Physics:
Symmetric equations of motion

Causal models:
Symmetric equations of motion
  Circumscription (in vs. out)
  Locality (autonomy of mechanisms)
  Intervention = surgery on mechanisms

39

include the entire universe in the model, causality disappears because interventions disappear – the manipulator and the manipulated lose their distinction. However, scientists rarely consider the entirety of the universe as an object of investigation. In most cases the scientist carves a piece from the universe and proclaims that piece *in* – namely, the *focus* of investigation. The rest of the universe is then considered *out* or *background* and is summarized by what we call *boundary conditions*. This choice of *ins* and *outs* creates asymmetry in the way we look at things, and it is this asymmetry that permits us to talk about "outside intervention" and hence about causality and cause–effect directionality.

This can be illustrated quite nicely using Descartes' classical drawing [36]. As a whole, this hand–eye system knows nothing about causation. It is merely a messy plasma of particles and photons trying their very best to obey Schroedinger's equation, which is symmetric.

However, carve a chunk from it – say, the object part [37] – and we can talk about the motion of the hand *causing* this light ray to change angle.

Carve it another way, focusing on the brain part [38], and lo and behold it is now the light ray that causes the hand to move – precisely the opposite direction. The lesson is that it is the way we carve up the universe that determines the directionality we associate with cause and effect. Such carving is tacitly assumed in every scientific investigation. In artificial intelligence it was called "circumscription" by J. McCarthy. In economics, circumscription amounts to deciding which variables are deemed endogenous and which exogenous, *in* the model or *external to* the model.

Let us summarize the essential differences between equational and causal models [39]. Both use a set of symmetric equations to describe normal conditions. The causal model, however, contains three additional ingredients: (i) a distinction between the *in* and the *out*; (ii) an assumption that each equation corresponds to an independent mechanism and hence must be preserved as a separate mathematical sentence; and (iii) interventions that are interpreted as surgeries over those mechanisms. This brings us closer to realizing the dream of making causality a friendly part of physics. But one ingredient is missing: *the algebra*. We discussed earlier how important the computational facility of algebra was to scientists and engineers in the Galilean era. Can we expect such algebraic facility to serve causality as well? Let me rephrase it differently: Scientific activity, as we know it, consists of two basic components:

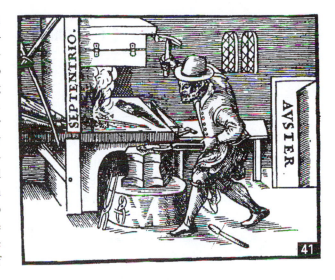

Observations [40] and interventions [41].

The combination of the two is what we call a *laboratory* [42], a place where we control some of the conditions and observe others. It so happened that standard algebras have served the observational component very well but thus far have not benefitted the interventional component. This is true for the algebra of equations, Boolean algebra, and probability calculus – all are geared to serve observational sentences but not interventional sentences.

Take, for example, probability theory. If we wish to find the chance that it rained, given that we see the grass wet, we can express our question in a formal sentence written like that: P (Rain | Wet), to be read: the probability of Rain, given Wet [43]. The vertical bar stands for the phrase: "given that we see." Not only can we express this question in a formal sentence, we can also use the machinery of probability theory and transform the sentence into other expressions. In our example, the sentence on the left can be transformed to the one on the right, if we find it more convenient or informative.

But suppose we ask a different question: "What is the chance it rained if we *make* the grass wet?" We cannot even express our query in the syntax of probability, because the vertical bar is already taken to mean "given that we see." We can invent a new symbol *do,* and each time we see a *do* after the bar we read it *given that we do* – but this does not help us compute the answer to our question, because the rules of probability do not apply to this new reading. We know intuitively what the answer should be: P (Rain), because making

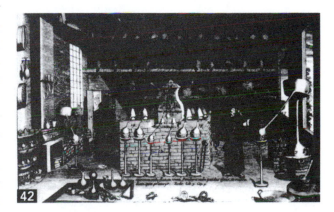

NEEDED: ALGEBRA OF DOING

Available: algebra of **seeing**

e.g., What is the chance it rained
if we **see** the grass wet?

$P\,(rain \mid wet) = ?$ $\{= P\,(wet \mid rain)\frac{P(rain)}{P(wet)}\}$

Needed: algebra of **doing**

e.g., What is the chance it rained
if we **make** the grass wet?

$P\,(rain \mid do(wet)) = ?$ $\{= P\,(rain)\}$

43

the grass wet does not change the chance of rain. But can this intuitive answer, and others like it, be derived mechanically, so as to comfort our thoughts when intuition fails?

The answer is *yes,* and it takes a new algebra. First, we assign a symbol to the new operator "given that we do." Second, we find the rules for manipulating sentences containing this new symbol. We do that by a process analogous to the way mathematicians found the rules of standard algebra.

Imagine that you are a mathematician in the sixteenth century, you are now an expert in the algebra of *addition,* and you feel an urgent need to introduce a new operator, *multiplication,* because you are tired of adding a number to itself all day long [44]. The first thing you do is assign the new operator a symbol: *multiply.* Then you go down to the meaning of the operator, from which you can deduce its rules of transformations. For example: the commutative law of multiplication can be deduced that way, the associative law, and so on. We now learn all this in high school.

In exactly the same fashion, we can deduce the rules that govern our new symbol: *do* (·). We have an algebra for seeing – namely, probability theory. We have a new operator, with a brand new outfit and a very clear meaning, given to us by the surgery procedure. The door is open for deduction, and the result is given in the next slide [45].

Please do not get alarmed, I do not expect you to read these equations right now, but I think you can still get the flavor of this new calculus. It consists of three rules that permit us to transform expressions involving actions and observations into other expressions of this type. The first allows us to ignore an irrelevant observation, the third to ignore an irrelevant action; the second allows us to exchange an action with an observation of the same fact. What are those symbols on the right? They are the "green lights" that the diagram gives us whenever the transformation is legal. We will see them in action on our next example.

This brings us to part three of the lecture, where I will demonstrate how the ideas presented thus far can be used to solve new problems of practical importance.

NEEDED: ALGEBRA OF **DOING** (Cont.)

Algebra of **Multiplication**	By Analogy
Available: algebra of addition	Available: algebra of seeing
e.g., $a+b = b+c,$ $a+(b+c) = (a+b)+c$	e.g., $P\,(x \mid y) = \dfrac{P\,(x,\,y)}{P(y)}$
New operation: $a \times b$	New operation: $do(z)$
Meaning: add a to itself b times	Meaning: surgery + substitution
New rules: $a \times b = b \times a,$ $a \times (b \times c) = (a \times b) \times c$ $a \times (b+c) = a \times b + a \times c$	New rules: $P\,(x \mid y,\,do(z)) = ?$

44

Consider the century-old debate concerning the effect of smoking on lung cancer [46]. In 1964, the Surgeon General issued a report linking cigarette smoking to death, cancer, and most particularly lung cancer. The report was based on nonexperimental studies in which a strong correlation was found between smoking and lung cancer, and the claim was that the correlation found is causal: If we ban smoking, then the rate of cancer cases will be roughly the same as the one we find today among non-smokers in the population.

RULES OF CAUSAL CALCULUS

Rule 1: Ignoring observations

$$P(y \mid do\{x\}, z, w) = P(y \mid do\{x\}, w)$$
$$\text{if } (Y \perp\!\!\!\perp Z \mid X, W)_{G_{\overline{X}}}$$

Rule 2: Action/observation exchange

$$P(y \mid do\{x\}, do\{z\}, w) = P(y \mid do\{x\}, z, w)$$
$$\text{if } (Y \perp\!\!\!\perp Z \mid X, W)_{G_{\overline{X}\underline{Z}}}$$

Rule 3: Ignoring actions

$$P(y \mid do\{x\}, do\{z\}, w) = P(y \mid do\{x\}, w)$$
$$\text{if } (Y \perp\!\!\!\perp Z \mid X, W)_{G_{\overline{X}, \overline{Z(W)}}}$$

45

These studies came under severe attacks from the tobacco industry, backed by some very prominent statisticians, among them Sir Ronald Fisher. The claim was that the observed correlations can also be explained by a model in which there is no causal connection between smoking and lung cancer. Instead, an unobserved genotype might exist that simultaneously causes cancer and produces an inborn craving for nicotine. Formally, this claim would be written in our notation as: $P(\text{Cancer} \mid do(\text{Smoke})) = P(\text{Cancer})$, meaning that making the population smoke or stop smoking would have no effect on the rate of cancer cases. Controlled experiments could decide between the two models, but these are impossible (and now also illegal) to conduct.

This is all history. Now we enter a hypothetical era where representatives of both sides decide to meet and iron out their differences. The tobacco industry concedes that there might be some weak causal link between smoking and cancer and representatives of the health group concede that there might be some weak links to genetic factors. Accordingly, they draw this combined model, and the question boils down to assessing, from the data, the strengths of the various links. They submit the query to a statistician and the answer comes back immediately: *impossible*. Meaning: there is no way to estimate the strength from the data, because any data whatsoever can perfectly fit either one of these two extreme models. So they give up and decide to continue the political battle as usual. Before parting, a suggestion comes up: perhaps we can resolve our differences if we measure some auxiliary factors. For example, since the

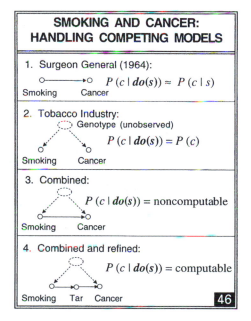

SMOKING AND CANCER: HANDLING COMPETING MODELS

1. Surgeon General (1964):

   Smoking ———→ Cancer $P(c \mid do(s)) \approx P(c \mid s)$

2. Tobacco Industry:

   Genotype (unobserved)

   Smoking   Cancer $P(c \mid do(s)) = P(c)$

3. Combined:

   Smoking ——→ Cancer $P(c \mid do(s)) = \text{noncomputable}$

4. Combined and refined:

   Smoking → Tar → Cancer $P(c \mid do(s)) = \text{computable}$

46

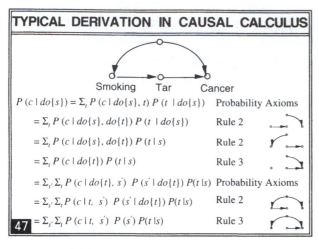

TYPICAL DERIVATION IN CAUSAL CALCULUS

Smoking   Tar   Cancer

$P(c \mid do\{s\}) = \Sigma_t P(c \mid do\{s\}, t) P(t \mid do\{s\})$ Probability Axioms

$\quad = \Sigma_t P(c \mid do\{s\}, do\{t\}) P(t \mid do\{s\})$ Rule 2

$\quad = \Sigma_t P(c \mid do\{s\}, do\{t\}) P(t \mid s)$ Rule 2

$\quad = \Sigma_t P(c \mid do\{t\}) P(t \mid s)$ Rule 3

$\quad = \Sigma_{s'} \Sigma_t P(c \mid do\{t\}, s') \; P(s' \mid do\{t\}) P(t \mid s)$ Probability Axioms

$\quad = \Sigma_{s'} \Sigma_t P(c \mid t, s') \; P(s' \mid do\{t\}) P(t \mid s)$ Rule 2

47 $\quad = \Sigma_{s'} \Sigma_t P(c \mid t, s') \; P(s') P(t \mid s)$ Rule 3

causal-link model is based on the understanding that smoking affects lung cancer through the accumulation of tar deposits in the lungs, perhaps we can measure the amount of tar deposits in the lungs of sampled individuals, and this might provide the necessary information for quantifying the links. Both sides agree that this is a reasonable suggestion, so they submit a new query to the statistician: Can we find the effect of smoking on cancer assuming that an intermediate measurement of tar deposits is available? The statistician comes back with good news: *it is computable* and, moreover, the solution is given in closed mathematical form. *How*?

SIMPSON'S PARADOX

(Pearson et al. 1899; Yule 1903; Simpson 1951)

- Any statistical relationship between two variables may be **reversed** by including additional factors in the analysis.

Application: The adjustment problem

- Which factors **should** be included in the analysis.

48

The statistician receives the problem and treats it as a problem in high school *algebra*: We need to compute P(Cancer), under hypothetical action, from nonexperimental data – namely, from expressions involving *no actions*. Or: We need to eliminate the "*do*" symbol from the initial expression. The elimination proceeds like ordinary solution of algebraic equations – in each stage [47], a new rule is applied, licensed by some subgraph of the diagram, eventually leading to a formula involving no "*do*" symbols, which denotes an expression that is computable from nonexperimental data.

49

You are probably wondering whether this derivation solves the smoking–cancer debate. The answer is *no*. Even if we could get the data on tar deposits, our model is quite simplistic, as it is based on certain assumptions that both parties might not agree to – for instance, that there is no direct link between smoking and lung cancer unmediated by tar deposits. The model would need to be refined then, and we might end up with a graph containing twenty variables or more. There is no need to panic when someone tells us: "you did not take this or that factor into account." On the contrary, the graph welcomes such new ideas, because it is so easy to add factors and measurements into the model. Simple tests are now available that permit an investigator to merely glance at the graph and decide if we can compute the effect of one variable on another.

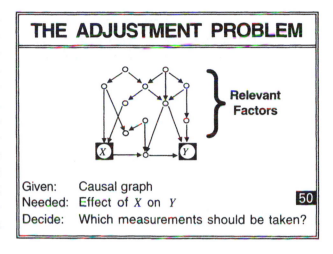

THE ADJUSTMENT PROBLEM

} Relevant Factors

Given: Causal graph
Needed: Effect of X on Y 50
Decide: Which measurements should be taken?

Our next example illustrates how a long-standing problem is solved by purely graphical means – proven by the new algebra. The problem is called *the adjustment problem* or "the covariate selection problem" and represents the practical side of Simpson's paradox [48].

Simpson's paradox, first noticed by Karl Pearson in 1899, concerns the disturbing observation that every statistical relationship between two variables may be *reversed* by including additional factors in the analysis. For example, you might run a study and find that

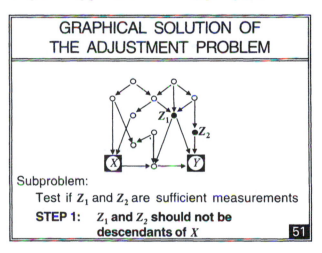

GRAPHICAL SOLUTION OF THE ADJUSTMENT PROBLEM

Z_1 Z_2

Subproblem:
   Test if Z_1 and Z_2 are sufficient measurements
STEP 1: Z_1 and Z_2 **should not be descendants of** X 51

students who smoke get higher grades; however, if you adjust for *age,* the opposite is true in every *age group,* that is, smoking predicts lower grades. If you further adjust for *parent income,* you find that smoking predicts higher grades again, in every *age–income* group, and so on.

Equally disturbing is the fact that no one has been able to tell us which factors *should* be included in the analysis. Such factors can now be identified by simple graphical means. The classical case demonstrating Simpson's paradox took place in 1975, when UC-Berkeley

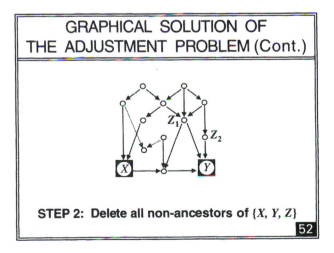

GRAPHICAL SOLUTION OF THE ADJUSTMENT PROBLEM (Cont.)

Z_1 Z_2

STEP 2: Delete all non-ancestors of $\{X, Y, Z\}$ 52

GRAPHICAL SOLUTION OF THE ADJUSTMENT PROBLEM (Cont.)

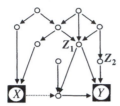

STEP 3: Delete all arcs emanating from X

53

was investigated for sex bias in graduate admission. In this study, overall data showed a higher rate of admission among male applicants; but, broken down by departments, data showed a slight bias in favor of admitting female applicants. The explanation is simple: female applicants tended to apply to more competitive departments than males, and in these departments, the rate of admission was low for both males and females.

To illustrate this point, imagine a fishing boat with two different nets, a large mesh and a small net [49]. A school of fish swim toward the boat and seek to pass it. The female fish try for the small-mesh challenge, while the male fish try for the easy route. The males go through and only females are caught. Judging by the final catch, preference toward females is clearly evident. However, if analyzed separately, each individual net would surely trap males more easily than females.

GRAPHICAL SOLUTION OF THE ADJUSTMENT PROBLEM (Cont.)

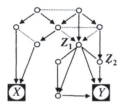

STEP 4: Connect any two parents sharing a common child

54

Another example involves a controversy called "reverse regression," which occupied the social science literature in the 1970s. Should we, in salary discrimination cases, compare salaries of equally qualified men and women or instead compare qualifications of equally paid men and women?

Remarkably, the two choices led to opposite conclusions. It turned out that men earned a higher salary than equally qualified women and, *simultaneously,* men were more qualified than equally paid women. The moral is that all conclusions are extremely sensitive to which variables we choose to hold constant when we are comparing,

GRAPHICAL SOLUTION OF THE ADJUSTMENT PROBLEM (Cont.)

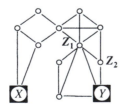

STEP 5: Strip arrow-heads from all edges

55

and that is why the adjustment problem is so critical in the analysis of observational studies.

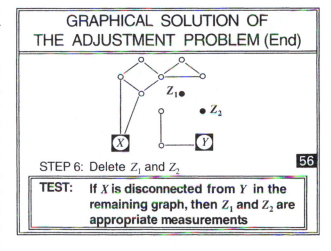

GRAPHICAL SOLUTION OF THE ADJUSTMENT PROBLEM (End)

STEP 6: Delete Z_1 and Z_2

56

TEST: **If X is disconnected from Y in the remaining graph, then Z_1 and Z_2 are appropriate measurements**

Consider an observational study where we wish to find the effect of X on Y, for example, treatment on response [50]. We can think of many factors that are relevant to the problem; some are affected by the treatment, some are affecting the treatment, and some are affecting both treatment and response. Some of these factors may be unmeasurable, such as genetic trait or life style; others are measurable, such as gender, age, and salary level. Our problem is to select a subset of these factors for measurement and adjustment so that, if we compare subjects under the same value of those measurements and average, we get the right result.

Let us follow together the steps that would be required to test if two candidate measurements, Z_1 and Z_2, would be sufficient [51]. The steps are rather simple, and can be performed manually even on large graphs. However, to give you the feel of their mechanizability, I will go through them rather quickly. Here we go [52–56].

At the end of these manipulations, we end up with the answer to our question: "If X is disconnected from Y, then Z_1 and Z_2 are appropriate measurements."

I now wish to summarize briefly the central message of this lecture. It is true that testing for cause and effect is difficult. Discovering causes of effects is even more difficult. But causality is not *mystical* or *metaphysical*. It can be understood in terms of simple processes, and it can be expressed in a friendly mathematical language, ready for computer analysis.

What I have presented to you today is a sort of pocket calculator, an *abacus* [57], to help us investigate certain problems of cause and effect with mathematical precision. This does not solve all the problems of causality, but the power of *symbols* and mathematics should not be underestimated [58].

Many scientific discoveries have been delayed over the centuries for the lack of a mathematical language that can amplify ideas and let scientists communicate results. I am convinced that many discoveries have been delayed in our century for lack of a mathematical language that can handle

Fig. 155 Little Johnny and his "calculating machine."

57

causation. For example, I am sure that Karl Pearson could have thought up the idea of *randomized experiment* in 1901 if he had allowed causal diagrams into his mathematics.

But the really challenging problems are still ahead: We still do not have a causal understanding of *poverty* and *cancer* and *intolerance,* and only the accumulation of data and the insight of great minds will eventually lead to such understanding.

The data is all over the place, the insight is yours, and now an abacus is at your disposal, too. I hope the combination amplifies each of these components.

Thank you.

Acknowledgments

Slide 1 (Dürer, *Adam and Eve,* 1504 engraving) courtesy of the Fogg Art Museum, Harvard University Art Museums, Gift of William Gray from the collection of Francis Calley Gray. Photo by Rick Stafford; image copyright © President and Fellows of Harvard College, Harvard University. Slide 2 (Doré, *The Flight of Lot*) copyright William H. Wise & Co. Slide 3 (Egyptian wall painting of Neferronpe playing a board game) courtesy of the Oriental Institute of the University of Chicago.

The following images were reproduced from antiquarian book catalogs, courtesy of Bernard Quaritch, Ltd. (London): slides 4, 5, 6, 7, 8, 9, 15, 27, 31, 36, 37, 38, 40, 42, and 58.

Slides 10 and 11 copyright The Courier Press. Slides 13 and 14 reprinted with the permission of Macmillan Library Reference USA, from *The Album of Science,* by I. Bernard Cohen. Copyright © 1980 Charles Scribner's Sons.

Slide 16 courtesy of the Library of California State University, Long Beach. Slides 20 and 22 reprinted with the permission of Cambridge University Press. Slide 25: copyright photograph by A. C. Barrington Brown, reproduced with permission.

Slide 30: from S. Wright (1920) in *Proceedings of the National Academy of Sciences,* vol. 6; reproduced with the permission of the American Philosophical Society and the University of Chicago Press. Slide 57 reprinted with the permission of Vandenhoeck & Ruprecht and The MIT Press.

NOTE: Color versions of slides 19, 26, 28–29, 32–35, and 43–56 may be downloaded from ⟨http://www.cs.ucla.edu/~judea/⟩.

Bibliography

Abbring, 2003 J.H. Abbring. Book reviews: Causality: Models, Reasoning, and Inference. *Economica,* 70:702–703, 2003.

Adams, 1975 E. Adams. *The Logic of Conditionals,* chapter 2. D. Reidel, Dordrecht, Netherlands, 1975.

Agresti, 1983 A. Agresti. Fallacies, statistical. In S. Kotz and N.L. Johnson, editors, *Encyclopedia of Statistical Science,* volume 3, pages 24–28. John Wiley, New York, 1983.

Aldrich, 1989 J. Aldrich. Autonomy. *Oxford Economic Papers,* 41:15–34, 1989.

Aldrich, 1993 J. Aldrich. Cowles' exogeneity and core exogeneity. Technical Report Discussion Paper 9308, Department of Economics, University of Southampton, England, 1993.

Aldrich, 1995 J. Aldrich. Correlations genuine and spurious in Pearson and Yule. *Statistical Science,* 10:364–376, 1995.

Andersson et al., 1997 S.A. Andersson, D. Madigan, and M.D. Perlman. A characterization of Markov equivalence classes for acyclic digraphs. *Annals of Statistics,* 24:505–541, 1997.

Andersson et al., 1998 S.A. Andersson, D. Madigan, M.D. Perlman, and T.S. Richardson. Graphical Markov models in multivariate analysis. In S. Ghosh, editor, *Multivariate Analysis, Design of Experiments and Survey Sampling,* pages 187–229. Marcel Dekker, Inc., New York, 1998.

Angrist and Imbens, 1991 J.D. Angrist and G.W. Imbens. Source of identifying information in evaluation models. Technical Report Discussion Paper 1568, Department of Economics, Harvard University, Cambridge, MA, 1991.

Angrist et al., 1996 J.D. Angrist, G.W. Imbens, and D.B Rubin. Identification of causal effects using instrumental variables (with comments). *Journal of the American Statistical Association,* 91(434):444–472, June 1996.

Angrist, 2004 J.D. Angrist. Treatment effect heterogeneity in theory and practice. *The Economic Journal,* 114:C52–C83, 2004.

Arah, 2008 O.A. Arah. The role of causal reasoning in understanding Simpson's paradox, Lord's paradox, and the suppression effect: Covariate selection in the analysis of observational studies. *Emerging Themes in Epidemiology,* 4:doi:10.1186/1742–7622–5–5, 2008. Online at http://www. ete-online.com/content/5/1/5.

Austin, 2008 P.C. Austin. A critical appraisal of propensity-score matching in the medical literature from 1996 to 2003. *Statistics in Medicine,* 27(12):2037–2049, 2008.

Avin et al., 2005 C. Avin, I. Shpitser, and J. Pearl. Identifiability of path-specific effects. In *Proceedings of the Nineteenth International Joint Conference on Artificial Intelligence (IJCAI-05),* pages 357–363, Edinburgh, UK, 2005.

Bagozzi and Burnkrant, 1979 R.P. Bagozzi and R.E. Burnkrant. Attitude organization and the attitude-behavior relationship. *Journal of Personality and Social Psychology,* 37:913–929, 1979.

Balke and Pearl, 1994a A. Balke and J. Pearl. Counterfactual probabilities: Computational methods, bounds, and applications. In R. Lopez de Mantaras and D. Poole, editors, *Uncertainty in Artificial Intelligence 10,* pages 46–54. Morgan Kaufmann, San Mateo, CA, 1994.

Balke and Pearl, 1994b A. Balke and J. Pearl. Probabilistic evaluation of counterfactual queries. In *Proceedings of the Twelfth National Conference on Artificial Intelligence,* volume I, pages 230–237. MIT Press, Menlo Park, CA, 1994.

Balke and Pearl, 1995a A. Balke and J. Pearl. Counterfactuals and policy analysis in structural models. In P. Besnard and S. Hanks, editors, *Uncertainty in Artificial Intelligence 11,* pages 11–18. Morgan Kaufmann, San Francisco, 1995.

Balke and Pearl, 1995b A. Balke and J. Pearl. Universal formulas for treatment effect from non-compliance data. In N.P. Jewell, A.C. Kimber, M.-L. Lee, and G.A. Whitmore, editors, *Lifetime Data: Models in Reliability and Survival Analysis,* pages 39–43. Kluwer Academic Publishers, Dordrecht, 1995.

Balke and Pearl, 1997 A. Balke and J. Pearl. Bounds on treatment effects from studies with imperfect compliance. *Journal of the American Statistical Association,* 92(439):1172–1176, 1997.

Balke, 1995 A. Balke. *Probabilistic Counterfactuals: Semantics, Computation, and Applications.* PhD thesis, Computer Science Department, University of California, Los Angeles, CA, November 1995.

Barigelli and Scozzafava, 1984 B. Barigelli and R. Scozzafava. Remarks on the role of conditional probability in data exploration. *Statistics and Probability Letters,* 2(1):15–18, January 1984.

Bayes,1763 T. Bayes. An essay towards solving a problem in the doctrine of chances. *Philosophical Transactions,* 53:370–418, 1763. Reproduced in W.E. Deming.

Becher, 1992 H. Becher. The concept of residual confounding in regression models and some applications. *Statistics in Medicine,* 11:1747–1758, 1992.

Berk and de Leeuw, 1999 R.A. Berk and J. de Leeuw. An evaluation of California's inmate classification system using a generalized regression discontinuity design. *Journal of the American Statistical Association,* 94:1045–1052, 1999.

Berk, 2004 R.A. Berk. *Regression Analysis: A Constructive Critique.* Sage, Thousand Oaks, CA, 2004.

Berkson, 1946 J. Berkson. Limitations of the application of fourfold table analysis to hospital data. *Biometrics Bulletin,* 2:47–53, 1946.

Bertsekas and Tsitsiklis, 1996 D.P. Bertsekas and J.M. Tsitsiklis. *Neuro-dynamic Programming.* Athena, Belmont, MA, 1996.

Bessler, 2002 D. Bessler. On world poverty: Its causes and effects, 2002. http://agecon2.tamu.edu/people/faculty/bessler-david/WebPage/poverty.pdf.

Bickel et al., 1975 P.J. Bickel, E.A. Hammel, and J.W. O'Connell. Sex bias in graduate admissions: Data from Berkeley. *Science,* 187:398–404, 1975.

Bishop et al., 1975 Y.M.M. Bishop, S.E. Fienberg, and P.W. Holland. *Discrete Multivariate Analysis: Theory and Practice.* MIT Press, Cambridge, MA, 1975.

Bishop, 1971 Y.M.M. Bishop. Effects of collapsing multidimensional contingency tables. *Biometrics,* 27:545–562, 1971.

Blalock, Jr., 1962 H.M. Blalock, Jr. Four-variable causal models and partial correlations. *American Journal of Sociology,* 68:182–194, 1962.

Bloom, 1984 H.S. Bloom. Accounting for no-shows in experimental evaluation designs. *Evaluation Review,* 8(2):225–246, April 1984.

Blumer et al., 1987 A. Blumer, A. Ehrenfeucht, D. Haussler, and M.K. Warmuth. Occam's razor. *Information Processing Letters,* 24, 1987.

Blyth, 1972 C.R. Blyth. On Simpson's paradox and the sure-thing principle. *Journal of the American Statistical Association,* 67:364–366, 1972.

Bollen, 1989 K.A. Bollen. *Structural Equations with Latent Variables.* John Wiley, New York, 1989.

Bonet, 2001 B. Bonet. A calculus for causal relevance. In *Proceedings of the Seventeenth Conference on Uncertainty in Artificial Intelligence,* pages 40–47. Morgan Kaufmann, San Francisco, CA, 2001.

Boumans, 2004 M. Boumans. Book reviews: Causality: Models, Reasoning, and Inference. *Review of Social Economy,* LXIII:129–135, 2004.

Bowden and Turkington, 1984 R.J. Bowden and D.A. Turkington. *Instrumental Variables.* Cambridge University Press, Cambridge, England, 1984.

Breckler, 1990 S.J. Breckler. Applications of covariance structure modeling in psychology: Cause for concern? *Psychological Bulletin,* 107(2):260–273, 1990.

Breslow and Day, 1980 N.E. Breslow and N.E. Day. *Statistical Methods in Cancer Research; Vol. 1, The Analysis of Case-Control Studies.* IARC, Lyon, 1980.

Brito and Pearl, 2002a C. Brito and J Pearl. Generalized instrumental variables. In A. Darwiche and N. Friedman, editors, *Uncertainty in Artificial Intelligence, Proceedings of the Eighteenth Conference,* pages 85–93. Morgan Kaufmann, San Francisco, 2002.

Brito and Pearl, 2002b C. Brito and J Pearl. A graphical criterion for the identification of causal effects in linear models. In *Proceedings of the Eighteenth National Conference on Artificial Intelligence,* pages 533–538. AAAI Press/The MIT Press, Menlo Park, CA, 2002.

Brito and Pearl, 2002c C. Brito and J Pearl. A new identification condition for recursive models with correlated errors. *Journal of Structural Equation Modeling,* 9(4):459–474, 2002.

Brito and Pearl, 2006 C. Brito and J Pearl. Graphical condition for identification in recursive SEM. In *Proceedings of the Twenty-Third Conference on Uncertainty in Artificial Intelligence,* pages 47–54. AUAI Press, Corvallis, OR, 2006.

Butler, 2002 S.F. Butler. Book review: A structural approach to the understanding of causes, effects, and judgment. *Journal of Mathematical Psychology,* 46:629–635, 2002.

Byerly, 2000 H.C. Byerly. Book reviews: Causality: Models, Reasoning, and Inference. *Choice,* 548, November 2000.

Cai and Kuroki, 2006 Z. Cai and M. Kuroki. Variance estimators for three 'probabilities of causation'. *Risk Analysis,* 25(6):1611–1620, 2006.

Cai et al., 2008 Z. Cai, M. Kuroki, J. Pearl, and J. Tian. Bounds on direct effect in the presence of confound intermediate variables. *Biometrics,* 64:695–701, 2008.

Campbell and Stanley, 1966 D.T. Campbell and J.C. Stanley. *Experimental and Quasi-Experimental Designs for Research.* R. McNally and Co., Chicago, IL, 1966.

Cartwright, 1983 N. Cartwright. *How the Laws of Physics Lie.* Clarendon Press, Oxford, 1983.

Cartwright, 1989 N. Cartwright. *Nature's Capacities and Their Measurement.* Clarendon Press, Oxford, 1989.

Cartwright, 1995a N. Cartwright. False idealisation: A philosophical threat to scientific method. *Philosophical Studies,* 77:339–352, 1995.

Cartwright, 1995b N. Cartwright. Probabilities and experiments. *Journal of Econometrics,* 67:47–59, 1995.

Cartwright, 1999 N. Cartwright. Causality: Independence and determinism. In A Gammerman, editor, *Causal Models and Intelligent Data Management,* pages 51–63. Springer-Verlag, Berlin, 1999.

Cartwright, 2007 N. Cartwright. *Hunting Causes and Using Them: Approaches in Philosophy and Economics.* Cambridge University Press, New York, NY, 2007.

Chajewska and Halpern, 1997 U. Chajewska and J.Y. Halpern. Defining explanation in probabilistic systems. In D. Geiger and P.P. Shenoy, editors, *Uncertainty in Artificial Intelligence 13,* pages 62–71. Morgan Kaufmann, San Francisco, CA, 1997.

Chakraborty, 2001 R. Chakraborty. A rooster crow does not cause the sun to rise: Review of Causality: Models, Reasoning, and Inference. *Human Biology,* 110(4):621–624, 2001.

Chalak and White, 2006 K. Chalak and H. White. An extended class of instrumental variables for the estimation of causal effects. Technical Report Discussion Paper, UCSD, Department of Economics, July 2006.

Cheng, 1992 P.W. Cheng. Separating causal laws from causal facts: Pressing the limits of statistical relevance. *Psychology of Learning and Motivation,* 30:215–264, 1992.

Cheng, 1997 P.W. Cheng. From covariation to causation: A causal power theory. *Psychological Review,* 104(2):367–405, 1997.

Chickering and Pearl, 1997 D.M. Chickering and J. Pearl. A clinician's tool for analyzing non-compliance. *Computing Science and Statistics,* 29(2):424–431, 1997.

Chickering, 1995 D.M. Chickering. A transformational characterization of Bayesian network struc-tures. In P. Besnard and S. Hanks, editors, *Uncertainty in Artificial Intelligence 11*, pages 87–98. Morgan Kaufmann, San Francisco, 1995.

Chou and Bentler, 1995 C.P. Chou and P. Bentler. Estimations and tests in structural equation mod-eling. In R.H. Hoyle, editor, *Structural Equation Modeling*, pages 37–55. Sage, Thousand Oaks, CA, 1995.

Christ, 1966 C. Christ. *Econometric Models and Methods*. John Wiley and Sons, Inc., New York, 1966.

Cliff, 1983 N. Cliff. Some cautions concerning the application of causal modeling methods. *Multivariate Behavioral Research*, 18:115–126, 1983.

Cohen and Nagel, 1934 M.R. Cohen and E. Nagel. *An Introduction to Logic and the Scientific Method*. Harcourt, Brace and Company, New York, 1934.

Cole and Hernán, 2002 S.R. Cole and M.A. Hernán. Fallibility in estimating direct effects. *International Journal of Epidemiology*, 31(1):163–165, 2002.

Cole, 1997 P. Cole. Causality in epidemiology, health policy, and law. *Journal of Marketing Research*, 27:10279–10285, 1997.

Cooper and Herskovits, 1991 G.F. Cooper and E. Herskovits. A Bayesian method for constructing Bayesian belief networks from databases. In B.D. D'Ambrosio, P. Smets, and P.P. Bonissone, editors, *Proceedings of Uncertainty in Artificial Intelligence Conference, 1991*, pages 86–94. Morgan Kaufmann, San Mateo, 1991.

Cooper, 1990 G.F. Cooper. Computational complexity of probabilistic inference using Bayesian belief networks. *Artificial Intelligence*, 42(2):393–405, 1990.

Cowell et al., 1999 R.G. Cowell, A.P. Dawid, S.L. Lauritzen, and D.J. Spielgelhalter. *Probabilistic Networks and Expert Systems*. Springer Verlag, New York, NY, 1999.

Cox and Wermuth, 1996 D.R. Cox and N. Wermuth. *Multivariate Dependencies – Models, Analysis and Interpretation*. Chapman and Hall, London, 1996.

Cox and Wermuth, 2003 D.R. Cox and N. Wermuth. A general condition for avoiding effect reversal after marginalization. *Journal of the Royal Statistical Society, Series B (Statistical Methodology)*, 65(4):937–941, 2003.

Cox and Wermuth, 2004 D.R. Cox and N. Wermuth. Causality: A statistical view. *International Statistical Review*, 72(3):285–305, 2004.

Cox, 1958 D.R. Cox. *The Planning of Experiments*. John Wiley and Sons, NY, 1958.

Cox, 1992 D.R. Cox. Causality: Some statistical aspects. *Journal of the Royal Statistical Society*, 155, Series A:291–301, 1992.

Crámer, 1946 H. Crámer. *Mathematical Methods of Statistics*. Princeton University Press, Princeton, NJ, 1946.

Cushing and McMullin, 1989 J.T. Cushing and E. McMullin (Eds.). *Philosophical Consequences of Quantum Theory: Reflections on Bell's Theorem*. University of Notre Dame Press, South Bend, IN, 1989.

Darlington, 1990 R.B. Darlington. *Regression and Linear Models*. McGraw-Hill, New York, 1990.

Darnell, 1994 A.C. Darnell. *A Dictionary of Econometrics*. Edward Elgar Publishing Limited, Brookfield, VT, 1994.

Darwiche, 2009 A. Darwiche. *Modeling and Reasoning with Bayesian Networks*. Cambridge University Press, New York, 2009.

Davidson and MacKinnon, 1993 R. Davidson and J.G. MacKinnon. *Estimation and Inference in Econometrics*. Oxford University Press, New York, 1993.

Dawid, 1979 A.P. Dawid. Conditional independence in statistical theory. *Journal of the Royal Statistical Society, Series B*, 41(1):1–31, 1979.

Dawid, 2000 A.P. Dawid. Causal inference without counterfactuals (with comments and rejoinder). *Journal of the American Statistical Association*, 95(450):407–448, June 2000.

Dawid, 2002 A.P. Dawid. Influence diagrams for causal modelling and inference. *International Statistical Review*, 70:161–189, 2002.

De Kleer and Brown, 1986 J. De Kleer and J.S. Brown. Theories of causal ordering. *Artificial Intelligence*, 29(1):33–62, 1986.

Dean and Wellman, 1991 T.L. Dean and M.P. Wellman. *Planning and Control*. Morgan Kaufmann, San Mateo, CA, 1991.

Dechter and Pearl, 1991 R. Dechter and J. Pearl. Directed constraint networks: A relational framework for casual modeling. In J. Mylopoulos and R. Reiter, editors, *Proceedings of the Twelfth International Joint Conference of Artificial Intelligence (IJCAI-91)*, pages 1164–1170. Morgan Kaufmann, San Mateo, CA, Sydney, Australia, 1991.

Dechter, 1996 R. Dechter. Topological parameters for time-space tradeoff. In E. Horvitz and F. Jensen, editors, *Proceedings of the Twelfth Conference on Uncertainty in Artificial Intelligence*, pages 220–227. Morgan Kaufmann, San Francisco, CA, 1996.

Decock, 2002 L. Decock. Bibliografische notities: Causality: Models, Reasoning, and Inference. *Tijdschrift voor Filosofie*, 64:201, 2002.

DeFinetti, 1974 B. DeFinetti. *Theory of Probability: A Critical Introductory Treatment*. Wiley, London, 1974. 2 volumes. Translated by A. Machi and A. Smith.

Dehejia and Wahba, 1999 R.H. Dehejia and S. Wahba. Causal effects in nonexperimental studies: Re-evaluating the evaluation of training programs. *Journal of the American Statistical Association*, 94:1053–1063, 1999.

Demiralp and Hoover, 2003 S. Demiralp and K. Hoover. Searching for the causal structure of a vector autoregression. *Oxford Bulletin of Economics*, 65:745–767, 2003.

Dempster, 1990 A.P. Dempster. Causality and statistics. *Journal of Statistics Planning and Inference*, 25:261–278, 1990.

Dhrymes, 1970 P.J. Dhrymes. *Econometrics*. Springer-Verlag, New York, 1970.

Didelez and Pigeot, 2001 V. Didelez and I. Pigeot. Discussions: Judea Pearl, Causality: Models, Reasoning, and Inference. *Politische Vierteljahresschrift*, 42(2):313–315, 2001.

Didelez, 2002 V. Didelez. Book reviews: Causality: Models, Reasoning, and Inference. *Statistics in Medicine*, 21:2292–2293, 2002.

Dong, 1998 J. Dong. Simpson's paradox. In P. Armitage and T. Colton, editors, *Encyclopedia of Biostatistics*, pages 4108–4110. J. Wiley, New York, 1998.

Dor and Tarsi, 1992 D. Dor and M. Tarsi. A simple algorithm to construct a consistent extension of a partially oriented graph. Technical Report R-185, UCLA, Computer Science Department, 1992.

Druzdzel and Simon, 1993 M.J. Druzdzel and H.A. Simon. Causality in Bayesian belief networks. In D. Heckerman and A. Mamdani, editors, *Proceedings of the Ninth Conference on Uncertainty in Artificial Intelligence*, pages 3–11. Morgan Kaufmann, San Mateo, CA, 1993.

Duncan, 1975 O.D. Duncan. *Introduction to Structural Equation Models*. Academic Press, New York, 1975.

Edwards, 2000 D. Edwards. *Introduction to Graphical Modelling*. Springer-Verlag, New York, 2nd edition, 2000.

Eells and Sober, 1983 E. Eells and E. Sober. Probabilistic causality and the question of transitivity. *Philosophy of Science*, 50:35–57, 1983.

Eells, 1991 E. Eells. *Probabilistic Causality*. Cambridge University Press, Cambridge, UK, 1991.

Efron and Feldman, 1991 B. Efron and D. Feldman. Compliance as an explanatory variable in clinical trials. *Journal of the American Statistical Association*, 86(413):9–26, March 1991.

Engle et al., 1983 R.F. Engle, D.F. Hendry, and J.F. Richard. Exogeneity. *Econometrica*, 51:277–304, 1983.

Epstein, 1987 R.J. Epstein. *A History of Econometrics*. Elsevier Science, New York, 1987.

Eshghi and Kowalski, 1989 K. Eshghi and R.A. Kowalski. Abduction compared with negation as failure. In G. Levi and M. Martelli, editors, *Proceedings of the Sixth International Conference on Logic Programming*, pages 234–254. MIT Press, Cambridge, MA, 1989.

Everitt, 1995 B. Everitt. Simpson's paradox. In B. Everitt, editor, *The Cambridge Dictionary of Statistics in the Medical Sciences*, page 237. Cambridge University Press, New York, 1995.

Feller, 1950 W. Feller. *Probability Theory and Its Applications*. Wiley, New York, 1950.

Fikes and Nilsson, 1971 R.E. Fikes and N.J. Nilsson. STIRPS: A new approach to the application of theorem proving to problem solving. *Artificial Intelligence*, 2(3/4):189–208, 1971.

Fine, 1975 K. Fine. Review of Lewis' counterfactuals. *Mind*, 84:451–458, 1975.

Fine, 1985 K. Fine. *Reasoning with Arbitrary Objects*. B. Blackwell, New York, 1985.

Finkelstein and Levin, 1990 M.O. Finkelstein and B. Levin. *Statistics for Lawyers*. Springer-Verlag, New York, 1990.

Fisher, 1926 R.A. Fisher. The arrangement of field experiments. *Journal of the Ministry of Agriculture of Great Britain,* 33:503–513, 1926. *Collected Papers,* 2, no. 48, and *Contributions,* paper 17.

Fisher, 1935 R.A. Fisher. *The Design of Experiments*. Oliver and Boyd, Edinburgh, 1935.

Fisher, 1970 F.M. Fisher. A correspondence principle for simultaneous equations models. *Econometrica,* 38(1):73–92, January 1970.

Fleiss, 1981 J.L. Fleiss. *Statistical Methods for Rates and Proportions*. John Wiley and Sons, New York, 2nd edition, 1981.

Frangakis and Rubin, 2002 C.E. Frangakis and D.B. Rubin. Principal stratification in causal inference. *Biometrics,* 1(58):21–29, 2002.

Freedman and Stark, 1999 D. A. Freedman and P. B. Stark. The swine flu vaccine and Guillain-Barré syndrome: A case study in relative risk and specific causation. *Evaluation Review,* 23(6):619–647, December 1999.

Freedman, 1987 D. Freedman. As others see us: A case study in path analysis (with discussion). *Journal of Educational Statistics,* 12(2):101–223, 1987.

Freedman, 1997 D.A. Freedman. From association to causation via regression. In V.R. McKim and S.P. Turner, editors, *Causality in Crisis?,* pages 113–161. University of Notre Dame Press, Notre Dame, IN, 1997.

Frisch, 1938 R. Frisch. Autonomy of economic relations. Reprinted [with Tinbergen's comments]. In D.F. Hendry and M.S. Morgan, editors, *The Foundations of Econometric Analysis,* pages 407–423. Cambridge University Press, 1938.

Frydenberg, 1990 M. Frydenberg. The chain graph Markov property. *Scandinavian Journal of Statistics,* 17:333–353, 1990.

Gail, 1986 M.H. Gail. Adjusting for covariates that have the same distribution in exposed and unexposed cohorts. In S.H. Moolgavkar and R.L. Prentice, editors, *Modern Statistical Methods in Chronic Disease Epidemiology,* pages 3–18. John Wiley and Sons, New York, 1986.

Galles and Pearl, 1995 D. Galles and J. Pearl. Testing identifiability of causal effects. In P. Besnard and S. Hanks, editors, *Uncertainty in Artificial Intelligence 11,* pages 185–195. Morgan Kaufmann, San Francisco, 1995.

Galles and Pearl, 1997 D. Galles and J. Pearl. Axioms of causal relevance. *Artificial Intelligence,* 97(1-2):9–43, 1997.

Galles and Pearl, 1998 D. Galles and J. Pearl. An axiomatic characterization of causal counterfactuals. *Foundation of Science,* 3(1):151–182, 1998.

Gardenfors, 1988 P. Gardenfors. Causation and the dynamics of belief. In W. Harper and B. Skyrms, editors, *Causation in Decision, Belief Change and Statistics II,* pages 85–104. Kluwer Academic Publishers, Dordrecht /Boston /London, 1988.

Geffner, 1992 H. Geffner. *Default Reasoning: Causal and Conditional Theories*. MIT Press, Cambridge, MA, 1992.

Geiger and Pearl, 1993 D. Geiger and J. Pearl. Logical and algorithmic properties of conditional independence. *The Annals of Statistics,* 21(4):2001–2021, 1993.

Geiger et al., 1990 D. Geiger, T.S. Verma, and J. Pearl. Identifying independence in Bayesian networks. *Networks,* 20: 507–534.

Geneletti, 2007 S. Geneletti. Identifying direct and indirect effects in a non-counterfactual framework. *Journal of the Royal Statistical Society, Series B (Methodological),* 69(2): 199–215, 2007.

Geng et al., 2002 Z. Geng, J. Guo, and W-K. Fung. Criteria for confounders in epidemiological studies. *Journal of the Royal Statistical Society, Series B,* 64(1):3–15, 2002.

Geng, 1992 Z. Geng. Collapsibility of relative risk in contingency tables with a response variable. *Journal of the Royal Statistical Society,* 54(2):585–593, 1992.

Gibbard and Harper, 1976 A. Gibbard and L. Harper. Counterfactuals and two kinds of expected utility. In W.L. Harper, R. Stalnaker, and G. Pearce (Eds.), *Ifs,* pages 153–169. D. Reidel, Dordrecht, 1976.

Gillies, 2001 D. Gillies. Critical notice: Judea Pearl, Causality: Models, Reasoning, and Inference. *British Journal of Science,* 52:613–622, 2001.

Ginsberg and Smith, 1987 M.L. Ginsberg and D.E. Smith. Reasoning about action I: A possible worlds approach. In Frank M. Brown, editor, *The Frame Problem in Artificial Intelligence,* pages 233–258. Morgan Kaufmann, Los Altos, CA, 1987.

Ginsberg, 1986 M.L. Ginsberg. Counterfactuals. *Artificial Intelligence,* 30(35–79), 1986.

Glymour and Cooper, 1999 C.N. Glymour and G.F. Cooper, editors. *Computation, Causation, and Discovery.* MIT Press, Cambridge, MA, 1999.

Glymour and Greenland, 2008 M.M. Glymour and S. Greenland. Causal diagrams. In K.J. Rothman, S. Greenland, and T.L. Lash, editors, *Modern Epidemiology.* Lippincott Williams & Wilkins, Philadelphia, PA, 3rd edition, 2008.

Glymour, 1998 C.N. Glymour. Psychological and normative theories of causal power and the probabilities of causes. In G.F. Cooper and S. Moral, editors, *Uncertainty in Artificial Intelligence,* pages 166–172. Morgan Kaufmann, San Francisco, CA, 1998.

Glymour, 2001 C.N. Glymour. *The Mind's Arrows: Bayes Nets and Graphical Causal Models in Psychology.* The MIT Press, Cambridge, MA, 2001.

Goldberger, 1972 A.S. Goldberger. Structural equation models in the social sciences. *Econometrica: Journal of the Econometric Society,* 40:979–1001, 1972.

Goldberger, 1973 A.S. Goldberger. Structural equation models: An overview. In A.S. Goldberger and O.D. Duncan, editors, *Structural Equation Models in the Social Sciences,* pages 1–18. Seminar Press, New York, NY, 1973.

Goldberger, 1991 A.S. Goldberger. *A Course of Econometrics.* Harvard University Press, Cambridge, MA, 1991.

Goldberger, 1992 A.S. Goldberger. Models of substance; comment on N. Wermuth, 'On block-recursive linear regression equations'. *Brazilian Journal of Probability and Statistics,* 6:1–56, 1992.

Goldszmidt and Pearl, 1992 M. Goldszmidt and J. Pearl. Rank-based systems: A simple approach to belief revision, belief update, and reasoning about evidence and actions. In B. Nebel, C. Rich, and W. Swartout, editors, *Proceedings of the Third International Conference on Knowledge Representation and Reasoning,* pages 661–672. Morgan Kaufmann, San Mateo, CA, 1992.

Good and Mittal, 1987 I.J. Good and Y. Mittal. The amalgamation and geometry of two-by-two contingency tables. *The Annals of Statistics,* 15(2):694–711, 1987.

Good, 1961 I.J. Good. A causal calculus, (I). *British Journal for the Philosophy of Science,* 11:305–318, 1961.

Good, 1962 I.J. Good. A causal calculus (II). *British Journal for the Philosophy of Science,* 12:43–51; 13:88, 1962.

Good, 1993 I.J. Good. A tentative measure of probabilistic causation relevant to the philosophy of the law. *Journal of Statistical Computation and Simulation,* 47:99–105, 1993.

Gopnik et al., 2004 A. Gopnik, C.N. Glymour, D.M. Sobel, L.E. Schulz, T. Kushnir, and D. Danks. A theory of causal learning in children: Causal maps and Bayes nets. *Psychological Review,* 111(1):3–32, 2004.

Granger, 1969 C.W.J. Granger. Investigating causal relations by econometric models and cross spectral methods. *Econometrica; Journal of the Econometric Society,* 37(3):424–438, July 1969.

Granger, 1988 C.W.J. Granger. Causality testing in a decision science. In W. Harper and B. Skyrms, editors, *Causation in Decision, Belief Change and Statistics I,* pages 1–20. Kluwer Academic Publishers, Dordrecht/Boston/London, 1988.

Grayson, 1987 D.A. Grayson. Confounding confounding. *American Journal of Epidemiology,* 126:546–553, 1987.

Greene, 1997 W.H. Greene. *Econometric Analysis.* Prentice Hall, Upper Saddle River, NJ, 1997.

Greenland and Brumback, 2002 S. Greenland and B. Brumback. An overview of relations among causal modelling methods. *International Journal of Epidemiology,* 31:1030–1037, 2002.

Greenland and Neutra, 1980 S. Greenland and R. Neutra. Control of confounding in the assessment of medical technology. *International Journal of Epidemiology,* 9(4):361–367, 1980.

Greenland and Robins, 1986 S. Greenland and J.M. Robins. Identifiability, exchangeability, and epidemiological confounding. *International Journal of Epidemiology,* 15(3):413–419, 1986.

Greenland and Robins, 1988 S. Greenland and J.M Robins. Conceptual problems in the definition and interpretation of attributable fractions. *American Journal of Epidemiology,* 128:1185–1197, 1988.

Greenland et al., 1989 S. Greenland, H. Morgenstern, C. Poole, and J.M. Robins. Re: 'Confounding confounding'. *American Journal of Epidemiology,* 129:1086–1089, 1989.

Greenland et al., 1999a S. Greenland, J. Pearl, and J.M Robins. Causal diagrams for epidemiologic research. *Epidemiology,* 10(1):37–48, 1999.

Greenland et al., 1999b S. Greenland, J.M. Robins, and J. Pearl. Confounding and collapsibility in causal inference. *Statistical Science,* 14(1):29–46, February 1999.

Greenland, 1998 S. Greenland. Confounding. In P. Armitage and T. Colton, editors, *Encyclopedia of Biostatistics,* page 905–6. J. Wiley, New York, 1998.

Gursoy, 2002 K. Gursoy. Book reviews: Causality: Models, Reasoning, and Inference. *IIE Transactions,* 34:583, 2002.

Guyon et al., 2008a I. Guyon, C. Aliferis, G.F. Cooper, A. Elisseeff, J.-P. Pellet, P. Spirtes, and A. Statnikov. Design and analysis of the causation and prediction challenge. *JMLR Workshop and Conference Proceedings,* volume 3: WCCI 2008 causality challenge, Hong Kong, June 3–4 2008.

Guyon et al., 2008b I. Guyon, C. Aliferis, G.F. Cooper, A. Elisseeff, J.-P. Pellet, P. Spirtes, and A. Statnikov. Design and analysis of the causality pot-luck challenge. *JMLR Workshop and Conference Proceedings,* volume 5: NIPS 2008 causality workshop, Whistler, Canada, December 12 2008.

Haavelmo, 1943 T. Haavelmo. The statistical implications of a system of simultaneous equations. *Econometrica,* 11:1–12, 1943. Reprinted in D.F. Hendry and M.S. Morgan (Eds.), *The Foundations of Econometric Analysis,* Cambridge University Press, 477–490, 1995.

Haavelmo, 1944 T. Haavelmo. The probability approach in econometrics (1944)*. Supplement to *Econometrica,* 12:12–17, 26–31, 33–39, 1944. Reprinted in D.F. Hendry and M.S. Morgan (Eds.), *The Foundations of Econometric Analysis,* Cambridge University Press, New York, 440–453, 1995.

Hadlock, 2005 C.R. Hadlock. Book reviews: Causality: Models, Reasoning, and Inference. *Journal of the American Statistical Association,* 100:1095–1096, 2005.

Hall, 2004 N. Hall. Two concepts of causation. In N. Hall, J. Collins, and L.A. Paul, editors, *Causation and Counterfactuals*, Chapter 9. MIT Press, Cambridge, MA, 2004.

Hall, 2007 N. Hall. Structural equations and causation. *Philosophical Studies,* 132:109–136, 2007.

Halpern and Pearl, 1999 J.Y. Halpern and J. Pearl. Actual causality. Technical Report R-266, University of California Los Angeles, Cognitive Systems Lab, Los Angeles, 1999.

Halpern and Pearl, 2000 J.Y. Halpern and J. Pearl. Causes and explanations. Technical Report R-266, Cognitive Systems Laboratory, Department of Computer Science, University of California, Los Angeles, CA, March 2000. Online at ⟨www.cs.ucla.edu/~judea/⟩.

Halpern and Pearl, 2001a J.Y. Halpern and J. Pearl. Causes and explanations: A structural-model approach—Part I: Causes. In *Proceedings of the Seventeenth Conference on Uncertainty in Artificial Intelligence,* pages 194–202. Morgan Kaufmann, San Francisco, CA, 2001.

Halpern and Pearl, 2001b J.Y. Halpern and J. Pearl. Causes and explanations: A structural-model approach—Part II: Explanations. In *Proceedings of the International Joint Conference on Artificial Intelligence,* pages 27–34. Morgan Kaufmann, CA, 2001.

Halpern and Pearl, 2005a J.Y. Halpern and J. Pearl. Causes and explanations: A structural-model approach—Part I: Causes. *British Journal of Philosophy of Science,* 56:843–887, 2005.

Halpern and Pearl, 2005b J.Y. Halpern and J. Pearl. Causes and explanations: A structural-model approach—Part II: Explanations. *British Journal of Philosophy of Science,* 56:843–887, 2005.

Halpern, 1998 J.Y. Halpern. Axiomatizing causal reasoning. In G.F. Cooper and S. Moral, editors, *Uncertainty in Artificial Intelligence,* pages 202–210. Morgan Kaufmann, San Francisco, CA, 1998. Also, *Journal of Artificial Intelligence Research* 12:3, 17–37, 2000.

Halpern, 2008 J.Y. Halpern. Defaults and normality in causal structures. In G. Brewka and J. Lang, editors, *Proceedings of the Eleventh International Conference on Principles of Knowledge Representation and Reasoning (KR 2008)*, page 198–208. Morgan Kaufmann, San Mateo, CA, 2008.

Hauck et al., 1991 W.W. Hauck, J.M. Heuhaus, J.D. Kalbfleisch, and S. Anderson. A consequence of omitted covariates when estimating odds ratios. *Journal of Clinical Epidemiology,* 44(1):77–81, 1991.

Hausman, 1998 D.M. Hausman. *Causal Asymmetries.* Cambridge University Press, New York, 1998.

Hayduk, 1987 L.A. Hayduk. *Structural Equation Modeling with LISREL, Essentials and Advances.* Johns Hopkins University Press, Baltimore, 1987.

Heckerman and Shachter, 1995 D. Heckerman and R. Shachter. Decision-theoretic foundations for causal reasoning. *Journal of Artificial Intelligence Research,* 3:405–430, 1995.

Heckerman et al., 1994 D. Heckerman, D. Geiger, and D. Chickering. Learning Bayesian networks: The combination of knowledge and statistical data. In R. Lopez de Mantaras and D. Poole, editors, *Uncertainty in Artificial Intelligence 10,* pages 293–301. Morgan Kaufmann, San Mateo, CA, 1994.

Heckerman et al., 1995 Guest Editors: David Heckerman, Abe Mamdani, and Michael P. Wellman. Real-world applications of Bayesian networks. *Communications of the ACM,* 38(3):24–68, March 1995.

Heckerman et al., 1999 D. Heckerman, C. Meek, and G.F. Cooper. A Bayesian approach to causal discovery. In C. Glymour and G. Cooper, editors, *Computation, Causation, and Discovery,* The MIT Press, Cambridge, MA, 143–167, 1999.

Heckman and Honoré, 1990 J.J. Heckman and B.E. Honoré. The empirical content of the Roy model. *Econometrica,* 58:1121–1149, 1990.

Heckman and Robb, 1986 J.J. Heckman and R.R. Robb. Alternative methods for solving the problem of selection bias in evaluating the impact of treatments on outcomes. In H. Wainer, editor, *Drawing Inference From Self Selected Samples,* pages 63–107. Springer-Verlag, New York, NY, 1986.

Heckman and Vytlacil, 1999 J.J. Heckman and E.J. Vytlacil. Local instrumental variables and latent variable models for identifying and bounding treatment effects. *Proceedings of the National Academy of Sciences, USA,* 96(8):4730–4734, April 1999.

Heckman and Vytlacil, 2007 J.J. Heckman and E.J. Vytlacil. *Handbook of Econometrics,* volume 6B, Econometric Evaluation of Social Programs, Part I: Causal Models, Structural Models and Econometric Policy Evaluation, pages 4779–4874. Elsevier B.V., 2007.

Heckman et al., 1998 J.J. Heckman, H. Ichimura, and P. Todd. Matching as an econometric evaluation estimator. *Review of Economic Studies,* 65:261–294, 1998.

Heckman, 1992 J.J. Heckman. Randomization and social policy evaluation. In C. Manski and I. Garfinkle, editors, *Evaluations: Welfare and Training Programs,* pages 201–230. Harvard University Press, Cambridge, MA, 1992.

Heckman, 1996 J.J. Heckman. Comment on 'Identification of causal effects using instrumental variables'. *Journal of the American Statistical Association,* 91(434):459–462, June 1996.

Heckman, 2000 J.J. Heckman. Causal parameters and policy analysis in economics: A twentieth century retrospective. *The Quarterly Journal of Economics,* 115(1):45–97, 2000.

Heckman, 2003 J.J. Heckman. Conditioning causality and policy analysis. *Journal of Econometrics,* 112(1):73–78, 2003.

Heckman, 2005 J.J. Heckman. The scientific model of causality. *Sociological Methodology,* 35:1–97, 2005.

Heise, 1975 D.R. Heise. *Causal Analysis.* John Wiley and Sons, New York, 1975.

Hendry and Morgan, 1995 D.F. Hendry and M.S. Morgan. *The Foundations of Econometric Analysis.* Cambridge University Press, Cambridge, 1995.

Hendry, 1995 David F. Hendry. *Dynamic Econometrics.* Oxford University Press, New York, 1995.

Hennekens and Buring, 1987 C.H. Hennekens and J.E. Buring. *Epidemiology in Medicine.* Little, Brown, Boston, 1987.

Hernán et al., 2002 M.A. Hernán, S. Hernández-Díaz, M.M. Werler, and A.A. Mitchell. Causal knowledge as a prerequisite for confounding evaluation: An application to birth defects epidemiology. *American Journal of Epidemiology,* 155(2):176–184, 2002.

Hernán et al., 2004 M.A. Hernán, S. Hernández-Díaz, and J.M. Robins. A structural approach to selection bias. *Epidemiology,* 15(5):615–625, 2004.

Hernández-Díaz et al., 2006 S. Hernández-Díaz, E.F. Schisterman, and Hernán M.A. The birth weight "paradox" uncovered? *American Journal of Epidemiology,* 164(11):1115–1120, 2006.

Hesslow, 1976 G. Hesslow. Discussion: Two notes on the probabilistic approach to causality. *Philosophy of Science,* 43:290–292, 1976.

Hiddleston, 2005 E. Hiddleston. Causal powers. *British Journal for Philosophy of Science,* 56:27–59, 2005.

Hitchcock, 1995 C. Hitchcock. The mishap of Reichenbach's fall: Singular vs. general causation. *Philosophical Studies,* 78:257–291, 1995.

Hitchcock, 1996 C.R. Hitchcock. Causal decision theory and decision theoretic causation. *Nous,* 30(4):508–526, 1996.

Hitchcock, 1997 C. Hitchcock. Causation, probabilistic, 1997. In *Stanford Encyclopedia of Philosophy,* online at: http://plato.stanford.edu/entries/causation-probabilistic.

Hitchcock, 2001 C. Hitchcock. Book reviews: Causality: Models, Reasoning, and Inference. *The Philosophical Review,* 110(4):639–641, 2001.

Hitchcock, 2007 C.R. Hitchcock. Prevention, preemption, and the principle of sufficient reason. *Philosophical Review,* 116:495–532, 2007.

Hitchcock, 2008 C.R. Hitchcock. Structural equations and causation: Six counterexamples. *Philosophical Studies,* page DOI 10.1007/s 11098–008–9216–2, 2008.

Hoel et al., 1971 P.G. Hoel, S.C. Port, and C.J. Stone. *Introduction to Probability Theory.* Houghton Mifflin Company, Boston, 1971.

Holland and Rubin, 1983 P.W. Holland and D.B. Rubin. On Lord's paradox. In H. Wainer and S. Messick, editors, *Principals of Modern Psychological Measurement,* pages 3–25. Lawrence Earlbaum, Hillsdale, NJ, 1983.

Holland, 1986 P.W. Holland. Statistics and causal inference. *Journal of the American Statistical Association,* 81(396):945–960, December 1986.

Holland, 1988 P.W. Holland. Causal inference, path analysis, and recursive structural equations models. In C. Clogg, editor, *Sociological Methodology,* pages 449–484. American Sociological Association, Washington, D.C., 1988.

Holland, 1995 P.W. Holland. Some reflections on Freedman's critiques. *Foundations of Science,* 1:50–57, 1995.

Holland, 2001 P.W. Holland. The false linking of race and causality: Lessons from standardized testing. *Race and Society,* 4(2): 219–233, 2001.

Hoover, 1990 K.D. Hoover. The logic of causal inference. *Economics and Philosophy,* 6:207–234, 1990.

Hoover, 2001 K. Hoover. *Causality in Macroeconomics.* Cambridge University Press, New York, 2001.

Hoover, 2003 K.D. Hoover. Book reviews: Causality: Models, Reasoning, and Inference. *Economic Journal,* 113:F411–F413, 2003.

Hoover, 2004 K.D. Hoover. Lost causes. *Journal of the History of Economic Thought,* 26(2):149–164, June 2004.

Hoover, 2008 K.D. Hoover. Causality in economics and econometrics. In S.N. Durlauf and L.E. Blume, editors, *From The New Palgrave Dictionary of Economics.* Palgrave Macmillan, New York, NY, 2nd edition, 2008.

Hopkins and Pearl, 2002 M. Hopkins and J Pearl. Strategies for determining causes of events. In *Proceedings of the Eighteenth National Conference on Artificial Intelligence,* pages 546–552. AAAI Press/ The MIT Press, Menlo Park, CA, 2002.

Howard and Matheson, 1981 R.A. Howard and J.E. Matheson. Influence diagrams. *Principles and Applications of Decision Analysis,* 1981. Strategic Decisions Group, Menlo Park, CA. Reprinted in *Decision Analysis* 2(3): 129–143, 2005.

Howard, 1960 R.A. Howard. *Dynamic Programming and Markov Processes.* MIT Press, Cambridge, MA, 1960.

Howard, 1990 R.A. Howard. From influence to relevance to knowledge. In R.M. Oliver and J.Q. Smith, editors, *Influence Diagrams, Belief Nets, and Decision Analysis,* pages 3–23. Wiley and Sons, Ltd., New York, NY, 1990.

Hoyer et al., 2006 P. Hoyer, S. Shimizu, and A.J. Kerminen. Estimation of linear, non-Gaussian causal models in presence of confounding latent variables. In *Proceedings of the Third European Workshop on Probabilistic Graphical Models (PGM'06),* pages 155–162. Institute of Information Theory and Automation, Prague, Czech Republic, 2006.

Huang and Valtorta, 2006 Y. Huang and M. Valtorta. Pearl's calculus of intervention is complete. In R. Dechter and T.S. Richardson, editors, *Proceedings of the Twenty-Second Conference on Uncertainty in Artificial Intelligence,* pages 217–224. AUAI Press, Corvallis, OR, 2006.

Hume,1739 D. Hume. *A Treatise of Human Nature*. Oxford University Press, Oxford, 1739. Reprinted 1888.

Hume, 1748 D. Hume. *An enquiry concerning human understanding*. Reprinted Open Court Press (1958), LaSalle, IL, 1748.

Humphreys and Freedman, 1996 P. Humphreys and D. Freedman. The grand leap. *British Journal for the Philosophy of Science*, 47:113–123, 1996.

Hurwicz, 1962 L. Hurwicz. On the structural form of interdependent systems. In E. Nagel, P. Suppes, and A. Tarski, editors, *Logic, Methodology, and Philosophy of Science,* pages 232–239. Stanford University Press, Stanford CA, 1962.

Imai et al., 2008 K. Imai, L. Keele, and T. Yamamoto. Identification, inference, and sensitivity analysis for causal mediation effects. Technical report, Department of Politics, Princeton University, December 2008.

Imbens and Angrist, 1994 G.W. Imbens and J.D. Angrist. Identification and estimation of local average treatment effects. *Econometrica,* 62(2):467–475, March 1994.

Imbens and Rubin, 1997 G.W. Imbens and D.R. Rubin. Bayesian inference for causal effects in randomized experiments with noncompliance. *Annals of Statistics,* 25:305–327, 1997.

Imbens, 1997 G.W. Imbens. Book reviews. *Journal of Applied Econometrics,* 12(1): 91–94, 1997.

Imbens, 2004 G.W. Imbens. Nonparametric estimation of average treatment effects under exogeneity: A review. *The Review of Economics and Statistics,* 86(1):4–29, 2004.

Intriligator et al., 1996 M.D. Intriligator, R.G. Bodkin, and C. Hsiao. *Econometric Models, Techniques, and Applications*. Prentice-Hall, Saddle River, NJ, 2nd edition, 1996.

Isham, 1981 V. Isham. An introduction to spatial point processes and Markov random fields. *International Statistical Review,* 49:21–43, 1981.

Iwasaki and Simon, 1986 Y. Iwasaki and H.A. Simon. Causality in device behavior. *Artificial Intelligence,* 29(1):3–32, 1986.

James et al., 1982 L.R. James, S.A. Mulaik, and J.M. Brett. *Causal Analysis: Assumptions, Models, and Data*. Studying Organizations, 1. Sage, Beverly Hills, 1982.

Jeffrey, 1965 R. Jeffrey. *The Logic of Decisions*. McGraw-Hill, New York, 1965.

Jensen, 1996 F.V. Jensen. *An Introduction to Bayesian Networks*. Springer, New York, 1996.

Jordan, 1998 M.I. Jordan. *Learning in Graphical Models*. Kluwer Academic Publishers, Dordrecht, series D: Behavioural and Social Sciences – vol. 89 edition, 1998.

Katsuno and Mendelzon, 1991 H. Katsuno and A.O. Mendelzon. On the difference between updating a knowledge base and revising it. In J.A. Allen, R. Fikes, and E. Sandewall, editors, *Principles of Knowledge Representation and Reasoning: Proceedings of the Second International Conference,* pages 387–394, Morgan Kaufmann, San Mateo, CA, 1991.

Kaufman et al., 2005 S. Kaufman, J.S. Kaufman, R.F. MacLenose, S. Greenland, and C. Poole. Improved estimation of controlled direct effects in the presence of unmeasured confounding of intermediate variables. *Statistics in Medicine,* 25:1683–1702, 2005.

Kennedy, 2003 P. Kennedy. *A Guide to Econometrics*. MIT Press, Cambridge, MA, 5th edition, 2003.

Khoury et al., 1989 M.J. Khoury, W.D Flanders, S. Greenland, and M.J. Adams. On the measurement of susceptibility in epidemiologic studies. *American Journal of Epidemiology,* 129(1):183–190, 1989.

Kiiveri et al., 1984 H. Kiiveri, T.P. Speed, and J.B. Carlin. Recursive causal models. *Journal of Australian Math Society,* 36:30–52, 1984.

Kim and Pearl, 1983 J.H. Kim and J. Pearl. A computational model for combined causal and diagnostic reasoning in inference systems. In *Proceedings of the Eighth International Joint Conference on Artificial Intelligence (IJCAI-83),* pages 190–193. Karlsruhe, Germany, 1983.

Kim, 1971 J. Kim. Causes and events: Mackie on causation. *Journal of Philosophy,* 68:426–471, 1971. Reprinted in E. Sosa and M. Tooley (Eds.), *Causation,* Oxford University Press, 1993.

King et al., 1994 G. King, R.O. Keohane, and S. Verba. *Designing Social Inquiry: Scientific Inference in Qualitative Research*. Princeton University Press, Princeton, NJ, 1994.

Kleinbaum et al., 1982 D.G. Kleinbaum, L.L. Kupper, and H. Morgenstern. *Epidemiologic Research*. Lifetime Learning Publications, Belmont, California, 1982.

Kline, 1998 R.B. Kline. *Principles and Practice of Structural Equation Modeling.* The Guilford Press, New York, 1998.

Koopmans et al., 1950 T.C. Koopmans, H. Rubin, and R.B. Leipnik. Measuring the equation systems of dynamic economics. In T.C. Koopmans, editor, *Statistical Inference in Dynamic Economic Models,* pages 53–237. John Wiley, New York, 1950.

Koopmans, 1950 T.C. Koopmans. When is an equation system complete for statistical purposes? In T.C. Koopmans, editor, *Statistical Inference in Dynamic Economic Models,* Cowles Commission, Monograph 10. Wiley, New York, 1950. Reprinted in D.F. Hendry and M.S. Morgan (Eds.), *The Foundations of Econometric Analysis,* pages 527–537. Cambridge University Press, 1995.

Koopmans, 1953 T.C. Koopmans. Identification problems in econometric model construction. In W.C. Hood and T.C. Koopmans, editors, *Studies in Econometric Method,* pages 27–48. Wiley, New York, 1953.

Korb and Wallace, 1997 K.B. Korb and C.S. Wallace. In search of the philosopher's stone: Remarks on Humphreys and Freedman's critique of causal discovery. *British Journal for the Philosophy of Science,* 48:543–553, 1997.

Koster, 1999 J.T.A. Koster. On the validity of the Markov interpretation of path diagrams of Gaussian structural equations systems with correlated errors. *Scandinavian Journal of Statistics,* 26:413–431, 1999.

Kramer and Shapiro, 1984 M.S. Kramer and S. Shapiro. Scientific challenges in the application of randomized trials. *Journal of the American Medical Association,* 252:2739–2745, November 1984.

Kraus et al., 1990 S. Kraus, D. Lehmann, and M. Magidor. Nonmonotonic reasoning, preferential models and cumulative logics. *Artificial Intelligence,* 44:167–207, 1990.

Kuroki and Cai, 2004 M. Kuroki and Z. Cai. Selection of identifiability criteria for total effects by using path diagrams. In M. Chickering and J. Halpern, editors, *Uncertainty in Artificial Intelligence, Proceedings of the Twentieth Conference,* pages 333–340. AUAI, Arlington, VA, 2004.

Kuroki and Miyakawa, 1999a M. Kuroki and M. Miyakawa. Estimation of causal effects in causal diagrams and its application to process analysis (in Japanese). *Journal of the Japanese Society for Quality Control,* 29:237–247, 1999.

Kuroki and Miyakawa, 1999b M. Kuroki and M. Miyakawa. Identifiability criteria for causal effects of joint interventions. *Journal of the Japan Statistical Society,* 29:105–117, 1999.

Kuroki and Miyakawa, 2003 M. Kuroki and M. Miyakawa. Covariate selection for estimating the causal effect of control plans using causal diagrams. *Journal of the Royal Statistical Society, Series B,* 65:209–222, 2003.

Kuroki et al., 2003 M. Kuroki, M. Miyakawa, and Z. Cai. Joint causal effect in linear structural equation model and its application to process analysis. *Artificial Intelligence and Statistics,* 9:70–77, 2003.

Kvart, 1986 I. Kvart. *A Theory of Counterfactuals.* Hackett Publishing, Co., Indianapolis, 1986.

Kyburg Jr., 2005 H.E. Kyburg Jr. Book review: Judea Pearl, Causality, Cambridge University Press, 2000. *Artificial Intelligence,* 169:174–179, 2005.

Laplace, 1814 P.S. Laplace. *Essai Philosophique sure les Probabilités.* Courcier, New York, 1814. English translation by F.W. Truscott and EL. Emory, Wiley, NY, 1902.

Lauritzen and Richardson, 2002 S.L. Lauritzen and T.S. Richardson. Chain graph models and their causal interpretations. *Royal Statistical Society,* 64(Part 2): 1–28, 2002.

Lauritzen and Spiegelhalter, 1988 S.L. Lauritzen and D.J. Spiegelhalter. Local computations with probabilities on graphical structures and their application to expert systems (with discussion). *Journal of the Royal Statistical Society, Series B,* 50(2): 157–224, 1988.

Lauritzen et al., 1990 S.L. Lauritzen, A.P. Dawid, B.N. Larsen, and H.G. Leimer. Independence properties of directed Markov fields. *Networks,* 20:491–505, 1990.

Lauritzen, 1982 S.L. Lauritzen. *Lectures on Contingency Tables.* University of Aalborg Press, Aalborg, Denmark, 2nd edition, 1982.

Lauritzen, 1996 S.L. Lauritzen. *Graphical Models.* Clarendon Press, Oxford, 1996.

Lauritzen, 2001 S.L. Lauritzen. Causal inference from graphical models. In D.R. Cox and C. Kluppelberg, editors, *Complex Stochastic Systems*, pages 63–107. Chapman and Hall/CRC Press, Boca Raton, FL, 2001.

Lauritzen, 2004 S.L. Lauritzen. Discussion on causality. *Scandinavian Journal of Statistics*, 31: 189–192, 2004.

Lawry, 2001 J. Lawry. Review: Judea Pearl, Causality: Models, Reasoning, and Inference. *MathSciNet, Mathematical Reviews on the Web*, MR1744773((2001d:68213)):http://www.ams.org/mathscinet–getitem?mr=1744773, 2001.

Leamer, 1985 E.E. Leamer. Vector autoregressions for causal inference? *Carnegie-Rochester Conference Series on Public Policy*, 22:255–304, 1985.

Lee and Hershberger, 1990 S. Lee and S.A. Hershberger. A simple rule for generating equivalent models in covariance structure modeling. *Multivariate Behavioral Research*, 25(3):313–334, 1990.

Lemmer, 1993 J.F. Lemmer. Causal modeling. In D. Heckerman and A. Mamdani, editors, *Proceedings of the Ninth Conference on Uncertainty in Artificial Intelligence*, pages 143–151. Morgan Kaufmann, San Mateo, CA, 1993.

Leroy, 1995 S.F. Leroy. Causal orderings. In K.D. Hoover, editor, *Macroeconometrics: Developments, Tensions, Prospects*, pages 211–227. Kluwer Academic, Boston, 1995.

Leroy, 2002 S.F. Leroy. A review of Judea Pearl's Causality. *Journal of Economic Methodology*, 9(1): 100–103, 2002.

Levi, 1988 I. Levi. Iteration of conditionals and the Ramsey test. *Synthese*, 76:49–81, 1988.

Lewis, 1973a D. Lewis. Causation. *Journal of Philosophy*, 70:556–567, 1973.

Lewis, 1973b D. Lewis. *Counterfactuals*. Harvard University Press, Cambridge, MA, 1973.

Lewis, 1973c D. Lewis. Counterfactuals and comparative possibility, 1973. In W.L. Harper, R. Stalnaker, and G. Pearce (Eds.), *Ifs*, pages 57–85, D. Reidel, Dordrecht, 1981.

Lewis, 1976 D. Lewis. Probabilities of conditionals and conditional probabilities. *Philosophical Review*, 85:297–315, 1976.

Lewis, 1979 D. Lewis. Counterfactual dependence and time's arrow. *Nous*, 13:418–446, 1979.

Lewis, 1986 D. Lewis. *Philosophical Papers*, volume II. Oxford University Press, New York, 1986.

Lin, 1995 F. Lin. Embracing causality in specifying the indeterminate effects of actions. In *Proceedings of the Fourteenth International Joint Conference on Artificial Intelligence (IJCAI-95)*, Montreal, Quebec, 1995.

Lindley and Novick, 1981 D.V. Lindley and M.R. Novick. The role of exchangeability in inference. *The Annals of Statistics*, 9(1):45–58, 1981.

Lindley, 2002 D.V. Lindley. Seeing and doing: The concept of causation. *International Statistical Review*, 70:191–214, 2002.

Lucas Jr., 1976 R.E. Lucas Jr. Econometric policy evaluation: A critique. In K. Brunner and A.H. Meltzer, editors, *The Phillips Curve and Labor Markets*, Vol. 1 of the Carnegie-Rochester Conferences on Public Policy, supplementary series to the *Journal of Monetary Economics*, pages 19–46. North-Holland, Amsterdam, 1976.

Luellen et al., 2005 J.K. Luellen, W.R. Shadish, and M.H. Clark. Propensity scores: An introduction and experimental test. *Evaluation Review*, 29(6):530–558, 2005.

MacCallum et al., 1993 R.C. MacCallum, D.T. Wegener, B.N. Uchino, and L.R. Fabrigar. The problem of equivalent models in applications of covariance structure analysis. *Psychological Bulletin*, 114(1): 185–199, 1993.

Mackie, 1965 J.L. Mackie. Causes and conditions. *American Philosophical Quarterly*, 2/4:261–264, 1965. Reprinted in E. Sosa and M. Tooley (Eds.), *Causation*, Oxford University Press, New York, 1993.

Mackie, 1980 J.L. Mackie. *The Cement of the Universe: A Study of Causation*. Clarendon Press, Oxford, 1980.

Maddala, 1992 G.S. Maddala. *Introduction to Econometrics*. McMillan, New York, NY, 1992.

Manski, 1990 C.F. Manski. Nonparametric bounds on treatment effects. *American Economic Review, Papers and Proceedings*, 80:319–323, 1990.

Manski, 1995 C.F. Manski. *Identification Problems in the Social Sciences*. Harvard University Press, Cambridge, MA, 1995.

Marschak, 1950 J. Marschak. Statistical inference in economics. In T. Koopmans, editor, *Statistical Inference in Dynamic Economic Models,* pages 1–50. Wiley, New York, 1950. Cowles Commission for Research in Economics, Monograph 10.

Maudlin, 1994 T. Maudlin. *Quantum Non-Locality and Relativity: Metaphysical Intimations of Modern Physics.* Blackwell, Oxford, UK, 1994.

McDonald, 1997 R.P. McDonald. Haldane's lungs: A case study in path analysis. *Multivariate Behavioral Research,* 32(1): 1–38, 1997.

McDonald, 2001 R.P. McDonald. Book reviews: Causality: Models, Reasoning, and Inference. *Chance,* 14(1):36–37, 2001.

McDonald, 2002a R.P. McDonald. What can we learn from the path equations?: Identifiability, constraints, equivalence. *Psychometrika,* 67(2):225–249, 2002.

McDonald, 2002b R.P. McDonald. Review: Judea Pearl, Causality: Models, Reasoning, and Inference. *Psychometrika,* 67(2):321–322, 2002.

McKim and Turner, 1997 V.R. McKim and S.P. Turner (Eds.). *Causality in Crisis?* University of Notre Dame Press, Notre Dame, IN, 1997.

Meek and Glymour, 1994 C. Meek and C.N. Glymour. Conditioning and intervening. *British Journal of Philosophy Science,* 45:1001–1021, 1994.

Meek, 1995 C. Meek. Causal inference and causal explanation with background knowledge. In P. Besnard and S. Hanks, editors, *Uncertainty in Artificial Intelligence 11,* pages 403–410. Morgan Kaufmann, San Francisco, 1995.

Mesarovic, 1969 M.D. Mesarovic. Mathematical theory of general systems and some economic problems. In H.W. Kuhn and G.P. Szego, editors, *Mathematical Systems and Economics I,* pages 93–116. Springer Verlag, Berlin, 1969.

Michie, 1999 D. Michie. Adapting Good's q theory to the causation of individual events. *Machine Intelligence,* 15:60–86, 1999.

Miettinen and Cook, 1981 O.S. Miettinen and E.F. Cook. Confounding essence and detection. *American Journal of Epidemiology,* 114:593–603, 1981.

Mill, 1843 J.S. Mill. *System of Logic,* volume 1. John W. Parker, London, 1843.

Mitchell, 1982 T.M. Mitchell. Generalization as search. *Artificial Intelligence,* 18:203–226, 1982.

Mittelhammer et al., 2000 R.C. Mittelhammer, G.G. Judge, and D.J. Miller. *Econometric Foundations.* Cambridge University Press, New York, NY, 2000.

Moneta and Spirtes, 2006 A. Moneta and P. Spirtes. Graphical models for identification of causal structures in multivariate time series models. In *Proceedings of the Ninth Joint Conference on Information Sciences,* Atlantis Press, Kaohsiung, Taiwan, 2006.

Moole, 1997 B.R. Moole. Parallel construction of Bayesian belief networks. Master's thesis, Department of Computer Science, University of South Carolina, Columbia, SC, 1997.

Moore and McCabe, 2005 D.S. Moore and G.P. McCabe. *Introduction to the Practice of Statistics.* W.H. Freeman and Co., Gordonsville, VA, 2005.

Morgan and Winship, 2007 S.L. Morgan and C. Winship. *Counterfactuals and Causal Inference: Methods and Principles for Social Research (Analytical Methods for Social Research).* Cambridge University Press, New York, NY, 2007.

Morgan, 2004 S.L. Morgan. Book reviews: Causality: Models, Reasoning, and Inference. *Sociological Methods and Research,* 32(3):411–416, 2004.

Mueller, 1996 R.O. Mueller. *Basic Principles of Structural Equation Modeling.* Springer, New York, 1996.

Muthen, 1987 B. Muthen. Response to Freedman's critique of path analysis: Improve credibility by better methodological training. *Journal of Educational Statistics,* 12(2): 178–184, 1987.

Nayak, 1994 P. Nayak. Causal approximations. *Artificial Intelligence,* 70:277–334, 1994.

Neuberg, 2003 L.G. Neuberg. Causality: Models, Reasoning, and Inference, reviewed by L.G. Neuberg. *Econometric Theory,* 19:675–685, 2003.

Neyman, 1923 J. Neyman. Sur les applications de la thar des probabilities aux experiences Agaricales: Essay des principle, 1923. English translation of excerpts (1990) by D. Dabrowska and T. Speed, in *Statistical Science,* 5:463–472.

Niles, 1922 H.E. Niles. Correlation, causation, and Wright's theory of "path coefficients." *Genetics,* 7:258–273, 1922.

Novick, 1983 M.R. Novick. The centrality of Lord's paradox and exchangeability for all statistical inference. In H. Wainer and S. Messick, editors, *Principals of Modern Psychological Measurement.* Earlbaum, Hillsdale, NJ, 1983.

Nozick, 1969 R. Nozick. Newcomb's problem and two principles of choice. In N. Rescher, editor, *Essays in Honor of Carl G. Hempel,* pages 114–146. D. Reidel, Dordrecht, 1969.

Orcutt, 1952 G.H. Orcutt. Toward a partial redirection of econometrics. *Review of Economics and Statistics,* 34:195–213, 1952.

O'Rourke, 2001 J. O'Rourke. Book reviews: Causality: Models, Reasoning, and Inference. *Intelligence,* 12(3):47–54, 2001.

Ortiz, Jr., 1999 C.L. Ortiz, Jr. Explanatory update theory: Applications of counterfactual reasoning to causation. *Artificial Intelligence,* 108(1–2): 125–178, 1999.

Otte, 1981 R. Otte. A critque of Suppes' theory of probabilistic causality. *Synthese,* 48:167–189, 1981.

Palca, 1989 J. Palca. Aids drug trials enter new age. *Science Magazine,* pages 19–21, October 1989.

Paul, 1998 L.A. Paul. Keeping track of the time: Emending the counterfactual analysis of causation. *Analysis,* 3:191–198, 1998.

Payson, 2001 S. Payson. Book review: Causality: Models, Reasoning, and Inference. *Technological Forecasting & Social Change,* 68:105–108, 2001.

Paz and Pearl, 1994 A. Paz and J. Pearl. Axiomatic characterization of directed graphs. Technical Report R-234, Department of Computer Science, University of California, Los Angeles, CA, 1994.

Paz et al., 1996 A. Paz, J. Pearl, and S. Ur. A new characterization of graphs based on interception relations. *Journal of Graph Theory,* 22(2): 125–136, 1996.

Pearl and Meshkat, 1999 J. Pearl and P. Meshkat. Testing regression models with fewer regressors. In D. Heckerman and J. Whittaker, editors, *Artificial Intelligence and Statistics 99,* pages 255–259. Morgan Kaufmann, San Francisco, CA, 1999.

Pearl and Paz, 1987 J. Pearl and A. Paz. Graphoids: A graph–based logic for reasoning about relevance relations. In B. Duboulay, D. Hogg, and L. Steels, editors, *Advances in Artificial Intelligence-II,* pages 357–363. North-Holland Publishing Co., Amsterdam, 1987.

Pearl and Paz, 2008 J. Pearl and A. Paz. Confounding equivalence in observational studies. Technical Report TR-343, University of California Los Angeles, Cognitive Systems Lab, Los Angeles, September 2008.

Pearl and Robins, 1995 J. Pearl and J.M. Robins. Probabilistic evaluation of sequential plans from causal models with hidden variables. In P. Besnard and S. Hanks, editors, *Uncertainty in Artificial Intelligence 11,* pages 444–453. Morgan Kaufmann, San Francisco, 1995.

Pearl and Verma, 1987 J. Pearl and T. Verma. The logic of representing dependencies by directed acyclic graphs. In *Proceedings of the Sixth National Conference on AI (AAAI-87),* pages 374–379, Seattle, WA, July 1987.

Pearl and Verma, 1991 J. Pearl and T. Verma. A theory of inferred causation. In J.A. Allen, R. Fikes, and E. Sandewall, editors, *Principles of Knowledge Representation and Reasoning: Proceedings of the Second International Conference,* pages 441–452. Morgan Kaufmann, San Mateo, CA, 1991.

Pearl, 1978 J. Pearl. On the connection between the complexity and credibility of inferred models. *International Journal of General Systems,* 4:255–264, 1978.

Pearl, 1982 J. Pearl. Reverend Bayes on inference engines: A distributed hierarchical approach. In *Proceedings AAAI National Conference on AI,* pages 133–136, Pittsburgh, PA, 1982.

Pearl, 1985 J. Pearl. Bayesian networks: A model of self-activated memory for evidential reasoning. In *Proceedings, Cognitive Science Society,* pages 329–334, Irvine, CA, 1985.

Pearl, 1988a J. Pearl. Embracing causality in formal reasoning. *Artificial Intelligence,* 35(2):259–271, 1988.

Pearl, 1988b J. Pearl. *Probabilistic Reasoning in Intelligent Systems.* Morgan Kaufmann, San Mateo, CA, 1988.

Pearl, 1990a J. Pearl. Probabilistic and qualitative abduction. In *Proceedings of AAAI Spring Symposium on Abduction,* pages 155–158, Stanford, CA, 1990.

Pearl, 1990b J. Pearl. System Z: A natural ordering of defaults with tractable applications to default reasoning. In R. Parikh, editor, *Proceedings of the Conference on Theoretical Aspects of Reasoning About Knowledge,* pages 121–135, San Mateo, CA, 1990. Morgan Kaufmann Publishers.

Pearl, 1993a J. Pearl. Belief networks revisited. *Artificial Intelligence,* 59:49–56, 1993.

Pearl, 1993b J. Pearl. Comment: Graphical models, causality, and intervention. *Statistical Science,* 8(3):266–269, 1993.

Pearl, 1993c J. Pearl. From conditional oughts to qualitative decision theory. In D. Heckerman and A. Mamdani, editors, *Proceedings of the Ninth Conference on Uncertainty in Artificial Intelligence,* pages 12–20, San Mateo, CA, July 1993. Morgan Kaufmann Publishers.

Pearl, 1994a J. Pearl. From Bayesian networks to causal networks. In A. Gammerman, editor, *Bayesian Networks and Probabilistic Reasoning,* pages 1–31. Alfred Walter Ltd., London, 1994.

Pearl, 1994b J. Pearl. A probabilistic calculus of actions. In R. Lopez de Mantaras and D. Poole, editors, *Uncertainty in Artificial Intelligence 10,* pages 454–462. Morgan Kaufmann, San Mateo, CA, 1994.

Pearl, 1995a J. Pearl. Causal diagrams for empirical research. *Biometrika,* 82(4):669–710, December 1995.

Pearl, 1995b J. Pearl. Causal inference from indirect experiments. *Artificial Intelligence in Medicine,* 7(6):561–582, 1995.

Pearl, 1995c J. Pearl. On the testability of causal models with latent and instrumental variables. In P. Besnard and S. Hanks, editors, *Uncertainty in Artificial Intelligence 11,* pages 435–443. Morgan Kaufmann, San Francisco, 1995.

Pearl, 1996 J. Pearl. Structural and probabilistic causality. In D.R. Shanks, K.J. Holyoak, and D.L. Medin, editors, *The Psychology of Learning and Motivation,* volume 34, pages 393–435. Academic Press, San Diego, CA, 1996.

Pearl, 1998a J. Pearl. Graphs, causality, and structural equation models. *Sociological Methods and Research,* 27(2):226–284, 1998.

Pearl, 1998b J. Pearl. On the definition of actual cause. Technical Report R-259, Department of Computer Science, University of California, Los Angeles, CA, 1998.

Pearl, 1999 J. Pearl. Probabilities of causation: Three counterfactual interpretations and their identification. *Synthese,* 121(1–2):93–149, November 1999.

Pearl, 2000 J. Pearl. Comment on A.P. Dawid's Causal inference without counterfactuals. *Journal of the American Statistical Association,* 95(450):428–431, June 2000.

Pearl, 2001a J. Pearl. Bayesianism and causality, or, why I am only a half-Bayesian. In D. Corfield and J. Williamson, editors, *Foundations of Bayesianism,* Applied Logic Series, Volume 24, pages 19–36. Kluwer Academic Publishers, the Netherlands, 2001.

Pearl, 2001b J. Pearl. Causal inference in the health sciences: A conceptual introduction. *Health Services and Outcomes Research Methodology,* 2:189–220, 2001. Special issue on Causal Inference.

Pearl, 2001c J. Pearl. Direct and indirect effects. In *Proceedings of the Seventeenth Conference on Uncertainty in Artificial Intelligence,* pages 411–420. Morgan Kaufmann, San Francisco, CA, 2001.

Pearl, 2003a J. Pearl. Comments on Neuberg's review of Causality. *Econometric Theory,* 19:686–689, 2003.

Pearl, 2003b J. Pearl. Reply to Woodward. *Economics and Philosophy,* 19:341–344, 2003.

Pearl, 2003c J. Pearl. Statistics and causal inference: A review. *Test Journal,* 12(2):281–345, December 2003.

Pearl, 2004 J. Pearl. Robustness of causal claims. In M. Chickering and J. Halpern, editors, *Proceedings of the Twentieth Conference Uncertainty in Artificial Intelligence,* pages 446–453. AUAI Press, Arlington, VA, 2004.

Pearl, 2005a J. Pearl. Direct and indirect effects. In *Proceedings of the American Statistical Association, Joint Statistical Meetings,* pages 1572–1581. MIRA Digital Publishing, Minneapolis, MN, 2005.

Pearl, 2005b J. Pearl. Influence diagrams – historical and personal perspectives. *Decision Analysis,* 2(4):232–234, 2005.

Pearl, 2008 J. Pearl. The mathematics of causal relations. Technical Report TR-338, http://ftp.cs.ucla.edu/pub/stat_ser/r338.pdf, Department of Computer Science, University of California, Los Angeles, CA, 2008. Presented at the American Psychopathological Association (APPA) Annual Meeting, NYC, March 6–8, 2008.

Pearl, 2009 J. Pearl. Remarks on the method of propensity scores. *Statistics in Medicine,* 28:1415–1416, 2009. See also <http://ftp.cs.ucla.edu/put/stat_ser/r345-sim.pdf>.

Pearson et al., 1899 K. Pearson, A. Lee, and L. Bramley-Moore. Genetic (reproductive) selection: Inheritance of fertility in man. *Philosophical Transactions of the Royal Society A,* 73:534–539, 1899.

Peikes et al., 2008 D.N. Peikes, L. Moreno, and S.M. Orzol. Propensity scores matching: A note of caution for evaluators of social programs. *The American Statistician,* 62(3):222–231, 2008.

Peng and Reggia, 1986 Y. Peng and J.A. Reggia. Plausibility of diagnostic hypotheses. In *Proceedings of the Fifth National Conference on AI (AAAI-86),* pages 140–145, Philadelphia, 1986.

Petersen et al., 2006 M.L. Petersen, S.E. Sinisi, and M.J. van der Laan. Estimation of direct causal effects. *Epidemiology,* 17(3):276–284, 2006.

Poole, 1985 D. Poole. On the comparison of theories: Preferring the most specific explanations. In *Proceedings of the Ninth International Conference on Artificial Intelligence (IJCAI-85),* pages 144–147, Los Angeles, CA, 1985.

Popper, 1959 K.R. Popper. *The Logic of Scientific Discovery.* Basic Books, New York, 1959.

Pratt and Schlaifer, 1988 J.W. Pratt and R. Schlaifer. On the interpretation and observation of laws. *Journal of Econometrics,* 39:23–52, 1988.

Price, 1991 H. Price. Agency and probabilistic causality. *British Journal for the Philosophy of Science,* 42:157–176, 1991.

Price, 1996 H. Price. *Time's arrow and Archimedes' point: New directions for the physics of time.* Oxford University Press, New York, 1996.

Program, 1984 Lipid Research Clinic Program. The Lipid Research Clinics Coronary Primary Prevention Trial results, parts I and II. *Journal of the American Medical Association,* 251(3):351–374, January 1984.

Rebane and Pearl, 1987 G. Rebane and J. Pearl. The recovery of causal poly-trees from statistical data. In *Proceedings of the Third Workshop on Uncertainty in AI,* pages 222–228, Seattle, WA, 1987.

Reichenbach, 1956 H. Reichenbach. *The Direction of Time.* University of California Press, Berkeley, 1956.

Reiter, 1987 R. Reiter. A theory of diagnosis from first principles. *Artificial Intelligence,* 32(l):57–95, 1987.

Richard, 1980 J.F. Richard. Models with several regimes and changes in exogeneity. *Review of Economic Studies,* 47:1–20, 1980.

Richardson, 1996 T. Richardson. A discovery algorithm for directed cyclic graphs. In E. Horvitz and F. Jensen, editors, *Proceedings of the Twelfth Conference on Uncertainty in Artificial Intelligence,* pages 454–461. Morgan Kaufmann, San Francisco, CA, 1996.

Rigdon, 2002 E.E. Rigdon. New books in review: Causality: Models, Reasoning, and Inference and Causation, Prediction, and Search. *Journal of Marketing Research,* XXXIX: 137–140, 2002.

Robert and Casella, 1999 C.P. Robert and G. Casella. *Monte Carlo Statistical Methods.* Springer Verlag, New York, NY, 1999.

Robertson, 1997 D.W. Robertson. The common sense of cause in fact. *Texas Law Review,* 75(7): 1765–1800, 1997.

Robins and Greenland, 1989 J.M. Robins and S. Greenland. The probability of causation under a stochastic model for individual risk. *Biometrics,* 45:1125–1138, 1989.

Robins and Greenland, 1991 J.M. Robins and S. Greenland. Estimability and estimation of expected years of life lost due to a hazardous exposure. *Statistics in Medicine,* 10:79–93, 1991.

Robins and Greenland, 1992 J.M. Robins and S. Greenland. Identifiability and exchangeability for direct and indirect effects. *Epidemiology,* 3(2): 143–155, 1992.

Robins and Wasserman, 1999 J.M. Robins and L. Wasserman. On the impossibility of inferring causation from association without background knowledge. In C.N. Glymour and G.F. Cooper, editors, *Computation, Causation, and Discovery,* pages 305–321. AAAI/MIT Press, Cambridge, MA, 1999.

Robins et al., 1992 J.M. Robins, D. Blevins, G. Ritter, and M. Wulfsohn. *g*-estimation of the effect of prophylaxis therapy for pneumocystis carinii pneumonia on the survival of AIDS patients. *Epidemiology,* 3:319–336, 1992.

Robins et al., 2003 J.M. Robins, R. Schemes, P. Spirtes, and L. Wasserman. Uniform consistency in causal inference. *Biometrika,* 90:491–512, 2003.

Robins, 1986 J.M. Robins. A new approach to causal inference in mortality studies with a sustained exposure period – applications to control of the healthy workers survivor effect. *Mathematical Modeling,* 7:1393–1512, 1986.

Robins, 1987 J. Robins. Addendum to "A new approach to causal inference in mortality studies with sustained exposure periods – application to control of the healthy worker survivor effect." *Computers and Mathematics, with Applications,* 14:923–45, 1987.

Robins, 1989 J.M.Robins. The analysis of randomized and non-randomized AIDS treatment trials using a new approach to causal inference in longitudinal studies. In L. Sechrest, H. Freeman, and A. Mulley, editors, *Health Service Research Methodology: A Focus on AIDS,* pages 113–159. NCHSR, U.S. Public Health Service, 1989.

Robins, 1993 J.M. Robins. Analytic methods for estimating HIV treatment and cofactors effects. In D.G. Ostrow and R. Kessler, editors, *Methodological Issues in AIDS Behavioral Research,* pages 213–290. Plenum Publishing, New York, 1993.

Robins, 1995 J.M. Robins. Discussion of "Causal diagrams for empirical research" by J. Pearl. *Biometrika,* 82(4):695–698, 1995.

Robins, 1997 J.M. Robins. Causal inference from complex longitudinal data. In M. Berkane, editor, *Latent Variable Modeling and Applications to Causality,* pages 69–117. Springer-Verlag, New York, 1997.

Robins, 1999 J.M. Robins. Testing and estimation of directed effects by reparameterizing directed acyclic with structural nested models. In C.N. Glymour and G.F. Cooper, editors, *Computation, Causation, and Discovery,* pages 349–405. AAAI/MIT Press, Cambridge, MA, 1999.

Robins, 2001 J.M. Robins. Data, design, and background knowledge in etiologic inference. *Epidemiology,* 12(3):313–320, 2001.

Rosenbaum and Rubin, 1983 P. Rosenbaum and D. Rubin. The central role of propensity score in observational studies for causal effects. *Biometrika,* 70:41–55, 1983.

Rosenbaum, 1984 P.R. Rosenbaum. The consequences of adjustment for a concomitant variable that has been affected by the treatment. *Journal of the Royal Statistical Society, Series A (General),* Part 5(147):656–666, 1984.

Rosenbaum, 1995 P.R. Rosenbaum. *Observational Studies.* Springer-Verlag, New York, 1995.

Rosenbaum, 2002 P.R. Rosenbaum. *Observational Studies.* Springer-Verlag, New York, 2nd edition, 2002.

Rothman and Greenland, 1998 K.J. Rothman and S. Greenland. *Modern Epidemiology.* Lippincott-Rawen, Philadelphia, 2nd edition, 1998.

Rothman, 1976 K.J. Rothman. Causes. *American Journal of Epidemiology,* 104:587–592, 1976.

Rothman, 1986 K.J. Rothman. *Modern Epidemiology.* Little, Brown, 1st edition, 1986.

Roy, 1951 A.D. Roy. Some thoughts on the distribution of earnings. *Oxford Economic Papers,* 3:135–146, 1951.

Rubin, 1974 D.B. Rubin. Estimating causal effects of treatments in randomized and nonrandomized studies. *Journal of Educational Psychology,* 66:688–701, 1974.

Rubin, 2004 D.B. Rubin. Direct and indirect causal effects via potential outcomes. *Scandinavian Journal of Statistics,* 31:161–170, 2004.

Rubin, 2005 D.B. Rubin. Causal inference using potential outcomes: Design, modeling, decisions. *Journal of the American Statistical Association,* 100(469):322–331, 2005.

Rubin, 2007 D.B. Rubin. The design *versus* the analysis of observational studies for causal effects: Parallels with the design of randomized trials. *Statistics in Medicine,* 26:20–36, 2007.

Rubin, 2008a D.B. Rubin. Author's reply (to Ian Shrier's Letter to the Editor). *Statistics in Medicine*, 27:2741–2742, 2008.

Rubin, 2008b D.B. Rubin. For objective causal inference, design trumps analysis. *The Annals of Applied Statistics*, 2:808–840, 2008.

Rubin, 2009 D.B. Rubin. Author's Reply: Should observational studies be designed to allow lack of balance in covariate distributions across treatment groups? *Statistics in Medicine*, 28:1420–1423, 2009.

Rücker and Schumacher, 2008 G. Rücker and M. Schumacher. Simpson's paradox visualized: The example of the Rosiglitazone meta-analysis. *BMC Medical Research Methodology*, 8(34):1–8, 2008.

Salmon, 1984 W.C. Salmon. *Scientific Explanation and the Causal Structure of the World*. Princeton University Press, Princeton, NJ, 1984.

Salmon, 1998 W.C. Salmon. *Causality and Explanation.* Oxford University Press, New York, NY, 1998.

Sandewall, 1994 E. Sandewall. *Features and Fluents,* volume 1. Clarendon Press, Oxford, 1994.

Savage, 1954 L.J. Savage. *The Foundations of Statistics.* John Wiley and Sons, Inc., New York, 1954.

Scheines, 2002 R. Schemes. Public administration and health care: Estimating latent causal influences: TETRAD III variable selection and bayesian parameter estimation. In W. Klosgen, J.M. Zytkow, and J. Zyt, editors, *Handbook of Data Mining and Knowledge Discovery,* pages 944–952. Oxford University Press, New York, 2002.

Schlesselman, 1982 J.J. Schlesselman. *Case-Control Studies: Design Conduct Analysis.* Oxford University Press, New York, 1982.

Schumaker and Lomax, 1996 R.E. Schumaker and R.G. Lomax. *A Beginner's Guide to Structural Equation Modeling.* Lawrence Erlbaum Associations, Mahwah, NJ, 1996.

Serrano and Gossard, 1987 D. Serrano and D.C. Gossard. Constraint management in conceptual design. In D. Sriram and R.A. Adey, editors, *Knowledge Based Expert Systems in Engineering: Planning and Design,* pages 211–224. Computational Mechanics Publications, 1987.

Shachter et al., 1994 R.D. Shachter, S.K. Andersen, and P. Szolovits. Global conditioning for probabilistic inference in belief networks. In R. Lopez de Mantaras and D. Poole, editors, *Uncertainty in Artificial Intelligence,* pages 514–524. Morgan Kaufmann, San Francisco, CA, 1994.

Shachter, 1986 R.D. Shachter. Evaluating influence diagrams. *Operations Research,* 34(6):871–882, 1986.

Shadish and Clark, 2006 W.R. Shadish and M.H. Clark. A randomized experiment comparing random to nonrandom assignment. Unpublished paper, University of California, Merced, 2006.

Shadish and Cook, 2009 W.R. Shadish and T.D. Cook. The renaissance of field experimentation in evaluating interventions. *Annual Review of Psychology,* 60:607–629, 2009.

Shafer, 1996 G. Shafer. *The Art of Causal Conjecture.* MIT Press, Cambridge, MA, 1996.

Shapiro, 1997 S.H. Shapiro. Confounding by indication? *Epidemiology,* 8:110–111, 1997.

Shep, 1958 M.C. Shep. Shall we count the living or the dead? *New England Journal of Medicine,* 259:1210–1214, 1958.

Shimizu et al., 2005 A. Shimizu, S. Hyvärinen, Y. Kano, and P.O. Hoyer. Discovery of non-Gaussian linear causal models using ICA. In R. Dechter and T.S. Richardson, editors, *Proceedings of the Twenty-First Conference on Uncertainty in Artificial Intelligence,* pages 525–533. AUAI Press, Edinburgh, Schotland, 2005.

Shimizu et al., 2006 S. Shimizu, P.O. Hoyer, Hyvärinen, and A.J. Kerminen. A linear non-Gaussian acyclic model for causal discovery. *Journal of the Machine Learning Research,* 7:2003–2030, 2006.

Shimony, 1991 S.E. Shimony. Explanation, irrelevance and statistical independence. In *Proceedings of the Ninth Conference on Artificial Intelligence (AAAI'91),* pages 482–487, 1991.

Shimony, 1993 S.E. Shimony. Relevant explanations: Allowing disjunctive assignments. In D. Heckerman and A. Mamdani, editors, *Proceedings of the Ninth Conference on Uncertainty in Artificial Intelligence,* pages 200–207, San Mateo, CA, July 1993. Morgan Kaufmann Publishers.

Shipley, 1997 B. Shipley. An inferential test for structural equation models based on directed acyclic graphs and its nonparametric equivalents. Technical report, Department of Biology, University of Sherbrooke, Canada, 1997. Also in *Structural Equation Modelling,* 7:206–218, 2000.

Shipley, 2000a B. Shipley. Book reviews: Causality: Models, Reasoning, and Inference. *Structural Equation Modeling,* 7(4):637–639, 2000.

Shipley, 2000b B. Shipley. *Cause and Correlation in Biology: A User's Guide to Path Analysis, Structural Equations and Causal Inference.* Cambridge University Press, New York, 2000.

Shoham, 1988 Y. Shoham. *Reasoning About Change: Time and Causation from the Standpoint of Artificial Intelligence.* MIT Press, Cambridge, MA, 1988.

Shpitser and Pearl, 2006a I. Shpitser and J Pearl. Identification of conditional interventional distributions. In R. Dechter and T.S. Richardson, editors, *Proceedings of the Twenty-Second Conference on Uncertainty in Artificial Intelligence,* pages 437–444. AUAI Press, Corvallis, OR, 2006.

Shpitser and Pearl, 2006b I. Shpitser and J Pearl. Identification of joint interventional distributions in recursive semi-Markovian causal models. In *Proceedings of the Twenty-First National Conference on Artificial Intelligence,* pages 1219–1226. AAAI Press, Menlo Park, CA, 2006.

Shpitser and Pearl, 2007 I. Shpitser and J Pearl. What counterfactuals can be tested. In *Proceedings of the Twenty-Third Conference on Uncertainty in Artificial Intelligence,* pages 352–359. AUAI Press, Vancouver, BC Canada, 2007. Also, *Journal of Machine Learning Research*, 9:1941–1979, 2008.

Shpitser and Pearl, 2008 I. Shpitser and J Pearl. Dormant independence. In *Proceedings of the Twenty-Third Conference on Artificial Intelligence*, pages 1081–1087. AAAI Press, Menlo Park, CA, 2008.

Shrier, 2009 I. Shrier. Letter to the Editor: Propensity scores. *Statistics in Medicine*, 28:1317–1318, 2009.

Simon and Rescher, 1966 H.A. Simon and N. Rescher. Cause and counterfactual. *Philosophy and Science,* 33:323–340, 1966.

Simon, 1953 H.A. Simon. Causal ordering and identifiability. In Wm. C. Hood and T.C. Koopmans, editors, *Studies in Econometric Method,* pages 49–74. Wiley and Sons, Inc., New York, NY, 1953.

Simpson, 1951 E.H. Simpson. The interpretation of interaction in contingency tables. *Journal of the Royal Statistical Society, Series B*, 13:238–241, 1951.

Sims, 1977 C.A. Sims. Exogeneity and causal ordering in macroeconomic models. In *New Methods in Business Cycle Research: Proceedings from a Conference, November 1975,* pages 23–43. Federal Reserve Bank, Minneapolis, 1977.

Singh and Valtorta, 1995 M. Singh and M. Valtorta. Construction of Bayesian network structures from data – a brief survey and an efficient algorithm. *International Journal of Approximate Reasoning,* 12(2): 111–131, 1995.

Sjölander, 2009 A. Sjölander. Letter to the Editor: Propensity scores and M-structures. *Statistics in Medicine*, 28:1416–1423, 2009.

Skyrms, 1980 B. Skyrms. *Causal Necessity.* Yale University Press, New Haven, 1980.

Smith and Todd, 2005 J. Smith and P. Todd. Does matching overcome LaLonde's critique of nonexperimental estimators? *Journal of Econometrics,* 125:305–353, 2005.

Sobel, 1990 M.E. Sobel. Effect analysis and causation in linear structural equation models. *Psychometrika,* 55(3):495–515, 1990.

Sober and Barrett, 1992 E. Sober and M. Barrett. Conjunctive forks and temporally asymmetric inference. *Australian Journal of Philosophy,* 70:1–23, 1992.

Sober, 1985 E. Sober. Two concepts of cause. In P. Asquith and P. Kitcher, editors, *PSA: Proceedings of the Biennial Meeting of the Philosophy of Science Association,* volume II, pages 405–424. Philosophy of Science Association, East Lansing, MI, 1985.

Sommer et al., 1986 A. Sommer, I. Tarwotjo, E. Djunaedi, K. P. West, A. A. Loeden, R. Tilden, and L. Mele. Impact of vitamin A supplementation on childhood mortality: A randomized controlled community trial. *The Lancet,* 327:1169–1173, 1986.

Sosa and Tooley, 1993 E. Sosa and M. Tooley (Eds.). *Causation.* Oxford readings in Philosophy. Oxford University Press, Oxford, 1993.

Spiegelhalter et al., 1993 D.J. Spiegelhalter, S.L. Lauritzen, P.A. Dawid, and R.G. Cowell. Bayesian analysis in expert systems (with discussion). *Statistical Science,* 8:219–283, 1993.

Spirtes and Glymour, 1991 P. Spirtes and C.N. Glymour. An algorithm for fast recovery of sparse causal graphs. *Social Science Computer Review,* 9(1):62–72, 1991.

Spirtes and Richardson, 1996 P. Spirtes and T. Richardson. A polynomial time algorithm for determinint DAG equivalence in the presence of latent variables and selection bias. *Proceedings of the Sixth International Workshop on Artificial Intelligence and Statistics,* 1996.

Spirtes and Verma, 1992 P. Spirtes and T. Verma. Equivalence of causal models with latent variables. Technical Report CMU-PHIL-33, Carnegie Mellon University, Pittsburgh, Pennsylvania, October 1992.

Spirtes et al., 1993 P. Spirtes, C.N. Glymour, and R. Scheines. *Causation, Prediction, and Search.* Springer-Verlag, New York, 1993.

Spirtes et al., 1995 P. Spirtes, C. Meek, and T. Richardson. Causal inference in the presence of latent variables and selection bias. In P. Besnard and S. Hanks, editors, *Uncertainty in Artificial Intelligence 11,* pages 499–506. Morgan Kaufmann, San Francisco, 1995.

Spirtes et al., 1996 P. Spirtes, T. Richardson, C. Meek, R. Scheines, and C.N. Glymour. Using *d*-separation to calculate zero partial correlations in linear models with correlated errors. Technical Report CMU-PHIL-72, Carnegie-Mellon University, Department of Philosophy, Pittsburgh, PA, 1996.

Spirtes et al., 1998 P. Spirtes, T. Richardson, C. Meek, R. Scheines, and C.N. Glymour. Using path diagrams as a structural equation modelling tool. *Sociological Methods and Research,* 27(2): 182–225, November 1998.

Spirtes et al., 2000 P. Spirtes, C.N. Glymour, and R. Scheines. *Causation, Prediction, and Search.* MIT Press, Cambridge, MA, 2nd edition, 2000.

Spirtes, 1995 P. Spirtes. Directed cyclic graphical representation of feedback. In P. Besnard and S. Hanks, editors, *Proceedings of the Eleventh Conference on Uncertainty in Artificial Intelligence,* pages 491–498. Morgan Kaufmann, San Mateo, CA, 1995.

Spohn, 1980 W. Spohn. Stochastic independence, causal independence, and shieldability. *Journal of Philosophical Logic,* 9:73–99, 1980.

Spohn, 1983 W. Spohn. Deterministic and probabilistic reasons and causes. *Erkenntnis,* 19:371–396, 1983.

Spohn, 1988 W. Spohn. A general non-probabilistic theory of inductive reasoning. In *Proceedings of the Fourth Workshop on Uncertainty in Artificial Intelligence,* pages 315–322, Minneapolis, MN, 1988.

Stalnaker, 1968 R.C. Stalnaker. A theory of conditionals. In N. Rescher, editor, *Studies in Logical Theory,* volume No. 2, American Philosophical Quarterly Monograph Series. Blackwell, Oxford, 1968. Reprinted in W.L. Harper, R. Stalnaker, and G. Pearce (Eds.), *Ifs,* D. Reidel, Dordrecht, pages 41–55, 1981.

Stalnaker, 1972 R.C. Stalnaker. Letter to David Lewis, 1972. In W.L. Harper, R. Stalnaker, and G. Pearce (Eds.), *Ifs,* D. Reidel, Dordrecht, pages 151–152, 1981.

Stelzl, 1986 I. Stelzl. Changing a causal hypothesis without changing the fit: Some rules for generating equivalent path models. *Multivariate Behavioral Research,* 21:309–331, 1986.

Steyer et al., 1996 R. Steyer, S. Gabler, and A.A. Rucai. Individual causal effects, average causal effects, and unconfoundedness in regression models. In F. Faulbaum and W. Bandilla, editors, *SoftStat'95, Advances in Statistical Software 5,* pages 203–210. Lucius & Lucius, Stuttgart, 1996.

Steyer et al., 1997 R. Steyer, A.A. von Davier, S. Gabler, and C. Schuster. Testing unconfoundedness in linear regression models with stochastic regressors. In W. Bandilla and F. Faulbaum, editors, *SoftStat'97, Advances in Statistical Software 6,* pages 377–384. Lucius & Lucius, Stuttgart, 1997.

Stone, 1993 R. Stone. The assumptions on which causal inferences rest. *Journal of the Royal Statistical Society,* 55(2):455–466, 1993.

Strotz and Wold, 1960 R.H. Strotz and H.O.A. Wold. Recursive versus nonrecursive systems: An attempt at synthesis. *Econometrica,* 28:417–427, 1960.

Suermondt and Cooper, 1993 H.J. Suermondt and G.F. Cooper. An evaluation of explanations of probabilistic inference. *Computers and Biomedical Research,* 26:242–254, 1993.

Suppes and Zaniotti, 1981 P. Suppes and M. Zaniotti. When are probabilistic explanations possible? *Synthese,* 48:191–199, 1981.

Suppes, 1970 P. Suppes. *A Probabilistic Theory of Causality.* North-Holland Publishing Co., Amsterdam, 1970.

Suppes, 1988 P. Suppes. Probabilistic causality in space and time. In B. Skyrms and W.L. Harper, editors, *Causation, Chance, and Credence.* Kluwer Academic Publishers, Dordrecht, The Netherlands, 1988.

Swanson and Granger, 1997 N.R. Swanson and C.W.J. Granger. Impulse response functions based on a causal approach to residual orthogonalization in vector autoregressions. *Journal of the American Statistical Association*, 92:357–367, 1997.

Swanson, 2002 N.R. Swanson. Book reviews: Causality: Models, Reasoning, and Inference. *Journal of Economic Literature*, XL:925–926, 2002.

Tian and Pearl, 2000 J. Tian and J. Pearl. Probabilities of causation: Bounds and identification. *Annals of Mathematics and Artificial Intelligence*, 28:287–313, 2000.

Tian and Pearl, 2001a J. Tian and J. Pearl. Causal discovery from changes. In *Proceedings of the Seventeenth Conference on Uncertainty in Artificial Intelligence,* pages 512–521. Morgan Kaufmann, San Francisco, CA, 2001.

Tian and Pearl, 2001b J. Tian and J. Pearl. Causal discovery from changes: A Bayesian approach. Technical Report R-285, Computer Science Department, UCLA, February 2001.

Tian and Pearl, 2002a J. Tian and J. Pearl. A general identification condition for causal effects. In *Proceedings of the Eighteenth National Conference on Artificial Intelligence,* pages 567–573. AAAI Press/The MIT Press, Menlo Park, CA, 2002.

Tian and Pearl, 2002b J. Tian and J Pearl. On the testable implications of causal models with hidden variables. In A. Darwiche and N. Friedman, editors, *Proceedings of the Eighteenth Conference on Uncertainty in Artificial Intelligence*, pages 519–527. Morgan Kaufmann, San Francisco, CA, 2002.

Tian et al., 1998 J. Tian, A. Paz, and J. Pearl. Finding minimal separating sets. Technical Report R-254, University of California, Los Angeles, CA, 1998.

Tian et al., 2006 J. Tian, C. Kang, and J. Pearl. A characterization of interventional distributions in semi-Markovian causal models. In *Proceedings of the Twenty-First National Conference on Artificial Intelligence,* pages 1239–1244. AAAI Press, Menlo Park, CA, 2006.

Tversky and Kahneman, 1980 A. Tversky and D. Kahneman. Causal schemas in judgments under uncertainty. In M. Fishbein, editor, *Progress in Social Psychology,* pages 49–72. Lawrence Erlbaum, Hillsdale, NJ, 1980.

VanderWeele and Robins, 2007 T.J. VanderWeele and J.M. Robins. Four types of effect modification: A classification based on directed acyclic graphs. *Epidemiology,* 18(5):561–568, 2007.

Verma and Pearl, 1988 T. Verma and J. Pearl. Causal networks: Semantics and expressiveness. In *Proceedings of the Fourth Workshop on Uncertainty in Artificial Intelligence,* pages 352–359, Mountain View, CA, 1988. Also in R. Shachter, T.S. Levitt, and L.N. Kanal (Eds.), *Uncertainty in AI 4,* Elesevier Science Publishers, 69–76, 1990.

Verma and Pearl, 1990 T. Verma and J. Pearl. Equivalence and synthesis of causal models. In *Proceedings of the Sixth Conference on Uncertainty in Artificial Intelligence,* pages 220–227, Cambridge, MA, July 1990. Also in P. Bonissone, M. Henrion, L.N. Kanal and J.F. Lemmer (Eds.), *Uncertainty in Artificial Intelligence 6,* Elsevier Science Publishers, B.V, 255–268, 1991.

Verma and Pearl, 1992 T. Verma and J. Pearl. An algorithm for deciding if a set of observed independencies has a causal explanation. In D. Dubois, M.P. Wellman, B. D'Ambrosio, and P. Smets, editors, *Proceedings of the Eighth Conference on Uncertainty in Artificial Intelligence,* pages 323–330. Morgan Kaufmann, Stanford, CA, 1992.

Verma, 1993 T.S. Verma. Graphical aspects of causal models. Technical Report R-191, UCLA, Computer Science Department, 1993.

Wainer, 1989 H. Wainer. Eelworms, bullet holes, and Geraldine Ferraro: Some problems with statistical adjustment and some solutions. *Journal of Educational Statistics,* 14:121–140, 1989.

Wang et al., 2009 X. Wang, Z. Geng, H. Chen, and X. Xie. Detecting multiple confounders. *Journal of Statistical Planning and Inference,* 139: 1073–1081, 2009.

Wasserman, 2004 L. Wasserman. *All of Statistics: A Concise Course in Statistical Inference.* Springer Science+Business Media, Inc., New York, NY, 2004.

Weinberg, 1993 C.R. Weinberg. Toward a clearer definition of confounding. *American Journal of Epidemiology,* 137:1–8, 1993.

Weinberg, 2007 C.R. Weinberg. Can DAGs clarify effect modification? *Epidemiology,* 18:569–572, 2007.

Wermuth and Lauritzen, 1983 N. Wermuth and S.L. Lauritzen. Graphical and recursive models for contingency tables. *Biometrika,* 70:537–552, 1983.

Wermuth and Lauritzen, 1990 N. Wermuth and S.L. Lauritzen. On substantive research hypotheses, conditional independence graphs and graphical chain models (with discussion). *Journal of the Royal Statistical Society, Series B,* 52:21–72, 1990.

Wermuth, 1987 N. Wermuth. Parametric collapsibility and the lack of moderating effects in contingency tables with a dichotomous response variable. *Journal of the Royal Statistical Society, Series B,* 49(3):353–364, 1987.

Wermuth, 1992 N. Wermuth. On block-recursive regression equations. *Brazilian Journal of Probability and Statistics (with discussion),* 6:1–56, 1992.

Whittaker, 1990 J. Whittaker. *Graphical Models in Applied Multivariate Statistics.* John Wiley, Chichester, England, 1990.

Whittemore, 1978 A.S. Whittemore. Collapsibility of multidimensional contingency tables. *Journal of the Royal Statistical Society, Series B,* 40(3):328–340, 1978.

Wickramaratne and Holford, 1987 P.J. Wickramaratne and T.R. Holford. Confounding in epidemiologic studies: The adequacy of the control group as a measure of confounding. *Biometrics,* 43:751–765, 1987.

Winship and Morgan, 1999 C. Winship and S.L. Morgan. The estimation of causal effects from observational data. *Annual Review of Sociology,* 25:659–706, 1999.

Winslett, 1988 M. Winslett. Reasoning about action using a possible worlds approach. In *Proceedings of the Seventh American Association for Artificial Intelligence Conference,* pages 89–93, 1988.

Woodward, 1990 J. Woodward. Supervenience and singular causal claims. In D. Knowles, editor, *Explanation and Its Limits,* pages 211–246. Cambridge University Press, New York, 1990.

Woodward, 1995 J. Woodward. Causation and explanation in econometrics. In D. Little, editor, *On the Reliability of Economic Models,* pages 9–61. Kluwer Academic, Boston, 1995.

Woodward, 1997 J. Woodward. Explanation, invariance and intervention. *Philosophy of Science,* 64(S):26–S41, 1997.

Woodward, 2003 J. Woodward. *Making Things Happen.* Oxford University Press, New York, NY, 2003.

Wright, 1921 S. Wright. Correlation and causation. *Journal of Agricultural Research,* 20:557–585, 1921.

Wright, 1923 S. Wright. The theory of path coefficients: A reply to Niles' criticism. *Genetics,* 8:239–255, 1923.

Wright, 1925 S. Wright. Corn and hog correlations. Technical Report 1300, U.S. Department of Agriculture, 1925.

Wright, 1928 P.O. Wright. *The Tariff on Animal and Vegetable Oils.* The MacMillan Company, New York, NY, 1928.

Wright, 1988 R.W. Wright. Causation, responsibility, risk, probability, naked statistics, and proof: Prunning the bramble bush by clarifying the concepts. *Iowa Law Review,* 73:1001–1077, 1988.

Wu, 1973 D.M. Wu. Alternative tests of independence between stochastic regressors and disturbances. *Econometrica,* 41:733–750, 1973.

Yanagawa, 1984 T. Yanagawa. Designing case-contol studies. *Environmental health perspectives,* 32:219–225, 1984.

Yule, 1903 G.U. Yule. Notes on the theory of association of attributes in statistics. *Biometrika,* 2:121–134, 1903.

Zelterman, 2001 D. Zelterman. Book reviews: Causality: Models, Reasoning, and Inference. *Technometrics,* 32(2):239, 2001.

Zidek, 1984 J. Zidek. Maximal Simpson disaggregations of 2×2 tables. *Biometrika,* 71:187–190, 1984.

Name Index

Subject Index

exchangeability
  causal understanding and, 179, 384
  confounding and, 196–9, 341n
  De Finetti's, 178
exclusion restrictions, 101, 232–3, 380
exogeneity, 97n, 165–70
  controversies regarding, 165, 167, 169–70,
    245–7
  counterfactual and graphical definitions,
    245–7
  definition, causal, 166, 289, 333
  error-based, 169–70, 247, 343
  general definition, 168
  hierarchy of definitions, 246
  use in policy analysis, 165–6
  see also confounding bias; ignorability
expectation, 9–10
  conditional, 9
  controlled vs. conditional, 97, 137n, 162
explaining away, 17
explanation, 25, 58, 221–3, 285, 308–9
  as attribution, 402–3
  purposive, 333–4

factorization
  Markov, 16
  truncated, 24
faithfulness, 48
  see also stability
family (in a graph), 13
front-door criterion, 81–3
  applications, 83–5, 106
functional models, 26, 203–20
  advantages, 32
  and counterfactuals, 33, 204–6
  intervention in, 32
  as joint distributions, 31
  nonparametric, 67, 69, 94, 154–7

G-estimation, 72, 102–4, 123, 352–3
Gibbs sampling
  in Bayesian inference, 21, 375–7
  for estimating attribution, 280
  for estimating effects, 275–7
graphical models
  in social science, 38–40, 97
  in statistics, 12–20
graphoids, 11–12, 234
graphs
  complete, 13
  cyclic, 12–13, 28, 95–6, 142
  directed, 12
  as models of intervention, 68–70
  mutilated, 23
  notation and probabilities, 12
  and relevance, 11

homomorphy, in imaging, 242
Hume's dilemma, 41, 238, 249, 406, 413

IC algorithm, 50
IC* algorithm, 52
identification, 77, 105, 366, 376
  of direct effects, 126–31
  by graphs, 89–94, 114–18
  of plans, 118–26, 354–5
identifying models, 91–2, 105, 114–15
ignorability, 19–80, 246, 248n, 289, 341–4
  and back-door criterion, 80, 100, 343, 350–2
  judgment of, 79, 100, 102, 350
  demystified, 341–4
  see also exogeneity
imaging, 112, 242–3
independence, 3
  conditional, 3, 11
  dormant, 64, 347n, 448
indirect effects, 132, 165, 355–8
inference
  causal, 22–3, 32, 85–9, 209
  counterfactual, 33–9, 210–13, 231–4
  probabilistic, 20, 30, 31
inferred causation, 44, 45
  algorithms for, 50, 52
  local conditions for, 54–7
influence diagrams, 111n, 382
instrumental variables, 90, 153, 168, 247–8,
    274–5, 366, 395
  definitions of, 247–8
  formula, 90, 153
  tests for, 274–5
intent to treat analysis, 261
intervention, 22–3, 332
  atomic, 70, 362
  calculus of, 85–9
  as conditionalization, 23, 72–4
  examples, 28–9, 32
  joint, 74–6, 91, 118–26
  notation, 67n, 70
  stochastic, 113–14
  as transformation, 72–4, 112, 242–3
  truncated factorization formula for, 24,
    72, 74
  as variable, 70–2, 111
  see also actions
intransitive dependence, 43, 57
INUS condition, 313–15, 321–2
invariance
  of conditional independence, 31, 48, 63
  of mechanisms, *see* autonomy
  of structural parameters, 63, 160–2, 332

join–tree propagation, 20

Laplacian models, 26, 257
latent structure, 45
  see also semi-Markovian models
  projection of, 52
  recovery of, 52
Lewis's counterfactuals, 238–40